From Artemisia annua *L.* *to Artemisinins*

From Artemisia annua *L.* *to Artemisinins*

The Discovery and Development of Artemisinins and Antimalarial Agents

Youyou Tu
Professor
China Academy of Chinese Medical Sciences
Beijing, China

ELSEVIER
Chemical Industry Press
ACADEMIC PRESS
An imprint of Elsevier

Academic Press is an imprint of Elsevier
125 London Wall, London EC2Y 5AS, United Kingdom
525 B Street, Suite 1800, San Diego, CA 92101-4495, United States
50 Hampshire Street, 5th Floor, Cambridge, MA 02139, United States
The Boulevard, Langford Lane, Kidlington, Oxford OX5 1GB, United Kingdom

Library of Congress Cataloging-in-Publication Data
A catalog record for this book is available from the Library of Congress

British Library Cataloguing-in-Publication Data
A catalogue record for this book is available from the British Library

ISBN: 978-0-12-811655-5

For information on all Academic Press publications visit our website at
https://www.elsevier.com/books-and-journals

Publisher: Glyn Jones
Acquisition Editor: Simon Tian
Editorial Project Manager: Naomi Robertson
Production Project Manager: Lucía Pérez
Designer: Vicky Pearson

Typeset by Thomson Digital

Contents

About the Author

Professor Youyou Tu graduated from the School of Pharmacy, Beijing Medical College, Beijing, China in 1955. She then studied and graduated from a training course on the theories of Traditional Chinese Medicine (TCM) for personnel with a background in Western Medicine.

Integrating her medical knowledge of Western Medicine and TCM, Professor Tu led a research group responsible for outstanding achievements in the chemical analysis and identification of over 200 herbs and 380 abstracts used in TCM. Her group and she eventually discovered a completely new antimalarial substance, artemisinin (*qinghaosu*) and its derivatives, including the extremely effective derivative dihydroartemisinin. For this breakthrough in the treatment of malaria, which has saved millions of lives, Professor Tu was awarded the Nobel Prize for Medicine in 2015.

Professor Tu has received much international recognition for her work, and has been honored with numerous prestigious awards (for full details of her Curriculum Vitae, please refer to the Appendix at the end of this volume).

List of Contributors

Original Authors

Youyou Tu Artemisinin Research Center, China Academy of Chinese Medical Sciences, Beijing, China

Jingyan Wang Academy of Military Medical Sciences, Beijing, China

Manyuan Wang Department of Pharmaceutics of TCM, School of Traditional Chinese Medicine, Capital Medical University, Beijing, China

Hechun Ye Key Laboratory of Plant Molecular Physiology (formerly known as Key Laboratory of Photosynthesis & Environmental Molecular Physiology), Institute of Botany, CAS, Beijing, China

Lixin Yang Institute of Chinese Materia Medica, China Academy of Chinese Medical Sciences, Beijing, China

Lan Yang Artemisinin Research Center, China Academy of Chinese Medical Sciences, Beijing, China

Tingliang Jiang Institute of Chinese Materia Medica, China Academy of Chinese Medical Sciences, Beijing, China

Muyun Ni Institute of Chinese Materia Medica, China Academy of Chinese Medical Sciences, Beijing, China

Xuanliang Gao China Medical Science Press, Beijing, China

Hong Wang Graduate University of Chinese Academy of Sciences, Beijing, China

Zhuchun Bei Academy of Military Medical Sciences, Beijing, China

Benye Liu Key Laboratory of Plant Molecular Physiology (formerly known as Key Laboratory of Photosynthesis & Environmental Molecular Physiology), Institute of Botany, CAS, Beijing, China

Fang Yan Key Laboratory of Plant Molecular Physiology (formerly known as Key Laboratory of Photosynthesis & Environmental Molecular Physiology), Institute of Botany, CAS, Beijing, China

Guofeng Li Key Laboratory of Plant Molecular Physiology (formerly known as Key Laboratory of Photosynthesis & Environmental Molecular Physiology), Institute of Botany, CAS, Beijing, China

Translators

Manyuan Wang School of Traditional Chinese Medicine, Capital Medical University, Beijing, China

Lan Yang Artemisinin Research Center, China Academy of Chinese Medical Sciences, Beijing, China

Jianying Shen Artemisinin Research Center, China Academy of Chinese Medical Sciences, Beijing, China

Hong Wang Graduate University of Chinese Academy of Sciences, Beijing, China

Junxian Yu Department of Pharmacy, Beijing Friendship Hospital, Capital Medical University, Beijing, China

Foreword

The forthcoming of this book is a very delightful thing. Up to now, artemisinin is the first original invention in China that has obtained international recognition, which has a profound influence both on the modern and the traditional medicine.

Malaria has traditionally been a serious epidemic disease that is harmful to human's health. In the 1820s, western scientists found an effective antimalarial ingredient named quinine from the bark of cinchona. And then, they developed a series of quinolines as antimalarial drugs. Among them, chloroquine, developed in the 1940s, showed the best curative effects. With the prevalence of malaria associated with the chloroquine-resistant *Plasmodium* parasites in 1960s, domestic and foreign scientists had done a lot of work to search for new antimalarial drugs, but without any promising progress. In the early 1970s, the team of Youyou Tu, at the Institute of Chinese Materia Medica of China Academy of Chinese Medical Sciences, carried out experiments to develop new antimalarial agents in the traditional Chinese medicine field. At that time, under the considerably underresourced research conditions, they went through some twists and turns, and inspired by the expression, that is, "*Artemisia annua* L. should be wrung into juice to treat malaria," recorded in *A Handbook of Prescriptions for Emergencies* written by Hong Ge (284–363) in the East Jin Dynasty, around more than 1600 years ago. Youyou Tu's team finally found a new antimalarial compound called artemisinin from *A. annua* L. with novel chemical structure. And so, they solved the threatening global resurgence malaria associated with the drug-resistant *Plasmodium* parasites. In the last 40 years, artemisinin and its derivatives have become the first choice of the WHO for the global malaria control, which saved millions of patients' lives. This is a huge breakthrough in the history of the antimalarial drug research. Therefore, I feel happy for the achievements of Chinese scientists who were engaged in the artemisinins research field. Also, it makes me feel proud of traditional Chinese medicine which contains great treasures.

In recent years, new progresses were made in the field of artemisinins research and are included in this book.

I hope that the publication of the book can make the readers get some beneficial experience and have some important enlightenment through the discovery and the development process of artemisinins, and can inspire the researchers to achieve much more progress in the modernization and internationalization of traditional Chinese medicine (TCM).

Jieping Wu

September 7, 2006

Malaria is a devastating worldwide epidemic disease caused by parasites and especially common in the tropical area of the world. The saying "One child dying of malaria every second" remains true to this day, and the disease still has a serious economic impact in Africa, retarding economic growth and development and perpetuating the vicious cycle of poverty. There were some antimalarial medicine that had been used internationally, such as cinchona, quinine, and quinoline derivatives, such as Pamaquine (Aminoquin), Mepacrine (Atabrine), Chloroquine, Primaquine (Pirmaquinum), and so on. Malaria was effectively treated and controlled by chloroquine and other quinoline drugs for a long period of time until the prevalence of chloroquine-resistant malaria parasites in the late 1960s, leading to a dramatically failure in a global attempt in eradicating malaria. Domestic and foreign researchers have done a lot of work to search for new antimalarial drugs, but without breakthrough findings of new effective antimalarial medicine. In 1972, the research team, led by Youyou Tu, found a highly antimalarial drug in TCM, that is, artemisinin, with novel chemical structure, and then developed another medicine, dihydroartemisinin, a drug that has higher potency and is more convenient to take with lower dosage. Successively, some domestic peers developed other artemisinin derivatives, such as artemether and artesunate. On the road of human fighting against malaria, Chinese scientists have achieved a considerable breakthrough. In recent years, WHO strongly recommended artemisinin-based drugs as the frontline antimalarial treatment for all *falciparum* malaria.

A. annua L. is a traditional Chinese herbal medicine and was first recorded in *Prescriptions for Fifty-two Diseases* (Wu Shi Er Bing Fang) written on a piece of silk treatise unearthed from the Mawangdui Han Dynasty tombs (168 BC), described it as a treatment for hemorrhoids. Later, *Shen Nong's Herbal Classic* (Shen Nong Ben Cao Jing) named it Cao Hao and also as Qing Hao. As *A Handbook of Prescriptions for Emergencies* (Zhou Hou Bei Ji Fang) described Qing Hao as an antimalarial herbal medicine, various related prescriptions were created, such as Qing Hao pills, Qing Hao decoction or drink, Qing Hao powder, Qing Hao apozem, Qing Hao dew, and Qing Hao wine. For example, there were Qing Hao decoction in *General Medical Collection of Royal Benevolence* (Sheng Ji Zong Lu, also translated as *General Records of Holy Universal Relief*) written in the Song Dynasty, 960–1279; Jie Nue Qing Hao pills in *Danxi's Mastery of Medicine* (Dan Xi Xin Fa) written in the Yuan Dynasty, 1271–1368; and Qing Hao powder and Qu Nue Shen Ying pills in *Prescriptions for Universal*

Relief (Pu Ji Fang) written in the Ming Dynasty, 1368–1644. In addition, Qing Hao was prescribed to treat the paroxysmal fever of malaria in *Compendium of Materia Medica* (Ben Cao Gang Mu) by Shizhen Li (1518–93) written in the Ming Dynasty, while it was used to cure chronic dysentery and SHAOYANG-NUE (a kind of malaria), respectively, in *Essentials of Materia Medica* (Ben Cao Bei Yao) and *Detailed Analysis of Epidemic Febrile Disease* (Wen Bing Tiao Bian) in the Qing Dynasty, 1644–1911. Qing Hao is a general name of a category of herbs in traditional Chinese medicine, but there are several types of herbs within this category (at least six types), and each contains different chemical components with various degrees of efficacy in treating malaria. Hong Ge's handbook did not specify which type of Qing Hao should be used for treating fever. Although there are different varieties of Qing Hao, yet only *A. annua* L. has effective antimalarial activity. Besides, the specie is also widely distributed and extensively utilized. Thus, *A. annua* L. is considered as the genuine herbal medicine. However, due to the uncertainties of appearance properties, quality control level, potency consistence, and methods of administration, there are obstacles for "original crude traditional Chinese medicine" to be accepted internationally. Therefore, the success of artemisinins is not only a major breakthrough of antimalarial medicine in the world, but also an expressive model in the process of modernization and internationalization of TCM.

This issue on artemisinin can spread scientific and technological knowledge and also provide people with some significant inspiration and motivation. First, TCM should be highly focused. During the Vietnam War (1955–75), a clandestine project was set out to develop antimalarial therapies since 1964 in China, in response to the North Vietnamese leader's request. In 1967, the 523 Office was established to organize dozens of institutes from seven provinces to tackle key research problems. It was not until January 21, 1969, that China's largest TCM research institute—Institute of Chinese Materia Medica of China Academy of Chinese Medical Sciences—was ordered to participate in the antimalarial research work. Believing in the great value of TCM, the group led by Youyou Tu bravely shouldered the responsibility and confronted the challenge with an innovative spirit of persistence and boldness. They made extractions from more than 200 prescriptions and 380 medicinal herbs. After going through a series of failures, in October 1971, they eventually obtained from *A. annua* L. an extract that inhibited malarial parasites by 100% rate, which was later purified into a novel compound that was named artemisinin. What a breakthrough of antimalaria therapies they achieved! We assume that, if there were no participation and multilateral collaboration of these medical research institutes, if they had not paid attention to *A Handbook of Prescriptions for Emergencies* (Zhou Hou Bei Ji Fang) and other TCM documents, it would be impossible to find artemisinin in such a short period and in such a poor research condition. Second, at present the drug export in China mainly depends on raw chemical material, which is not advisable to be a long-term choice. Based on the pharmaceutical research and development level in China and the advantage of TCM, to realize the modernization and internationalization of TCM is China's key route to meet the world's medical needs. To comply with demands at home and abroad, it is necessary not only to inherit "the traditional model," but also to develop "the modern model." Only in this way we are able to make a greater contribution to the world. Third, although Chinese have advantages in both rich resource of *A. annua* L. and related antimalarial applications, we still need to learn advanced international experience in many aspects to continuously promote social and economic benefits. These aspects could be considered carefully, such as intellectual property protection, large-scale cultivation of natural resources, enterprise reform, enterprise internationalization, long-term cooperation between research and industry, and so on.

Finally, in a very delighted mood, I extol the dear hope that artemisinins have brought to mankind. I wish this book could offer domestic and foreign colleagues some beneficial knowledge and inspiration.

Ruixiang Zhang

June 8, 2006

After more than 30 years, a belated report of artemisinins is finally going to be published. People may be familiar with artemisinin, but for the research history of artemisinin few people really know about it. Therefore, the detailed introduction of the process of artemisinins research in this issue is particularly remarkable and exciting. Now, the work is going to come out, isn't it a comforting and amazing event!

The author, Youyou Tu, is a lifelong chief professor at the China Academy of Chinese Medical Sciences and a chemist in TCM. In December 1930, she was born in Ningbo, Zhejiang province, a city on the east coast of China. In 1951, she was admitted to the Department of Pharmacy in Peking University (the department was once adjusted to known as the Department of Pharmacy of Beijing Medical College in 1952) which now is the School of Pharmaceutical Sciences of Peking University. Graduated in 1955, she was assigned to the Institute of Chinese Materia Medica of Academy of Traditional Chinese Medicine affiliated to the Ministry of Health (now is China Academy of Chinese Medical Sciences). In 1959, she attended the third phase of national TCM full-time courses for the ones having the western medicine background which is hosted by the Ministry of Health, and she took 2.5 years to learn the knowledge of TCM systematically. In more than 60 years of TCM research, Youyou Tu has been working in the first line, and has successively engaged in the pharmacognosy, processing, chemistry of Chinese Materia Medica, and the development of new TCM drugs. Hardworking as she is, she has won plenty of awards. In 1958, she won the youth socialist construction activist prize held by the ministry of health. And in 1977 she was selected as the advanced worker held by the Institute of Chinese Materia Medica of China Academy of Chinese Medical Sciences. In 1978, she attended the National Science Congress, and the artemisinin research team she led was awarded the National Science Congress award team. In

1979, the findings of antimalarial drug artemisinin won the second national invention award. In the same year she attended the national awards ceremony of March-Eighth Red-Banner Pacesetter and Red-Banner Groups and won the prize flag. In 1982, as the only representative of the project of artemisinin invention, she attended the national science and technology awards ceremony and received the certificates of invention and the invention medal as the first inventor. In 1984, she was awarded the title of "young and middle-aged expert with outstanding contributions" by the National Ministry of Personnel. Since 1990, she has received the first special allowances from government by the state council.

Her achievement of the "dihydroartemisinin and its tablets" won the national top 10 scientific and technological achievements in 1992. Then, the project was selected to be the "new China top ten health achievements" by the Ministry of Health. In the same year, "the study of the quality of the commonly used Chinese Materia Medica" was awarded the first prize of the national prize for the progress in science and technology, "the study of the species and the quality of *Artemisia annua* L." led by Youyou Tu was a part of the awarded project. Since 1992, she has been awarded the lifelong professor of China Academy of Chinese Medical Sciences. In 1995, she was selected to be the advanced worker by the state council. In 2002, she was awarded "the first new century female inventors" by the state intellectual property office. In conclusion, she has gained remarkable achievements for over 60 years, and has a very high attainment in academic credentials.

This book is based on the author's research experience, a large amount of experiment records, notes, and related information that have been cloaked for more than 30 years, and shows us the TCM malaria research, the discovery of artemisinin, and the whole process of the development and application of the artemisinins in detail. This book almost contains little magnificent language, and has no vigorous scene description, but once one starts reading the book carefully, it can be drawn that the work done by the scientists across China is thought-provoking and admirable.

During the Vietnam War, North Vietnamese leaders, suffering equally heavy losses of soldiers due to malaria as from warfare, asked the government of China to urgently find malaria remedies. In China, the military institutes started confidential antimalarial research in 1964. In 1967, Chinese leadership set up a national leading group office for malaria control (abbreviated as national 523 office) to coordinate nationwide research. Several thousands of compounds were screened between 1967 and 1969, but no useful medicines were found. In January 1969, two directors and another member from the national 523 office visited the Academy of Traditional Chinese Medicine and the Institute of Chinese Materia Medica, seeking help for searching novel remedies from TCM. After the mobilization, Youyou Tu was assigned to accept the task in this urgent time. Reviewing all the medical books in the ancient China systematically, she built the research team and screened more than 200 Chinese Materia Medica. The team finally found that the extract of the *A. annua* L. has

100% inhibition rate of the malaria parasites in October 1971, which means a breakthrough of treating malaria by using *A. annua* L. By purifying the extract, they got the antimalarial compound, artemisinin, in November 8, 1972. Behind the invention process of artemisinin, there is the wisdom of the researchers to solve questions of varying difficulty. And the fact proved that the invention of artemisinin is the pride of Chinese people. Youyou Tu is the main creator of artemisinin. The Institute of Chinese Materia Medica of China Academy of Chinese Medical Sciences is the main institute of the invention of artemisinin. Applications of artemisinin and its derivatives as antimalarial drugs are the results of the joint efforts from many organizations and plenty of researchers.

It is reported that the invention of artemisinin is one of the major breakthrough in the field of medicine and health care in the 20th century, and is a major original innovation of China and even the world. It should be said that it is worthy of such evaluation. For more than 40 years, artemisinins have made a great contribution to save thousands of lives by treating malaria. In the 7th Chinese International Conference on the Technology History, the Nobel Prize Laureate, Dr. Zhenning Yang, said that "such an invention, should be awarded internationally" after listening to the introduction of Youyou Tu and her findings from Heng Wu in 1996. Jeffrey D. Sachs, a professor in Columbia University, the director of United Nations Millennium Development Goals, and the special consultant of Kofi Annan (the former UN secretary general), made a report named "the investment for the development, a practical plan for realizing the Millennium Development Goals" in the United Nations Millennium Development Goals forum in 2003, Beijing. And after the meeting, he was interviewed by the People's Daily, saying that artemisinin, a new breakthrough in antimalarial technologies, was one of the most important scientific developments in the history of China. He noted that in every 30 s in Africa, there is a child who dies due to malaria, and each year there are about 3 million people who died of malaria in the world. Therefore, he suggested that China should add 8000 ha for the cultivation of *A. annua* L. and improve the extraction methods to increase the production of artemisinin, so as to save millions of lives. He stressed that "this will make historic contributions to the Millennium Development Goals."

In Africa, the new generation of antimalarial drug, dihydroartemisinin (trade name is Cotecxin), developed by Youyou Tu, was widely used in antimalarial treatments, and known as the miracle treatment. Some people even named her newborn child as "Cotecxin."

The finding of artemisinin brings people a new structure of antimalarial drug, which has solved the problem of the resistance against quinolines for the malaria treatment which has puzzled the field of medicine for a long time. And it also has saved the lives of patients with antichloroquine malaria patients infected by *Plasmodium falciparum*. This is a great contribution we have made to the global health.

TCM has a history of thousands of years, and has made an indelible contribution for the living and breeding of Chinese people. And it still plays a vital role for Chinese. Development and modernization of TCM should be given more attention. Chairman Mao gave an instruction that TCM is a great treasure house that should be explored and improved. How to discover? How to improve? The invention of artemisinin may give some ideas to the researchers. We need to use modern science and technology, and study TCM through a variety of ways and all kinds of methods. But at the same time, we should also pay great importance to the guidance of TCM theory and the rich TCM experience in clinical practice.

We feel grateful that Youyou Tu offered us this vital and readable work in her 70s! And also thank her and all colleagues in China for their outstanding contributions in the discovery, research, and clinical application of artemisinins! The older scientists' spirit of seeking truth from facts and working hard which can be persevered will certainly inspire us to work hard and make greater contributions to the global health care.

Institute of Chinese Materia Medica,
China Academy of Chinese Medical Sciences

May, 2008

Preface

For the malaria treatments, Chinese physicians traditionally used saxifrage herbal plants known as *dichodra* (Chang Shan); European used rubiaceae herbal plants known as cinchona bark in the 17th century. From those herbal medicines, people got febrifugines and cinchona alkaloids, respectively, which are nitrogen-containing heterocyclic natural products. The former is hard to popularize to use for its stronger side effects. In 1820, the latter was isolated and purified to get an effective antimalarial compound, quinine, which subsequently became widely used as antimalarial therapies. And by the synthesis and structure modification, people successively developed the pamaquine (plasmoquine), mepacrine (atabrine), pyrimethamine, primaquine, and so on, and especially chloroquine that came out in 1948 became the first choice of quinoline drugs.

In the early 1960s, the chloroquine-resistant *Plasmodium falciparum* strains that first emerged in South America quickly spread, and became a serious global problem of controlling, preventing, and treating malaria. To find malaria remedies, Chinese scientists had made a lot of structural modification of febrifugine but failed. The academy of military medical sciences in China developed benflumetol and naphthoquine, which are still quinoline drugs, and their capability of inducing drug resistance of malaria parasites has been confirmed, which means they are inappropriate to use alone. Therefore, national 523 office started to pay much attention on the natural medicines, especially TCM. Youyou Tu regarded the study of TCM antimalarial drugs by using modern scientific research as her own duty. She believed that there must be an essence which deserves inheritance and development in TCM that has 5000 years of history. She successfully led the team to find the modern antimalarial drug, artemisinin, which is excavated from Chinese herbal medicine *Artemisia annua* L., and to solve the problem of chloroquine-resistant malaria.

After the discovery of artemisinin, the team at the Institute of Chinese Materia Medica of China Academy of Chinese Medical Sciences continued to research the structure–activity relationship; the research results illustrated that the peroxide bridge in artemisinin is the main antimalarial activity group. On the basis of peroxide bridge, it will enhance the antimalarial activity many times after changing the C_{12} carbonyl group to hydroxyl group or bringing acetyl group to the chemical structure. In 1975, at the national internal conference on antimalarial drugs, the progress of artemisinin's structure–activity relationship research result was reported in public, which has led the further studies on the derivative across the nation. In 1976 and 1977, Guangxi Guilin pharmaceutical factory and Shanghai Institute of Materia Medica appointed some scientists to develop artemisinin derivatives, respectively, and in 1987, they developed artesunate and artemether. In 1992, the research team at the academy of military medical sciences in China developed coartem (artemether + lumefantrine) and the combination of naphthoquine and artemisinin. Mefloquine was once combined with artesunate, even some antimalarial drugs that had developed the parasites resistance, such as piperaquine, also combined with artemisinins to be reused, and that is also some contribution of the artemisinins to promote the new therapies. After the finding of artemisinin, the antimalarial drugs team at Institute of Chinese Materia Medica of China Academy of Chinese Medical Sciences recreated dihydroartemisinin and its tablets as antimalarial medicine. Dihydroartemisinin was discovered in 1973 and its tablet was named as cotecxin in 1992. The antimalarial efficacy of cotecxin is 10 times higher than artemisinin, and the recrudescence rate is lower to 1.95%, which makes it won the "national top ten science and technology achievements award." During the period of the 7th Five Year Plan of China, Youyou Tu's team finished the research of "the study of the species and the quality of *A. annua* L." as a part project of "the study of the quality of the commonly used Chinese Materia Medica" which was awarded the first prize of the national prize for the progress in science and technology. And successively Youyou Tu's team achieved six new drug certificates, one new drug clinical research approval documents, and two Chinese invention patents. The name lists are as follows:

	Items	Approval Number	Issued Date
1	Artemisinin bulk drug	(86) New drug certificate X-01	October 1986
2	Artemisinin suppository	(86) New drug certificate X-04	October 1986
3	Artemisinin tablet	New drug H20030144	June 2003
4	Dihydroartemisinin bulk drug	(92) New drug certificate X-66	July 1992
5	Dihydroartemisinin tablet	(92) New drug certificate X-67	July 1992
6	Dihydroartemisinin suppository	New drug certificate H20030341	June 2003
7	Dihydroartemisinin tablet (indication of lupus erythematosus)	Drug clinical research certificate 2004L02089 and 2016L02562	June 2004 February 2016
8	Dihydroartemisinin tablet (indication of lupus erythematosus)	Invention patent Zl99103346.9	July 2003
9	The antimalarial treatment using dihydroartemisinin and pyronaridine	Invention patent ZL99109669.X	February 2004

Ever since artemisinin and its derivatives came into being, which have played a key role in saving people from malaria. Owing to various reasons, the related research conducted at the Institute of Chinese Materia Medica of China Academy of Chinese Medical Sciences has not published systematically. WHO has begun to recommend artemisinin-based combination therapies for global malaria control in 2001. Artemisinin and its derivatives are extremely concerned once again in recent years. There were lots of interviewers from domestic and foreign visited the Institute of Chinese Materia Medica of China Academy of Chinese Medical Sciences, and they were particularly interested in the topic of the invention history of artemisinin. It also reflects the attention and the demand of those related personnel.

Since the advent of quinine in 1820, after almost 180 years of looking and searching, researchers desired to get a novel chemical entity as antimalarial drugs to solve the problem of the drug resistance malaria, which caused the millions of deaths annually. *A. annua* L., the ordinary and humble grass, which is widely dispersed throughout the temperate region in China, shocked the world by the finding of artemisinin. This fully shows that the TCM is a great treasure house; it both has the systematic theory and valuable medical practice, and contains up to 10,000 kinds of Chinese herbal medicine. In terms of "Qing Hao can treat malaria," it has been recorded more than 1000 years ago; however, researchers failed to find its antimalarial essence timely, which has caused artemisinin cannot become a modern antimalarial therapy to save millions of people died of malaria for many years, and it is how much regretful. Authors hope that introducing the discovery history of artemisinin can arouse people to innovate, thus can find much more essence of TCM. Artemisinin against human malaria has proven effective and rapid, which proves that the TCM treatment can directly kill pathogenic factor quickly, in addition to adjusting the body's Yin and Yang imbalance. And so, the TCM research field to explore is very broad and extensive.

The application of artemisinins' indication is still expanding. Except for the parasitic diseases, it also has the effect of immune regulation to treat immunity disease, even cancer. An innovative drug has a strong vitality.

It is hoped that reader will find this book useful and helpful.

The Discovery History of Artemisinin

Malaria has been a life-threatening disease in the human history. Even though tremendous progress has been made in the prevention and treatment of the disease, it remains an acute public health problem in many regions. In 2015 alone, the most recent year for which data are available, around 2 billion people in more than 50 countries with an annual incidence of approximately 200 million and nearly 400,000 of those patients succumbed to their infections.

1 Discovery of Antimalarial Effect of Artemisinin

Malaria was effectively treated and controlled by chloroquine and quinolines for a long period of time until the development of drug-resistant malaria in the late 1960s following a catastrophic failure in a global attempt in eradicating malaria. Resurgence of malaria and rapidly increased mortality due to loss of effective treatment presented a serious global challenge, in particular, in the regions with the prevalence of malaria associated with the drug-resistant *Plasmodium* parasites, especially *Plasmodium falciparum*. Southern East Asia was one of the most severe endemic areas in the late 1960s.

During the Vietnam War, malaria caused by chloroquine-resistant *P. falciparum* was a major problem that spurred research efforts on both sides of the battlefield. As reported, the casualty to the US military force caused by medical disability due to the full seasonal prevalence of malaria reached 4–5 times higher than the actual casualty from the direct combat in 1964. Malaria infected nearly half of the total military individuals or around 500,000 US troops in 1965. Fighting against malaria became one of the top medical priorities, as well as challenges for the US army in Vietnam. A program coordinated through the Division of Experimental Therapeutics at the Walter Reed Army Institute of Research (WRAIR) in Washington, DC, was launched in searching for new antimalarial drugs. The program involved numerous research institutes and a vast investment. Up to 1972, over 214,000 compounds were screened by the Walter Reed Army Institute of Research which, however, ended up with no breakthrough findings and discovery of novel antimalarial medicines. Vietnamese, suffering equally heavy losses of soldiers due to malaria as from warfare, lacked a research infrastructure and thus turned to China for help. North Vietnamese leaders asked the government of China to urgently find malaria remedies. Chairman Mao and Premier Zhou called for an urgent effort to find solutions. Research on novel antimalarial medicines became an important political assignment for the medical researchers in the Chinese army. In China, the military institutes started confidential antimalarial research in 1964. In 1967, Chinese leadership set up a national leading group office for malaria control (abbreviated as national 523 office) to organize and coordinate antimalarial drug research activities of dozens of joint units in seven provinces and cities across the country. For the confidential purpose, the project was named by the date of 23rd May when the project was initiated. Several thousands of compounds and traditional Chinese medicine (TCM) drugs were screened between 1967 and 1969, but no satisfactory therapies were identified.

The outlook is serious. To overcome the difficulties of antimalarial drug research in China, on January 21, 1969, three representatives (Bingqiu Bai, Jianfang Zhang, and Xin Tian) from the national 523 office, two directors and one member, visited academy of TCM affiliated to the Ministry of Health (now renamed as China Academy of Chinese Medical Sciences, abbreviated as CACMS), seeking help for searching new antimalarial therapies from TCM. A reception and consultative meeting was convened by the vice-president of CACMS, Henian Gao, and the deputy director of the Institute of Chinese Materia Medica of CACMS (abbreviated as ICMM), Guozhen Zhang. The representatives of national 523 office introduced the research background of antimalarial drugs, which was concerned closely by the Central Government of China. "A lot of work about TCM antimalarial research has been conducted, but the promising outcome is still awaited due to little knowledge and experience in TCM study. National 523 office hope the scientists at CACMS can participate in the secret antimalarial research" was stated at the meeting. Both the leaders of CACMS and ICMM noted, "although it is still during the Cultural Revolution, all research work at CACMS mostly stalled, the greatest efforts will be made to undertake the covert mission."

Youyou Tu was appointed by the leadership teams both at CACMS and ICMM to build and head the project 523 research group at ICMM (abbreviated as 523 group at ICMM), responsible for the overall work. The reason she was appointed as the head of the group was partially due to the fact that she received training on both modern pharmaceutical sciences and traditional medicine and due to her hard work.

Youyou Tu started by collecting and combing information on the relevant TCM antimalarial practice. Within 3 months, she gathered over 2000 herbal, animal, and mineral medicine prescriptions for either internal or external uses through reviewing ancient TCM literatures, folk recipes, and interviewing those well-known experienced Chinese medical doctors, such as Fuzhou Pu and Meizhong Yue, for potential prescriptions and herbal recipes. She then narrowed down and compiled the prescriptions from 2000 to 640 and summarized the recipes in a brochure named *Antimalarial Collections of Recipes and Prescriptions* (Kang Nue Dan

Mi Yan Fang Ji). It was the information collection and deciphering that laid sound foundation for the discovery of artemisinin. This also differentiates the approaches taken by TCM and general photochemistry in searching for novel drugs. In April 1969, she circulated copies of the brochure to other research groups outside the institute for reference through the national project 523 office. And then, she started with experiments on dichroine using the animal models. The study was soon given up due to its severe side effects.

From May 1969, the 523 group at ICMM tested aqueous and ethanol extracts of over 100 herbal medicines in a rodent malaria model with few promising results found up to June 1971, which they then assessed to see whether the extracts could clear *Plasmodium* parasites from the bloodstream of infected mice. It was a very laborious and tedious job, in particular when they faced on failure and another. The screening result of pepper extract was the best, with 84% inhibition rate of *Plasmodium* parasites. The extract alleviated the symptoms of fever, but was unable to clear parasites apparently in clinical trials. Among the few promising screening results, extracts from Qing Hao were shown to inhibit parasite growth by 68%. Follow-up studies, however, only achieved 12%–40% inhibition. This puzzled members of the 523 group at ICMM. Qing Hao would be a potential candidate only if the original promising observations (above the 68% inhibition rate) could be repeated.

Malaria was one of the epidemic diseases with the most comprehensive records in traditional Chinese medical literature. For example, *Prescriptions for Universal Relief* (Pu Ji Fang), one of the most comprehensive Chinese medicine prescription texts in the Ming Dynasty, contained at least four chapters entitled “Zhu Nue Men” on malaria. Qing Hao is a traditional Chinese Herbal medicine and was first recorded in *Prescriptions for Fifty-two Diseases* (Wu Shi Er Bing Fang) written on a piece of silk treatise unearthed from the Mawangdui Han Dynasty tombs (168 BC), described it as a treatment for hemorrhoids. Later, *Shen Nong's Herbal Classic* (Shen Nong Ben Cao Jing), the oldest herbal classic existed in China, named it Cao Hao and also as Qing Hao. As *A Handbook of Prescriptions for Emergencies* (Zhou Hou Bei Ji Fang) described Qing Hao as an antimalarial herbal medicine, various related prescriptions were recorded. Qing Hao was also prescribed to treat the paroxysmal fever of malaria in *Compendium of Materia Medica* (Ben Cao Gang Mu), one of the most famous and comprehensive herbal classic in China, by Shizhen Li (1518–93) written in the Ming Dynasty. No doubt, clinical practice in alleviating malaria symptoms utilizing Qing Hao was true and consistent; although countless ways of treatments were tested over the thousands of years, success was occasional and failure was common.

It was the most challenging process. The 523 group at ICMM started to focus on herb Qing Hao and work hard in multiple attempts, but still received no promising results. Progress was not smooth, and no significant results emerged easily. After the confirmatory experiments appeared to be contradictory to what was recorded in the literature, Youyou Tu started reviewing the literature again. Fortunately, she found one sentence in the Chinese medical archive that caught her attention. In reading Hong Ge's *A Handbook of Prescriptions for Emergencies*, she further digested the sentence "a handful of Qing Hao immersed in two liters of water, wring out the juice and drink it all" when Qing Hao was mentioned for alleviating the malaria symptoms. This reminded her that the heating might need to be avoided during extraction; therefore the method was modified by using the solvent with a low boiling point. In September 1971, a modified procedure was designed to reduce the extraction temperature by immersing or distilling Qing Hao using ethyl ether. The obtained extracts were then treated with an alkaline solution to retain the neutral portion by removing acidic impurities.

In the experiments carried out on October 4, 1971, sample No. 191, that is, the neutral portion of the Qinghao ethyl ether extract, was found 100% effective on rodent malaria when administered orally at a dose of 1.0 g/kg for 3 consecutive days. The same results were observed when tested in malaria-infected monkeys between December 1971 and January 1972. This breakthrough finding became a critical step in the discovery of artemisinin. In the same studies, extracts from Huang Yao Zi (Dioscoreae bulbiferae Tuber, *Dioscorea bulbifera* L.), Ya Dan Zi (Bruceae Fructus, *Brucea javanica* (L.) Merr.), Xian He Cao (Agrimoniae Herba, *Agrimonia pilosa* Ledeb), Tu Fu Ling (Smilacis glabrae Rhizoma, *Smilax glabra* Roxb.), and extract of Qing Hao using other solvents were also tested with negative or no comparable results.

Youyou Tu reported the novel findings at the nationwide Project 523 meeting held in Nanjing on March 8, 1972, saying that "We have screened over a hundred types of single and combination herbal recipes since July 1971 and found that Qing Hao ethyl ether extract showed 95%–100% inhibition of rodent malaria. We performed further purification to retain the effective neutral portion by removing the noneffective toxic acidic portion. We observed the same efficacy when we tested the Qing Hao ether extract and the neutral portion on the monkey malaria model in late December." This report attracted overwhelming interests and triggered nationwide concern about Qing Hao and the extracts.

The 523 group at ICMM subsequently carried out a clinical trial (21 cases) between August and October 1972 in Changjiang County (Hainan province, an island off the southern coast of China) and simultaneously at 302 hospital (9 cases) in Beijing. This was the first time the neutral portion of ethyl ether Qing Hao extract was tested in humans. In the trial carried out in Hainan province, all the patients quickly returned to normal temperature from 40°C. A total of 21 local and migrant malaria patients, 9 infected by *P. falciparum* and 11 infected by *Plasmodium vivax*, were treated in three dose groups and all of them recovered from the fevers with full clearance of malaria parasites. All nine patients were successfully treated at Beijing 302 hospital. The results from the first clinical trial in Hainan and Beijing were reported in the nationwide project 523 meeting held in Beijing in November 1972. The national office for malaria control issued a communication on malaria control research recording the clinical findings on November 5, 1972. It was confirmed "in the expedited clinical trial on the 21 cases of local and migrant malaria patients in August, the Qing Hao extract from the Beijing (research) district showed relatively good efficacy (over 90%) against *Plasmodium vivax* and *Plasmodium falciparum*. This is a promising antimalarial drug with potential for further improvement." Her achievement and experience had inspired other researchers. Proving the efficacy of the neutral portion of ethyl ether Qing Hao extract in the experiments on rodent and monkey malaria models in October 1971 and the subsequent clinical trial between August and October 1972 triggered and steered nationwide antimalarial research on Qing Hao for malaria treatment.

The next step was to isolate the active ingredients, as at the time, the antimalarial active substance had not yet been isolated in its pure form, nor was its structure known. For this purpose, polyamide was first added into this concentrated extract, being shaken by votexing, before percolated with 47% alcohol. The diluted alcohol percolated extract was then subject to another round of ethyl ether extraction, with the extracted components being separated by silica gel column chromatography. The fraction eluted with 10% acetyl ester from the silicon column was able to form needle-like crystals. On November 8, 1972, the 523 group at ICMM isolated a colorless, crystalline substance. Lately, the compound was identified with a melting point of 156–157°C, and exhibited the full antimalaria activity when tested with mice of malaria infected by *Plasmodium berghei* in 50–100 mg/kg dose regimes, while those eluted with other petroleum ether or 15% acetyl ester did not show any antimalaria activity. The active compound was named as Qinghaosu (it means the "active principle" of *Artemisia annua* L.; *qinghao* is the TCM name of *A. annua* L.; *su* means "basic element" in Chinese; now artemisinin is the popular name).

At the beginning of 1973, the 523 group at ICMM received multiple letters from other institutes requesting that they further clarify and share information on the findings and experience in the Qing Hao antimalarial research field, to which they responded with thorough explanations. And then, the colleagues from Shandong Provincial Institute of Chinese Medicine and Yunnan Provincial Institute of Materia Medica, respectively, conducted antimalarial research on different Qing Hao resources. In 1973, both the institutes sent letters to ICMM that expressed gratitude for sharing information on the Qing Hao antimalarial research.

The first formulation the 523 group at ICMM tested in patients was tablets, which yielded unsatisfactory results. The first human trial of eight cases using this preparation did not achieve expected results, with only three cases being cured after 3 weeks and two cases with some side effects. Then, the group found out in subsequent work that this was due to the poor disintegration of an inappropriately formulated tablet produced in an old compressing machine. The 523 group at ICMM shifted to a new preparation (a capsule of raw powder of pure artemisinin without any excipients). To conduct clinical observation once again, Guozhen Zhang, the deputy director of ICMM, brought the new preparation to established field site in malaria endemic areas (in Changjiang County, Hainan province). All trial of three patients achieved satisfactory clinical efficacy. Patients received a total dose of 3–3.5 g oral artemisinin capsule, and recovered quickly after taking the medicine. The average parasite clearance time was 18.5 h, and the average fever clearance time was 30 h. The road leading toward the creation of a new antimalarial drug opened again. In October 1973, the 523 group at ICMM returned to Beijing from Changjiang County and timely reported the clinical studies to the national 523 office. On November 2, 1973, the national 523 office sent an official letter to CACMS requesting for convening a relevant malaria prevention and treatment conference to discuss how to conduct the future studies of the new antimalarial treatment based on Qing Hao. According to the requirement, Youyou Tu presented the comprehensive research progress with Qing Hao, as well as artemisinin.

Between February 28 and March 1, 1974, an antimalarial treatment collaborative meeting was convened at ICMM to share and exchange information obtained and to plan for future work. The critical information of research progress and the laboratories at ICMM was introduced in detail to the colleagues across the nation. After the meeting, much more teams conducted researches on Qing Hao, and the national cooperation group of antimalarial research teams was expanded. Until 1978, a total of 529 malaria cases clinical trials were conducted by CACMS. According to the outward information of the artemisinin achievements assessment meeting held in 1978, over 40 units from 10 provinces, cities, and municipalities participated in the clinical trials. Between 1972 and 1978, taking crude Qing Hao extracts or artemisinin, 6555 malaria patients were tested from all over China covering all types of malaria (some of whom had chloroquine-resisitant or brain malaria), all were effective except about 10 people who died due to other complications. Among them, 2099 cases were tested with artemisinin.

According to the requirements of new drug approval, the 523 group at ICMM completed development activities, application, and received “New Drug Certificate for Artemisinin” [(86)New drug certificate X-01] from the Chinese Ministry of Health in 1986, which was the first modern “New Drug Certificate” in China since 1949.

It was not a simple and easy journey in the discovery of the artemisinin from Qing Hao, a Chinese herbal medicine with over 2000 years of clinical application, especially in the 1970s when research was significantly underresourced in China. Although Qing Hao was widely documented in TCM literature, however, few details were given on either the species or effective parts of the herbal medicine when clinical application was mentioned. According to plant taxonomy, there are at least six species in the *Artemisia* family: *A. annua* L., *Artemisia apiacea* Hance (*Artemisia carvifolia*), *Artemisia scoparia* Waldst. et kit., *Artemisia capillaris* Thunb., *Artemisia japonica* Thunb., and *Artemisia eriopoda* Bunge. However, no clear classification was given for the Qing Hao (it means a category of green herbs and is the general name of the Artemisia family in Chinese) regardless of numerous mentions of the name Qing Hao in the literature, nor did the texts specify the effective parts of the plant. All the species in Qing Hao (Artemisia) family were used. By the time that research on artemisinin was carried out, two Qing Hao (Artemisia) species were listed in the Chinese Pharmacopoeia and four others were also being prescribed. Youyou Tu's studies confirmed that only *A. annua* L. (sweet wormwood) contains meaningful quantity of artemisinin. They subsequently carried out a thorough study on the herbal medicine Qing Hao. In addition to the confusion in finding the right plant, variables, such as the part and origin of the plant, its harvest season, low artemisinin content in the plant, extraction and purification process, and so on added extra difficult in the discovery of artemisinin. All the factors mentioned earlier are integrated, make the probability of the discovery of artemisinin small and small, and the discovery for malaria therapy by Youyou Tu and her colleagues represents an outstanding contribution to human health. As a result, they demonstrated that the effective components are mainly found in the leaves of mature plants and can be effectively extracted with ethyl ether or alcohol as solvent, but have to operate at a temperature below 60°C. Qing Hao extract, only when prepared in a proper way, was effective in treating malaria; success in identifying effectiveness of neutral ethyl ether extract of *A. annua* L. was not a simple and easy win.

2 Structural Identification and Structure–Activity Relationship Research of Artemisinin and Derivatives

The isolation of the crystal of artemisinin allowed further investigation of its chemical structure. The 523 group at ICMM proceed to determine the chemical structure of artemisinin in December 1972 through elemental analysis, spectrophotometry, mass spectrum, polarimetric analysis, and other techniques. First, elemental analysis showed that structure of artemisinin only contains carbon, hydrogen, oxygen, and no nitrogen atoms which previous quinoline antimalarials contained. The compound was further purified with different recrystallization processes and tested at the Department of Analytical Chemistry of the Institute of Materia Medica, China Academy of Medical Sciences. Based on the elemental analysis and results from other studies, colleagues at the Institute of Material Medica verified that the compound contained no nitrogen and had a potential formula of $C_{15}H_{22}O_5$ on April 27, 1973. And then, knowing from the literature that Professor Zhujin Liu of the Shanghai Institute of Organic Chemistry of Chinese Academy of Sciences had more experience in the research of sesquiterpenes natural products, they started collaboration with the Shanghai Institute of Organic Chemistry and the Institute of Biophysics of Chinese Academy of Sciences on artemisinin chemical structure analysis in 1974. Through comprehensive chemical and physical analyses, it was demonstrated that artemisinin is a new type of sesquiterpene lactone. The stereo-structure was finally determined by X-ray crystallography in 1975, which verified that artemisinin was a new sesquiterpene lactone containing a peroxyl group. The work was one of the first applications reported in China in determining an absolute molecular configuration utilizing the scattering effects of oxygen atoms by X-ray diffraction technique.

The stereo-structure of artemisinin was published in 1977 and cited by *Chemical Abstracts*. The publication application was reported to the Chinese Ministry of Health. And after approval, the paper was submitted to Kexue Tongbao on February 20, 1976. Then the paper (A new sesquiterpene lactone-Qinghaosu) was published in issue No. 3 in 1977. Considering the confidentiality requirement of Chinese military force missions, to prevent further development of medical use by other countries, the paper was published in the name of "the Collaborating Group of Qinghaosu Structure." The paper only reported the chemical structure and partial physical and chemical properties of artemisinin, but its excellent antimalaria activity was not reported. After the publication, the related contents have been indexed by *Chemical Abstracts* (C.A. 1977, 87, 98788g), which promoted the process that artemisinin was concerned by international peers. Structural identification was coordinated with scientists in the Shanghai Institute of Organic Chemistry and the Institute of Biophysics of Chinese Academy of Sciences and the 523 group at ICMM played a role in determining the structure using the crystal they obtained.

In 1973, to determine the functional groups in the artemisinin molecule, the 523 group at ICMM chemically modified the peroxyl bridge and carboxyl groups of the molecule. They produced deoxyartemisinin through reduction of the peroxyl group to an epoxy group by subjecting the artemisinin in the palladium and calcium carbonate methanol solution under room temperature and pressure, and then treating it using an acetone and *n*-hexane mixture. They synthesized dihyaroartemisinin by reducing the carboxyl group to a hydroxyl group using sodium borohydride ($NaBH_4$), to confirm that the chemical structure of the active substance had a carbonyl ketone group, and determined the molecular formula of dihydroartemisinin is $C_{15}H_{24}O_5$. They were unaware that the chemical substance they had produced would later be found to be more effective than the natural compound. Dihydroartemisinin was further reduced to dihydro-deoxyartemisinin by reacting in a palladium and calcium carbonate methanol solution. Some new compounds were obtained by derivatizing through the hydroxyl group of dihydroartemisinin.

The structure–activity relationship research showed the effective doses and observation in clearance of malaria parasites when the structure-modified compounds were administered. The results showed that the dose was reduced from 50 to 100 mg/kg/day for artemisinin to 12.5 and 6 mg/kg/day for dihydroartemisinin and acetate of dihydroartemisinin. The dose was similar between deoxyartemisinin and artemisinin. However, deoxyartemisinin was unable to clear malaria parasites. This study verified that the peroxyl group in the artemisnin molecule was critical for its antimalarial function, while reducing the carboxyl ketone group to hydroxyl group improved the efficacy, as well as allowed derivatization of artemisinin to form new compounds. This led to the development of dihydroartemisinin and other derivatives, such as artemether, artesunate, and arteether into new antimalarial drugs. Up to now, no clinical application has been reported with other artemisinin derivatives except for the aforementioned factors.

In 1975, the national 523 office convened the national antimalarial cooperation conference in Henan province. The 523 group at ICMM reported the results of structure–activity relationship research of artemisinin at the meeting. In 1976, researches of derivatives of artemisinin were fully carried out, and other units/researchers had synthesized and developed artesunate and artemether as antimalarial drugs.

Since 1986, Youyou Tu's team carried out a series of development activities on the chemistry, pharmacology, pharmacokinetics, and stability of dihydroartemisinin, which was created in 1973, and performed clinical trials according to regulatory requirements. The Chinese Ministry of Health granted "New Drug Certificate for Dihydroartemisinin" [(92)New drug certificate X-66, trademarked as Cotecxin] to ICMM in 1992. Dihydroartemisinin is 10 times more potent than artemisinin clinically, again demonstrating the "high efficacy, rapid action and low toxicity" of the drugs in the artemisinins category. More importantly, there was much less disease recurrence during treatment with this derivative. And successively Youyou Tu's team achieved three new drug certificates and two Chinese invention patents for the treatment of malaria and other diseases.

3 Artemisinin—A Gift From Traditional Chinese Medicine to the World

Artemisinin represents a new class of antimalarial drugs. Artemisinin and its derivatives, first discovered, tested, and marketed in China, produce more resolution of fever and parasitaemia than all known antimalarial agents. Due to these remarkable properties, there were many concerns among the Chinese scientists regarding the most effective way to transform an ancient Chinese method of using the herbal medicine Qing Hao into artemisinin.

Research conditions were relatively poor in China in the 1970s. To produce sufficient quantities of Qing Hao extract for clinical trials, the 523 group at ICMM carried out extraction using several household water vats. Some team members' health deteriorated due to long-term exposure to large quantities of organic solvents and insufficient ventilation equipment. To launch clinical trials sooner while not compromising patient safety, based on the limited safety data from the animal study, the team members and Youyou Tu volunteered to take Qing Hao extract themselves to assure its safety. In 1973, unsatisfactory results were observed in the clinical trial using artemisinin tablets, so the team carried out a thorough investigation and verified poor disintegration of the tablets as the root cause, which allowed them to quickly resume the trial using capsules and confirmed artemisinin's clinical efficacy in time.

In the unlikely atmosphere of the Culture Revolution in China, few scientific papers were published. With the change of social environment, the research results of *A. annua* L. and artemisinin were further arranged and issued. In November 1978, based on the early preliminary meeting (Xiangshan, Beijing), the national 523 office held a conference about artemisinin assessment in Yangzhou, Jiangsu Province. Among early-stage artemisinin-related conferences, this conference first invited media representatives, promoting the publicity of the related achievement. After the conference, all relevant units and researchers published the related research achievement to make the research progress of artemisinin against malaria known to the public.

In October 1981, in the 4th meeting of the scientific working group on the chemotherapy of malaria, an international conference in the name of "special conference on qinghaosu" was held by the United Nations Development Programme/the World Bank/the World Health Organization (WHO) special program for research and training in malaria chemotherapy group of tropical diseases in Beijing. The Chinese artemisinin research and development group made a systematic introduction of the related research at the meeting. A series of presentations on artemisinin and its clinical application, including Youyou Tu's report "Studies on the Chemistry of Qinghaosu," received positive and enthusiastic responses. Related papers of Chinese specialists were published at the beginning of 1982 in the name of "China cooperative research group on qinghaosu and its derivatives as antimalarials" in *Journal of Traditional Chinese Medicine*, which included six papers: "Chemical studies on qinghaosu (artemisinine)," "The chemistry and synthesis of qinghaosu derivatives," "Antimalarial efficacy and mode of action of qinghaosu and its derivatives in experimental models," "Metabolism and pharmacokinetics of qinghaosu and its derivatives," "Studies on the toxicity of qinghaosu and its derivatives," and "Clinical studies on the treatment of malaria with qinghaosu and its derivatives."

Artemisinins represent a new class of antimalarial drugs, which lead to two paradigm shifts in antimalarial research and therapy. The first one is the change from quinoline-based antimalarial drugs to artemisinin-based therapies due to the emergence of parasites resistant to quinolone drugs. Currently, artemisinin-based combination therapies are the drugs recommended by WHO for treating the deadly *P. falciparum* infections and are being used worldwide. The second paradigm shift is the change of research direction in antimalarial drug development. Artemisinin and its derivatives are in a new class of antimalarial agents with unique structure (1,2,4-trioxane peroxide pharmacophore), which has become a new direction for antimalarial drug development.

There is no doubt that the new birth of these drugs with their novel structure, activated much research in this field and related subjects, leading to more scientific discovery and application. Since 1972, artemisinin has been extracted from *A. annua* L., which quickly evoked worldwide research interest.

"Traditional Chinese medicine is a great treasure-house. We should explore it and raise it to a higher level." Artemisinin was explored using this resource. From the research experience in discovering artemisinin, strengths can be learned by both Chinese and Western medicine scientists. There is great potential for future advances if these strengths can be fully integrated. A substantial amount of TCM resources can be developed as modern medicines in the future. Since "Tasting a hundred herbs by Shen Nong," Chinese have accumulated substantial experience in clinical practice, integrated, and summarized the medical application of most nature resources over the past several thousand years. Adapting, exploring, developing, and advancing these practices would allow researchers to discover more novel medicines beneficial to global health care.

(Supported by the Institute of Chinese Materia Medica of China Academy of Chinese Medical Sciences)

PART 1

Artemisia annua L.

CHAPTER 1

Introduction

Chapter Outline

Artemisia annua L. is commonly used in traditional Chinese medicine (TCM) and it has a history of more than 2000 years. *A. annua* L. was firstly recorded in *Prescriptions for Fifty-two Diseases* (**Wu Shi Er Bing Fang**) written on a piece of silk treatise unearthed from the Mawangdui Han Dynasty tombs (168 BC), described it as a treatment for hemorrhoids. Later, *Shen Nong's Herbal Classic* (**Shen Nong Ben Cao Jing**), *Grand Materia Medica* (**Da guan Ben Cao**) and *Compendium of Materia Medica* (**Ben Cao Gang Mu**), etc. all shared the involvement of *A. annua* L. as a herbal medicine. And it is in *A Handbook of Prescriptions for Emergencies* (**Zhou Hou Bei Ji Fang**) by Hong Ge, a physician of the Eastern Jin Dynasty, that *A. annua* L. was first recorded as a herbal medicine that could cure malaria by "administering the juice wrung out of fresh *A. annua* L."

In 1971, the 523 group at ICMM, headed by Youyou Tu, isolated from the extract of *A. annua* L. an effective part that turned out to have a 100% inhibition of malarial parasite in mice. Later, a novel structured antimalarial monomer compound was separated from that part and was named Qinghaosu (*qinghao* is the TCM name of *A. annua* L.; *su* means "basic element" in Chinese; now artemisinin is the popular name). The Discovery of artemisinin was groundbreaking in history of antimalaria agents, which provides artemisinin with worldwide recognition as China's first innovative medicine and also helps China build a solid foundation for further research on artemisinins. Consequently, attention from various related fields was attracted to this herbal medicine. In this part, the authors are going to introduce knowledge about *A. annua* L. in aspects such as herbal medicine textual study, plant taxonomy, original plant resources, manufacturing techniques, chemistry, pharmacology, clinical application, and others.

The success of artemisinin has brought opportunities to plant resources development. *A. annua* L. is a kind of annual herb, growing up in every spring. In some areas of China, the natural resources are very rich and local people benefit profitably from *A. annua* L., which can be considered as a major content of Technology-Alleviate-Poverty Project. WHO strongly recommended artemisinin-based drugs as the frontline antimalarial treatment for all *falciparum* malaria. And so, the demand for *A. annua* L. in international

From *Artemisia annua* L. to Artemisinins
http://dx.doi.org/10.1016/B978-0-12-811655-5.00001-5

market is dramatically increasing. Natural resources are now far from enough. Therefore, *A. annua* L. becomes an economic plant in some areas that yield high quality ones and is cultivated in large scale in large number of fields. However, many problems still existed to be solved, such as the occupancy of crop fields. Besides, the quantity of artemisinin content in crude herbal medicine especially needs to be improved. Therefore, scientists have carried out bioengineering research to resolve the resource problem. For this reason, Authors specially invited experts from the Institute of Botany of Chinese Academy of Science to give an overview on this field in the final two chapters of this part.

1 *The History of* Artemisia annua *L. and Its Antimalarial Efficacy*

Formed during the Western Zhou Dynasty to the Spring and Autumn (around 1100 BC to 600 BC), **Shi Jing** (the *Chinese Book of Odes*) records that "**You You Lu Ming, Shi Ye Zhi Hao**" [1], which translated as "'Youyou' the deer bleating; wild Hao they are eating." This shows that the name of Qing Hao has already been recorded, indicating that the herb shares the same long history with China. **Yanzi** (578 BC to 500 BC) once explained "Hao (note: the Chinese word for the plant of *Artemisia* species) is the plant taller than normal grass" [2]. And it was also recorded in *Er Ya* (the book synthesized natural study of Chinese national culture in the Pre-Qin Dynasty Period) that "Hao (蒿) is also known as Qin (鼓) [3], which is consistent with the illumination by **Shuo Wen Jie Zi** (*Origin of Chinese Characters*) that "Qin (鼓) refers to Hao (蒿) [4]. Then, the relation between Hao (蒿) and Qin (鼓) was further clarified in later times. In the period of the Three Kingdoms, Ji Lu pointed it out that "Hao (蒿) specially means Qing Hao (青蒿) (the Chinese word for several plants of *Artemisia* species including *A. annua* L.), which was called as Qin (鼓) in the border of **Jing** and **Yu**, in the area of **Ru'nan** and **Ru'yin**." And Yan Sun also said that "at the border of **Jing** and **Chu**, Hao (蒿) is called as Qin (鼓). Thus Qin (鼓) is Hao (蒿), Hao (蒿) means Qing Hao (青蒿) and Qing Hao (青蒿) means Cao Hao (草蒿) [5]. Xi Zhu (the southern song dynasty poet, philosopher, and educator) once annotated for the *Chinese Book of Odes* that "Hao (蒿) refers to Qing Hao (青蒿). The mentioned location such as **Jing**, **Yu**, **Ru'nan**, **Ru'yin**, and **Chu** nowadays belong to Hunan, Henan, and Hubei Province in China. Besides, in 1973, in Changsha of Hunan Province, a silk manuscript treatise *Prescriptions for Fifty-two Diseases* (**Wu Shi Er Bing Fang**) was unearthed from Mawangdui Han Tomb III (168 BC) and is thought to be the oldest surviving Chinese medical book. In the book, in the *Male Haemorrhoids Item* it reads that "Qin (鼓) is the alternative name of Qing Hao (青蒿) in the area of **Jing**," clearly telling the ancient name of several plants of *Artemisia* species used in the areas now located in Hunan and Hubei province in China. The word "Qin" (鼓) in this book was originally missing, but managers put on the word "Di" (荻) instead. According to the original text of *Er Ya* and the commentaries by Lu and Sun, now we can surely infer that it is Qin (鼓) instead of Di (荻).

Shen Nong's Herbal Classic (**Shen Nong Ben Cao Jing**, also translated as *Canon of the Divine Husbandman's Materia Medica*) [6] says that "Cao Hao is another name of Qing Hao." Hongjing Tao added that "Qing Hao is everywhere. People take and eat it together with some aromatic spicy vegetables, such as cilantro" [7]. And *Commentaries on the Illustrations* (**Ben Cao Tu Jing**, also translated as *Illustrated Pharmacopoeia*) even gives more detail that "Cao Hao means Qing Hao. The seedling grows in spring. When leaves become extremely slim and fresh, people take and eat it with some aromatic spicy vegetables, such as cilantro. When it comes to summer, the plant grows to five-feet height. In autumn, fine yellowish flowers blooms and the mature fruits are as big as millets" in 1070. In addition, another book *Dream Stream Essays* (**Meng Xi Bi Tan**, also translated as *Dream Pool Essays*) has noticed something important and it tells that "there are numerous species of *Artemisia* using the name Hao. For the instance of *Artemisia annua* L. and *Artemisia apiacea* Hance (*Artemisia carvifolia*), one yellow and the other green. So the Qing Hao mentioned in Chinese herbal classics might have some differences among themselves. There are several Qing Hao species growing in border of the area of **Sui** and **Yin** in Shaanxi province. In the group of Qing Hao, there will sometimes grow some individuals of special green color, called by local people as Xiang Hao (literally means aromatic *Artemisia* specie). Their leaves are same as normal Hao (literally means common *Artemisia* species), but different in color. The common ones are of common green while these ones are of emerald green, as deep as pines' color. Amid a population of yellow ones, these special *Artemisia* species have an exclusively green color and slightly fragrant perfume. I am afraid that they might be the most utilized ones by our ancestors." From the aforementioned, we know that "Cao Hao" in *Shen Nong's Herbal Classic* still actually refers to several *Artemisia* species in the Song Dynasty.

In the Ming Dynasty, "when people recite the saying '**Yin Chen** in March, **Hao** in April', they just believe that the two are actually of one thing and the reason why they are named differently is simply due to their maturity," says *Materia Medica Companion* (**Ben Cao Meng Quan**), "However, although the two are so close to each other in leaf shape, they are two radically different species." **Yin Chen** mainly refers to *Artemisia capillaris*; **Hao** (**Qing Hao**) mainly refers to *A. annua*. And the subtle distinction between *A. capillaris* and *A. annua* is further illuminated in *Compendium of Materia Medica* (**Ben Cao Gang Mu**), which tells that "**Qing Hao** sprouts in February. The caulicle is soft, as thick as a finger. Both the caulicle and the young leaf are of deep green. The leaf is colored on both sides and its shape is slightly similar to that of **Yin Chen**. And the root is white and hard. During July and August, fine yellow flowers open and fruit, very fragrant. The fruit is as big as linseed, hiding little seeds inside." With the advent of taxonomy by Linnaeus in the 18th century and by using divergent taxonomy many species have been named and renamed over the year. One of Qing Hao original plants was later described by Linnaeus and given the scientific name of *A. annua* L. The famous Chinese pharmacist Mr. Yuhuang Zhao wrote an article in 1936, saying that "Qing Hao (Cao Hao) is ranked low-grade in *Shen Nong's Herbal Classic*......most of them are *Artemisia annua* L." [8].

According to the classic document and herbal literature mentioned previously, we may know one of the herb's earliest official names is Qing Hao. Even though *Shen Nong's Herbal Classic* originally puts Cao Hao as genuine medical herb, it also gives Qing Hao as an alternative name. In the book *Newly Revised Materia Medica* (**Xin Xiu Ben Cao**), the item of Bai Hao (the Chinese word for the plant of white *Artemisia* species) reads "the leave of Bai Hao is thicker than that of Qing Hao" [7], indicating that Qing Hao is the official name in the Tang Dynasty. On the other hand, it can be inferred from the attached figures in the book *Syndrome of Materia Medica* (**Zheng Lei Ben Cao**) that there are more than one type of Qing Hao in the Song Dynasty [7]. Among them, there is one type very similar to *A. annua* L. So *A. annua* L. could be identified as the genuine herbal medicine at that time. In fact, according to the book *Amplification on Materia Medica* (**Ben Cao Yan YI**) by Zongshi Kou, the book *Revised Classified Materia Medica from Historical Classics for Emergency* (**Chong Xiu Jing Shi Zheng Lei Bei Ji Ben Cao**) by Cunhui Zhang in the southern Song Dynasty [7] and other historical classics [9] of the Yuan Dynasty, we may infer that Qing Hao was still used as the genuine medical herb in Dynasty of Song and Yuan. Even though in the time of the book *Compendium of Materia Medica* (**Ben Cao Gang Mu**) people might confuse Qing Hao with Yin Chen, yet the distinction between them still could be figured out in 1525.

The application of Qing Hao was originally recorded in *Prescriptions for Fifty-two Diseases* (**Wu Shi Er Bing Fang**) as a herbal medicine for male hemorrhoids [10]. During 200–300, Qing Hao was included in *Shen Nong's Herbal Classic*, in which Qing Hao is classified as low-grade and its function are described as "able to treat scabs, itching and sores, to kill lice and to extinguish bone steaming, heat vexation and fever, as well as to improve eyesight" [6]. It was not until 281–340 that Qing Hao was recorded to be applied to malaria. *A Handbook of Prescriptions for Emergencies* (**Zhou Hou Bei Ji Fang**) by Hong Ge of the Eastern Jin Dynasty is the first book to record the antimalarial effect of Qing Hao in history, because it says that "a handful of Qing Hao immersed in two liters of water, wring out the juice and drink it all" [11]. However, the specie of Qing Hao is not clearly mentioned in the aforementioned documents. There comes a question: whether the Qing Hao that is included in *Shen Nong's Herbal Classic* and the one that is applied to malaria in *A Handbook of Prescriptions for Emergencies* are the same one or of same specie? To answer this question, we may refer to the book *Compendium of Materia Medica*. From 1552 to 1578, Shizhen Li, the author of that book, connected those two "Qing Hao" with the one that was proved by his own clinical practice to be effective to malaria, concluding that the previous Qing Hao written in classic documents and the present Qing Hao applied in his medical practice was actually the same one and of same species. Additionally, he further introduced the antimalaria function of Qing Hao. Between the Eastern Jin Dynasty and Ming Dynasty, prescriptions for malaria all involve Qing Hao, such as the Qing Hao decoction recorded in *General Medical Collection of Royal Benevolence* (**Sheng Ji Zong Lu**) [12] in the Song Dynasty, Jie Nue Qing Hao pills in *Danxi's Mastery of Medicine* (**Dan Xi Xin Fa**) [13] in the Yuan Dynasty, Qing

Hao powder and Qu Nue Shen Ying pills in *Prescriptions for Universal Relief* (**Pu Ji Fang**) [14] in the Ming Dynasty, and so on. After the time of *Compendium of Materia Medica*, in the Qing Dynasty, another two books also recorded the antimalarial effect of Qing Hao, one being *Essentials of Materia Medica* (**Ben Cao Bei Yao**) [15] and the other *Detailed Analysis of Epidemic Febrile Disease* (**Wen Bing Tiao Bian**) [16]. The TCM herbal medicine Qing Hao has a highly special efficacy of antimalaria. Based on that character, we may chain all the historical literature recordings of Qing Hao and conclude that there is only one herb of *Artemisia* species that can be considered genuine Qing Hao and the genuine herbal medicine fights against malaria.

In 1830s, Japanese scholars translated and annotated the book *Compendium of Materia Medica* [17], adding botanical name to related herbal medicine. However, they mistakenly added "*Artemisia apiacea* Hance" to Qing Hao and "*Artemisia annua* L." to Huang Hua Hao. So was it followed by our colleagues in China. However, scientific practice has proved it to be an error and for the scientific spirit of responsibility, the authors recommend correction of the mistake.

References

[1] Jing S, 诗经 Chinese Book of Odes (annotated by Zhu Xi 朱熹). Shanghai: Shanghai Classics Publishing House; 1987.

[2] Mu BCG, 本草纲目 (Compendium of Materia Medica). 1596. Li Shizhen 李时珍. Beijing: People's Medical Publishing House; 1975, p. 943–46.

[3] Er Ya 尔雅 annotated by Guo Pu 郭璞. Shanghai: Tong Wen Shu Ju Publishing House; 1985.

[4] Shuo Wen Jie Zi 说文解字 (Origin of Chinese Characters), Xu Shen 许慎 annotated by Duan Yucai 段玉裁. Shanghai: Shanghai Classics Publishing House; 1981.

[5] Er Ya Yi Shu 尔雅义疏 Hao Yixing 郝懿行, Zhong Guo Shu Dian Publishing House, Beijing, 1982.

[6] Shen Nong Ben Cao Jing 神农本草经 (Shen Nong's Herbal Classic), annotated by Wu Pu 吴谱 and Sun Xingyan 孙星衍. Beijing: Shang Wu Yin Shu Guan Publishing House; 1955, p. 103.

[7] Chong Xiu Jing Shi Zheng Lei Bei Ji Ben Cao 重修政和经史证类备急本草 (Revised Classified Materia Medica from Historical Classics for Emergency), 1082. Tang Shen Wei 唐慎微 annotated in 1249 by Zhang Cunhui 张存惠. Beijing: People's Medical Publishing House; 1957.

[8] Qi Zhou Yao Zhi 祁州药志 1936. Zhao Yuhuang 赵燏黄, vol. 1. 58.

[9] Chong Kan Jing Shi Zheng Lei Da Quan Ben Cao 重刊经史证类大全本草 (Reprint Classified Materia Medica from Historical Classics for Emergency), 1082. Tang Shen Wei 唐慎微 Reprint and annotated by Wang Qiu 王秋 in 1577.

[10] Cooperative research group on the archaeological finds of Mawangdui Han Tomb. Chinese Cultural Relics. Beijing: Cultural Relics Publishing House; 1975; 9:42.

[11] Zhou Hou Bei Ji Fang 肘后备急方 (A Handbook of Prescriptions for Emergencies). 281–340. Ge Hong 葛洪. Beijing: People's Medical Publishing House; 1956. p. 44.

[12] Sheng Ji Zong Lu 圣济总录 (General Medical Collection of Royal Benevolence). 1111–1118. Beijing: People's Medical Publishing House; 1962. p. 724.

[13] Dan Xi Xin Fa 丹溪心法 (Danxi's Mastery of Medicine). 1481. Shanghai: Shanghai Science and Technology Press; 1959. p. 47.

[14] Pu Ji Fang 普济方 (Prescriptions for Universal Relief). 1406. Vol. 197-198 Zhu Su 朱橚. Beijing: People's Medical Publishing House; 1959.

[15] Ben Cao Bei Yao 本草备要 (Essentials of Materia Medica). 1694. Wang Ang 汪昂. Beijing: Shang Wu Yin Shu Guan Publishing House; 1954. p. 102.
[16] Wen Bing Tiao Bian 温病条辨 (Detailed Analysis of Epidemic Febrile Disease). 1798. Wu Tang 吴瑭. Beijing: People's Medical Publishing House; 1964. p. 108.
[17] Shirai Kōtarō', et al. Li Shizhen's Compendium of Materia Medica. Tokyo, Japan: Shun'yodo Bookstore Co. Ltd; 1933.

Further Reading

Meng Xi Bi Tan 梦溪笔谈 (Dream Stream Essays). 1086. Shen Kuo 沈括. Beijing: Cultural Relics Publishing House; 1975.
Ben Cao Meng Quan 本草蒙筌 (Materia Medica Companion). 1565. Chen Jiamo 陈嘉谟. Beijing: People's Medical Publishing House; 1988.

CHAPTER 2

Studies of the Certified and Confused Herba Artemisiae Annuae

Chapter Outline

1 A Revision of the Chinese Name of the Sweet Wormwood Plant

Since artemisinin, a novel antimalarial, had been studied and developed from the Herba Artemisiae Annuae (*Artemisia annua* L.), it had been revised in the *Pharmacopoeia of the People's Republic of China* of version 1985 to remove an error that *A. annua* L. was described as Herba Artemisiae Annuae same as *Artemisia apiacea* Hance in previous versions (1963 and 1977 versions). However, Herba Artemisiae Annuae had always been known as *A. annua* L. from version 1985 to version 2005 of the *Pharmacopoeia of the People's Republic of China* [1]. Particularly in the *Modern Chinese Material Medica* (version 2002) [2], the original plant of Herba Artemisiae Annuae, defined in an article of Herba Artemisiae Annuae

From *Artemisia annua* L. to Artemisinins
http://dx.doi.org/10.1016/B978-0-12-811655-5.00002-7

in the *Chinese Material Medica* (version 1988), was accordingly changed into *A. annua* L. again, claiming a consistency with the *Pharmacopoeia of the People's Republic of China* (version 2000) (hereinafter referred to as the *Pharmacopoeia*), also known as "medicinal sweet wormwood."

For this purpose, the author considers that it is necessary to make textual research [3,4] on the certified Herba Artemisiae Annuae and revise the original plant of Herba Artemisiae Annuae to *A. annua* L.

According to the *Modern Chinese Material Medica*, in the plant kingdom, *A. apiacea* Hance and *A. annua* L. are two different species which are known as "*Qinghao* (sweet wormwood or sweet annie)" and "*Huanghuahao*" in Chinese, respectively. Now, we are to discuss the origins of synonyms of Herba Artemisiae Annuae from the initial scientific names by Carl Linnaeus and H. F. Hance, both of whom were botanical authority figures.

A. annua Linn. was firstly named by Linnaeus in 1753 [4], whereas *A. apiacea* Hance was first named by Hance et al. in 1852, which were not linked to Chinese names. In 1856, *A. apiacea* Hance and *A. annua* L. were firstly defined as scientific names of "*Qinghao*" and "*Huanghuahao*" in the *Illustrated Flora of Japan* by Yokusai Iinuma, respectively. They were accepted by Ninzo Matsumura when he compiled the *Nomenclature of Japanese Plants in Latin, Japanese, and Chinese* from 1884 to 1915 [7]. After that, they were cited in the *Makino's Illustrated Flora of Japan* (by Tomitaro Makino) [8] and the *Chinese Materia Medica: Vegetable Kingdom* (by G. A. Stuart) [9] in 1911 in succession. In 1918, they were still in use in the *Botanical Nomenclature, A Complete Dictionary of Botanical Terms* by Kong Qinglai [10]. *Qinghao* and *Huanghuahao* were not scientifically named as *A. apiacea* Hance and *A. annua* L. until a Japanese version of the *Compendium of Materia Medica* [11] added scientific names item by item for each Chinese materia medica recorded in Li Shizhen's *Compendium of Materia Medica*. Thus, their reprints and citations led to a case of mistaken identity between names of Chinese materia medica and scientific names. However, people did not note that Li simultaneously listed both "*Qinghao*" and "*Huanghuahao*" in his *Compendium of Materia Medica* [12], which were two herbs of different natures, flavors, and indications; particularly under the term of "*Huanghuahao*" (absolutely found by Li himself), it was noted as "acrid and bitter in taste, cool in nature, nontoxic, and indicated for children with convulsion and fever due to wind-cold." However, "*Qinghao*" was noted as "bitter in taste and cold in nature," consistent with those used conventionally in the *Shen Nong's Herbal Classic*; particularly, under the term of "treating malaria with alternate attacks of chill and fever" in its attached formula, there were records of such treatment confirmed by the *Handbook of Prescriptions for Emergencies* and Li's own clinical practice. Visibly, this is the certified Herba Artemisiae Annuae with antimalarial effect today. Therefore, Herba Artemisiae Annuae should be derived from the plant *A. annua* L., and errors made by

Japanese scholars should be corrected. Correction should be made for the *Pharmacopoeia of the People's Republic of China* correspondingly. What on earth is the "*Huanghuahao*" found by Li for "children with convulsion and fever" is a different matter, and is not associated with the certified Herba Artemisiae Annuae at all. As for *A. apiacea* Hance, C. M. Chen et al. [13] had established it as *Seseli libanotis* Kock. Hu Shilin also suggested it as the "correct name" of Herba Artemisiae Annuae [14].

2 Market Research on the Commercial Herba Artemisiae Annuae [15]

Through research on the commercial Herba Artemisiae Annuae in China, there are totally six *Artemisia* congeners (Table 2.1), five of which are confused ones.

3 Original Plants of Artemisia annua L. and Its Five Confused Congeners [15]

3.1 Artemisia annua L.

A. annua L. is also known as sweet wormwood, sweet annie, sweet sagewort, annual mugwort, or annual wormwood (Fig. 2.1).

Herbs, annual, 40–150 cm tall, specifically aromatic. Stem erect, terete, much branched, 6 mm in diameter. Surface shallowly longitudinally sulcate, green when young and yellowish-brown when old, glabrous; lowermost stem slightly ligneous. Basal leaves decumbent, cauline leaves alternate, often tripinnatisect, 4–7 cm long, 1.5–3 cm wide. Lobes or lobules oblong or obovate, apex acute. Lowermost stem leaves: leaf blade slightly ampliate and amplexicaul, upper green and lower yellowish green, bifacially puberulous or mealy glandular-punctate. Uppermost stem leaves: compound, attenuating and more finely divided upward. Synflorescence globose, 1.5 mm in diameter, a panicle produced from most nodes to form broad, conical compound panicle. Capitula many, 2 mm shortly pedunculate, closely subtended by whorl of bracteoles. Phyllaries two or three, glabrous, forming an involucre; outer bract long narrow, middle, and inner elliptic; abaxially, middle green, margins light yellow, scarious, and transparent. Receptacle oblong. Floret tubiform; female florets less, often at outer layer; only 1 pistil in the center, stigma bifid, long furcate and divaricate; c. 20 bisexual florets internally, c. 1-mm long, yellow. Corolla tube retracted into the upper part of ovary, slightly ampliate upward; apex quinquefid, lobules triangular, with acute apex. Stamens five; anther adnate and surrounding the stigma, apex caudate-acuminate; filaments fine and short, adnate in the middle of the interior surface of the corolla tube; one pistil present centrally. Ovary elliptic; style filiform; stigma bifid and bifurcate. Achenes ovoid, tiny, and light brown. Distributed throughout China. Wildernesses, hillsides, waysides, and river banks.

Table 2.1: Six species as Herba Artemisiae Annuae all over China (individual regions)

Original Plant					
Family	**Scientific Name**	**Place of Origin**	**Area Used**	**Medicinal Parts**	**Notes**
Compositae	*A. annua* L.	All over China	Beijing, Hebei, Shandong, Henan, Anhui, Hubei, Hunan, Sichuan, Yunnan, Guangdong, Guangxi, Qinghai, etc.	Herba	The variant *A. annua* L.F. macrocephala Pamp is produced in Yunnan and used as Herba Artemisiae Annuae as the stock
Compositae	*A. apiacea* Hance	Liaoning, Hebei, Shandong, Jiangsu, Anhui, Hubei, Shaanxi, Zhejiang, Jiangxi, Fujian, and Guangdong	Used by some regions in Hebei, Shandong, Henan, Jiangsu, Anhui, Shaanxi, Hunan, Jiangxi, Zhejiang, Yunnan, etc.	Herba	
Compositae	*A. scoparia* Waldst. et Kit.	All over China	Used by some regions in Shandong, Hubei, Sichuan, Jilin, Shaanxi, etc.	Herba	It is known as *Yuzi Qinghao* in Hubei and widely used in Hubei and Shandong
Compositae	*A. capillaris* Thunb.	Liaoning, Hebei, Shandong, Jiangsu, Fujian, and Guangdong	Used by some regions in Liaoning, Hebei, Shandong, Jiangsu, Fujian, Guangdong, etc.	Herba	
Compositae	*A. japonica* Thunb.	Yunnan, Sichuan, Henan, Hebei, Zhejiang, Jiangsu, Shandong, and Northeast China	Used by individual regions in Yunnan, Sichuan, Henan, Zhejiang, Jiangsu, Shanghai, etc.	Herba	
Compositae	*A. eriopoda* Bunge	Yunnan, Sichuan, Henan, Hebei, Zhejiang, Jiangsu, Shandong, and Northeast China	Used by individual regions in Yunnan, Sichuan, Henan, Zhejiang, Jiangsu, Shanghai, etc.	Herba	

3.2 Artemisia apiacea *Hance*

A. apiacea Hance is also known as sweet wormwood and sweet annie (Fig. 2.2).

Herbs, annual or perennial, 40–150 cm tall, glabrous, fragrant. Stem terete, dark green when young. Surface finely shallowly longitudinally sulcate; lowermost stem slightly ligneous; uppermost stem axillarily branched. Leaves alternate, bipinnatisect; the first segment anisophyllous, pinnatisect again; the second segment linear, entire or 1- to 3-pinnatilobate,

Figure 2.1: *A. annua* L.

Figure 2.2: *A. apiacea* Hance.

also anisophyllous; apex acute, soft, dark green; rachis ctenoid; petioleshort and narrow; base slightly ampliate and amplexicaul. Heads arranged in paniculiform cymes, each capitulum lateral, slightly nodding, c. 6 mm in diameter, short-pedicelled, axillary. Involucre hemispherical, phyllaries in 3–4 series; outer phyllaries elongated in different sizes, inner ones oval, 2–3 mm long, c.1-mm wide; abaxially, middle grass-green, margins scarious and transparent. Receptacle flat, subtended by whorl of female florets, infertile; each floret tubular, with only 1 pistil in the center, stigma bifid, divaricate; internally, bisexual florets 60–80, green-yellow; corolla tubular, c. 2-mm long. Upper corolla tube retracted, gradually ampliate upward; outer surface of the middle microrugulate, outer surface of the upper smooth, apex quinquefid; lobes triangular-lanceolate, apex acute, patulose or revolute. Stamens five; anther adnate and surrounding the stigma, apex caudate-acuminate, bilateral base decurrens mucronate; filaments fine and short, adnate in the middle of the interior surface of the corolla tube; one pistil present centrally. Ovary elliptic; style filiform; stigma bifid and bifurcate. Achenes oblong to elliptic, tiny, and brown; surface prominently longitudinally striate.

River banks, sandy lands and seaside. Distributed in Northeast, South, and Southwest China.

3.3 Artemisia scoparia *Waldst. et Kit.*

Herbs, perennial, biennial, or annual. Roots fusiform or conical, mostly vertical. The whole plant off-white sericeous when young, 45–100 cm tall later. Stem usually single, occasionally 2–4; base usually ligneous. Surface purple or yellow green, longitudinally striate, much branched; old branches glabrescent, twigs off-white sericeous, sometimes with larger and densely foliferous sterile branches. Leaves dense. Lower stem leaves isomorphous with those of sterile branches, long petiolate; leaf blade oblong, 1.5–5 cm long, 2- or 3-pinnatisect; finally segments oblanceolate or filiform; apex acute, usually sericeous or sparsely puberulent superiorly. Middle stem leaves 1–2 cm long, 2-pinnatisect; base amplexicaul; lobules linear or filiform, piliferous or glabrous. Uppermost leaves sessile, 3-cleft or entire; lobules short, filiform. Capitula many, pedunculate, arranged in a synflorescence cauline on the lateral branches. Involucre ovate or subglobose, 1–2 mm in diameter; phyllaries 3 in 3–5 series, imbricately arranged, ovate, elliptic, oblong or obovate; apex obtuse; outer phyllaries short, inner ones large; margins broad scarious, abaxially green, subglabrous. Florets polygamous, tubiform; outer, female florets 5–15, mostly 10–12, fertile, stigma bifid and bifurcate, exserted outside of the corolla; inner, bisexual florets 3–9, apex slightly ampliate, quinquefid, lobes triangular, sometimes purple, inferiorly included, obovate, with abortive and infertile ovaries. Achenes small, oblong or obovoid, c. 0.7 mm, longitudinally striate, glabrous. Florescence August to September, fruit period September to October (Fig. 2.3).

Ditches, slopes, gravel lands, and saline lands. Distributed throughout China.

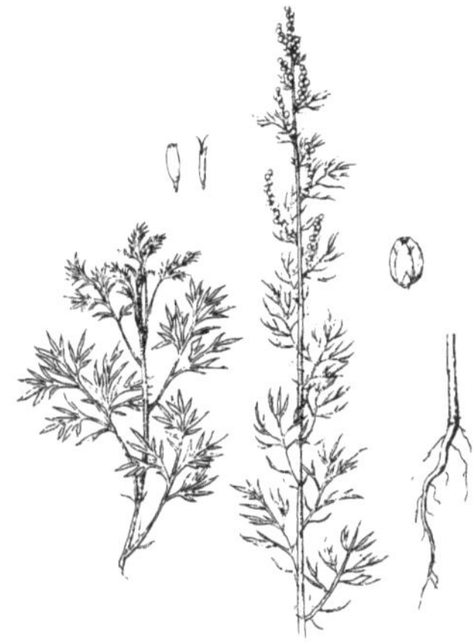

Figure 2.3: ***A. scoparia*** **Waldst. et Kit.**

3.4 Artemisia capillaris *Thunb*

Herbs, perennial; root branched, usually plagiotropic or mostly conically orthotropic, but not fusiform. Stems usually several and caespitose, assurgent, usually solitary during its first year; base robust, and woodier than *A. scoparia* Waldst. et Kit. Sometimes central filiform lobules more slender, straighter, and longer than previous species, up to 2.5 cm. Marginal female florets 4–12, usually 7 or so. Achenes large, up to 1 mm. The remaining similar to *A. scoparia* Waldst. et Kit. (Fig. 2.4).

Beach and coastal or riverine sandy soils; sparsely coastwise slopes. Distributed in East and South China, including Shandong, Jiangsu, Zhejiang, and Fujian.

Figure 2.4: ***A. capillaris*** **Thunb.**

Figure 2.5: *A. japonica* Thunb.

3.5 Artemisia japonica *Thunb*

Herbs, perennial; rhizomes thick. Stem erect, usually caespitose, 50–150 cm tall, upper parts with patulous or erect branches. Lower stem leaves withered at anthesis, spatulate, 3–8 cm long, 1–2.5 cm wide, linearly pseudostipulate, serrate or lobate apically. Middle stem leaves: leaf blade cuneate, serrate or almost palmately divided apically. Uppermost leaves almost linear, 3-cleft or entire. Capitula many, arranged in a synflorescence, shortly pedunculate or linearly bracteate. Involucre subglobose, 1–2 mm in diameter; phyllaries in c. 4 series, with broad scarious margins. Marginal female florets c. 10, fertile. Disk florets bisexual, sterile. Achenes up to 1-mm long (Fig. 2.5).

3.6 Artemisia eriopoda *Bunge*

Herbs, perennial. Stem erect, 30–70 cm tall, solitary or several, caespitose, glabrescent but lanuginous at base, with branches of the inflorescence uppermost or from the lower part. Basal and lower stem leaves: petiole 5–10 long, blade 2–5 cm wide, sometimes spatulate, or serrate or cleft at margin, lobes 5–7, broadly obovate, cuneate at base, palmately divided apically. All leaves: adaxial leaf surface glabrous, abaxial leaf surface puberulous. Uppermost leaves 3-pinnatisect or entire; lobules linear. Capitula many, arranged in a synflorescence at the stem or branch end, sessile or shortly pedunculate, with linear bracts. Involucre ovoid, c. 2-mm long, inner phyllaries oblong. Marginal female florets fertile. Disk florets bisexual, sterile. Achenes oblong, tiny, glabrous (Fig. 2.6).

4 *Pharmacognostical and Physicochemical Studies of* Artemisia annua L. *and Its Five Confused Congeners*

Based on the previous plant identification, comparative studies were conducted on Material Medica derived from these six *Artemisia* species with respect to their pharmacognostical descriptions, histology, microscopic characteristics, thin-layer chromatography (TLC), and chemical constituents.

Figure 2.6: ***A. eriopoda*** **Bunge.**

Properties and Histological Identification of Material Medica [15,16]: The materials used in the experiment included commercial medicinal materials, fresh materials or exsiccata, in which original plants had been identified. Their sources are listed in Table 2.2.

Table 2.2: Sources of experimental materials

Material Medica	Original Plant	Sample From		
		Commodity	Fresh Materials	Exsiccata
Herba Artemisiae Annuae	*A. annua* L.	Dexing, Jiangxi; Tong'an, Fujian; Hainan	Beijing	
Herba Artemisiae Apiaceae	*A. apiacea* Hance	Qichun, Hubei; Tai'an, Shandong	Changli, Hebei	
Herba Artemisiae Scopariae	*A. scoparia* Waldst. et Kit.	Henan	Beijing	
Herba Artemisiae Capillariae	*A. capillaris* Thunb.		Qingdao, Shandong; Zhejiang	
Herba Artemisiae Eriopodae	*A. eriopoda* Bunge		Beidaihe	Laoshan Mountain, Shandong
Herba Artemisiae Japonicae	*A. japonica* Thunb.		Beijing	Laoshan Mountain, Shandong

Figure 2.7: The Plant of *A. annua* L. (Preflowering).

4.1 Artemisia annua *L.*

1. Description (Figs. 2.7–2.9)

   Seventy to hundred centimeters long. Dark green when shed-dried, and tan when sundried. Stems terete, surfaces longitudinally canaliculate and angulate, light, easily broken; a large white pith present in the middle of the fracture, with pistachio margins. Uppermost stems much branched. Rachis closely subtended by many yellowish green capitula, c. 1 mm in diameter, caducous. Leaves alternate, leaf blade crenate, green or dark green, texture fragile, easily broken. Leaves, flowers, and fruits specifically aromatic; taste, slightly bitter.

Figure 2.8: The Plant of *A. annua* L. (Flowering and Fruit-Bearing).

Figure 2.9: *A. annua* L.

2. Microscopic characteristics
   a. Surface view of leaf
   Epidermal cells irregular. Anticlinal walls sinuately curved, 18-41-80 μm long. Epidermal cells narrowly rectangular on the vein rib, stomata anomocytic. Densely covered with medifixed hairs, especially on the midrib, with a pluricellular stalk and a slightly asymmetrical unicellular head; stalk cells short, 3–10, mostly 4–5, basal cells usually ampliate, apical cell walls thin and crenated; head cell 240-480-816 μm, with thin cell wall (Figs. 2.10–2.12). Apical stalk cells frequently withered near the midrib, with their head caducous and the pluricellular stalk left only (Fig. 2.13). Densely covered with pluricellular glandular hairs; head elliptic, full of flavescent

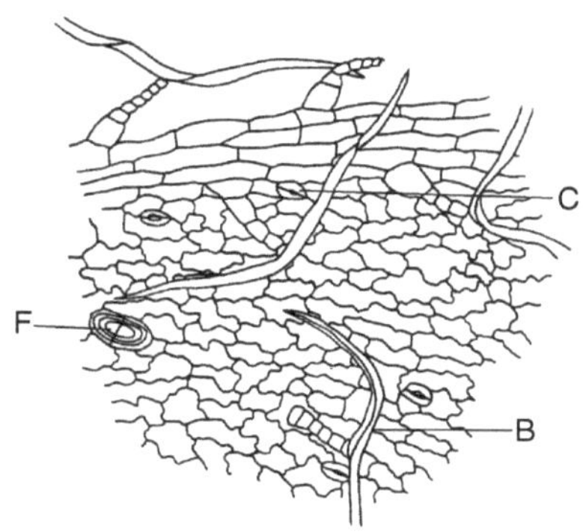

Figure 2.10: Leaf Epidermal Tissue of *A. annua* L. (×206).
B, Medifixed hairs or their stalks; C, stomata; F, glandular hairs.

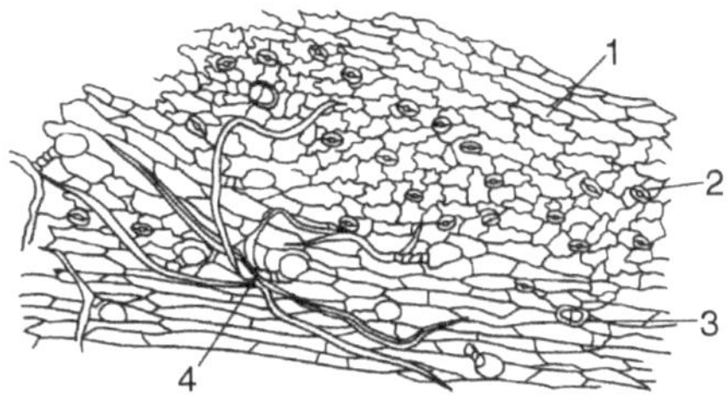

Figure 2.11: Leaf Epidermal Tissue of *A. annua* L. (×130).
1, Epidermal cells; 2, stomata; 3, glandular hairs; 4, medifixed hairs.

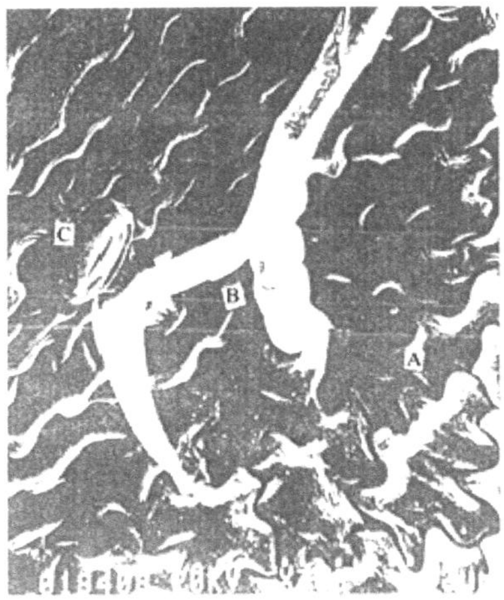

Figure 2.12: SEM Shows Medifixed Hairs of the Leaf Epidermis of *A. annua* L.
A, Epidermis or epidermal cells; B, medifixed hairs or their stalks; C, stomata.

volatile oil; two semicircular secreting cells usually parallel to the midrib (Fig. 2.14). There is no significance between upper and lower epidermal tissues, but only more hairs and stomata are present in the upper epidermis than in the lower one.

b. Transection of ultimate lobe
Transversely, epidermal cells arranged closely, superiorly with stalks of medifixed hairs, glandular hairs and stomata; glandular hairs concaved in the epidermis, mostly two to three cells arranged in a single row; longitudinally, mostly two to three pairs of cells arranged in two rows. Mesophyll tissue isobilateral, with palisade tissues closely arranged superiorly and inferiorly. Midrib prominently raised above and below the surface; upper palisade tissue through the midrib, arranged in two to three layers; upper one disrupted by collenchyma at the midrib; collateral vascular bundle quasicircular, without secretory cavity (Figs. 2.15 and 2.16).

c. Leaf powder is shown in Fig. 2.17

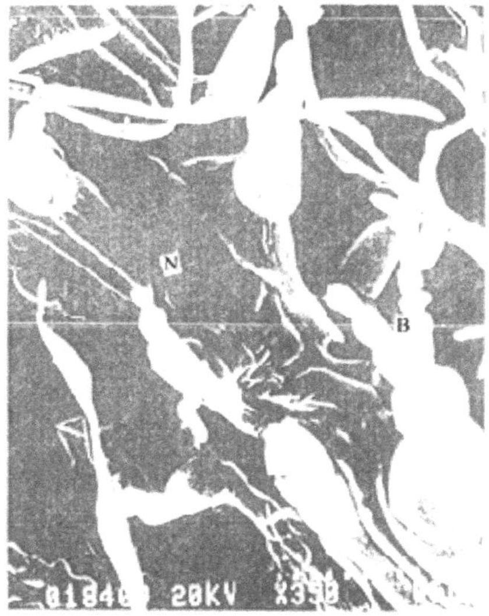

Figure 2.13: SEM Shows Medifixed Hairs of the Midrib of the Leaf of *A. annua* L.
B, Medifixed hairs or their stalks; N, midrib.

d. Surface view of stem
Epidermal cells narrow rectangular, 35-69-112 μm long, 13-18-25 μm wide; epidermal hairs far less than those of leaf surface; others basically consistent with leaves.

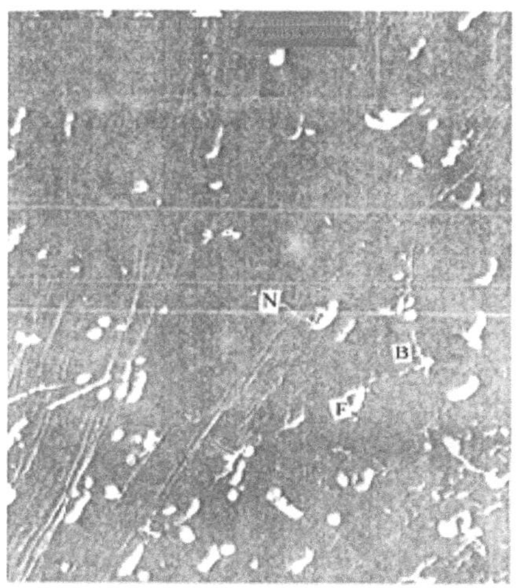

Figure 2.14: SEM Shows the Orientation of Glandular Hair Cells of the Leaf of *A. annua* L. (×100).
B, Medifixed hairs or their stalks; F, glandular hairs; N, midrib.

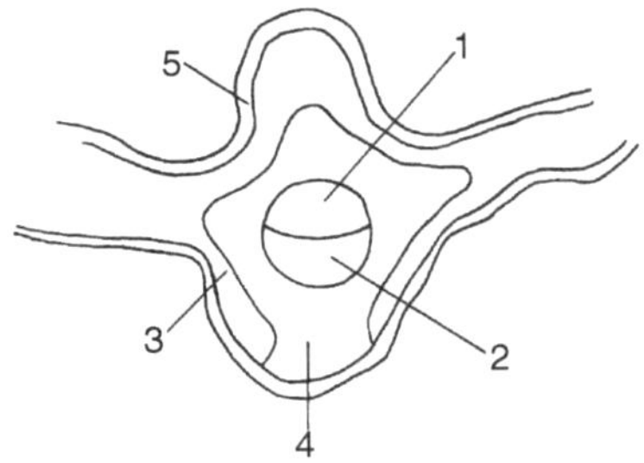

Figure 2.15: Schematic of Transection of Leaf Midrib of *A. anrua* L. (×28).
1, Secretory cavity; 2, xylem of vascular bundle; 3, phloem of vascular bundle; 4, mesophyll tissue; 5, fundamental tissue; 6, epidermal tissue.

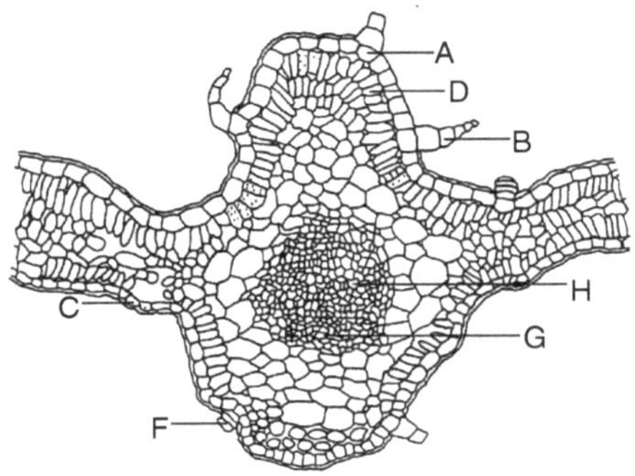

Figure 2.16: Detail of Transection of the Leaf of *A. annua* L. (×160).
A, Epidermis or epidermal cells; B, medifixed hairs or their stalks; C, stomata; D, palisade tissue; F, glandular hairs; G, phloem; H, xylem.

e. Transversely

Transversely, stems (0.5 cm in diameter) quasicircular, usually with more than 10 round horns. Epidermal cells squarish or rectangular; sometimes with stomata, glandular hairs, and stalks of medifixed hairs present. Cortical cells 3–5 layers, quasicircular or oblong, 15-25-45 μm in diameter, up to 12 layers at prominent horns; layers of collenchyma close to the subepidermis; innermost cortical cells long and narrow, without Casparian dots and starch granules. Starch granules present in the endodermis of the caulicle and its adjacent cortical cells. Collateral vascular bundles arranged in circles, primary phloem fibers present outside each vascular bundle. The diameter of pith is 1/2 to 2/3 of that of stem, without secretory cavity in the cortex and pith (Figs. 2.18 and 2.19). Transversely, horns of branchlets become more prominent, almost polygonous; when differentiated secondary tissue is present, the transverse section of the stem tends to become round.

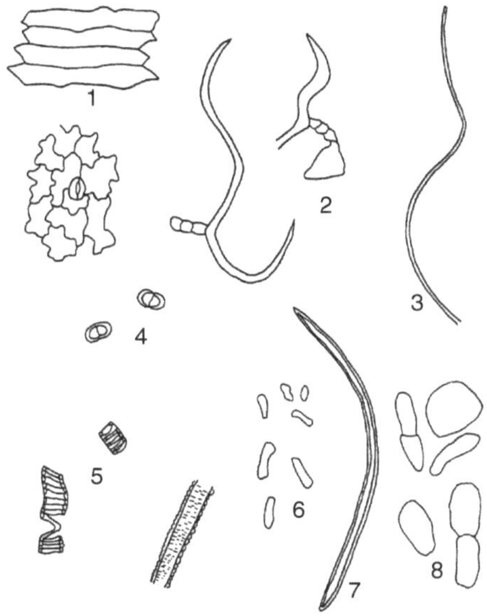

Figure 2.17: Left Powder of *A. annua* L. (×260).
1, Epidermal cells and stomata; 2, medifixed hairs; 3, linear hairs; 4, glandular hairs; 5, vessels; 6, mesophyll cells; 7, fibers; 8, parenchyma cells.

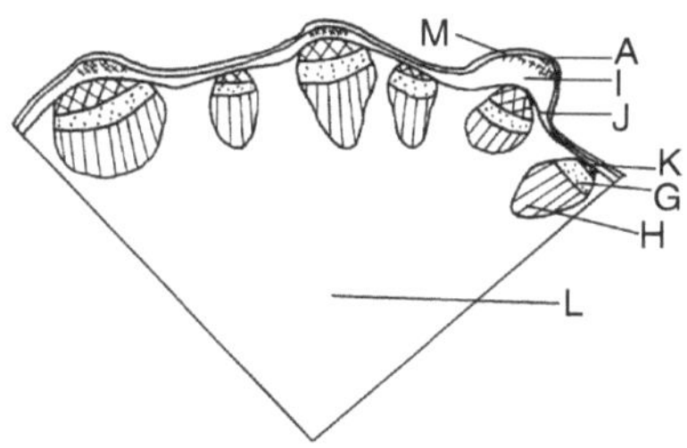

Figure 2.18: Schematic of Transection of Stem of *A. annua* L. (×24).
A, Epidermis or epidermal cells; G, phloem; H, xylem; I, cortex; J, endodermis; K, fiber bundle; L, pith; M, collenchyma.

4.2 Artemisia apiacea *Hance*

1. Description (Fig. 2.20)
   Similar to Herba Artemisiae Annuae, but cauline cortical fibers flexible, not easily fractured; fracture fibrous; leaves dark green; odor, not as strong as *A. annua* L.
2. Microscopic characteristics
   a. Surface view of leaf
      Epidermal cells larger as compared to *A. annua* L., 25-45-85 μm long; stalk cells of medifixed hairs 1–4, mostly 2–3; head cell 93-186-300 μm (Fig. 2.21).

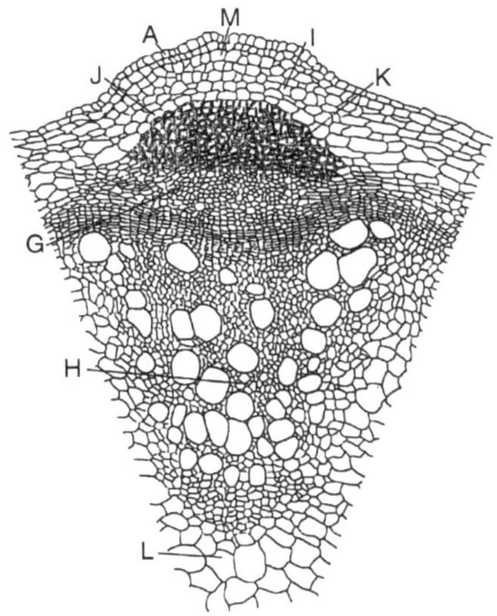

Figure 2.19: Detail of Transection of the Stem of *A. annua* L. (×190).
A, Epidermis or epidermal cells; G, phloem; H, xylem; I, cortex; J, endodermis; K, fiber bundle; L, pith; M, collenchyma.

Figure 2.20: *A. apiacea* Hance.

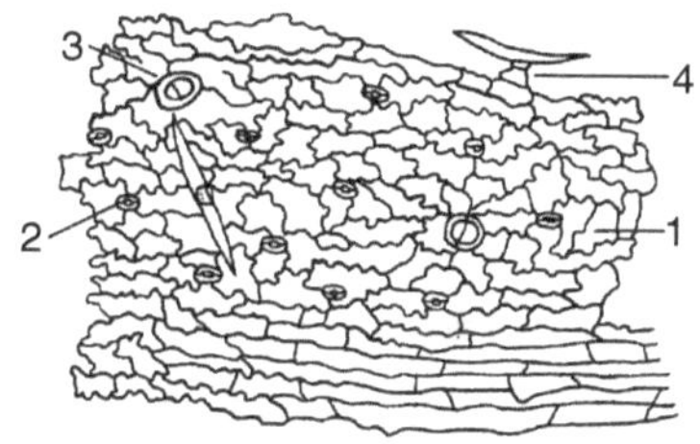

Figure 2.21: Leaf Epidermal Tissue of *A. apiacea* Hance (×130).
1, Epidermal cells; 2, stomata; 3, glandular hairs; 4, medifixed hairs.

b. Transection of ultimate lobe
Transversely, palisade cells arranged loosely, mesophyll cells long and large. Midrib not prominent or lower surface slightly prominent; 1–2 oblate secretory cavities present above or below the vascular bundle of the midrib, 20-26-35 μm in diameter, with 5–7 secretory cells (Figs. 2.22 and 2.23).

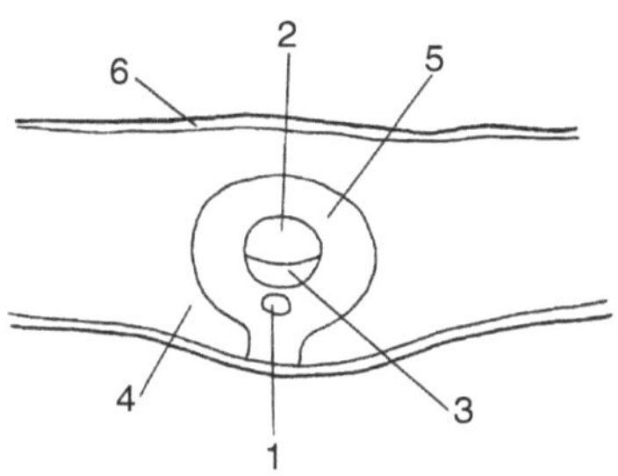

Figure 2.22: Schematic of Transection of Leaf Midrib of *A. apiacea* Hance (×28).
1, Secretory cavity; 2, xylem of vascular bundle; 3, phloem of vascular bundle; 4, mesophyll tissue; 5, fundamental tissue; 6, epidermal tissue.

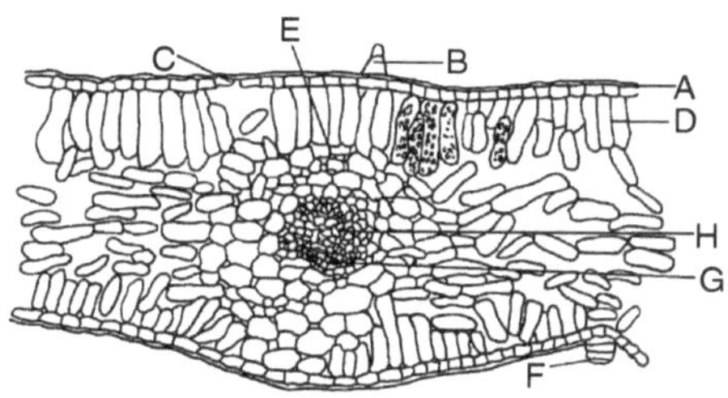

Figure 2.23: Detail of Transection of the Leaf of *A. apiacea* Hance (×160).
A, Epidermis or epidermal cells; B, medifixed hairs or their stalks; C, stomata; D, palisade tissue; E, secretory cavity; F, glandular hairs; G, phloem; H, xylem.

c. Leaf powder is shown in Fig. 2.24
d. Surface view of stem
Epidermal cells narrow rectangular, 40-79-120 μm long, 15-22-30 μm wide; epidermal hairs sparse; others basically consistent with leaves.
e. Transversely
Transversely, stems (0.5 cm in diameter) quasicircular; cortical cells 8–12 layers, quasicircular, 30-50-90 μm in diameter. The diameter of pith is c. 1/2 of that of stem. Oblate secretory cavities are observed in both pith and cortex (Fig. 2.25).

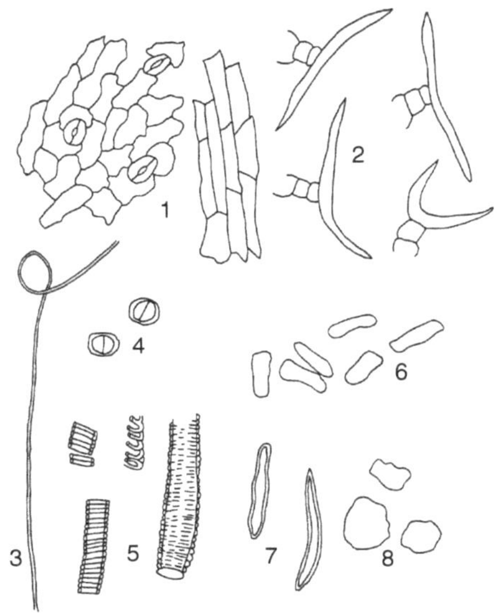

Figure 2.24: Left Powder of *A. apiacea* Hance (×260).
1, Epidermal cells and stomata; 2, medifixed hairs; 3, linear hairs; 4, glandular hairs; 5, vessels; 6, mesophyll cells; 7, fibers; 8, parenchyma cells.

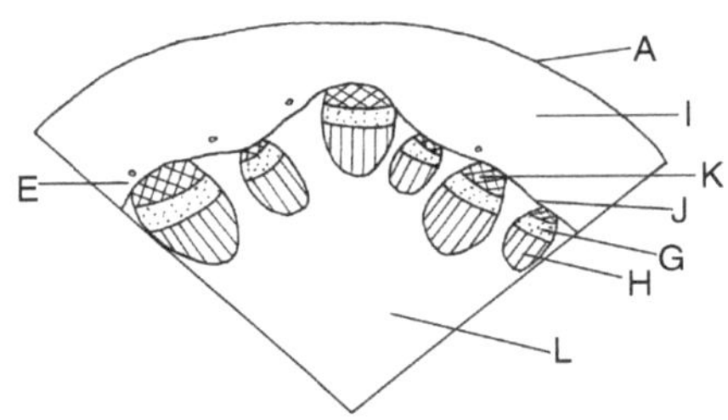

Figure 2.25: Schematic of Transection of Stem of *A. apiacea* Hance (×24).
A, Epidermis or epidermal cells; E, secretory cavity; G, phloem; H, xylem; I, cortex; J, endodermis; K, fiber bundle; L, pith.

Figure 2.26: ***A. scoparia*** **Waldst. et Kit.**

4.3 Artemisia scoparia *Waldst. et Kit.*

1. Description (Fig. 2.26)

   Seedling frizzy, off-white or celadon, densely off-white sericeous, downy. Stems slender, 1.5–2.5 cm long, 1–2 μm in diameter, prominently longitudinally striate after removing surficial sericeous hair; texture fragile, easily broken. Leaves petiolate; flattened leaves 2-3-pinnatifid or palmatifid; blade 1–3 cm long, c. 1-cm wide; lobules ovate, obovate or oblanceolate, linear; apex pungent. Odor, fragrant; taste, slightly bitter.

2. Microscopic characteristics

   a. Surface view of leaf

      Epidermal cells of linear leaves mostly similar, narrow rectangular, with flat or slightly curved walls; pinnate leaves irregular. Glandular hairs sparse; stalk cells of the medifixed hair 1–2, head cell extremely long, thick; hairs dense on the pinnate leaf and sparse on the linear leaf (Fig. 2.27).

   b. Transection of ultimate lobe

      Lower surface of pinnate leaf prominent, mesophyll of linear leaf cyclocytic (Fig. 2.28).

   c. Leaf powder is shown in Fig. 2.29

   d. Surface view of stem

      Epidermal cells narrow rectangular; tender stem frequently with medifixed hairs; head cells mostly caducous, stalk cells 1–2.

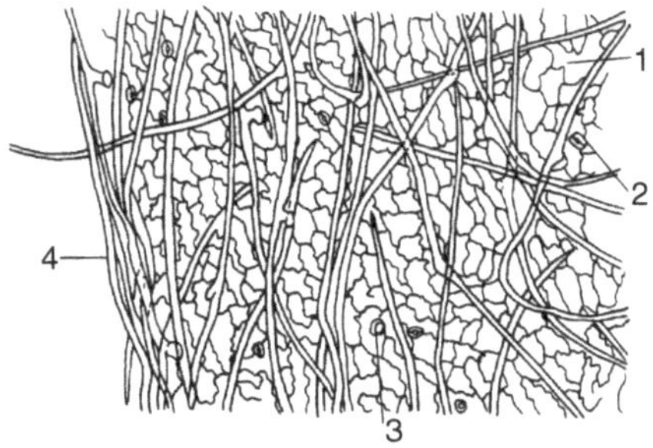

Figure 2.27: Leaf Epidermal Tissue of *A. scoparia* Waldst. et Kit. (×130).
1, Epidermal cells; 2, stomata; 3, glandular hairs; 4, medifixed hairs.

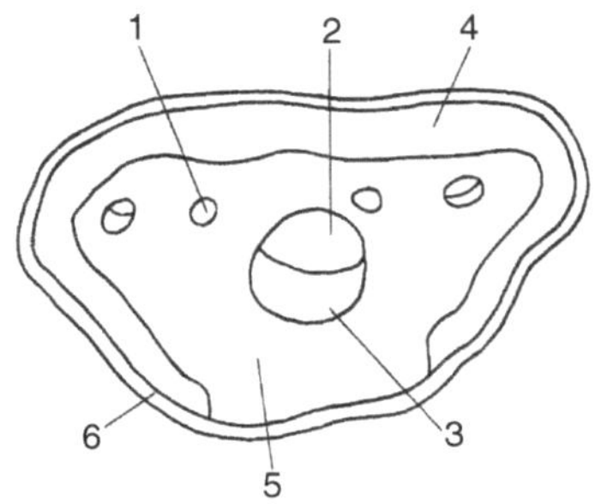

Figure 2.28: Schematic of Transection of Leaf Midrib of *A. scoparia* Waldst. et Kit. (×28).
1, Secretory cavity; 2, xylem of vascular bundle; 3, phloem of vascular bundle; 4, mesophyll tissue; 5, fundamental tissue; 6, epidermal tissue.

e. Transection of stem (0.5 cm in diameter)
Cortical cells 2–6, arranged densely, with secretory cavities. The diameter of pith is only c. 1/3 of that of stem.

4.4 Artemisia capillaris *Thunb*

1. Description (Fig. 2.30)
Dried seedlings mostly kneaded, celadon, white piliferous *en bloc*, downy. Stems small, 6–10 cm long, mostly curved or broken; branches thin, base thick, 1.5 mm in diameter, prominently longitudinally striate after removing surficial white hairs. Entire leaves mostly petiolate, connecting with thin stems; blade laciniate. Odor, specifically aromatic; taste, slightly bitter.
2. Microscopic characteristics
There were no significant differences in histological characteristics of stems and leaves between this species and *A. scoparia* Waldst. et Kit., except for the following two points.
a. Thicker
Thicker stalk cell walls of medifixed hairs of *A. capillaris* Thunb. Stalk cell walls of medifixed hairs are 1.3-3.4-5 μm and 2.5-4.7-7.5 μm thick for *A. scoparia* Waldst. et Kit. and *A. capillaris* Thunb., respectively.

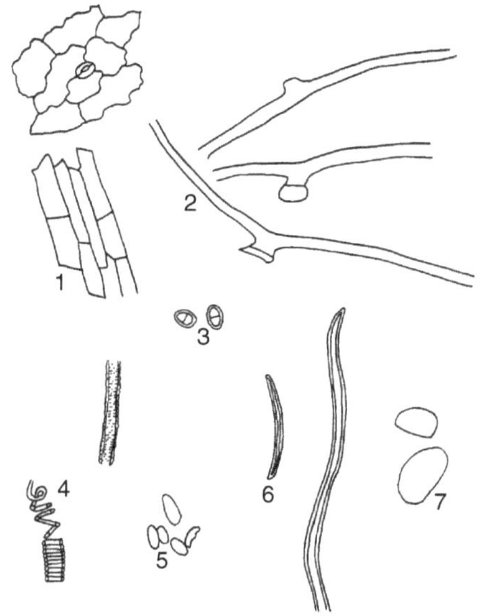

Figure 2.29: Leaf Powder of *A. scoparia* Waldst. et Kit. (×260).
1, Epidermal cells and stomata; 2, medifixed hairs; 3, glandular hairs; 4, vessels; 5, mesophyll cells; 6, fibers; 7, parenchyma cells.

Figure 2.30: *A. capillaris* Thunb.

b. Longer
   Longer head cells of medifixed hairs of *A. capillaris* Thunb. The full length of the head cell is 1030-1240-1700 μm for *A. scoparia* Waldst. et Kit., whereas sometimes more than 2200 μm for *A. capillaris* Thunb. (Fig. 2.31F).

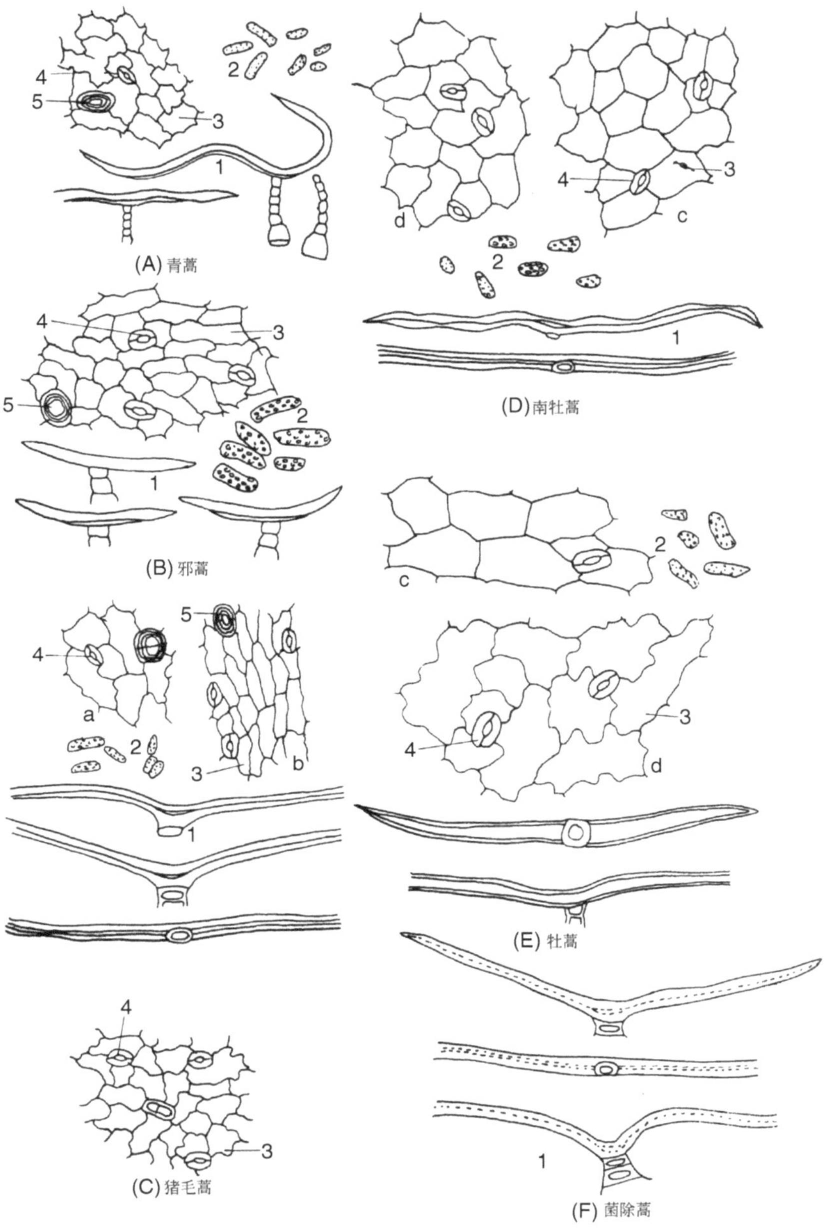

Figure 2.31: Microscopic Characteristics of Leaf (×165).
(A) *Artemisia annua* L.; (B) *Artemisia apiacea* Hance; (C) *Artenisia scoparia* Waldst. et Kit; (D) *Artemisia eriopoda* Bunge; (E) *Artemisia japonica* Thunb ; (F) *Artemisia capillaris* Thunb; 1, medifixed hairs; 2, mesophyll cells; 3, epidermal cells; 4, stomata; 5, glandular hairs.

Figure 2.32: ***A. eriopoda*** **Bunge.**

4.5 Artemisia eriopoda *Bunge*

1. Description (Fig. 2.32)
   Differences between *A. eriopoda* Bunge and *A. japonica* Thunb. are: for *A. eriopoda* Bunge, uppermost stem leaves 1-2-pinnatiparted, lobes lanceolate, middle and lowermost stem leaves pinnatifid, lobe ends usually 3-5-pinnatifid.
2. Microscopic characteristics
   a. Surface view of leaf
      Epidermal cell walls usually flat; head cell walls of medifixed hairs thick, some of which are undulately curved (Fig. 2.33).

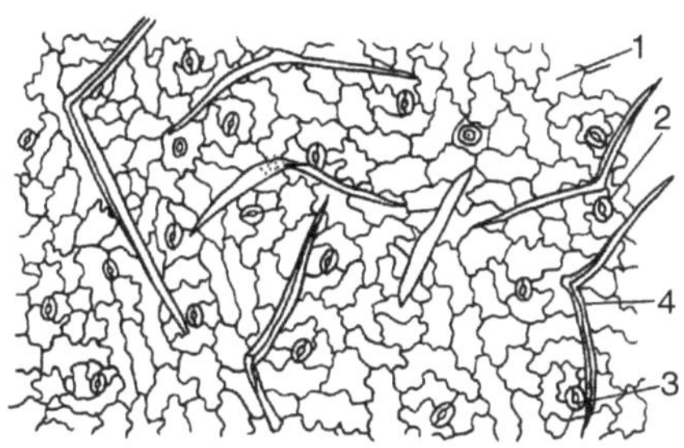

Figure 2.33: Leaf Epidermal Tissue of *A. eriopoda* Bunge (×130).
1, Epidermal cells; 2, stomata; 3, glandular hairs; 4, medifixed hairs.

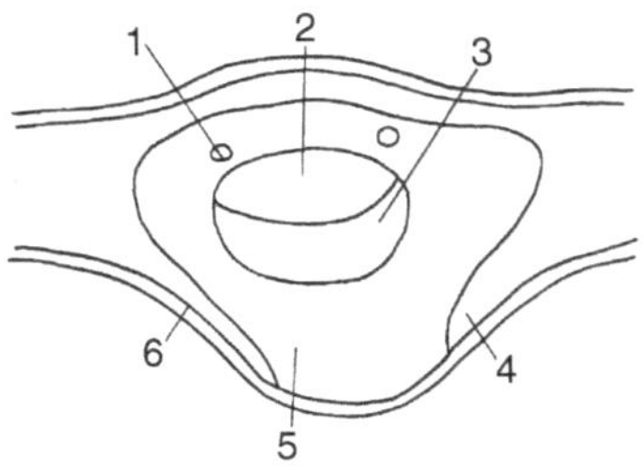

Figure 2.34: Schematic of Transection of Leaf Midrib of *A. eriopoda* Bunge (×28).
1, Secretory cavity; 2, xylem of vascular bundle; 3, phloem of vascular bundle; 4, mesophyll tissue; 5, fundamental tissue; 6, epidermal tissue.

b. Transection of ultimate lobe
   Lower surface of midrib prominent, mesophyll dorsiventral, with secretory cavities (Fig. 2.34).
c. Leaf powder is shown in Fig. 2.35
d. Surface view of stem
   As described in *A. scoparia* Waldst. et Kit.
e. Transection of stem (0.5 cm in diameter)
   Cortical cells 3–10 layers; secretory cavities present in the cortex, occasionally in the pith. The diameter of pith is c. 1/3 of that of stem.

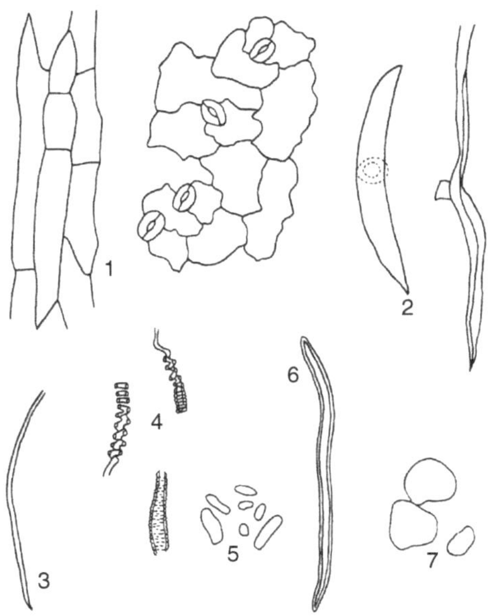

Figure 2.35: Leaf Powder of *A. eriopoda* Bunge (×260).
1, Epidermal cells and stomata; 2, medifixed hairs; 3, linear hairs; 4, vessels; 5, mesophyll cells; 6, fibers; 7, parenchyma cells.

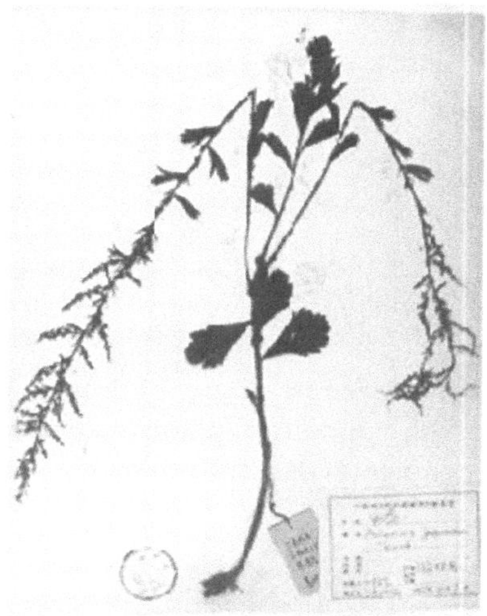

Figure 2.36: ***A. japonica*** **Thunb.**

4.6 Artemisia japonica *Thunb*

1. Description (Fig. 2.36)
   Dried herbs: stems terete, 1–3 mm in diameter, surface fuscous or brown; texture hard, fracture fibriform, yellowish white, centrally with a white and loose pith. Residual leaves yellow-green to fuscous, mostly fragmentary, crinkly and frizzy; texture fragile. Inflorescence yellow-green; stem slice showing several brown oblong seeds; odor, aromatic; taste, slightly bitter.
2. Microscopic characteristics
   a. Surface view of leaf
      Upper epidermal cell wall usually flat, not undulately curved (Fig. 2.37).
   b. Transection of ultimate lobe
      Lower surface of midrib prominent, mesophyll cells long and large, with secretory cavities (Fig. 2.38).

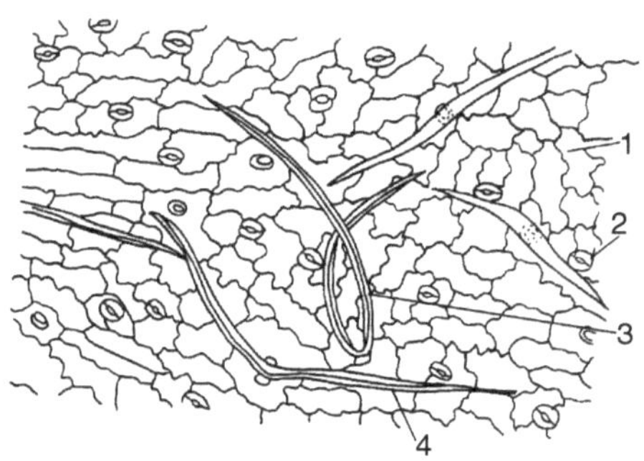

Figure 2.37: Leaf Epidermal Tissue of ***A. japonica*** **Thunb. (×130).**
1, Epidermal cells; 2, stomata; 3, glandular hairs; 4, medifixed hairs.

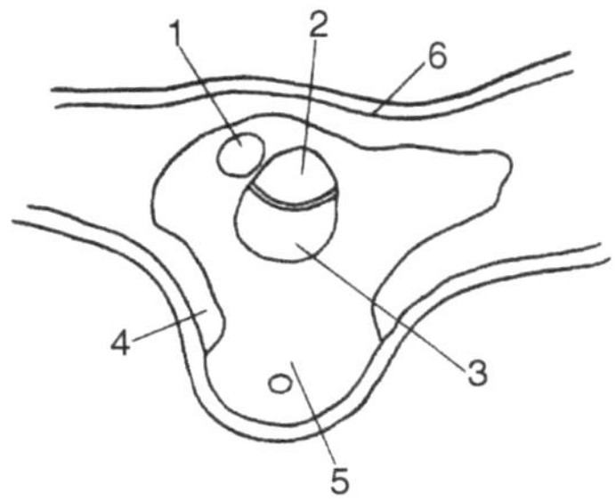

Figure 2.38: Schematic of Transection of Leaf Midrib of *A. japonica* Thunb. (×28).
1, Secretory cavity; 2, xylem of vascular bundle; 3, phloem of vascular bundle; 4, mesophyll tissue; 5, fundamental tissue; 6, epidermal tissue.

c. Leaf powder is shown in Fig. 2.39
d. Surface view of stem
As described in *A. scoparia* Waldst. et Kit.
e. Transection of stem
Secretory cavities are common in both cortex and pith. The pith is of a large size, with c. 2/3 of the whole stem in diameter.

Comparisons of characteristics of leaf tissues of all species aforementioned are shown in Table 2.3 and Fig. 2.31; comparison of transections of stems is shown in Table 2.4.

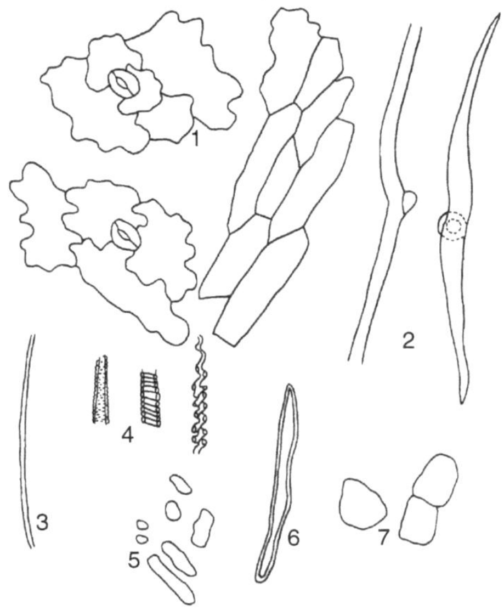

Figure 2.39: Leaf Powder of *A. japonica* Thunb. (×260).
1, Epidermal cells and stomata; 2, medifixed hairs; 3, linear hairs; 4, vessels; 5, mesophyll cells; 6, fibers; 7, parenchyma cells.

Table 2.3: Comparison of leaf tissues

Original Plant	Surface View				Transection				
	Medifixed Hair		Epidermal Cell			Mesophyll Cell		Secretory Cavity	
	No. of Stalk Cells (*N*)	Length of Head Cell (μm)	Shape	Length (μm)	Midrib	Type	Cell Length (μm)	Shape	Diameter (μm)
A. annua L.	3–10	240~480~816, cell wall thin	Irregular shape; anticlinal wall sinuately curved	18~41~80	Both upper and lower surfaces prominent	Isobilateral	20~30~55	None	None
A. apiacea Hance	1–4	93~186~300, cell wall thin	Irregular shape; anticlinal wall sinuately curved	25~45~85	Lower surface slightly prominent	Isobilateral	35~68~100	Oblate	20~26~35
A. scoparia Waldst. et Kit.	1–2 (cell wall 1–3~3–4~5 μm thick)	960~1236~1572, cell wall thick	Linear leaves almost narrow rectangular, wall flat or slightly curved; pinnate leaves like *A. annua* L.	30~58~130	Lower surface prominent (pinnate leaf)	Cyclocytic (linear leaf), isobilateral (pinnate leaf)	20~42~80	Oblate	25~37~50
A. capillaris Thunb.	1~2 (cell wall 2.5~4.7~7.5 μm thick)	960~1236~1572, cell wall thick	Linear leaves almost narrow rectangular, wall flat or slightly curved; pinnate leaves like *A. annua* L.	25~58~112	Lower surface prominent (pinnate leaf)	Cyclocytic (linear leaf), isobilateral (pinnate leaf)	20~42~80	Oblate	25~37~50
A. eriopoda Bunge	1–2	425~665~1,020, cell wall thick, some sinuately curved	Irregular shape; anticlinal wall sinuately curved; wall of upper epidermis slightly sinuately curved	42~60~135	Lower surface prominent	Dorsiventral	15~30~60	Oblate	27~34~42
A. japonica Thunb.	1–2	235~340~672, cell wall thin	Irregular shape; anticlinal wall sinuately curved; wall of upper epidermis slightly sinuately curved	60~119~215	Lower surface prominent	Dorsiventral	30~59~95	Quasicircular	30~48~60

Table 2.4: Comparison of transections of stem tissues

Original Plant	Shape of Transection		No. of Stalk Cells of the Medifixed Hair (*N*)	Secretory Cavities in the Cortex and the Pith
	Trunk	Twig		
A. annua L.	Quasicircular, with more than 10 round horns	Acutely polygonal	3–10	No
A. apiacea Hance	Quasicircular	Acutely polygonal	1–4	Yes
A. scoparia Waldst. et Kit.	Quasicircular	Acutely polygonal	1–2	Yes
A. capillaris Thunb.	Quasicircular	Acutely polygonal	1–2	Yes
A. eriopoda Bunge	Quasicircular	Acutely polygonal	1–2	Yes
A. japonica Thunb.	Quasicircular	Acutely polygonal	1–2	Yes

Discussion

Many sagebrushes have very prominent heterophylls, *A. scoparia* Waldst. et Kit., *A. eriopoda* Bunge, and *A. japonica* Thunb. fall into this category, but for *A. annua* L. and *A. apiacea* Hance, their uppermost stem leaves are more slender and narrow than their lowermost and basal leaves. Due to the practical needs of identification of medicinal materials, observations and comparisons have been made on the tissue between heterophylls. Such difference in leaf shape will generally not lead to prominent changes in cellular tissue; however, only *A. scoparia* Waldst. et Kit., from pinnate to linear leaves, shapes of epidermal cells lead to a gradual transition of irregular shape to oblong shape and mesophyll tissues change from isobilateral to centric mesophylls in surface view. Epidermal cells on the surface of the midrib of the pinnate leaf are also narrowly rectangular, while the epidermis of the linear leaf basically surrounds the surface of the midrib. From this perspective, it is indeed consistent with the pinnate leaf. Additionally, epidermal cells of basal leaves of different species are slightly larger than those of stem leaves.

In this article, all observations and comparisons of stem tissues focus on the primary structure; because of inconsistent differentiation and priority of secondary tissues, they are difficult to compare and will no longer be explained. As the main antimalarial part of *A. annua* L. is its leaves, the tissue identification key listed is dominated by microscopic characteristics of leaf powder materials, and description of Material Medica powders is omitted.

Whether the secretory cavity is present in *A. annua* L. is a different characteristic from other five species. Its shape of transection of stem, composition and morphology of the medifixed hair are also special. In addition, the author has also observed the fluorescence of water and ethanol extracts of leaves, showing azure (water extract) and violet (ethanol extract) colors, which significantly differ from green and red colors of other five species.

Notes:

1. **Identification key to materia medica description**
   a. Identification key to materia medica description is shown in Table 2.5.
2. **Identification key to microscopic characteristics**
   a. Tissue identification key is shown in Table 2.6.
   b. Leaf tissue identification key is shown in Table 2.7.
   c. Leaf powder identification key is shown in Table 2.8.

5 Thin-Layer Chromatography (TLC) [17]

As previously mentioned, *A. annua* L. and its confused congeners are totally six sagebrushes. Except for morphology and histology, their main difference is that certified *A. annua* L. contains artemisinin, the active antimalarial ingredient, whereas other congeners do not. For this, optimized TLC conditions for artemisinin are selected to use a TLC scanner for TLC fingerprinting and comparison, providing a new basis for the physical and chemical identification of *A. annua* L. and its confused congeners.

Table 2.5: Identification key to materia medica description

Stems terete, longitudinally angulate; texture slightly hard, easily broken; pith present in the middle of the fracture, 1–2 mm in diameter.
Leaves dark green or breen, apex pungent, frizzy, and fragile; flattened entire leaves 3-pinnatipartite.
Capitula minutely globose, c. 2 mm in diameter; taste, slightly bitter; odor, specifically aromatic ... *A. annua* L.
Large pith present in the middle of the fracture, 3–5 mm in diameter.
Leaf apex very pungent, flattened entire leaves 2-pinnatipartite.
Capitula semiglobose, large, c. 5 mm in diameter; taste, mild; odor, not specifically aromatic ... *A. apiacea* Hance
Southern and northern Herba Artemisiae Capillariae homomorphic, commercially known as capillary wormwood, white lanuginous throughout, frizzy, off-white or celadon, downy; flattened leaves 1-3-pinnatifid; taste, slightly bitter; odor, fragrant ... *A. capillaris* Thunb. and *A. scoparia* Waldst. et Kit.
Flattened leaves cuneate, dentate apically, flatly palmatifid ... *A. japonica* Thunb.
Flattened leaves spatulate, margins dentate, lobate ... *A. eriopoda* Bunge

Table 2.6: Tissue identification key

No secretory cavity in either leaves and stems; stalk cells of the medifixed hair 3–10 ... *A. annua* L.
Secretory cavities present in the fundamental tissue of the vein, the cortex of the stem, and the pith.
Leaf epidermal cells mostly narrow rectangular, walls flat or curved; stalk cells of the medifixed hair 1–2, head cell extremely long (960~1236~1572 μm) ... *A. scoparia* Waldst. et Kit.
Leaf epidermal cells irregular, anticlinal wall sinuately curved.
Stalk cells of the medifixed hair 1–4, head cell short (93~186~300 μm) ... *A. apiacea* Hance
Stalk cells of the medifixed hair 1–2, head cell long.
Stalk cells of the medifixed hair 425~665~1020 μm long, cell walls thick; epidermal cells 42~60~135 μm long ... *A. eriopoda* Bunge
Stalk cells of the medifixed hair 245~340~672 μm long, cell walls thin; epidermal cells 60~119~125 μm long ... *A. japonica* Thunb.

Table 2.7: Leaf tissue identification key

Epidermal cells almost narrow rectangular, walls flat or curved ... *A. scoparia* Waldst. et Kit. (northern Herba Artemisiae Capillariae) (linear leaf)
Epidermal cells mostly irregular
Midrib of the ultimate lobe basically not prominent; or lower surface slightly prominent; head cell of the medifixed hair short (93∼186∼300 μm), stalk cells mostly 2–3; glandular hairs sparse ... *A. apiacea* Hance
Midrib of the ultimate lobe prominent.
Stalk cells of the medifixed hair mostly 4–5, densely glandular-pubescent ... *A. annua* L.
Stalk cells of the medifixed hair mostly 1–2, rarely glandular-pubescent
Medifixed hair very dense, head cell extremely long (960∼1572 μm) ... *A. scoparia* Waldst. et Kit. (northern Herba Artemisiae Capillariae) (pinnate leaf)
Medifixed hair sparse, head cell extremely short (206∼816 μm)
Epidermal cells c. 95 μm in mean long diameter, mesophyll cells 20∼44∼65 μm long; secretory cavity c. 30 μm in diameter ... *A. eriopoda* Bunge
Epidermal cells c. 119 μm in mean long diameter; mesophyll cells 30∼59∼95 μm long; secretory cavity large, c. 45 μm in diameter ... *A. japonica* Thunb.

Table 2.8: Leaf powder identification key

Epidermal cells almost narrow rectangular, rarely irregular; stalk cells of the medifixed hair 1–2, head cell extremely long (960∼1236∼1572 μm) ... *A. scoparia* Waldst. et Kit. (northern Herba Artemisiae Capillariae)
Epidermal cells mostly irregular, rarely narrow rectangular.
Epidermal cells usually less than 65 μm in long diameter.
Stalk cells of the medifixed hair mostly 4–5, minute, head cell long (240∼480∼816 μm); mesophyll cells minute, 20∼30∼55 μm ... *A. annua* L.
Stalk cells of the medifixed hair mostly 2–3, thick, head cell short (93∼186∼300 μm); mesophyll cells large, 35∼75∼100 μm long ... *A. apiacea* Hance
Epidermal cells usually greater than 65 μm in long diameter.
Epidermal cells c. 95 μm in mean long diameter, mesophyll cells 20∼45∼65 μm long ... *A. eriopoda* Bunge
Epidermal cells c. 119 μm in mean long diameter, mesophyll cells 30∼59∼95 μm long ... *A. japonica* Thunb.

Experiments

1. **Instrument and drugs**
   a. Shimadzu CS-910 dual wavelength TLC scanner (Shimadzu, Japan), Kieselgel 60 F254 (Merck, Germany). All chemical reagents are analytically pure (AP).
   b. Sources and origins of medicinal materials are as follows: (1) *A. annua* Linn. (Hainan Island); (2) *A. annua* L. (Beijing); (3) *A. scoparia* Waldst. et Kit. (northern Herba Artemisiae Capillariae) (Beijing); (4) *A. eriopoda* Bunge (Beijing); (5) *A. capillaris* Thunb. (Beijing); (6) *A. japonica* Thunb. (Shanghai); and (7) *A. apiacea* Hance (Beijing).
   c. Chemical reference substance (CRS): artemisinin (qinghaosu).

2. **Experimental methods**
   a. Sample preparation: According to the extraction method for artemisinin from A. annua L., smash each of the aforementioned 7 samples into the powder and pass through a 40-mesh screen, fill 2 g of each sample in a dry filtration paper cylinder and place it in a sorbitic extractor, then extract in the petroleum ether for 2 h; concentrate the extract to dryness, and transfer into 1 mL of chloroform (CCl4) for sample application purpose.
   b. TLC: Apply 5 μL of each test sample to the Kieselgel 60 F254 (E. Merck) (20 × 20 cm). With petroleum ether-ethyl acetate (85:15) as developing solvent, in a distance of 15 cm, transfer the evaporated solvent, spray with anisaldehyde, develop (in a dryer at 80 ± 1°C for 30 min).
   c. Selection of scanned wavelength: Spots of artemisinin CRS are scanned in the ultraviolet spectra using absorption curves of different wavelengths (400–800 nm), demonstrating a maximum absorption wavelength (λ_{max}) of 525 nm (Fig. 2.40).
   d. Scanning parameters: Dual wavelength reflection zigzag scanning, λs = 525 nm, λR = 700 nm, scanning speed: 20 nm/min, chart speed: 20 nm/min, slit: 1.25 × 1.25 mm, sensitivity × 2, zigzag scan range: ± 5 mm.
   e. Mobile phase: petroleum ether-ethyl acetate (85:15). The thin-layer chromatogram obtained is shown in Fig. 2.41

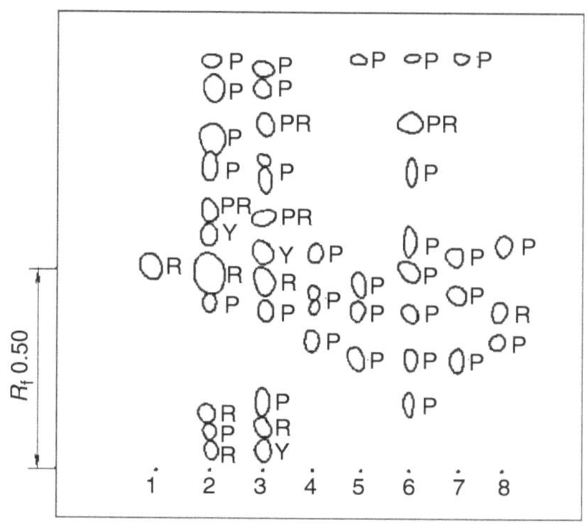

Figure 2.40: Thin-Layer Chromatograms of *A. annua* L. (Southern and Northern) and Its Confused Congeners.
Mobile phase: petroleum ether-ethyl acetate (85:15). 1, Artemisinin CRS; 2, *A. annua* L. (Hainan Island); 3, *A. annua* L. (Beijing); 4, *A. scoparia* Waldst. et Kit.; 5, *A. eriopoda* Bunge; 6, *A. capillaris* Thunb.; 7, *A. japonica* Thunb.; 8, *A. apiacea* Hance. P, Purple; R, red; Y, yellow; PR, purplish red.

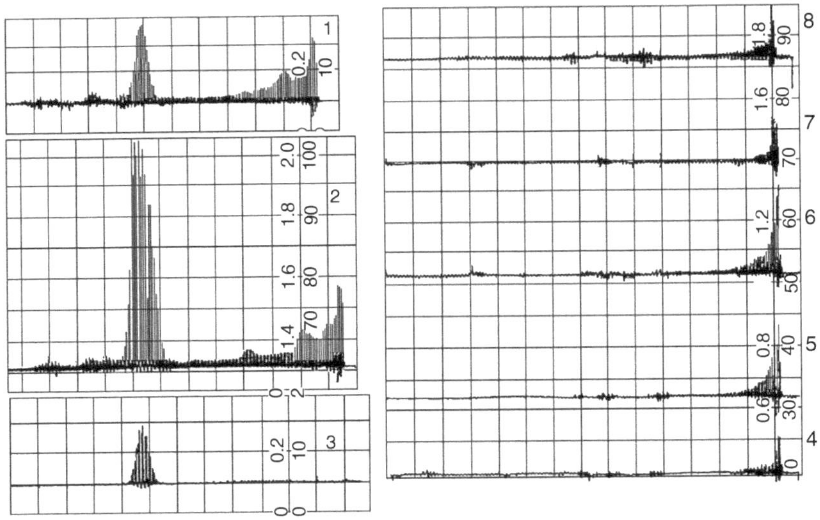

Figure 2.41: Thin-Layer Chromatograms of *A. annua* L. (Southern and Northern) and Confused Congeners.

Mobile phase: petroleum ether-ethyl acetate (85:15). 1, Artemisinin CRS; 2, *A. annua* L. (Hainan Island); 3, *A. annua* L. (Beijing); 4, *A. scoparia* Waldst. et Kit.; 5, *A. eriopoda* Bunge; 6, *A. capillaris* Thunb.; 7, *A. japonica* Thunb.; 8, *A. apiacea* Hance.

Discussion

1. Perfect extraction methods and TLC conditions for artemisinin should be well selected for comparisons because the main difference between certified *A. annua* L. and several other confused congeners is that *A. annua* L. contains artemisinin while other species do not. Experiments have shown that the following conditions are also appropriate for each of other species, with many spots identified: extraction of crude drugs with petroleum ether, a mixture of petroleum ether and ethyl acetate (85:15) as the mobile phase for TLC.
2. Selection of chromogenic agent: Artemisinin can be developed with vanillin/sulfuric acid, para-Dimethylaminobenzaldehyde (PDAB), anisaldehyde, etc. Vanillin is more commonly used because it is prominently characteristic. Artemisinin can change from yellow to green and blue, successively, but TLC requires good stability. Thus, anisaldehyde is selected due to its prominent red spots and stability within 2 h; in contrast, PDAB shows indefinite spots and poor chromogenic effect.
3. Sweet wormwood (*A. annua* L.) is widely distributed throughout China, but there is great difference in artemisinin content: some produced in certain regions are suitable for artemisinin production in industry, but some are not. Some southern provinces and municipalities are superior to northern ones. It can be seen from thin-layer chromatograms that the artemisinin content is significantly higher in *A. annua* L. from Hainan Island than from Beijing, whereas each of other confused congeners does not contain artemisinin. L. and its confused congeners.

6 Study of Chemical Constituents of *A. annua L.* and Its Confused Congeners

- Comparison of Chemical Constituents
  Comparison of chemical constituents of *A. annua* L. and its five confused congeners is listed in Table 2.9 [18–26].
- Structural Formulae of Chemical Constituents
  - Structural formulae of chemical constituents of *A. annua* L. are illustrated in Fig. 2.42 [19–22]
  - Structural formulae of chemical constituents of *A. apiacea* Hance are illustrated in Fig. 2.43 [23]
  - Structural formulae of chemical constituents of *A. eriopoda* Bunge are illustrated in Fig. 2.44 [24]
  - Structural formulae of chemical constituents of *A. japonica* Thunb. are illustrated in Fig. 2.45 [25]

7 Dynamic Research on Chemical Constituents of *A. annua L.* [26]

It has been found in antimalarial studies that premature *A. annua* L. and a large amount of stalks have no antimalarial activity. Thus, further discussion is needed on active antimalarial parts and harvest periods of *A. annua* L., to provide scientific basis for harvesting, processing, and rational use of sources of *A. annua* L.

- Chemical Research on Young (From Beijing in Early May) and Adult Plants (Leaves Harvested in Autumn) of *A. annua* L.
  - Comparison between neutral and acidic fractions
    According to antimalarial studies, the neutral extract from the leaves of *A. annua* L. is the active antimalarial part, from which artemisinin is extracted. The acidic fraction has no antimalarial activity, but the artemisinic acid contained is the precursor of sesquiterpene lactones, including artemisinin (Table 2.10).
  - Comparison of chemical constituents
    Comparison of chemical constituents between young and adult plants of *A. annua* L. is illustrated in Table 2.11.
- Comparison of Artemisinin Content With Respect to Different Harvesting Seasons, Medicinal Parts and Storage Time of *A. annua* L.
  - Beijing area [15]. Comparison of artemisinin content with respect to different harvest time, medicinal parts, and storage time of *A. annua* L. from Beijing area is illustrated in Table 2.12
  - Shandong area [27]: Comparison of artemisinin content with respect to different harvest time and medicinal parts of *A. annua* L. from Shandong area is illustrated in Table 2.13

Table 2.9: Comparison of chemical constituents of *A. annua* L. and its five confused congeners

Identification Point	*A. annua* L.	*A. apiacea* Hance	*A. japonica* Thunb.	*A. eriopoda* Bunge	*A. capillaris* Thunb.	*A. scoparia* Waldst. et Kit.
Volatile oil	Myrcene, limonene, 1,8-cineole, α-pinene, terpinen-4-ol, borneol, camphene, camphor, β-bisabolene, aromadendrene, γ-cadinene, γ-muurolene, alkene, artemisia ketone, benzyl isovalerate, trimethyl bicyclo-heptene	α-Farnesene, *trans*-β-farnesene, lavandulol, myrcene, terpinolene, α-curcumen, δ-cadinene, α-cadinene, methyleugenol	Myrcene, *p*-cymene, limonene, perillene, α-pinene, β-pinene, α-terpilenol, bomyl acetate, camphene, calamenene, alkene, methyleugenol, naphthalene	α-Curcumene, δ-elemene, γ-cadinene, calamenene, γ-muurolene, alkene	Myrcene, 1,8-cineole, borneol, camphene, α-curcumene, benzaldehyde, davanone, capillin, eugenol, isoeugenol, naphthalene	Carvone, *p*-cymene, limonene, perillene, α-phellandrene, thymol, α-pinene, β-pinene, terpilenol, verbanol, methyleugenol, naphthalene, ar-methyl acetophenone
Sesquiterpenoids	Artemisinin, arteannuin A, arteannuin B, arteannuin C, arteannuin D, arteannuin E, artemisinic acid					
Coumarins	Scopoletin	7-Methoxycoumarin, 7-prenyloxy-8-methoxycoumarin, 7,8-(methylenedioxy) coumarin, 7,8-dimethoxycoumarin	7,8-Dimethoxy coumarin, 6,7-dimethoxy coumarin		6,7-Dimethoxy coumarin, scopoletin, capilarin, etc.	6,7-Dimethoxycoumarin, scopoletin, 6-hydroxy-7-methoxycoumarin, capilarin, etc.
Flavonoids	3,5-Dihydroxy-6,7,3′,4′-tetramethoxy flavonol, 5-hydroxy-3,6,7,3′,4′-pentamethoxy flavone, 5-hydroxy-3,6,7,4′-tetramethoxy flavone, 3,5,3′-trihydroxy-6,7,4′-trimethoxy flavone		8,4′-dihydroxy-3,7,2′-trimethoxy flavone, 3,5-dihydroxy-6,7,3′,4′-tetramethoxy flavone		Rhamnocitrin, cirsimaritin, eupalitin, quercetin, arcapillin, isoarcapillin, etc.	Rhamnocitrin, cirsimaritin, eupalitin, quercetin, etc.
Organic acid	Artemisinic acid		Cinnamic acid, *p*-methoxybenzoic acid, ferulic acid, melissic acid		Capillartemisin, etc.	Chlorogenic acid, salicylic acid, azelaic acid
Other	Octacosanol, β-sitosterol, stigmasterol, aurantiamide acetate	Stigmasterol	β-Amyrin, β-sitosterol, stigmasterol	α-Amyrin, δ-amyrin, β-sitosterol, triacontanol	Capillarisin, etc.	*p*-Hydroxyacetophenone choline

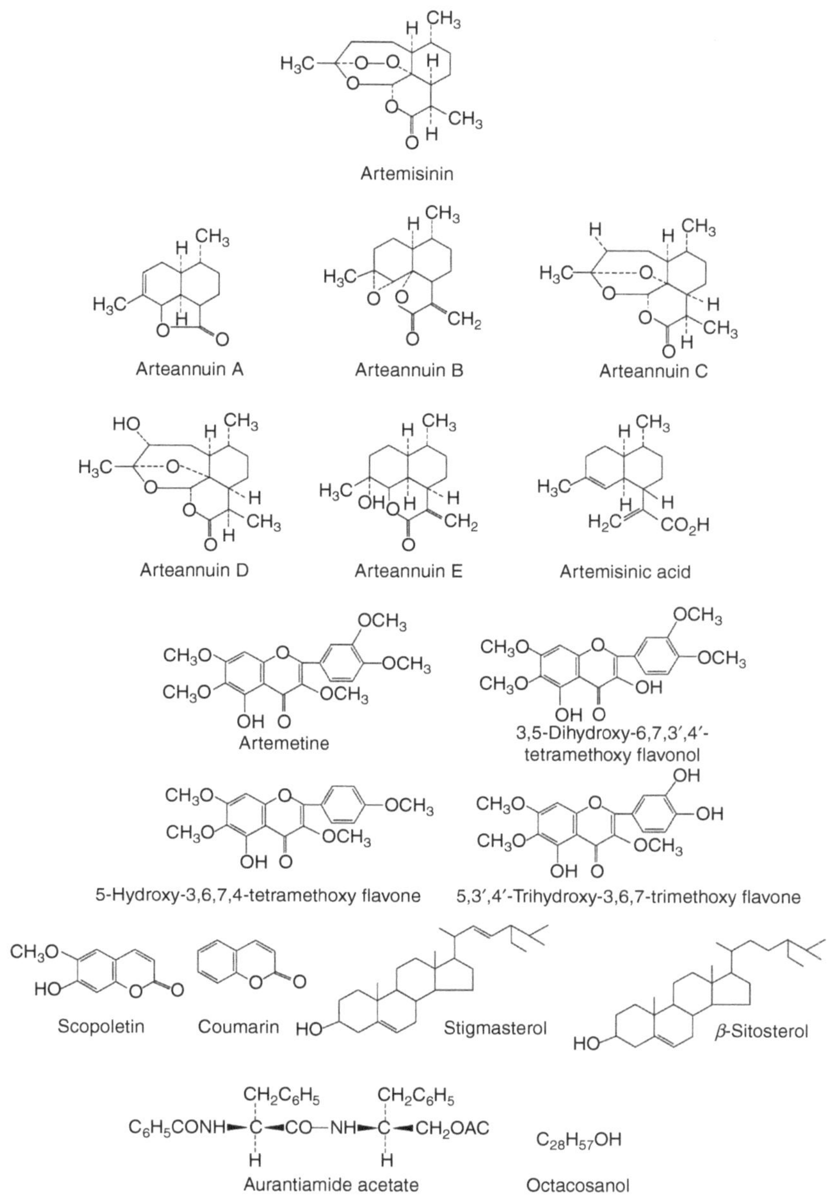

Figure 2.42: Structural Formulae of Chemical Constituents of *A. annua* L.

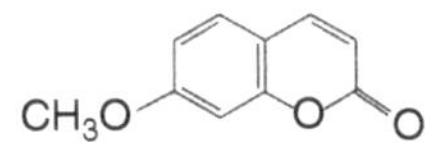

7-Methoxycoumarin

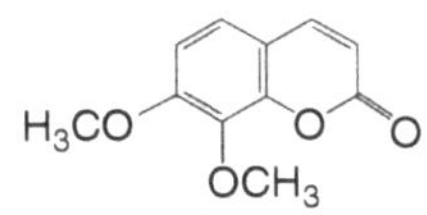

7,8-Dimethoxycoumarin

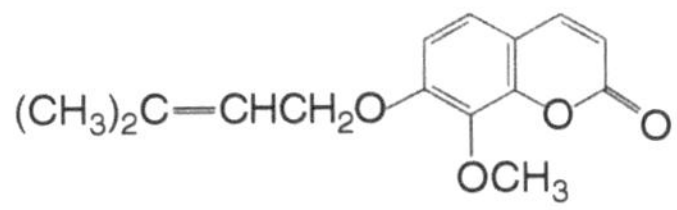

7-Prenyloxy-8-methoxycoumarin

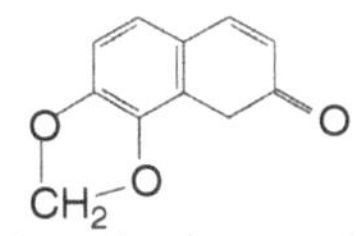

7,8-(Methylenedioxy)coumarin

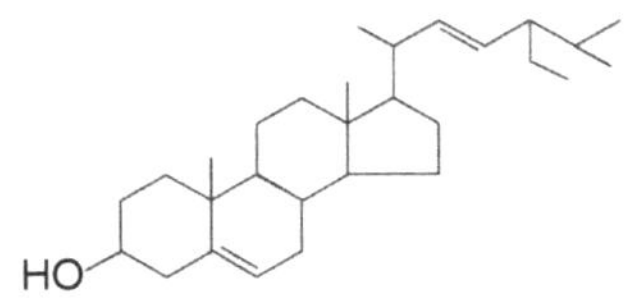

Stigmasterol

Figure 2.43: Structural Formulae of Chemical Constituents of *A. apiacea* Hance.

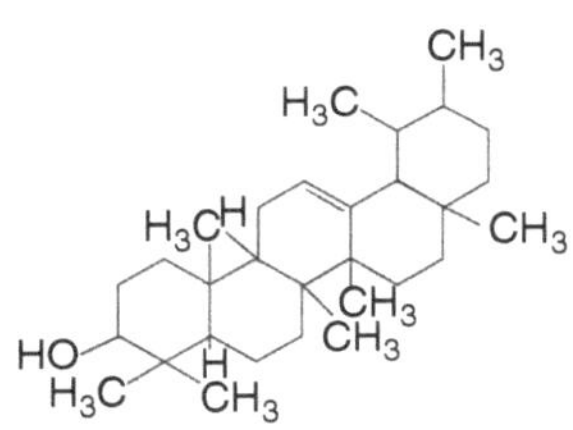

α-Amyrin

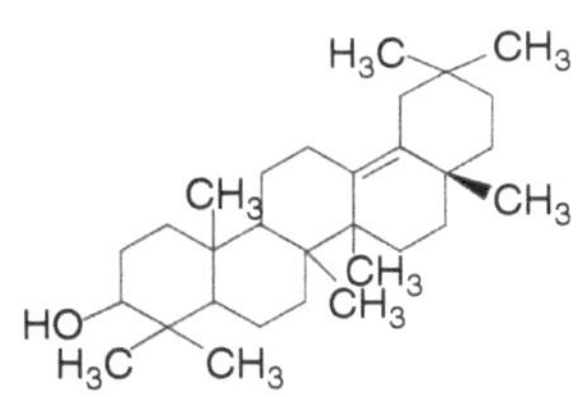

δ-Amyrin

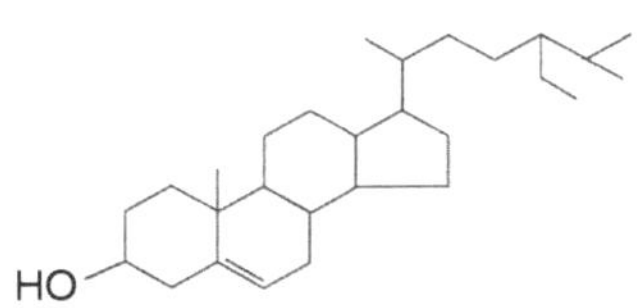

β-Sitosterol

$CH_3{\cdot}[CH_2]_{28}{\cdot}CH_2OH$

Triacontanol

Figure 2.44: Structural Formulae of Chemical Constituents of *A. eriopoda* Bunge.

- Sichuan area [28]
  - Different harvest time: Comparison of artemisinin content of leaves of *A. annua* L. from Sichuan area at different harvest time is illustrated in Table 2.14
  - Storage time: Comparison of artemisinin content of leaves of *A. annua* L. from Sichuan area at different storage time is illustrated in Table 2.15

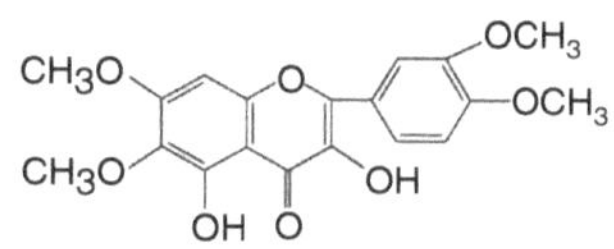

3,5-Dihydroxy-6,7,3′4′-tetramethoxyflavone

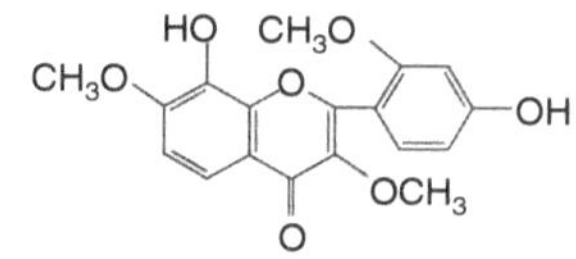

8,4-Dihydroxy-3,7,2′-trimethoxyflavone

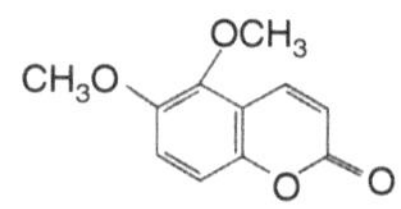

7,8-Dimethoxycoumarin

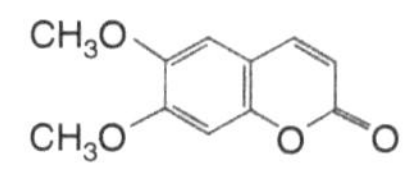

6,7-Dimethoxycoumarin

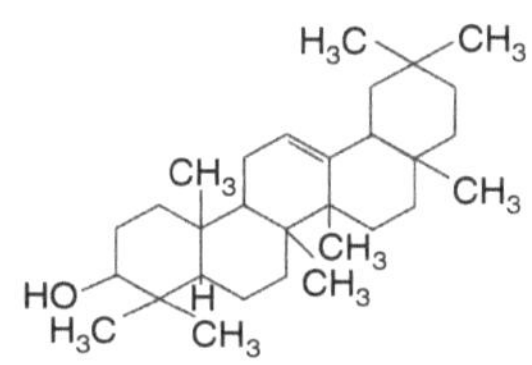

β-Amyrin

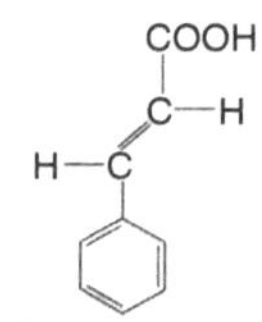

4-hydroxycinnamic acid

$CH_3 \cdot [CH_2]_{28} \cdot COOH$

Triacontanoic acid

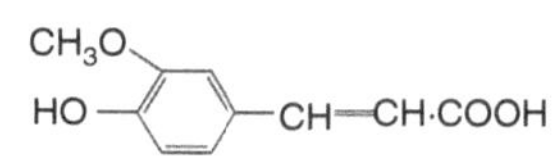

Ferulaic acid

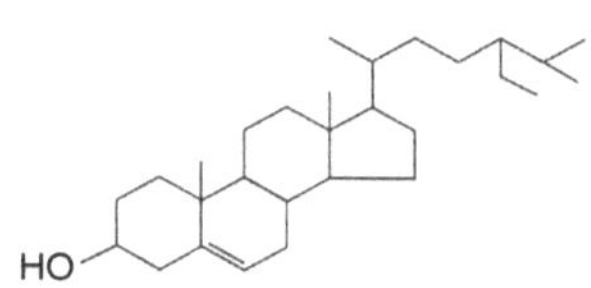

β-Sitosterol

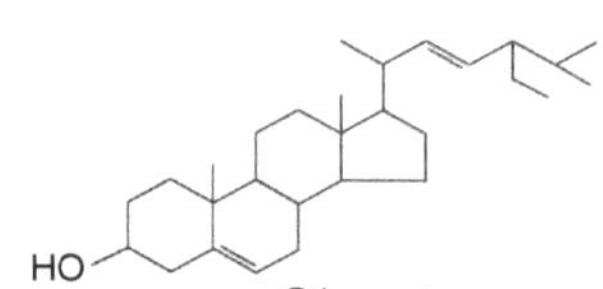

Stigmaster

Figure 2.45: Structural Formulae of Chemical Constituents of *A. japonica* Thunb.

Table 2.10: Comparison between neutral and acidic fractions of young and adult plants of *A. annua* L.

Harvest Period	Acidic Fraction (%)	Neutral Fraction (%)
Young plant	3.7	1.4
Adult plant	1.3	2.3

Table 2.11: Comparison of chemical constituents between young and adult plants of *A. annua* L.

Harvest Period	Sesquiterpene Lactone	Organic Acid	Flavonoids	Other
Young plant	None	Artemisinic acid	3,5-dihydroxy-6,7,3′,4′-tetramethoxy flavonol	Octacosanol, β-sitosterol, stigmasterol
Adult plant	Artemisinin, arteannuin A, arteannuin B, arteannuin C, arteannuin D, arteannuin E	Artemisinic acid	3,5-dihydroxy-6,7,3′,4′-tetramethoxy flavonol, 5-hydroxy-3,6,7,4′-tetramethoxy flavone, 5-hydroxy-3,6,7,3′,4′-*p*entamethoxy flavone, 5,3′,4′-trihydroxy-3,6,7-trimethoxy flavone	Octacosanol, β-sitosterol, stigmasterol, coumarin, scopoletin, aurantiamide acetate

Table 2.12: Comparison of artemisinin content with respect to different harvest time, medicinal parts, and storage time of *A. annua* L. from Beijing area

		Artemisinin (%)		
Harvest Time	**Medicinal Part**	**First**	**Second**	**Mean**
Beijing, July 12	Leaves	0.11	0.09	0.10 ± 0.01
	Stems	No	No	
Beijing, August 10 (before flowering)	Leaves	0.13	0.11	0.12 ± 0.01
	Stems	No	No	
Beijing, August 14 (bud time)	Leaves	0.15	0.16	0.16 ± 0.01
	Stems	No	No	
	Flowers	0.20	0.23	0.22 ± 0.02
Beijing, August 21 (anthesis)	Leaves	0.23	0.25	0.24 ± 0.01
	Stems	No	No	
	Flowers	0.52	0.56	0.54 ± 0.02
Beijing, August 29 (anthesis)	Leaves	0.12	0.11	0.12 ± 0.01
	Stems	No	No	
	Flowers	0.14	0.27	0.21 ± 0.07

Table 2.13: Comparison of artemisinin content with respect to different harvest time and medicinal parts of *A. annua* L. from Shandong area

Harvest Time	Medicinal Part	Artemisinin Content
Early June	Leaves	Trace
Early July	Leaves	0.02%
	Stems	None
Early August	Leaves	0.08%
	Stems	None
	Roots	None
Early September	Leaves (mixed with flower buds)	0.12%
	Stems	None
	Roots	None
Early September	Flower buds	0.20%
Late September	Flowers and fruits	0.075%

Table 2.14: Comparison of artemisinin content of leaves of *A. annua* L. from Sichuan area at different harvest time

Harvest Time	Medicinal Part	Artemisinin Content (%)	Note
May 30	Leaves	0.38	
June 15	Leaves	0.38	
June 30	Leaves	0.43	
July 15	Leaves	0.65	
July 30	Leaves	0.61	
August 15	Leaves	0.58	
August 30	Leaves	0.56	
	Spears of the upper plant	0.65	Suggestive of higher content in spears than old leaves
	Leaves of the middle plant	0.55	
	Leaves of the lower plant	0.36	

Table 2.15: Comparison of artemisinin content of leaves of *A. annua* L. from Sichuan area at different storage time

Medicinal Part	Storage Time	Content	Note
Leaves	Newly harvested Stored for 9 months	0.56%, 0.61% 0.51%, 0.51%	Suggestive of certain effect of storage time on content

Discussion

1. Young plants of *A. annua* L. contain no artemisinin, so there is no antimalarial activity. A quantity of artemisinic acid will be gradually converted into various sesquiterpenoids over time. Therefore, *A. annua* L. is inadvisable for premature harvesting and young plants are inappropriate for pharmaceutical use, especially for raw materials for antimalarial.
2. The active antimalarial part of *A. annua* L. is mainly leaves, but neither stem nor root is active. However, studies have shown that both leaves and stems are active in conventional antipyretic effect, and leaves are more potent than stems. Thus all things considered, aerial parts, including leaves and stems, can still be used for pharmaceutical use, but it is advisable for harvesting when leafy.
3. As for the harvest season, experiments of samples from Beijing, Shandong, and Sichuan have indicated that the preanthesis and leafy stage in autumn (July and August) is preferable. Flower buds contain high artemisinin content, but in full bud stage, leaves have been scarce; after all, the ratio of flower bud is small, so *A. annua* L. is preferable to be harvested in the preanthesis and leafy stage.
4. The artemisinin content is higher in spears or leaf apexes than in stem leaves, which is consistent with the record in ancient literature that leaf apexes are used for the treatment of malaria.
5. Content decreases as storage time rises. Thus, it is advisable to harvest and use in the same year.

References

[1] 国家药典委员会．中华人民共和国药典．北京：化学工业出版社, 2005. [Available only in Chinese: National Pharmacopoeia Committee. Pharmacopoeia of the People's Republic of China. Beijing: Chemical Industry Press; 2005.]

[2] 肖培根．新编中药志．北京：化学工业出版社, 2002. [Available only in Chinese: Xiao PG. Modern chinese material medica. Beijing: Chemical Industry Press; 2002.]

[3] 屠呦呦．中药青蒿的正品研究．中药通报 12 (4) (1987) 194–197.[Available only in Chinese: Tu YY. Authentic research on traditional Chinese medicine of Artemisia annua. Bull Chin Mater Med 1987;12(4):194–7.]

[4] 屠呦呦，张衍箴，张定媛等. "青蒿"中药志.第四册.北京：人民卫生出版社, 1988. [Available only in Chinese: Tu YY, Zhang YZ, Zhang DY, et al. "Qinghao". Chinese medicinal herbal, book 4. Beijing: People's Medical Publishing House; 1988.]

[5] Deleted in review

[6] Deleted in review

[7] ［日］松村任三.改正增补植物名汇.日本东丸善株式会社, 1895.30. [Available only in Chinese: Matsumura N. Correct supplement plant name. Japan Higashi Shanzhu Association; 1895 [p. 30].]

[8] ［日］牧野富太郎.植物图鉴.参文舍第, 2 版.东京北隆馆，明治 42 年 (1909年). [Available only in Chinese: Makino T. Botanical illustrations, Canwen Association version 2. Tokyo Beilong Gallery; 1909.]

[9] Stuart GA. Chinese Materia Medica: Vegetable Kingdom. American Presbyterian Mission Press, Shanghai, 1911, p. 50. [Available only in Chinese: Stuart GA. Chinese materia medica, vegetable kingdom. Shanghai: American Presbyterian Mission Press; 1911 [p. 50].]

[10] 孔庆莱等.植物学大辞典.第, 7 版.北京：商务印书馆，民国17年 (1928年). [Available only in Chinese: Kong QL. Botany dictionary. Version 2. Beijing: Commercial Press; 1928.]

[11] 白井光太郎等.头注图译本草纲目.第 5 册.日本东京：春阳堂藏版，昭和 8 年 (1933 年). [Available only in Chinese: Shirai FK, et al. Figure version of compendium of materia medica, vol. 5. Tokyo, Japan: Chun Yang-Tang Haidishijie; 1933.]

[12] 李时珍.本草纲目.北京：人民卫生出版社 15 (1975) 13, 15. [Available only in Chinese: Li SZ. Compendium of materia medica. Beijing: People's Medical Publishing House; 1975 [pp. 13,15].]

[13] 陈重明等.邪蒿的原植物考证和 Seseli L□属中名的订正.植物分类学报 19 (4) (1981) 534. [Available only in Chinese: Chen CM., et al. Original plants of array research and name of the correction in *Seseli* L. genus. Acta Phytotaxon Sin 1981;19(4):534.]

[14] 胡世林.为青蒿"正名".健康报, 2005.11,10. [Available only in Chinese: Huang MY. Shilin hunamedrectification for "qinghaosu". Health News 2005;11:10.]

[15] 中国中医研究院中药研究所.青蒿抗疟研究专辑(1971~1978 年). 中药研究资料, 1978. [Available only in Chinese: Traditional Chinese Medicine Research Institution. Artemisinin-based antimalarial album (1971–1978). Traditional Chinese Medicine Research Data; 1978.]

[16] 张宝媛，付桂芳.中药青蒿的组织鉴定研究.药物分析杂志 4 (1985) 199–203. [Available only in Chinese: Zhang BY, Fu GF. The organization of the Chinese medicine artemisinin identification research. Chin J Pharmaceut Anal 1985;4:199–203.]

[17] 屠呦呦,黄珉珉.青蒿品种整理和质量研究——青蒿.内部资料, 1990, 63–66. [Available only in Chinese: Tu YY, Huang MM. *Artemisia annua* species systematics and quality study—*Artemisia annua*. Int Mater; 1990:63–6.]

[18] 钟裕蓉，崔淑莲.青蒿挥发油化学成分研究.中药通报 8 (6) (1983) 31–32. [Available only in Chinese: Zhong YR, Cui SL. Study on chemical constituents of volatile oil from *Artemisia annua* L. Bull Chin Mater Med 1983;8(6):31–2.]

[19] 屠呦呦，倪慕云，钟裕蓉等.中药青蒿化学成分的研究□.药学学报 16 (5) (1981) 366–370. [Available only in Chinese: Tu YY, Ni MY, Zhong YR, et al. Study on the constituents of *Artemisia annua* L. Acta Pharmaceut Sin 1981;16(5):366–70.]

[20] 屠呦呦，倪慕云，钟裕蓉等.中药青蒿的化学成分和青蒿素衍生物的研究.中药通报 6 (2) (1981) 31. [Available only in Chinese: Tu YY, Ni MY, Zhong YR, et al. Studies on the constituents of *Artemisia annua* and derivatives of artemisinin. Bull Chin Mater Med 1981;6(2):31.]

[21] Tu YY, Ni MY, Zhong YR, et al. Studies on the constituents of *Artemisia annua*: part II. Planta Medica 1982;44:143–5.

[22] 屠呦呦，倪慕云，钟裕蓉等.中药青蒿的化学成分和青蒿素衍生物的研究.中药通报(III). 屠呦呦，尹建平，吉力等.中药青蒿化学成分的研究(□).中草药1985, 16(5):200~201. [Available only in Chinese: Tu YY, Yin JP, Ji L, et al. Studies on the constituents of *Artemisia annua*, part III. Chin Herbal Med 1985;16(5):200–1.]

[23] 吴崇明，屠呦呦.邪蒿化学成分的分离鉴定.中草药 16 (6) (1985) 2–3. [Available only in Chinese: Wu CM, Tu YY. The separation of array chemical identification. Chin Herbal Med 1985;16(6):2–3.]

[24] 尹建平，屠呦呦.南牡蒿化学成分的研究.中草药 20 (4) (1989) 5–6. [Available only in Chinese: Yin JP, Tu YY. Studies on the constituents of *Artemisia eriopoda* Bge. Chin Herbal Med 1989;20(4):5–6.]

[25] 顾玉诚，屠呦呦.牡蒿的化学成分研究.中草药 24 (3) (1993) 122–124. [Available only in Chinese: Gu YC, Tu YY. Studies on the constituents of *Artemisia japonica*. Chin Herbal Med 1993;24(3):122–4.]

[26] 屠呦呦，朱启聪，沈星.中药青蒿幼株的化学成分研究.中药通报 10 (9) (1985) 419–420. [Available only in Chinese: Tu YY, Zhu QC, Shen X. Studies on the constituents of *Artemisia annua* young plants. Bull Chin Mater Med 1985;10(9):419–20.]

[27] 山东中医药研究所.中药青蒿不同采收期及药用部位青蒿素含量测定.山东医药工业 2 (1983) 7–8. [Available only in Chinese: Shandong Institute of Traditional Chinese Medicine. Different harvest time and medicinal parts of artemisinin content determination. Shandong Pharmaceut Ind 1983;2:7–8.]

[28] 四川中药研究所抗疟小组.抗疟药青蒿的研究.中草药通讯 1 (1979) 5–12. [Available only in Chinese: Sichuan Institute of Traditional Chinese Medicine of Antimalarial Group. Research of antimalarial drug artemisinin. Chin Herbal Med 1979;1:5–12.]

CHAPTER 3

Research on Artemisia annua *L. for the Treatment of Malaria*

Chapter Outline

On January 21, 1969 Youyou Tu was appointed as the head of the research group to undertake the task of exploring new drug for treatment of malaria by the Institute of Chinese Material Medica, China Academy of Chinese Medical Sciences after being mobilized by the director and deputy director of national project "523" office (formerly known as State Leading Group Office of Treatment and Prevention of Malaria). Starting with systematic collection of herbal, animal, and mineral prescriptions, Youyou Tu summarized numerous prescriptions in a brochure *Antimalarial Collections of Recipes and Prescriptions*. After that, she obeyed "523" arrangement and was sent to work in malaria-endemic area in Hainan until the National "523" General Assembly was convened in Guangzhou on May 22, 1971. As the task was difficult to be completed, the leadership role of three ministries (Ministry of Health, Medical Department of General Logistics Department and Ministry of Chemical Industry) and one academy (Chinese Academy of Sciences) was established to promote the work. Premier Zhou also gave instructions via telegram, thus reflecting the importance of this task. Youyou Tu was assigned

From *Artemisia annua* L. to Artemisinins
http://dx.doi.org/10.1016/B978-0-12-811655-5.00003-9

by the leader of the Institute to attend the Assembly. She built up the antimalaria research group to act in the spirit of the 1971 State Council Release (29) Document after the Assembly. The first job was to conduct efficacy screening studies. At that time, Comrade Lang Linfu was responsible for screening of *Plasmodium berghei* and *Plasmodium inui*, *Plasmodium cynomolgi*. Although only four members, all of them worked very hard and often examined two lots of samples once a week, whose practical action really implemented the fighting spirit "with laboratory as battlefield" that was needed for military project. After screening the efficacy of more than 380 extracted samples from more than 200 kinds of traditional Chinese medicine, they finally focused on "A Handful of Qinghao Immersed in Two Liters of Water, Wring out the Juice and Drink It All" described in *Recipe 16 of Malaria Treatment* in *A Handbook of Prescriptions for Emergencies* (East Jin Dynasty). Through comprehensive study on the resources of varieties, medicinal parts, collection season, and extraction methods, they made a breakthrough that *Artemisia annua* neutral extract achieved 100% inhibition of *Plasmodium* in *P. berghei*, *P. inui*, *P. cynomolgi* on October 04, 1971. Based on clinical studies, they successfully treated first 30 cases with artemisinin in 1972. On November 08, 1972, an artemisinin monomer was separated from neutral extract, which was identified as the active substance in *A. annua* in 1973. Since then, a new antimalaria drug, Artemisinin, was available. Five hundred twenty-nine cases were validated for several successive years. In 1986, Artemisinin was granted with *New Drug Certificate*. This chapter will describe the active part of *A. annua* as well as the discovery and identification of effective artemisinin monomer.

1 Researches on Active Antimalarial Ingredients in Herbs

1.1 Exploration and Identification

Leaves of *A. annua* L. (origin: Beijing) at the adult stage were harvested in autumn, and then extracted with water and 95% ethanol, but the extracted volatile oil had no effect on *Plasmodium berghei*. After being cold dipped in ethanol, the extract exhibited higher potency on *P. berghei* if it was concentrated at the temperature of 60°C, but lost its potency if the temperature was higher than 60°C. The active ingredients extracted by cold dipping or ether reflux method had significantly higher and stable potency on *P. berghei* (Table 3.1). Notably, the temperature that was controlled to be below 60°C was critical for the extraction of *A. annua* L. crude drug. The antimalarial effect of extracted artemisinin monomer that had been boiled in water for half an hour or refluxed in ethanol for 4 h remained stable. This suggested the antimalarial effect of artemisinin deteriorated in the presence of certain substances only when the temperature rose in the process of the crude extract. In the practice of exploring active antimalarial ingredients from *A. annua* L., researchers have profoundly understood that Chinese medicine research work must be seriously conducted to "discard the dross and select the essential, eliminate the false and retain the true." Only through carefully repeated practice, success can be achieved by grasping and following the inherent law.

Ether extracts had high and stable efficacy, but its toxicity was higher with increasing dose. Thus, it was necessary to refine the active ingredients by discarding the dross and selecting the essence,

Table 3.1: Titer of different extracts for rodent malaria

Extract	Dosage (g/kg)	Inhibition Rate of Rodent Malaria (%)
Water extract	Crude drug: 50 equivalently	6
	Crude drug: 120 equivalently	12
Volatile oils	Crude drug: 750 equivalently	−27
Hot ethanol extract	Crude drug: 30–40 equivalently	33
	Crude drug: 60–90 equivalently	36
Low-temperature ethanol-quenched concentrate	Crude drug: 50 equivalently	95
Ether extract	Crude drug: 10–20 equivalently	71
	Crude drug: 25–35 equivalently	95–100

Table 3.2: The titer and toxicity of neutral versus acidic part for rodent malaria

Extract	Dosage (g/kg)	Inhibition Rate of Rodent Malaria (%)	Mortality (%)
Neutral part	0.6	99	0
	0.9-1.0	100	0
Acidic part	2.2	−11	30
	3.2	−6	73

so as to facilitate in-depth research. After further investigation, researchers found that the ether extracts can be divided into neutral and acidic moieties, of which the acidic moiety accounting for two-thirds exhibited no efficacy but had high toxicity, and neutral moiety showed significant antimalarial effect at an oral dose of 1 g/kg × 3 days, with 100% inhibition rate of *P. berghei* and low toxicity (Table 3.2). Its extraction process and efficacy results are shown as follows.

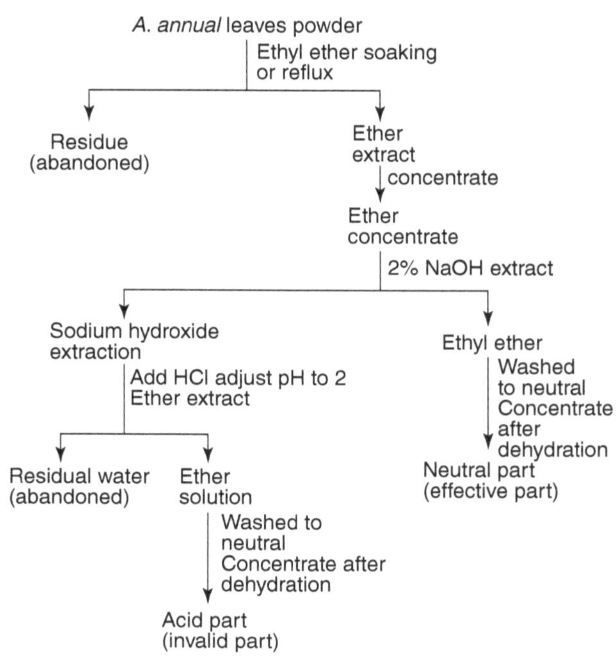

After succeeding in inhibition of *P. berghei*, the experiments were carried out in *P. inui* (*P. cynomolgi*), and the results were consistent with those obtained in *P. berghei* experiments, with inhibition rate of 100%.

The key breakthrough point in achieving antimalarial success is to find active antimalarial ingredients with stable 100% inhibition rate of *Plasmodium* from *A. annua* L. On March 08, 1972, Youyou Tu gave a research report titled "Exploration of Antimalarial Herbs Guided by Mao Zedong Thought" regarding the aforementioned finding under the arrangement of "523" Office at the National Research Council on Antimalarial Medicines held in Nanjing, thus greatly inspiring the attendees and attracting their attention. The Institute of Chinese Material Medica, China Academy of Chinese Medical Sciences was assigned by "523" Office to observe the clinical antimalarial efficacy in malaria endemic areas of Hainan in the same year.

In 1972, the neutral moiety in ether extracts were prepared into capsules which showed significant antimalarial efficacy in the first 30 clinical cases, thereby proving that the neutral moiety in *A. annua* L. had antimalarial action.

1.2 Research on Pharmacodynamics of Active Neutral Moiety in Artemisia annua L.

1.2.1 Plasmodium berghei

P. berghei model was established in the experiment. Healthy mice, weighing 18–22 g, were randomly assigned into groups. Blood samples were collected from eyeball removed from malaria infected mice, diluted with saline containing anticoagulant, and each healthy mouse was inoculated intraperitoneally 1×10^7 *Plasmodium*. All animals were treated with drugs since 24 h after inoculation for 3 consecutive days. Blood smears were collected from tails of animals at 24 h after the last administration for observation of inhibition of *Plasmodium* and negative returning and calculating inhibition rate and median effective dose (ED50).

The neutral moiety of *A. annua* L. that was administered via oral gavage at a dose of 1 g/kg × 3 days was able to achieve negative conversion of *Plasmodium*, with ED50 of 2646 mg/kg.

1.2.2 Plasmodium inui (Plasmodium cynomolgi)

P. inui (*P. cynomolgi*) model was established in the experiment. After blood transmission, when the parasitic rate on Day 8 following inoculation of *Plasmodium* reached more than 10‰, animals were orally given the drug once a day for 8–9 consecutive times. Blood smears were collected on the day of discontinuation and examined every 24 h. Alternatively, after sporozoite inoculation, when the parasitemia was detected, animals were treated via oral gavage once a day for 3 consecutive days, to observe the antimalarial effect on asexual parasite in *P. inui* (*P. cynomolgi*) during erythrocytic stage.

Findings: The neutral moiety had significant antimalarial effect on *P. inui* (*P. cynomolgi*) (Table 3.3). Two-dose groups successfully achieved negative conversion in simian plasmodia.

Table 3.3: The antimalarial effect of neutral part on simian malaria

Number of Animals	Dosage (g/kg)	Negative Conversion in Simian Plasmodia	Reappearance of Simian Plasmodia
2	0.32, once a day, 9 times	48 h	4–10 days
1	1.3, once a day, 8 times	29 h	None reappeared within 6 months

In the experiment, the neutral moiety of *A. annua* L. has shown the ability to kill *Plasmodium* in *P. berghei* and *P. inui* (*P. cynomolgi*), achieving 100% negative conversion rate.

1.3 Researches on the Safety of Neutral Moiety in Artemisia annua *L.*

1.3.1 Acute toxicity

Healthy mice were given a single dose via oral gavage and then observed for 3 days. The median lethal dose (LD50) was calculated with the mortality, and the therapeutic index was calculated with LD50/ED50 (Table 3.4).

1.3.2 Effect on the heart

In the experiments, the effect of neutral moiety on the heart was observed in mice, cats, and dogs respectively. The effect on the hearts was investigated in both infected and uninfected mice, and in both anesthetized and unanesthetized cats.

1. The effect on the heart rate was observed in 11 cats at doses of 0.5, 1.0, 2.5, and 7.1 g/kg, successively. No significant change was seen in the heart rate among the majority of cats after oral administration when compared with the heart rate of 100–160 bpm before administration, except that individual experienced accelerated heart rate.
2. The effect on the heart was observed in 13 dogs successively. Of them, 10 dogs were assigned into the administration group to receive the total dose of 0.5–4 g/kg, respectively. There were no significant changes in the heart rate, blood pressure, ECG before and after administration.

1.3.3 Effect on the liver

Liver function tests were conducted in 13 healthy dogs ($n = 10$ in the administration group and $n = 3$ in the control group). All animals were divided into three batches in the experiment. The first batch was given a single dose via oral gavage and observed for changes in liver function before and after administration; the second batch (normal control before administration) was tested for liver function 7, 5, and 3 d prior to administration and on the day of dosing. The

Table 3.4: Acute toxicity of neutral part in *A. annua* L.

Drug Extract	LD_{50} (mg/kg)	ED_{50} (mg/kg)	Therapeutic Index
Neutral active ingredient in *A. annua* L.	7425	2646	2.81

Table 3.5: Effect of neutral part on dog liver

Experimental Lot	Animal No.	Dose (g/kg)	GPT (X + SD)		T·T·T		T·F·T	
			Predose	Postdose	Predose	Postdose	Predose	Postdose
1	1	0.5	162	148				
	2	2.0	148	192				
	3	4.0	136	172				
2	4	1.0 × 4 d	115.80 ± 10.50	189.00 ± 12.70	Normal	Normal	Normal	Normal
	5	1.0 × 4 d	84.50 ± 16.30	78.00 ± 25.50	Normal	Normal	Normal	Normal
	6	1.0 × 4 d	64.30 ± 8.10	89.00 ± 26.90	Normal	Normal	Normal	Normal
3	7	0.5 × 4 d	76.00 ± 17.70	104.00 ± 8.49	Normal	Normal	Normal	Normal
	8	0.5 × 4 d	68.00 ± 14.80	97.00 ± 36.06	Normal	Normal	Normal	Normal
	9	0.5 × 4 d	128.30 ± 4.74	159.00 ± 14.10	Normal	Normal	Normal	Normal
	10	1.0 × 4 d	126.30 ± 13.65	139.00	Normal	Normal	Normal	Normal
	11	Control	103.00 ± 11.50	113.00 ± 4.90	Normal	Normal	Normal	Normal
	12	Control	98.00 ± 57.10	104.00 ± 8.50	Normal	Normal	Normal	Normal
	13	Control	103.00 ± 18.10	91.00 ± 44.60	Normal	Normal	Normal	Normal

animals received a dose of 1.0 g/kg for 4 consecutive days, and underwent liver function tests 2 and 5 d after the administration; and the third batch also received different doses for 4 consecutive days, while the control group was set up separately. The results showed that 4 out of 10 animals had higher-than-normal transaminase activity at the dose of up to 2 g/kg, of which two animals had a decline in transaminase activity in the second determination, and the other two animals failed to reflect such dynamic change as they only underwent determination once. The aforementioned findings suggested that neutral moiety exerted effect on the liver leading to elevated transaminase when it increased to a higher dose (Table 3.5).

1.3.4 Effect on kidneys

The changes in renal function were observed in 10 healthy dogs that received neutral moiety with different doses via oral gavage. The first batch received a dose of 1.0 g/kg for 4 consecutive days. Animals (normal control before administration) were tested for kidney function 7, 5, and 3 d prior to administration and on the day of dosing. Kidney function tests were conducted 2 and 5 d after administration, respectively, and compared to measurements before administration. The second batch received doses of 0.5 and 1.0 g/kg for 4 consecutive days. Kidney function tests were conducted in the morning of the fourth dosing and the following day, respectively, and compared to measurements before administration and in the control group.

Findings: No significant effect on renal function in each batch was found in the experiment (Table 3.6).

1.3.5 Pathological examination

All animals were divided into two batches in the experiment. Eight dogs received neutral moiety via oral gavage at a dose of 0.5 g/kg for 4 consecutive days and 1.0 g/kg for 4

Table 3.6: Effect of neutral part on kidney function

Lot No.	Animal No.	Dose (g/kg)	Predose					Postdose		
			Measured Value				Mean	1	2	Mean
1	1	0.5 × 4 d	34.30	29.20	34.90		32.80 ± 3.13	33.00	36.80	34.90 ± 2.69
	2	0.5 × 4 d	54.20	35.40	31.60		40.40 ± 12.10	31.00	38.90	34.95 ± 5.59
	3	0.5 × 4 d	40.00	32.70	32.50		35.10 ± 4.30	35.90	36.00	35.90 ± 0.00
	4	1.0 × 4 d	37.60	36.50	38.20		37.40 ± 0.86	34.50	37.50	36.00 ± 2.14
	5	Control	37.10	35.00	32.50		34.90 ± 2.30	35.00	34.00	34.50 ± 0.70
	6	Control	45.00	36.30	32.50		37.80 ± 6.40	36.50	32.70	34.60 ± 2.70
	7	Control	44.40	28.50			36.50 ± 11.24	35.00	34.00	34.50 ± 0.70
Lot No.	**Animal No.**	**Dose (g/kg)**	**Predose**					**Postdose**		
			Measured Value				**Mean**	**1**	**2**	**Mean**
2	8	1.0 × 4 d	37.80	38.70	36.50	36.00	37.30 ± 1.20	42.00	38.50	40.30 ± 2.50
	9	1.0 × 4 d	40.00	35.10	40.80	33.50	37.40 ± 3.60	36.00	37.50	36.80 ± 1.05
	10	1.0 × 4 d	31.20	34.40	43.00		36.20 ± 6.10	35.40	35.00	35.20 ± 0.28
			36.63 ± 5.17							

consecutive days. After being observed for changes in liver and kidney function, ECG and so on starting from 5 d after administration, all animals were sacrificed for histopathological examination, while three dogs were observed as controls.

The results of two batch animals showed that no acute toxicity was seen in major organs responsible for absorption, detoxification, and excretion of the drug (i.e., liver, stomach, intestine, kidney, and so on).

Based on previous animal experiments, the preclinical trials were conducted to ensure the safety of the drug in human body.

1.3.6 Human trials

Two trials were conducted. Led by Youyou Tu, three researchers with the approval from the leadership launched the first trial (known as "Pathfinder") in Dongzhimen Hospital Affiliated to Beijing University of Chinese Medicine, in which the initial dose was 0.35 g per patient, and then the dose was sequentially escalated to 0.5, 1.0, 2.0, 3.0, 4.0, and 5.0 g once a day for 7 consecutive days. To be in time for malaria season, Youyou Tu, as a member of the medical team of China Academy of Chinese Medical Sciences led by Dai Shaode rushed to malaria endemic area in Changjiang district of Hainan for the purpose of conducting clinical research on the drugs in August 1972. Considering the flexibility of clinical regimens, the dose was increased in the second trial to fully display their antimalarial efficacy and win the first battle. A total of five patients received the drug at a dose of 3.0 g/time twice daily for 3 consecutive days. ECG, liver function, kidney function, chest X-ray, blood, urine, and stool examination were conducted before, during, and after administration, respectively.

Findings: After administration, no abnormalities were found in routine blood and urine tests, and renal function was within the normal range. No abnormalities were found in BNU test, chest X-ray (−), and ECG during and after administration. No significant changes were seen in blood pressure. Fundus was normal, and there were no significant changes in visual field before and after administration. No abnormalities were found in body temperature and pulse. No clinical symptoms of respiratory system, urinary system, and central nervous system were found. Adverse reactions were mild gastrointestinal symptoms. Two cases of abdominal pain 1 h after administration were reported, but mild in severity and healed without treatment.

The observation of animal toxicity experiment and human trial suggested: neutral moiety of *A. annua* L. had no significant effect on internal organs except a mild or transient increase in liver transaminase activity in a few animals.

1.4 First Research on Clinical Efficacy of Active Neutral Moiety in Artemisia annua L.

From August to October in 1972, 11 cases of tertian malaria, 9 cases of falciparum malaria and 1 case of mixed infection (all of the cases were counted as tertian malaria since the majority of cases was tertian malaria) selected from local residents and migrants in Changjiang district in Hainan received neutral moiety capsule, while 4 cases received chloroquine as control. Nine cases of tertian malaria were also verified in 302 Military Hospital of China. A total of 30 cases were included for clinical observation. All of three dosage regimens (3 g/time, the low-dose group received the drug twice daily; median dose group 3 times daily, high dose group 4 times daily) were effective, of which the high-dose group showed the highest efficacy. The average fever subsidence time was 19.06 h for tertian malaria and 35.09 h for falciparum malaria in migrants. Negative conversion of *Plasmodium* was achieved in all cases, but *Plasmodium* reemerged in some cases in the short term. The observation results of three dose groups showed that the high-dose group had a lower recrudescence rate of *Plasmodium* than the low dose group. This suggested that the antimalarial efficacy may be improved by further purifying drugs and increasing the effective dose, which also put forward new requirements to the laboratory work. No significant adverse effect of drugs on the gastrointestinal tract as well as liver and kidney function were seen clinically. Only individual patients experienced vomiting or diarrhea. Two cases with elevated transaminase in 302 Military Hospital of China experienced continuous increase in transaminase, but they returned to normal in 1–2 weeks after administration (Tables 3.7 and 3.8).

The first trial was carried out with extreme prudence. The dose delivery was sequentially escalated from low to high (3 dose groups); in terms of subject selection, local residents with better immunity took precedence over migrants; in terms of types of malaria, tertian malaria was firstly selected followed by falciparum malaria. In other words, the trial started from local population to migrants with tertian malaria, and then conducted in local population with

Table 3.7: Efficacy of tertian malaria and falciparum malaria in Changjiang district in Hainan

Type of Malaria	Total dose (g)	No. of Cases	Mean Defervescence time (h)	Mean Duration of Negative Conversion in Plasmodia (days)	Efficacy		Recrudescence (Within 10 days)/Cases
					Cured/Cases	Effective/Case	
Tertian malaria	18	3	36.20	4	2	1	1
	27	4	11.23	2.25	4		2
	36	5	19.09	2	5		1
Falciparum malaria	18	1 (Local population)	39.50	5	1		
	18	1 (Population from low malaria area)				1	
	27	1 (Local population)	24.00	4	1		
	27	1 (Foreign population)				1	
	36	5 (Foreign population)	35.09	1.7	4	1	4

Table 3.8: Efficacy of tertian malaria and falciparum malaria in the 302 Military Hospital of China

No. of Cases/ Cases	Mean Defervescence Duration (days)	Mean Duration of Negative Conversion in Plasmodia (days)	Efficacy			Recrudescence (within 6–14 days)/Cases
			Cured/ Cases	Effective/ Cases	Ineffective/ Cases	
9	1.375	3.99	8	1	0	2

falciparum malaria followed by 5 migrants with falciparum malaria, while 4 cases received chloroquine as controls. The antimalarial effect of neutral moiety extracted from *A. annua* L. was proved in a total of 21 patients. The results demonstrated that the neutral moiety extracted from *A. annua* L. had a better efficacy than chloroquine. As observed by 302 Military Hospital of China, all of 9 cases with tertian malaria had responded to the treatment.

In the research and development of antimalarial drugs, the exploration of the active ingredient in *A. annua* L. was the key problem, although it tended to be a process. Particularly, due to lack of fruits over years, whether the pharmacological models of *P. berghei*, *P. inui* (*P. cynomolgi*) were consistent with the clinical manifestation urgently needed to be further confirmed. For this reason, the efficacy results of the first clinical study conducted in 1972 was of special importance. The consistency between laboratory results and clinical practice results demonstrated that the 100% inhibition of *Plasmodium* in *P. berghei* and *P. inui* (*P. cynomolgi*) was equivalent to 100% human clinical effectiveness. Convincingly, the relationship between preclinical trials and clinical efficacy was perfectly parallel.

On November 17, 1972, Youyou Tu firstly reported the summary of efficacy in 30 cases of malaria that were cured with *A. annua* L. at the General Assembly of "523" held in Beijing, arousing great concern again. In early 1973, the Institute of Chinese Material Medica, China Academy of Chinese Medical Sciences carried out antimalarial work with *A. annua* L. successively on the basis of the reply to the letters from Yunnan and Shandong after in-depth consultation. Afterwards, with the collaboration of Shandong Province, Shandong Research Institute of TCM that took over the work of Shandong Institute of Parasitic Diseases became the second inventor (Fig. 3.1), and the third inventor was Yunnan Institute of Materia Medica (Fig. 3.2). In 1975, Guangzhou University of Chinese Medicine which had always been the main force responsible for acupuncture treatment of malaria in "523" task also abandoned acupuncture for drugs, to further extend clinical validation of artemisinin-like drugs (became the sixth inventor later).

2 Separation and Purification of the Active Ingredient (Artemisinin) and Determination of Its Antimalarial Efficacy and Chemical Structure

2.1 Discovery of Artemisinin

Since *A. annua* L. that was harvested from Beijing had poor quality and limited content of artemisinin, its separation and extraction process were relatively complex. After repeated

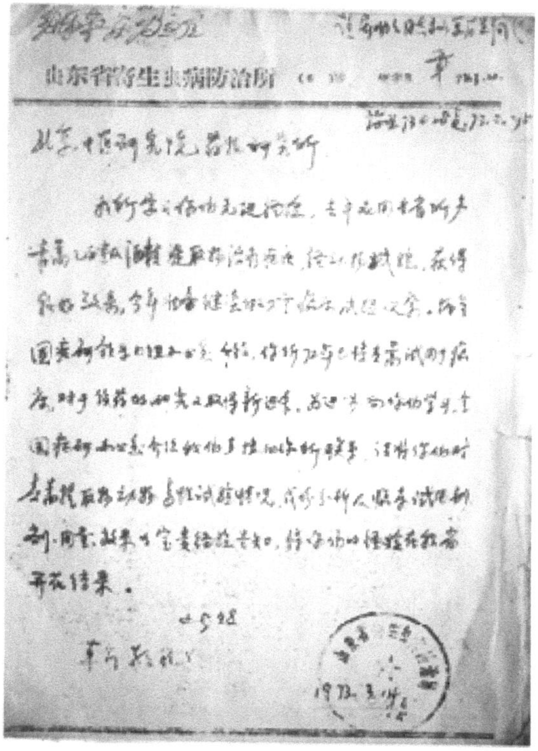
山东省寄生虫病防治所

Figure 3.1: A Letter From the Shandong Institute of Parasitic Diseases.

exploration, three crystallographic components were successfully isolated, which were Arteannuin A, artemisinin and Arteannuin B on November 08, 1972. After seeking for appropriate color development reagent and chromatoplate, the Rf values were determined to facilitate initial identification. In combination with the experiment in *P. berghei*, in which 100% negative conversion of rodent *Plasmodium* was achieved at a dose of 50–100 mg/kg of artemisinin, the active antimalarial monomer in *A. annua* L. was proved to be artemisinin.

Its separation processes and effect on *P. berghei* were as follows.

The neutral active ingredient in ether extract was mixed with polyamide, diafiltrated with 47% ethanol, and then concentrated under vacuum. Afterwards, the concentrate was extracted with ether, and then purified by silica gel column chromatography and finally eluted with petroleum ether, 10% ethyl acetate—petroleum ether and 15% ethyl acetate—petroleum ether in sequence to yield yellow oil, orange oil, Arteannuin A, artemisinin, Arteannuin B, black oil, and other ingredients, successively. The efficacy in the treatment of *P. berghei* was shown in Table 3.9. The separation process was as follows.

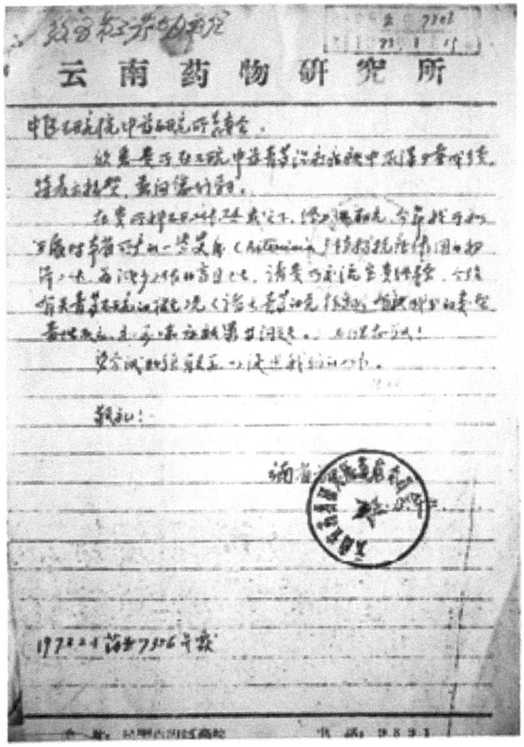
云南药物研究所

Figure 3.2: A Letter From the Yunnan Institute of Materia Medica.

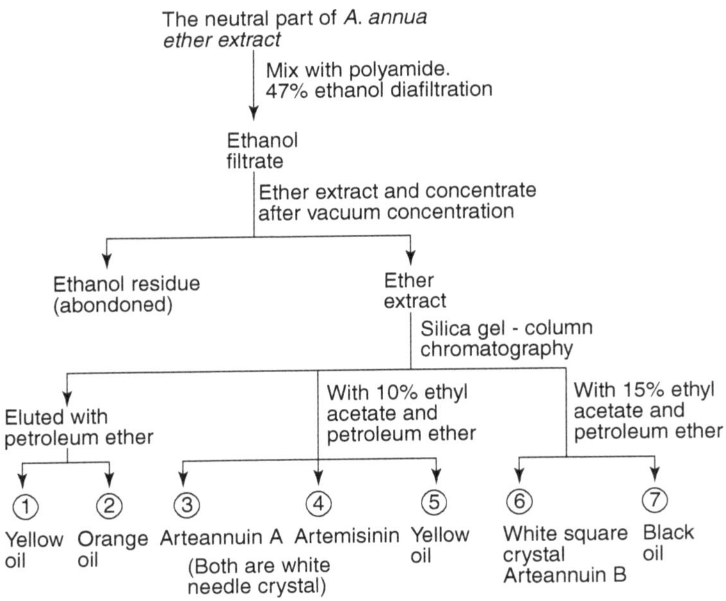

Table 3.9: Titers of different ingredients for rodent malaria

Isolate	Dose (mg/kg)	Titer for Rodent Malaria
Yellow oily matter	300	Ineffective
Arteannuin A	100	Ineffective
Artemisinin	50~100	All plasmodia became negative
Arteannuin B	400	Ineffective

The separated ingredients were tested with *P. berghei* model to determine their antimalarial effect.

Artemisinin (Chinese pinyin: Qinghaosu) that had been isolated and purified from neutral moiety of *A. annua* L. by the Institute of Chinese Material Medica, China Academy of Chinese Medical Sciences in 1972 was proved to be the first active antimalarial monomer derived from *A. annua* L. To confirm its clinical efficacy and identify its chemical structure as soon as possible, the research group devoted every effort to isolating more than 100 g extracts of artemisinin for future use. In 1973, the researchers went to the malaria endemic area in Changjiang district of Hainan with artemisinin monomer capsules for the purpose of validating the clinical efficacy once again, and they found that the artemisinin was the active antimalarial ingredient in *A. annua* L.

2.2 Researches on Safety of Artemisinin and Determination of Clinical Antimalarial Effect

In 1973, to gain time for initiating the observation of clinical efficacy, the acute toxicity test was conducted, in which histopathological examination was performed on major organs of animals, such as heart, liver, spleen, lung, kidney, brain, and so on. Meanwhile, researchers tried the drugs themselves to ensure the safety of drugs at that time.

2.2.1 Acute toxicity test

Kunming mice, weighing 12–22 g, half of them male and half of them female were included. The mice were fasted for about 20 h before the experiment, and then randomly assigned into five groups ($n = 10$) to receive artemisinin suspended in 1% Tween saline via oral gavage. All animals were observed for 1 week since 30 min after the administration, and any case of poisoning and death was recorded to calculate LD50.

The results are shown in Table 3.10.

The LD50 for oral gavage was calculated to be 4721.6 mg/kg with simplified probit method (for details, see Chapter 14, Pharmacological Studies on Artemisinin).

2.2.2 Effect on heart

a. The effect on the heart of mice: observe the effect of artemisinin and chloroquine on the ECG of infected and noninfected healthy mice which had not been anaesthetized. Infected group was intraperitoneally inoculated with *Plasmodium* 1×10^7 1 day before

Table 3.10: Results of acute toxicity test

Dose (mg/kg)	Logarithmic Dose	Mortality/Deaths	Mortality (%)	Probit (g)
10000	4.000	10	100	7.40
7000	3.8451	8	80	5.84
4900	3.6902	4	60	5.85
3430	3.5353	3	30	4.48
2401	3.3804	0	0	2.60

administration. ECG was traced before administration and 2 h after oral administration, respectively. The results showed that artemisinin slowed the heart rate with dose-independent relationship in infected and noninfected mice and the difference in infected mice at the dose of up to 2 g/kg was statistically significant when compared with preadministration.

b. The effect on the heart of cats: a total of 10 healthy cats were included, of which 7 cats without anesthesia received artemisinin via oral gavage at doses of 100 mg/kg ($n = 3$), 500 mg/kg ($n = 3$), and 1600 mg/kg ($n = 1$). Blood pressure and 2–3 h ECG were recorded after administration and then compared with those before administration. Three urethane-anesthetized cats were given artemisinin at a dose of 800 mg/kg. Blood pressure and 4 h ECG were recorded after administration and compared with those before administration and anesthesia. The results showed that no significant effect was seen in any dose group except a decline in the heart rate with increasing dose, no matter whether they were anesthetized or not.

2.2.3 Effect on liver

The effects on liver function were observed in mice. Artemisinin was administered via oral gavage for 3 consecutive days. ALT activity was measured on Day 1, 4, 8, 15, and 22, respectively, after administration and compared to the control group and the first measurement in the same group.

Findings: Except that the 800 mg/kg × 3 days group had a transient increase in transaminases on 4 and 8 d after administration, other groups showed no significant effect (Table 3.11).

Table 3.11: Effect of artemisinin on liver function in mice (GPT activity)

Group	Dose (mg/kg) × 3 days	Day 1 After Administration	Day 4 After Administration	Day 8 After Administration	Day 15 After Administration	Day 22 After Administration
Control		20.4 ± 1.3	17.0 ± 2.1	28.3 ± 3.7		27.9 ± 5.4
Administration	100	17.9 ± 1.4	14.2 ± 1.7	27.7 ± 3.5	17.3 ± 0.4	36.3 ± 9.3
	200	28.4 ± 2.3	16.8 ± 2.5	30.8 ± 8.2	16.0 ± 3.9	23.0 ± 3.6
	400	25.5 ± 1.3	32.3 ± 7.5	23.8 ± 3.3	19.5 ± 1.4	29.1 ± 5.1
	800	21.7 ± 1.6	36.9 ± 7.6	34.6 ± 3.7		23.3 ± 2.8

2.2.4 Histopathological examination

The mice whose ALT activity was affected by artemisinin in the foregoing experiment were treated for 3 days and sacrificed 1 month later. Their heart, liver, spleen, lung, kidney, brain, and other tissues were collected for observing pathological and histological changes. Another group of mice was sacrificed 3 or 6 days immediately after administration, and then tissue sections were collected. The results showed that the effect on the organs and tissues started from 200 mg/kg × 3 d dose group, mainly including cloudy swelling of liver cells and brain bleeding (Table 3.12). The number of animals who experienced changes also increased with increasing dose. No significant changes were seen in other organs.

2.2.5 Observation of toxic and side effect in human trials

Based on previous animal experiments, the researchers received the drug themselves to ensure the safety of the drug in patients.

Three researchers received artemisinin orally at doses of 3.5, 5, and 5 g once a day for 3 consecutive days. No significant changes were seen in ECG, EEG, liver function, kidney function, chest X-ray, routine blood and urine tests, and so on after administration. One of them experienced limb numbness but returned to normal 12 h later; and another researcher experienced accelerated heart rate (112 bpm) but returned to normal 30 min later.

The aforementioned animal toxicity tests and human trials showed that artemisinin had no significant effect on the major organs in animal and human except mild or transient increase in ALT activity in few animals or individuals, thus proving that artemisinin resulted in less toxic and side effect.

From September to October in 1973, the medical team of China Academy of Chinese Medical Sciences, composed of Li Chuanjie, Liu Jufu, etc., went to malaria endemic area in Changjiang district of Hainan again for the purpose of clinical validation. They firstly treated patients with artemisinin tablets (the preparation workroom was shut down at that time of Cultural Revolution, so the compression of tablets was outsourced). In the first trial, although three out of five patients had responded, the efficacy was not ideal. After prompt

Table 3.12: Relationships of different doses of artemisinin with cloudy swelling of hepatocytes and cerebral hemorrhage in mice

		Cloudy Swelling of Hepatocytes		Cerebral Hemorrhage	
Drug Dose	No. of Animals	No. of Animals	Percentage (%)	No. of Animals	Percentage (%)
800 mg/kg × 3 d	10	10	100.0	5	50.0
400 mg/kg × 3 d	9	7	77.8	0	0
200 mg/kg × 3 d	10	2	20.0	1	10.0
100 mg/kg × 3 d	10	0	0	0	0
Control	10	0	0	0	0

investigation, the tablet disintegration was found to be the cause of low efficacy. Researchers changed the formulation of artemisinin from tablets to capsules soon. The Deputy Director, Zhang Guozhen, went to malaria endemic area in Changjiang district of Hainan to investigate the efficacy. Three cases who were treated with artemisinin capsules at a dose of 3–3.5 g achieved satisfactory remission, among whom all *Plasmodium* was killed quickly and the average time of negative conversion was 18.5 h and the average fever subsidence time was 30 h. The previous finding proved that artemisinin was the active ingredient in *A. annua* L. The number of accumulative cases for clinical validation of artemisinin by the institute of Chinese Material Medica reached 529 as of 1978 when the appraising meeting was held.

In October 1973, the research group reported the results to "523" Office upon completion of the first clinical work of artemisinin. Subsequently, "523" Office notified China Academy of Chinese Medical Sciences by letter on November 02, 1973 (Fig. 3.3) to convene a national professional meeting of antimalaria drugs (including chemical synthesis) research and discuss "Looking for New Drug Through Combination of Chinese and Western Medicine." In this letter, "523" Office especially proposed that "*Artemisia annua* L. was priority medicine so that the relevant information should be exchanged at the meeting." Ycuyou Tu made a comprehensive report upon request at the meeting. From January 10, 1974 to January 17, 1974, the seminar for principals of "523" Offices in various regions was also held in Beijing.

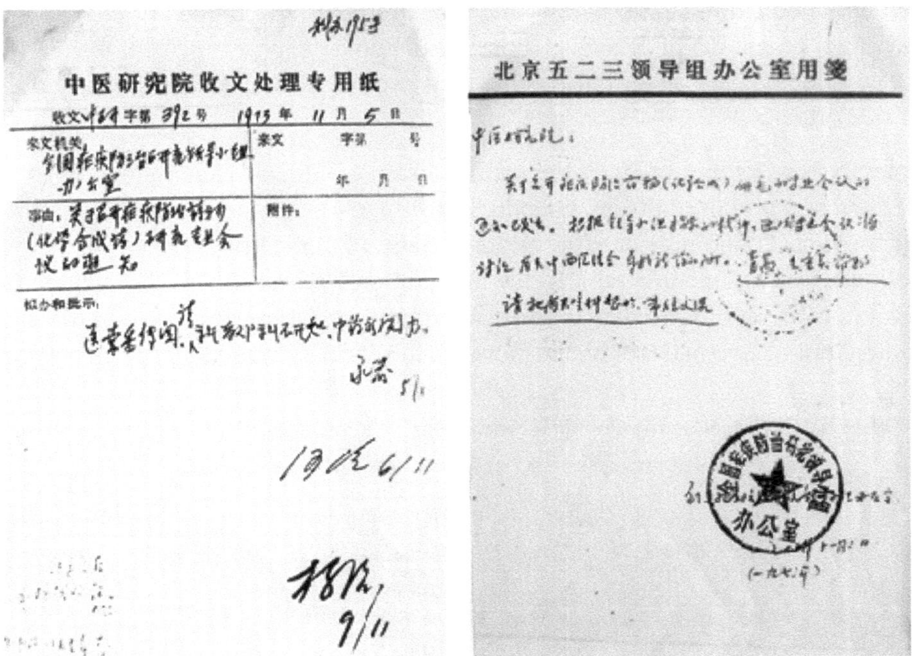
中医研究院收文处理专用纸

北京五二三领导组办公室用笺

Figure 3.3: Letters From the National Professional Conference on Research on Drugs for Malaria Control (November 2, 1973).

At this seminar, the further research work on artemisinin was assigned to China Academy of Chinese Medical Sciences under the coordination of Yunnan, Shandong, and other regions. Literally entrusted by "523" Office, China Academy of Chinese Medical Sciences convened the "Research Seminar of *Artemisia annua* L." from February 28 to March 1 in 1974, and firstly opened its laboratories involving artemisinin researches together for all attendees while offering detailed introduction and site visits, thereby forming a nationwide collaboration. With the ongoing participation of Sichuan, Guangxi, Guangdong, and other regions—the data of artemisinin appraisal meeting in 1978 showed that the number of participants exceeded more than 40, involving 6555 cases that had been "clinically" validated.

Institute of Chinese Material Medica completed the overall work according to the provisions on new drug approval and obtained the first *New Drug Certificate* (86) the Ministry of Health No. X-01, since the implementation of the approach of new drug approval in 1986.

2.3 Identification of the Chemical Structure of Artemisinin

On November 08, 1972, the antimalarial research group of the Institute of Chinese Material Medica, China Academy of Chinese Medical Sciences proceed with the identification of the chemical structure of artemisinin which was active monomer isolated from neutral moiety.

Artemisinin was white needle-like crystal, mp: 156–157°C, optical rotation: [a] 17D = + 66.3 (c = 1.64, chloroform). The absence of nitrogen element and double bonds in its structure was confirmed by chemical reaction. The elemental analysis results showed that C element accounted for 63.72% and H element accounted for 7.86 %. In combination with the four spectral data, its formula was determined as $C_{15}H_{22}O_5$, and molecular weight was 282. The stereo-structure of artemisinin, a sesquiterpene lactone, was finally determined with the assistance of Professor Lin Qishou (School of Pharmacy, Beijing Medical College), belonging to a new type of antimalarial drug.

During the studies on derivatives that were carried out in 1973 aiming to determine the functional groups, Youyou Tu found that the carbonyl peak disappeared when artemisinin was reduced by sodium borohydride together with the research team, thus confirming the presence of C12 carbonyl group in its molecular structure. Based on previous findings, they firstly yielded the derivatives of artemisinin, dihydroartemisinin, with molecular formula of $C_{15}H_{24}O_5$ and molecular weight of 284. Furthermore, they tried to improve the molecular structure of dihydroartemisinin via acetylation. Thanks to the introduction of the carbonyl group in the artemisinin structure, a variety of derivatives were generated so as to lay a foundation for analysis of structure-activity relationships of artemisinin. At the national "523" conference held in Henan Province in 1975, Institute of Chinese Material Medica gave a report on the law of the structure–activity relationship, thus facilitating launch of the studies in the field of artemisinin derivatives in 1976 and 1977.

Upon determination of the structure of artemisinin as sesquiterpene lactone in 1973, Institute of Chinese Material Medica that was eager to meet the needs of military industrial task considered to seek for collaboration to determine its stereo-structure as soon as possible. After learning from the literatures that Professor Liu Zhujin in Shanghai Institute of Organic Chemistry had rich experience in sesquiterpene compound, Youyou Tu contacted Shanghai Institute of Organic Chemistry with relevant information and greeted by Comrade Chen Yuqun. In January 1974, Chen agreed in his reply that a person was to be sent by Institute of Chinese Material Medica in his reply letter to work with Shanghai Institute. In February, Comrade Ni Muyun was sent by Institute of Chinese Material Medica to go to Shanghai Institute with large amount of information about artemisinin, dihydroartemisinin as well as relevant chromatographic data, who would coordinate with Comrade Wu Zhaohua in Room 1 of the institute where Zhou Weishan acted as the director to conduct further research on the basis of repeating the work in Beijing. Meanwhile, the coordination with the Institute of Biophysics, Chinese Academy of Sciences (the fourth inventor) that cultured desired crystal and provided relevant data were hosted by Youyou Tu in Beijing in 1974. After the chemical structure of artemisinin was determined with domestically advanced X-ray diffraction method at that time, and the absolute configuration was detected based on the accurate measurements of anomalous scattering intensity, the chemical structure of artemisinin was finally determined on the November 30, 1975. To ensure its accuracy, Professor Liang Xiaotian was invited from the Institute of Materia Medica, Chinese Academy of Medical Sciences by Institute of Chinese Material Medica and the Institute of Biophysics to participate in the discussion and confirm its structure from January 14, 1976 to January 18, 1976 (Fig. 3.4). On January 26, 1976, Youyou Tu went to Shanghai Institute of Organic Chemistry with Comrade Li Pengfei from the Institute of Biophysics to report the identification progress of artemisinin structure. On the following day, Li Pengfei gave a report on the identification of the chemical structure of artemisinin with X-ray diffraction method, which was confirmed by attendees, Zhou Weishan, Wu Yulin, Wu Zhaohua, etc. Subsequently, China Academy of Chinese Medical Sciences reported the findings to the Ministry of Health. With the approval of the Ministry of Health, an article titled "*A Novel Sesquiterpene Lactones–Artemisinin*" was submitted to *Chinese Science Bulletin* on February 20, 1976 and published in Volume 3

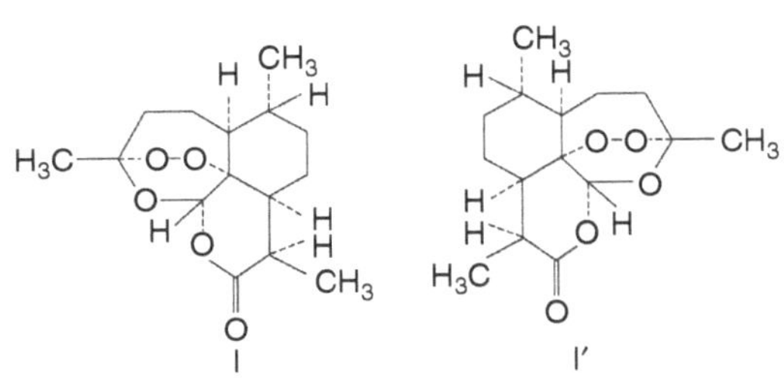

Figure 3.4: The Absolute Configuration of Artemisinin.

of 1997 and immediately included by C.A. As a confidential mission for combat readiness, "to avoid externally disclosure of study dynamic and medical production," the author of the article was the "Artemisinin Structure Collaborative Group" rather than the Institute of Chinese Material Medica, China Academy of Chinese Medical Sciences. In 1975, on the premise of clarifying artemisinin structure, the Institute of Chinese Material Medica continued the cooperation (another two comrades were later sent by the Institute of Chinese Material Medica to work till the end of the project) with Shanghai Institute of Organic Chemistry (the fifth inventor later) to conduct further research on the artemisinin structure and reaction, and published an article titled "Artemisia Annua L Absolute Configuration Structure and Reaction of Artemisinin" (Fig. 3.4) in *Acta Chimica Sinica* in May 1979.

Notes: Due to the special molecular structure of the originator product artemisinin, containing only carbon, hydrogen, and oxygen, it was difficult to determine how oxygen atom joined the carbon skeleton by chemical means. For this reason, X-ray diffraction was employed to measure the molecular structure in the early 1970s, while its absolute configuration was measured by using anomalous scattering of oxygen atoms. The efforts and contribution of the Artemisinin Structure Collaboration Group of Institute of Biophysics, Chinese Academy of Sciences will be described in this book to express our respect.

3 First Measurement of Absolute Configuration of Artemisinin With Anomalous Scattering of Oxygen Atom in China

Artemisinin (molecular formula: $C_{15}H_{22}O_5$), which is a colorless needle-like crystal, is active antimalarial ingredient extracted from traditional Chinese medicine, *A. annua* L. Its structure was finally proved to be a new sesquiterpene lactone with oxy groups by the Institute of Chinese Material Medica, China Academy of Chinese Medical Sciences with chemical and spectroscopic methods. However, due to the special structure of artemisinin, it was difficult to exactly determine how oxygen atom joined the carbon skeleton in a molecular unit relying on only the previously described methods. Thus, X-ray diffraction method was employed to determine the stereo-structure of artemisinin ultimately. On the basis of anomalous scattering intensity data, its absolute configuration was also established (Fig. 3.5).

3.1 Experiment and Data Processing

The morphology of the crystal belonged to the orthorhombic crystal system after being determined by the Weissenberg photographical technique, with absent reflection of h00, $h = 2n + 1$; 0k0, $k = 2n + 1$; 00l, $l = 2n + 1$. Thereby, the space group was uniquely determined as D42-P212121. The crystal density was measured with the specific gravity liquid column method. Cell parameters were accurately measured with $0.17 \times 0.17 \times 0.42$ mm^3 crystals on automatic Phillip PW-1100 four-circle diffractometer. When the peak and its foreground and background were scanned in equal time to collect all the intensity at

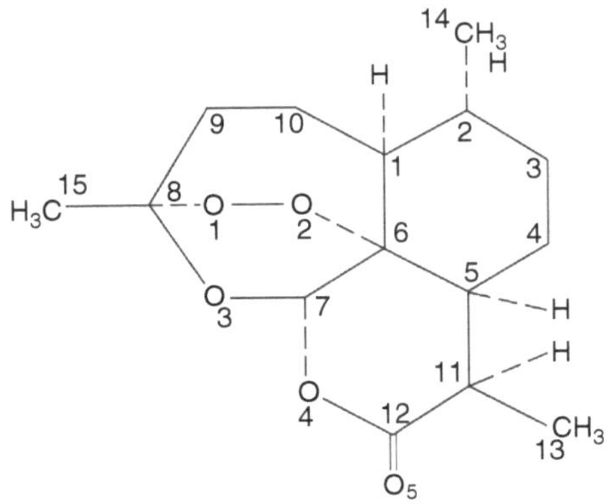

Figure 3.5: The Absolute Configuration of Artemisinin.

scan angle θ between 3 and 68° by using graphite monochromator ($2\theta M = 26.6°$) CuKα radiation ($\lambda = 1.5418$) with θ–2θ scan mode, at the scanning speed of 0.1° per second, 1553 independent reflections were obtained, of which the number of unobserved reflection was 254 [$I \leq 3\sigma(I)$], allowing $I = \sigma(I)/2$. The standard deviation of the diffraction intensity under the author's experimental conditions was calculated with the following equation.

$$\sigma 2(I) = N\,\text{Peak} + N\,\text{Foreground} + N\,\text{Background} + [0.03 \times (N\,\text{Peak} - N\,\text{Foreground} - N\,\text{Background})]2 \tag{3.1}$$

where N was the count value of counter tubes. During the data collection process, the repeated measurement was conducted once an hour with 811 ($\theta = 17.2°$) reflection as reference, to verify the decomposition of crystals and possible movement. Throughout the course of the experiment, artemisinin monocrystal was stable to X-ray diffraction. The crystallographic and physical parameters of artemisinin are listed in Table 3.13.

Table 3.13: Crystallographic and physical parameters of artemisinin

Crystal system: Orthorhombic
Space group: D_2^4-$P2_12_12_1$
Cell parameters: a = 24.077(6) Å b = 9.443(3) c = 6.356(1) V = 1445.095 Å^3 $F(000)$ = 608
Number of intracellular molecules: Z = 4
Relative molecular weight (M. W.): 282.1472
Melting point (m. p.): 156–157°C
Density measured by experiment: d_e = 1.30 g/cm^3
Density calculated: d_c = 1.296 g/cm^3
Linear absorption coefficient (CuKα): μ = 8.1023 cm^{-1}
Optical rotation: $[a]_D^{17}$ =+66.3° (c = 1.66, CCl_4)

Table 3.14: Statistical properties of normalized structure factor modulus

Item	<|*E*|>	<|*E*|²>	<||*E*|²–1|>	Fraction of Normalized Structure Factor >1	>2	>3
Center of symmetry Theory Group	0.798	1.000	0.968	0.32	0.05	0.003
Center of asymmetry Theory Group	0.886	1.000	0.736	0.37	0.02	0.001
Artemisinin Observed value	0.867	1.025	0.823	0.36	0.03	0.001

The experimental data was corrected with Lorentz factor and polarization factor.

$$\mathrm{Lp} = 1\sin 2\theta \cos 22\theta \mathrm{M} + \cos 22\theta 1 + \cos 22\theta \mathrm{M} \tag{3.2}$$

After the data was restored with the temperature factor and scale factor derived by Wilson method, the observed value of structure amplitude $|F|$ and normalized structure factor module $|E|$ were obtained. No absorption correction was conducted. The statistical average of normalized structure factor mode is shown in Table 3.14.

3.2 Determination of Structure

In the absence of heavy atoms, the structure of artemisinin molecule was determined directly. With symbolic addition method, the problem of phase angle was solved.

The reflection of three specific original points, of a specific enantiomer, and of seven additional symbols introduced during the derivation of phase angle (Table 3.15).

Table 3.15: A starting reflexion set of phase angles

h	|E|	Φ_h	
1901	3.07	$\frac{\pi}{2}$	Specified origin
870	2.74	0	
045	2.23	0	
1120	2.73	$\frac{\pi}{2}$	Designated enantiomer
210	2.89	m	$(0, \pi)$
1220	2.56	n	
2010	2.49	l	
2201	2.27	r	$\left(\pm\frac{\pi}{2}\right)$
011	2.78	p	
2101	2.67	q	
342	3.22	a	

The phase angle was measured with tangent equation f.

where h, k, and h–k—index at reflection point;

Φ—phase angle at reflection point.

Tangent formula

$$\Phi k \sim \Phi k + \Phi h\text{-}k \tag{3.3}$$

$$\tan < \Phi h > \sum | EkEh\text{-}k | \sin(\Phi k + \Phi h\text{-}k) \quad k \sum | EkEh\text{-}k | \cos(\Phi k + \Phi h\text{-}k) \quad k \tag{3.4}$$

Eq. (3.3) was applied starting from the diffraction point of the starting set to derive phase angles of 101 independent reflections with $|E| \geq 1.60$. In the derivation process, the additional letter symbols were introduced as the phase angle of seven reflection points to extend the phase angle, which were numerically and uniquely determined in the late measurement process of phase angle: $m = n = l = 0$, $r = p = q = \pi/2$ and $a \sim \pi/2$. With the help of the seven symbols, the initial values Φ of 100 reflection phase angles were derived (Table 3.16). By substituting this set of phase angles into Eq. (3.4), convergence of <Φh> occurred after 17 iterative correction, as shown in Table 3.16, which can be compared to the final phase angle (Φc) whose structure was refined by least square.

Figure E was derived from above 100 independent reflections with $|E| \geq 1.60$ by Fourier synthesis. The overlap diagram of Figure E along direction C is shown in Fig. 3.6.

As 18 separate peaks with value greater than 140 (in any proportion) were shown in Figure E, the coordinates of 17 nonhydrogen atoms were obtained considering chemical information, and remaining three atoms in molecule were also reflected in Figure E. In Fig. 3.6, the pseudopeak was indicated in black and white.

The electron cloud density function ρ1 (xyz) of 17 atoms derived from Figure E was calculated regardless of categories (tentatively considering as carbon atoms). With the aid of ρ1 (xyz) map, not only the accurate selection of 17 independent atoms was confirmed, but also the coordinates of the remaining 3 atoms were obtained. When the second round of the electron cloud density function, ρ2 (xyz) was calculated with 20 atoms, the reliability factor R was 0.3273. Based on the peak overlap and peak shape on the Figure ρ2 (xyz), 15 carbon and 5 oxygen atoms were identified with reference to the chemical information. Thus, the third round of the electron cloud density function, ρ3 (xyz) was calculated while R factor was reduced to 0.2202. Fig. 3.6 shows the overlap diagram of Figure E along direction C.

Figure ρ3 (xyz) reveals the chemical structure of artemisinin molecule as well as spatial distribution of carbon and oxygen atoms. Fig. 3.7 shows the electron cloud intensity overlap diagram of artemisinin molecule along direction C.

Table 3.16: $\lvert E \rvert$, Φ, and Φ_h values of 100 reflection points and their phase angles (Φ_c) after structure refinement

Even-Even-Even Type					**Even-Even-Odd Type**				
1220	2.56	0	0	0	1021	2.52	180	136	156
1060	2.46	180	180	180	2201	2.27	90	90	270
002	2.39	180	180	180	045	2.23	0	0	0
1020	2.20	180	180	180	203	2.18	90	90	90
402	2.20	0	0	0	041	2.13	180	180	180
2200	2.13	180	180	180	201	2.07	270	270	270
242	2.09	90	88	61	423	2.05	180	157	169
600	1.95	180	0	0	825	1.96	180	170	147
244	1.92	270	262	234	1643	1.86	180	170	184
644	1.88	0	345	341	1205	1.80	90	90	90
680	1.87	180	180	180	1605	1.80	270	270	270
260	1.86	0	0	0	403	1.79	90	272	90
620	1.77	0	0	0	463	1.79	180	153	116
2000	1.69	180	180	0	861	1.72	180	244	257
					245	1.70	90	164	75
					2001	1.62	270	270	270
					2203	1.60	270	270	90
Even-Odd-Even Type					**Even-Odd-Odd Type**				
210	2.89	0	0	0	011	2.78	90	90	90
870	2.74	0	0	0	1031	2.68	180	197	212
212	2.53	180	166	144	1231	2.58	180	146	152
2010	2.49	0	0	0	211	2.34	0	356	41
032	2.16	90	90	90	235	2.22	180	165	169
2210	2.02	180	180	0	1033	2.03	0	22	37
1212	1.82	0	53	10	1415	1.84	90	121	79
1272	1.72	0	325	325	413	1.78	90	56	52
034	1.72	270	270	270	013	1.76	270	270	270
654	1.66	0	27	23	213	1.73	180	146	190
					1233	1.71	0	316	318
					435	1.61	180	231	231
Even-Odd-Even Type					**Odd-Even-Odd Type**				
hkl	**$\lvert E \rvert$**	**Φ**	**$\langle\Phi h\rangle$**	**Φ_c**	***hkl***	**$\lvert E \rvert$**	**Φ**	**$\langle\Phi h\rangle$**	**Φ_c**
754	2.98	270	213	196	1901	3.07	90	90	90
534	2.22	90	128	130	2101	2.67	90	90	90
1134	2.14	270	262	276	125	2.27	90	122	125
154	1.94	90	136	145	523	2.07	90	114	122
184	1.88	270	221	205	165	2.06	270	247	288
934	1.76	270	266	298	1168	2.00	270	203	211
1354	1.75	180	198	147	705	1.99	90	90	90
952	1.72	180	199	154	701	1.97	270	270	270
932	1.70	90	96	111	961	1.96	270	270	249
570	1.68	90	270	270	1543	1.87	270	230	314
1154	1.67	90	106	82	1505	1.85	270	270	270
					1305	1.79	90	90	90
					1343	1.75	90	13	0
					525	1.75	270	16	6
					1161	1.66	90	30	315

(*Continued*)

Table 3.16: |E|, Φ, and Φ_h values of 100 reflection points and their phase angles (Φ_c) after structure refinement (*cont.*)

Odd-Even-Even Type					Odd-Odd-Odd Type				
hkl	|*E*|	Φ	<Φ*h*>	Φ_c	*hkl*	|*E*|	Φ	<Φh>	Φ_c
342	8.22	90	98	76	1711	2.60	270	287	308
1120	2.73	90	90	90	1171	2.15	270	271	259
1162	2.51	270	185	180	311	2.04	270	318	345
744	2.36	90	79	109	155	1.96	90	120	128
944	2.28	90	63	52	771	1.87	90	110	103
924	2.24	270	289	299	933	1.78	0	351	121
720	1.98	270	270	270	1911	1.70	270	243	270
1702	1.79	180	180	180	1951	1.65	90	90	85
1502	1.78	0	358	180	1813	1.62	270	289	259
544	1.70	270	218	199					
580	1.60	90	90	90					
1524	160	270	330	339					

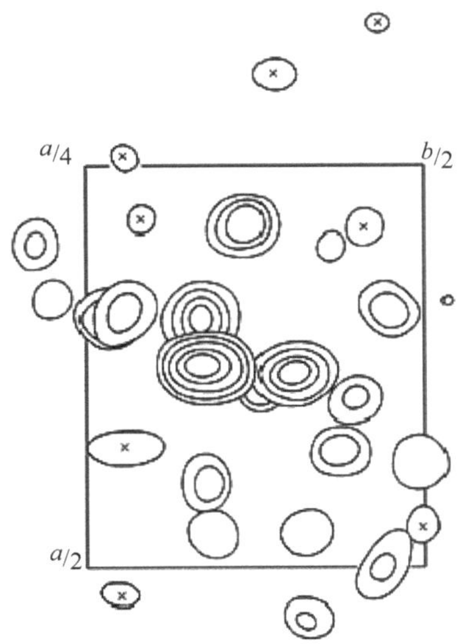

Figure 3.6: E Graphical Congruence in the C Direction.

3.3 Amendment of Structure

We developed a computer program for full-matrix least-squares-refinement of the structure by which artemisinin structure was amended. A total of 1553 independent reflections were refined. The scattering factors of carbon and oxygen atoms given by Cromer and Waber were employed, and the data of hydrogen atom was derived from scattering factors given by Eiland and Pepinsky.

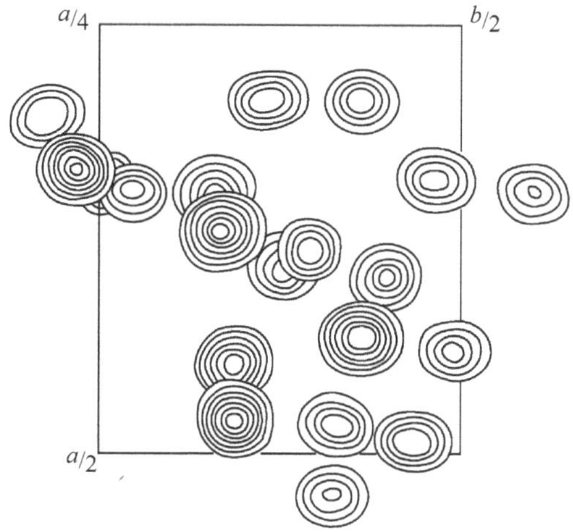

Figure 3.7: The Congruence of Electron Cloud Density of Artemisinin Molecules in the C Direction.

Based on the refinement of the coordinates of 20 nonhydrogen atoms, scale factors and the average temperature factor of the whole crystal, the thermal motion of atoms was further refined, including refinement of the isotropic and anisotropic temperature factor, and R factor decreased to 0.1129. The following weighting program was selected: where $|F0| > 16$, $W = 1\ |F0|$; otherwise $W = 1$. After another round of weighted full-matrix least-square correction, R factor was 0.1070.

On such basis, the electron density difference function was calculated. Fig. 3.8 showed the overlap diagram of electron density difference. Supported by the difference map, the

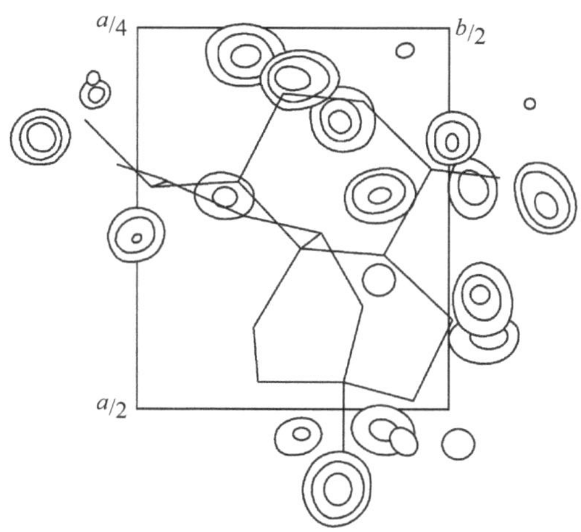

Figure 3.8: Differential Electron Cloud Density in the C Direction Shows Congruence of Location of Hydrogen Atoms.

location of all of the hydrogen atoms in the molecule was identified with structural chemistry knowledge. The coordinates of the hydrogen atoms were adjusted according to the tetrahedral angle of carbon atom, carbon–hydrogen bond length and other structural chemistry knowledge. The coordinates and thermal motion parameters of the atoms in artemisinin molecule are shown in Tables 3.17 and 3.18.

Assuming that the mean of isotropic thermal parameters in all directions was 3.56 for all nonhydrogen atoms in crystals wherein the hydrogen atoms linked to carbon ring were refined by the least square, and the mean of isotropic thermal parameters in all directions was 5.20 if the thermal parameters for methyl hydrogen atoms were carbon atoms of three methyl group, the results of structure factor of artemisinin were calculated: $R = 0.085$, R observed value = 0.0745.

3.4 Determination of Absolute Configuration

The absolute configuration of artemisinin was determined based on the refinement of its crystal structure.

The reflection intensity of the Bijvoet pairs which were sensitive to enantiomers was measured directly based on the Eqs. (3.5) and (3.6):

$$B = QH - 112(QH + 1) \tag{3.5}$$

$$QH = | Fhkl | / | Fhkl | \tag{3.6}$$

The values of $\Delta f'$ and $\Delta f''$ in experiment B were calculated based on the structural parameters of artemisinin molecule and anomalous scattering of carbon and oxygen atoms. And Theory B was calculated. If symbols between Experiment B and Theory B were consistent, the absolute configuration was the proposed structure model of the molecule based on electron cloud density function, and if symbols between them were opposite, the absolute configuration was actually the mirror of the structure model.

Due to the smaller anomalous scattering effect of light atoms, the effect of absorption must be considered. An absorption pair neighboring to major Bijvoet pairs was selected for relative correction of absorption effect. The corrected Bijvoet ratio was expressed as BA [Eq. (3.7)]:

$$BA = QH1QH2\text{-}112QH1QH2 + 1 \tag{3.7}$$

where QH1—ratio between $|F|$ values of major Bijvoet pairs;

QH2—ratio between $|F|$ values of neighboring absorption pair.

We chose those with theory B ≥ 0.006 and appropriate intensity from all Bijvoet pairs and 15 neighboring Bijvoet pairs to collect highly accurate reflection intensity data. The

Table 3.17: Coordinates of carbon and oxygen atoms (×10[4]) and their isotropic and anisotropic parameters of thermal motion

	x/a	*y/b*	*z/c*
C_1	3988(3)	3963(6)	7582(10)
C_2	3421(3)	4693(7)	7258(11)
C_3	2973(3)	3667(9)	6620(11)
C_4	2981(2)	2371(8)	8054(11)
C_5	3498(2)	1608(6)	8096(9)
C_6	3951(2)	2618(6)	8881(8)
C_7	3864(2)	2933(7)	1218(9)
C_8	4840(2)	3294(7)	1165(11)
C_9	4956(3)	4355(7)	9405(12)
C_{10}	4417(3)	5032(7)	8492(11)
C_{11}	3514(3)	222(7)	9393(9)
C_{12}	3508(2)	508(7)	1729(8)
C_{13}	1988(3)	849(8)	3779(13)
C_{14}	1529(4)	4104(8)	646(15)
C_{15}	277(8)	1695(8)	7079(13)
O_1	4823(2)	1892(4)	437(8)
O_2	4469(2)	1857(4)	8539(8)
O_3	4317(2)	3587(5)	2165(6)
O_4	3782(2)	1701(5)	2487(7)
O_5	1649(2)	885(6)	807(8)

	β_{11}	β_{22}	β_{33}	β_{12}	β_{23}	β_{13}
C_1	17(2)	24(11)	185(24)	19(10)	-34(33)	-6(14)
C_2	18(2)	120(14)	201(28)	-2(12)	-6(40)	7(15)
C_3	18(2)	165(15)	264(3)	0(12)	-86(44)	17(17)
C_4	15(2)	181(16)	254(30)	-11(12)	-144(42)	13(25)
C_5	12(2)	114(13)	128(21)	1(10)	-41(85)	21(13)
C_6	18(2)	108(12)	105(20)	-2(9)	18(31)	-9(12)
C_7	16(2)	158(14)	118(22)	-23(10)	-60(34)	6(13)
C_8	16(2)	101(13)	236(28)	4(11)	1(40)	-17(15)
C_9	16(2)	131(13)	340(34)	-2(10)	-57(45)	-19(16)
C_{10}	26(2)	149(14)	266(30)	-88(12)	-16(42)	-25(16)
C_{11}	15(2)	144(14)	172(26)	5(11)	-91(39)	9(15)
C_{12}	11(2)	94(12)	115(24)	21(10)	58(37)	-8(18)
C_{13}	22(2)	127(16)	215(35)	10(13)	23(48)	22(19)
C_{14}	32(2)	150(16)	369(40)	-6(15)	-126(52)	-9(22)
C_{15}	18(2)	173(18)	425(42)	21(13)	-209(54)	-15(19)
O_1	15(1)	107(8)	258(20)	12(7)	-32(27)	-14(10)
O_2	18(1)	122(9)	193(16)	2(7)	-38(24)	-16(9)
O_3	19(1)	117(8)	146(15)	-7(7)	27(23)	-16(9)
O_4	20(2)	139(11)	96(15)	-2(9)	14(26)	21(9)
O_5	27(2)	170(12)	292(23)	10(9)	-77(34)	35(13)

Table 3.18: Coordinates of hydrogen atom (×10⁴)

	x/a	*y/b*	*z/c*
H_1	4142	3681	6056
H_2	1704	4844	3593
H_3	3143	3280	5171
H'_3	2567	4090	6415
H_4	2841	2771	9596
H'_4	2656	1718	7593
H_5	3610	1450	6473
H_7	3570	3775	1275
H_9	149	1195	1900
H'_9	4800	200	4995
H_{10}	469	4375	2281
H'_{10}	770	4400	4631
H_{11}	1080	222	3980
H_{13}	2100	550	2240
H'_{13}	1750	1718	3750
H''_{13}	2266	781	4700
H_{14}	1500	4600	9150
H'_{14}	1900	3500	625
H''_{14}	1205	3400	1100
H_{15}	650	1800	7954
H'_{15}	234	750	6250
H''_{15}	200	2600	6200

measurements of hkl and hkl reflections were alternately repeated during the data collection and the peak cumulative count met certain requirements, namely the probability was 90% when the relative error of count fluctuation was 0.8%, so that the cumulative count value *N* could be estimated according to Eq. (3.8):

$$0.008\,(N\,\text{Peak} - N\,\text{Background}) \sim \pm 1.655 N\,\text{Peak} + N\,\text{Background} \qquad (3.8)$$

The experimental conditions for intensity collection: scanning mode: θ~2θ, scan width: 1.5°, scanning speed: 0.02°/s, crystal size: 0.33 × 0.21 × 0.66 mm^3.

The arithmetic mean and standard deviation of each reflection intensity are listed in Table 3.19. The formula for calculating the standard deviation (3.9):

$$\sigma = \sum ni = 1(I\text{i-}I)2 / n(n-1)12 \qquad (3.9)$$

where *I*—mean of multiple measurements; *I*i—the *i*th measurement; *n*—number of measurements.

The experiment B, experiment BA, and theory B of all reflection are shown in Table 3.20. As can be seen from Table 3.20, except two symbols (651, 251) in

Table 3.19: Strengths and standard deviations of all reflection points

hkl	<*I*>	*n*	$\sigma(\bar{I})$	$\sigma(I)$/<*I*>	*hkl*	<*I*>	*n*	$\sigma(\bar{I})$	$\sigma(I)$/<*I*>
621	9203	8	52.09	0.006	841	18543	5	72.72	0.004
621	8956	8	62.05	0.007	741	64186	4	186.8	0.003
421	107561	3	247	0.002	741	63329	4	76.61	0.001
421	108201	3	65.24	0.001	251	11915	6	66.25	0.006
641	6876	6	46.84	0.007	251	12155	6	60.30	0.005
641	7256	6	45.30	0.006	151	16615	5	65.11	0.004
541	52159	2	157	0.003	151	17147	5	52.87	0.003
541	52908	2	243	0.005	651	3809	18	26.26	0.007
841	18391	5	45.67	0.002	651	3883	18	18.40	0.005
751	5265	12	31.68	0.006	943	14909	4	48.07	0.003
751	5459	12	28.56	0.005	653	4168	12	28.49	0.007
113	11378	5	79.90	0.007	653	4304	12	24.63	0.006
113	11741	5	39.19	0.003	553	2938	12	22.76	0.008
013	149948	3	326	0.002	553	2963	12	21.81	0.007
013	152380	3	296	0.002	514	11519	4	104.2	0.009
813	17317	4	98.19	0.006	514	11702	4	91.52	0.008
813	17637	4	51.72	0.003	414	16836	5	73.99	0.004
913	14483	3	119.8	0.008	414	16993	5	74.26	0.004
913	14420	3	73.30	0.005	424	8733	6	45.60	0.005
1223	6764	10	28.38	0.004	424	9159	6	39.48	0.004
1228	6871	10	42.66	0.006	224	15016	7	67.07	0.004
1323	17378	3	88.15	0.005	224	14858	7	73.65	0.005
1823	17418	3	57.59	0.003	234	19296	3	162.8	0.008
543	8373	8	30.80	0.004	234	19706	3	34.85	0.002
543	8683	8	50.85	0.006	034	27222	3	72.88	0.003
643	1201	12	15.33	0.013	034	27306	3	193.7	0.007
643	1385	12	16.20	0.012	634	3709	8	22.64	0.006
1043	8348	6	45.99	0.006	634	3836	8	15.22	0.004
1043	8490	6	37.96	0.005	534	36357	5	120.3	0.003
943	14827	4	92.09	0.006	534	36701	5	61.22	0.002

Table 3.20: $B_{experiment}$, $BA_{experiment}$, and B_{theory} of each reflection point

hkl	$B_{实验}$	$BA_{实验}$	$B_{实验}$	*hkl*	$B_{实验}$	$BA_{实验}$	$B_{实验}$
621	0.014	0.016	−0.006	543	−0.018	0.053	0.008
641	−0.027	−0.020	0.016	1043	−0.008	−0.006	0.008
841	−0.004	−0.011	0.008	653	−0.016	−0.012	0.009
251	−0.010	0.006	−0.006	514	−0.008	−0.003	0.009
651	−0.010	0.008	−0.011	424	−0.024	−0.029	0.011
113	−−0.016	−0.008	0.016	234	−0.011	−0.009	0.008
813	−0.009	−0.011	0.007	634	−0.017	−0.012	0.011
1223	−0.008	−0.007	0.010				

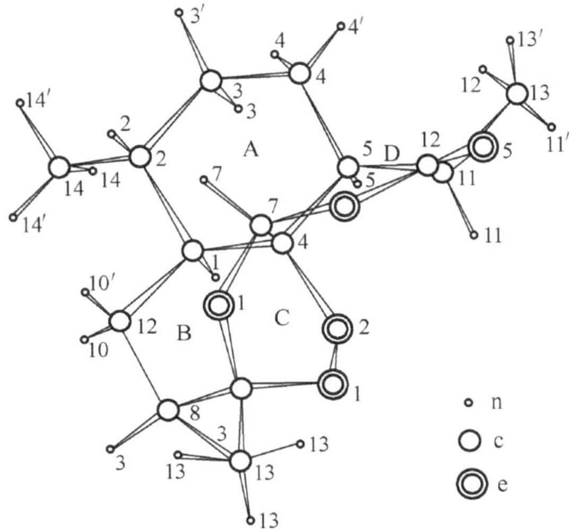

Figure 3.9: The Scenograph of the Artemisinin Molecule.

experiment B which were consistent with those in theory B in 15 reflection before absorption correction, all symbols between them were opposite; after absorption correction, except the symbol of only one reflection (543) in experiment BA which was consistent with that in theory B, all symbols between them were opposite. Therefore, it can be concluded that the absolute configuration of the molecule was a mirror image of the structure model obtained from the electron density map, as shown in Fig. 3.9. Moreover, the aforementioned conclusion was supported by absorption correction. The error indication of reflection 543 was caused by experimental error of its neighboring absorption Pair 643. Table 3.19 showed that Pair 643 had the weakest reflective strength and maximum relative standard deviation (more than 1%), so large absorption error was introduced when it was used to correct the absorption of Pair 543, thus overwhelming the anomalous scattering effect.

Discussion

Fig. 3.9 shows four interlinked rings of twelve carbon atoms with five oxygen atoms in artemisinin molecule: Rings A, B (C6, C7, O3, C8, C9, C10 and C1), C (C6, C7, O2, C8, O1 and O2), and D, of which Ring A is a chair-shaped cyclohexane, Ring D is a δ- lactone, distorted chair arrangement, Rings B and C are saturated oxetane. Based on structural parameters that are measured by X-ray diffraction, A/D, A/B, and C/D are *cis*-bonded, with the exception of Ring D and Ring B which are *trans*-bonded. Intramolecular and intermolecular hydrogen bonds are not found.

The bond length and bond angles as well as the accumulation of molecules in crystals are shown in Figs. 3.10–3.12. The bond length and bond angle associated with hydrogen atom are listed in Tables 3.21 and 3.22.

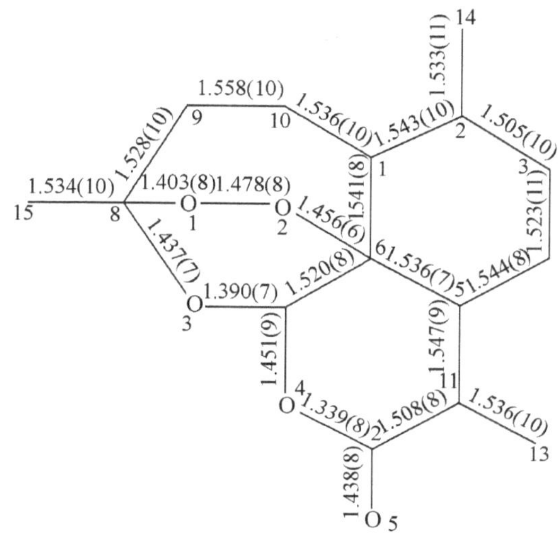

Figure 3.10: The C—O Bond Length of the Artemisinin Molecule.

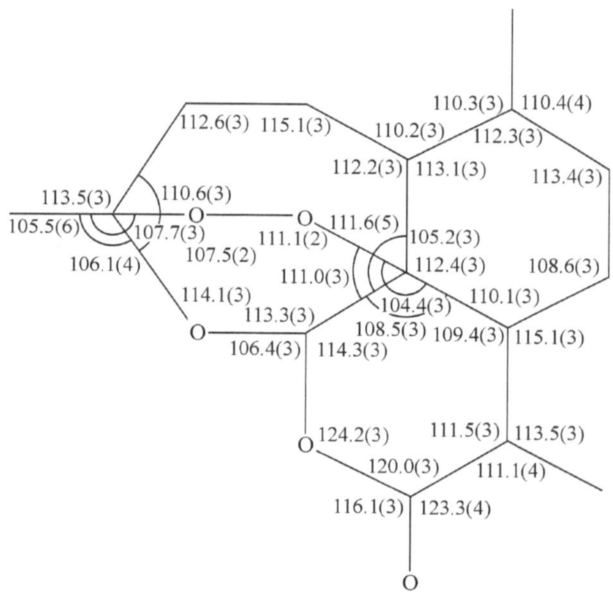

Figure 3.11: The C—O Bond Angle of the Artemisinin Molecule.

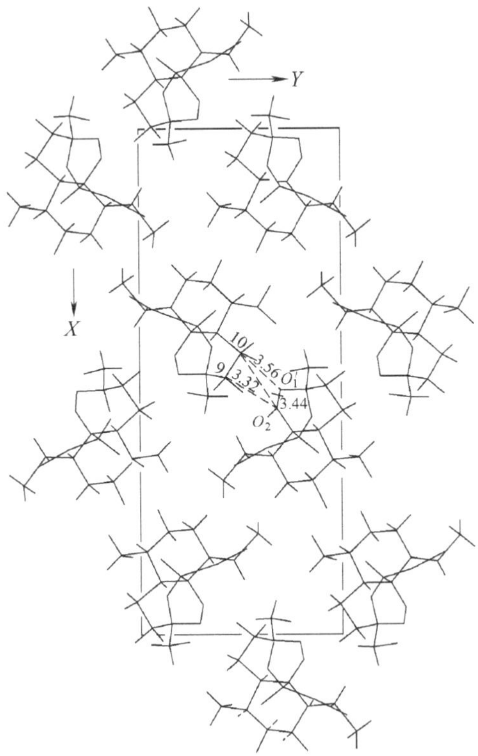

Figure 3.12: The Stacked Plot of Artemisinin Molecules in the Crystal.

As can be seen from Fig. 3.10, the length of two C—C single bonds (C1–C10 and C10–C9) is slightly longer than normal, where intermolecular distance is the nearest from the perspective of the crystal structure (Fig. 3.12). This can be possibly explained by interaction between adjacent molecules in crystals, resulting in the variation in the bond lengths.

The six-membered Ring C consisting of oxygen and peroxide bridges contained in the molecular structure of artemisinin was rarely seen, while similar six-membered ring existed only in 3-iso-propyl-6, 6-dimethyl-5-(1-naphthylamino)-1, 2, 4-trioxan.

With reference to some heterocyclic compounds containing peroxide group and oxygen bridges, the bond lengths and bond angles were compared (Table 3.23).

The first five peroxy bonds formed a ring, while the later four peroxide bonds did not form a ring. Thus, the O—O bond in artemisinin was within the normal range. Furthermore, the bond angles between C_6—O_2—O_1 and C_8—O_1—O_2 in artemisinin were 111.1 and 107.5°, respectively, which were similar to the bond angles in triacetone triperoxice compounds.

Table 3.21: Bond lengths related to the hydrogen atom

Bond	Bond Length (Å)	Bond	Bond Length (Å)
$C_1—H_1$	1.08	$C_{10}—H'_{10}$	1.01
$C_2—H_2$	1.01	$C_{11}—H_{11}$	1.10
$C_3—H_3$	1.08	$C_{13}—H_{13}$	1.10
$C_3—H'_3$	1.06	$C_{13}—H'_{13}$	1.02
$C_4—H_4$	1.08	$C_{13}—H''_{13}$	0.99
$C_4—H'_4$	0.95	$C_{14}—H_{14}$	1.07
$C_5—H_5$	1.08	$C_{14}—H'_{14}$	1.06
$C_7—H_7$	1.07	$C_{14}—H''_{14}$	1.07
$C_9—H_9$	1.09	$C_{15}—H_{15}$	1.06
$C_9—H'_9$	1.06	$C_{15}—H'_{15}$	1.04
$C_{10}—H_{10}$	1.00	$C_{15}—H''_{15}$	1.04

Table 3.22: Bond angles related to the hydrogen atom

Bond Angle	Degree (°)	Bond Angle	Degree (°)
$H_1C_1C_2$	107.4	$H_{10}C_{10}C_1$	104.9
$H_1C_1C_6$	108.4	$H_{10}C_{10}C_9$	107.4
$H_1C_1C_{10}$	105.8	$H_{10}C_{10}H'_{10}$	112.8
$H_2C_2C_1$	110.5	$H'_{10}C_{10}C_1$	108.7
$H_2C_2C_3$	107.5	$H'_{10}C_{10}C_9$	108.7
$H_2C_2C_{14}$	105.9	$H_{11}C_{11}C_5$	102.7
$H_3C_3C_2$	100.6	$H_{11}C_{11}C_{12}$	108.6
$H_3C_3C_4$	105.8	$H_{11}C_{11}C_{13}$	108.7
$H_3C_3H_3$	110.8	$H_{13}C_{13}H_{11}$	108.5
$H'_3C_3C_2$	116.8	$H_{13}C_{13}H'_{13}$	110.5
$H'_3C_3C_4$	108.5	$H_{13}C_{13}H''_{13}$	103.9
$H_4C_4C_3$	106.6	$H'_{13}C_{13}H''_{13}$	115.5
$H_4C_4C_5$	109.5	$H'_{13}C_{13}C_{11}$	106.8
$H_4C_4H'_4$	110.8	$H''_{13}C_{13}C_{11}$	112.7
$H'_4C_4C_3$	112.5	$H_{14}C_{14}C_2$	106.6
$H'_4C_4C_5$	108.7	$H_{14}C_{14}H'_{14}$	110.6
$H_5C_5C_4$	105.7	$H_{14}C_{14}H''_{14}$	118.4
$H_5C_5C_6$	102.9	$H'_{14}C_{14}C_2$	106.9
$H_5C_5C_{11}$	113.4	$h''_{14}C_{14}C_2$	109.7
$H_7C_7C_6$	105.7	$H'_{14}C_{14}H''_{14}$	105.6
$H_7C_7C_3$	100.4	$H_{15}C_{15}H'_5$	115.7
$H_7C_7C_4$	115.8	$H_{15}C_{15}C_8$	101.5
$H_9C_9C_8$	109.6	$H'_{15}C_{15}C_8$	108.4
$H_9C_9C_{10}$	105.6	$H_{15}C_{15}H''_{15}$	111.4
$H_9C_9H'_9$	113.7	$H''_{15}C_{15}C_8$	105.7
$H'_9C_9C_8$	109.4	$H'_{15}C_{15}H''_{15}$	114.5
$H_9C_9C_{10}$	106.8		

Table 3.23: Comparisons of bond lengths and angles of some compounds

Compound	O—O Bond Length (Å)
Artemisinin	1.478(6)
Dimeric cyclohexanone peroxide	1.482(2)
Dimeric cycloheptanone peroxide	1.472(3)
Dimeric cyclooctanone peroxide	1.474(2)
Trimeric acetone peroxide	1.483(2)
1,1-Dihydroperoxy cyclohexanyl-peroxide-1,1	1.477(4)
1,1′-Dihydroperoxy cyclododecanyl-peroxide-1,1′	1.474(3)
1,1-Dihydroperoxy cyclododecane	1.462
	1.467
ABP	1.445

However, it was worth mentioning that the dihedral angle between C—O—O—C of −47.4° (dihedral angle of absolute configuration was 47.4°) in artemisinin was less than that in the normal heterocyclic ring. This may be due to the fact that peroxide bridge was a six-membered ring formed by two tertiary carbon atoms C_6 and C_8 attached to the whole skeletal structure of molecule.

C—O bonds on Ring C were divided into two types: one was formed by oxygen bridge; the other was formed by peroxide bridge. However, the average of four C—O bonds was 1.423 regardless of formation type, which was consistent with the bond length of C—O in saturated heterocyclic ring (1.426 ± 0.005). The aforementioned phenomenon can also be seen in the structure of Phragmalin. Nevertheless, five oxygen atoms were allocated on one side of the molecule of artemisinin, and there was a carbon-oxygen chain, O_5—$C_{12}O_4$—C_7—O_3—C_8—O_1—O_2—C_6, similar to the "ether chain" from O_5, of which the length of carbon-oxygen bond from C_{12}—O_4 was short-long-short-long-short in order and intermediated in length between normal single and double bonds. Perhaps the lone pair of electrons that was not limited within each oxygen atom experienced variation in bond types. This helped to stabilize the whole molecule, which was consistent with stability to heat and light manifested by artemisinin molecules or crystals. This was a chain of carbon and oxygen similar to the "ether chain," which might be the site with specific performance in artemisinin molecule.

CHAPTER 4

Advancement in Researches on Chemical Ingredients in Artemisia annua *L.*

Chapter Outline

In the early 1970s, the research group of the Institute of Chinese Material Medica, China Academy of Chinese Medical Sciences led by Youyou Tu first conducted the systematic studies on chemical components of *Artemisia annua* L. Among 17 compounds (including artemisinin) isolated from *A. annua* L. (for chemical structure, see Fig. 2.42) (excluding gas chromatography analysis of volatile oils), six were oxygenated sesquiterpenes with unique structure that were rarely seen in nature [1–8]. The new structure and outstanding efficacy of artemisinin aroused great interest in chemical composition of *A. annua* L. at home and abroad. So far, more than 200 compounds have been isolated. Meanwhile, the researches on volatile components of *A. annua* L. by means of gas chromatography have also been reported. One excellent review on the phytochemistry of *Artemisia annua* L. was published in 2010 in the open access journal *Molecules*. The article by Geoffrey D. Brown, titled as "The biosynthesis of Artemisinin (Qinghaosu) and the phytochemistry of *Artemisia annua* L. (Qinghao)," reviewed almost 600 compounds in total and belonged to the special issue *Artemisinin (Qinghaosu): Commemorative Issue in Honor of Professor Youyou Tu on the Occasion of her 80th Anniversary*. Now, the chemical composition of *A. annua* L. is summarized as follows.

1 Terpenoids

Terpenoids are an important biologically active component of natural medicine, which are characterized by variable skeleton and huge quantity and extensive biological activity. It is one of the major physiologically active compounds in Chinese herbal medicine and

From *Artemisia annua* L. to Artemisinins
http://dx.doi.org/10.1016/B978-0-12-811655-5.00004-0

widely distributed in nature. From the perspective of the chemical structure, its skeleton is generally consisted of five carbon units, most of which are polyisoprene or isoprene derivatives. Terpenoids are often classified into monoterpene, sesquiterpenes, diterpene, etc. according to the number of isoprene units in the molecular structure, and further divided into chain, monocyclic, bicyclic, tricyclic, tetracyclic, and other terpenes depending on the absence or presence and the number of carbon ring in molecular structure of terpenes, such as chain diterpene, monocyclic diterpene, bicyclic diterpene, tricyclic diterpenoid, and tetracyclic diterpene. Most of the terpenoids are oxygenated derivatives, which can be divided into alcohols, aldehydes, ketones, carboxylic acids, esters, glycosides, etc.

Artemisinin is one of typical terpenoids in Chinese medicine-*A. annua* L. which has aroused great concern in the research community. Now, about 84 terpenoids have been isolated, including monoterpene, sesquiterpenes, diterpenes and triterpenoids, and other types.

1.1 Sesquiterpenes-Like Compounds

Domestic and foreign scholars have conducted extensive researches on artemisinin-based sesquiterpenes compounds which are the most active portion in *A. annua* L. There are about 61 isolated sesquiterpenes compounds to date (Table 4.1 and Fig. 4.1).

Table 4.1: Sesquiterpenoids isolated from *Artemisia annua* L.

No.	Compound	Part of the Plant	References
1	8[7→6]-Abeoamorphane,3α-hydroxy-4α,5α-epoxy-7-oxo	Seed	[9]
2	Abscisic acid	Aerial part	[10]
3	Abscisic acid methyl ester	Aerial part	[10]
4	Amorphan-12-oic acid,4α,5α-epoxy-6α-hydroxy	Seed	[9]
5	Amorphan-12-ol,4α,5α-epoxy-6α-hydroxy	Seed	[9]
6	Annuic acid,nor	Aerial part	[11]
7	Annuic acid formyl ester,nor	Seed	[9]
8	Annulide	Aerial part	[12]
9	Annulide,iso	Leaf	[12]
10	Arteannuic acid (artemisinic acid)	Leaf	[4,5]
11	Arteannuic acid,6,7-dehydro	Leaf	[13]
12	Arteannuic acid,11(*R*)-Dihydro	Flower	[14]
13	Arteannuic acid,epoxy	Herb	[15]
14	Arteannuic acid,epoxy-dihydro	Seed	[9]
15	Arteannuic acid methyl ester	Herb	[16]
16	Arteannuic alcohol	Root	[10]
17	Arteannuin A	Leaf	[3,5]
18	Arteannuin B	Leaf	[1,3,5,17,18]
19	Arteannuin B,deoxy	Herb	[18]
20	Arteannuin B,dihydro-deoxy	Herb	[18]
21	Arteannuin B,epi-deoxy	Herb	[19]
22	Arteannuin B,dihydro	Leaf	[20]
23	Arteannuin B,dihydro-epi-deoxy	Aerial part	[21]

Table 4.1: Sesquiterpenoids isolated from *Artemisia annua* L. (*cont.*)

No.	Compound	Part of the Plant	References
24	Arteannuin C	Aerial part	[22]
25	Arteannuin D	Leaf	[5,6]
26	Arteannuin E	Leaf	[5,6]
27	Arteannuin F(artemisilactone)	Aerial part	[23]
28	Arteannuin G	Leaf	[24]
29	Arteannuin H	Leaf	[20]
30	Arteannuin I	Leaf	[20]
31	Arteannuin J	Leaf	[20]
32	Arteannuin K	Leaf	[20,25]
33	Arteannuin L	Leaf	[20,25]
34	Arteannuin M	Leaf	[20,25]
35	Arteannuin N	Leaf	[20]
36	Arteannuin O	Leaf	[25]
37	Artemisia dihydroxycadinolide	Aerial part	[26]
38	Artemisinin	Herb	[10]
39	Artemisinin (Qinghaosu)	Leaf	[1,2,5]
40	Artemisinin,dehydro(artemisitene)	Aerial part	[27]
41	Artemisinin, deoxy (hydroartemisinin)	Leaf	[3,5]
42	Artemisinol	Herb	[16]
43	Cadina-4-en-11-ol,3-iso-butyryl	Aerial part	[28]
44	Cadina-4-ene,3α,7α-dihydroxy	Leaf	[29]
45	Cadina-4-ene,3α,7α-dihydroxy,3-acetate(amorpha-4-ene,3α, 7α-dihydroxy,3-acetate)	Seed	[9]
46	Cadina-4 (15),11-dien-9-one	Aerial part	[28]
47	Cadina-4 (7),11-dien-12-al	Aerial part	[28]
48	Cedrane,3α,15-dihydroxy	Seed	[9]
49	Cyclohexane,lα-aldehyde-2β-[3-butanone]-3α-methyl-6β-[2-propenoic acid](artemisia secocadinane)	Aerial part	[26]
50	Cyclohexane,lα-aldehyde-2β-[3-butanone]-3α-methyl-6β-[2-propanoic acid]	Seed	[9]
51	Cyclohexane,1-oxo-2β-[3-butanone]-3α-methyl-6β-[2-propanol formyl ester]	Seed	[9]
52	Cyclohexane,1-oxo-2β-[3-butanone]-3α-methyl-6β-[2-propanoic acid]	Seed	[9]
53	Eudesma-4(15),11-diene	Seed	[9]
54	Eudesma-4(15),11-diene,5α-hydroperoxy	Seed	[9]
55	Eudesma-4(15),11-diene,5α-hydroxy	Leaf	[28]
56	Eudesma-4(15),5(6)-diene, 1β-hydroxy	Seed	[9]
57	Eudesma-4(15),7(8)-diene, 1β-hydroxy	Seed	[9]
58	Eudesma-4(15)-ene,lβ,6α-dihydroxy	Seed	[9]
59	Germacrane-4(15),5*E*,10(14)-triene, 1β-hydroxy	Seed	[9]
60	(+)-Nortaylorione	Herb	[30]
61	Oplopan-4-oic acid,15-nor-10-hydroxy	Seed	[9]

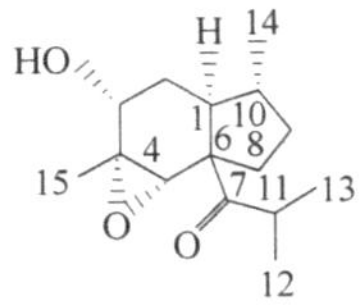

8[7→6]-Abeoamorphane,3α-hydroxy-4α,5α-epoxy-7-oxo

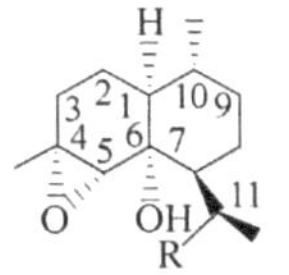

Amorphan-12-oic acid,4α,5α-epoxy-6α-hydroxy R=COOH

Amorphan-12-ol,4α,5α-epoxy-6α-hydroxy R=CH_2OH

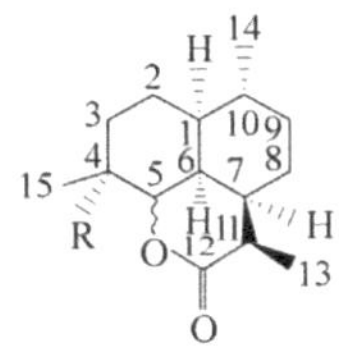

Annulide

5α-O,4(15),11(13)-diene R=H

Annulide, iso

5α-O,3(4),11(13)-diene R=H

Arteannuin E

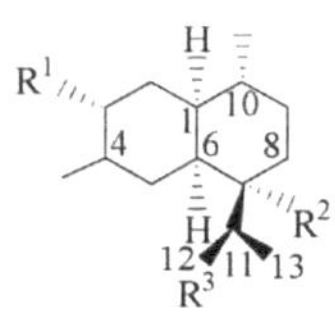

Arteannuic alcohol

4(5),11(13)-diene R^1,R^2=H,R^3=CH_2OH

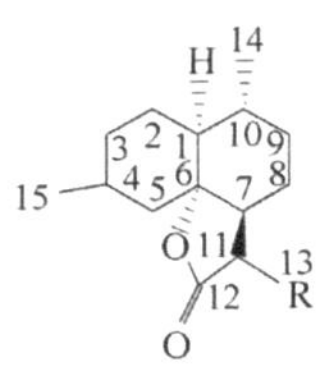

Arteannuin B

4,5α-epoxy R=CH_2

Arteannuin B,deoxy

4(5)-ene R=CH_2

Arteannuin B,dihydro-deoxy

4(5)-ene R=CH_3

Arteannuin B,dihydro

4,5α-epoxy R=CH_3

Arteannuin C

4,5β-epoxy R=CH_2

Artemisia dihydroxycadinolide

4-OH,5-OH R=CH_2

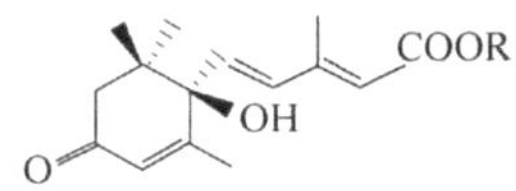

Abscisic acid R=H

Abscisic acid methyl ester R=CH_3

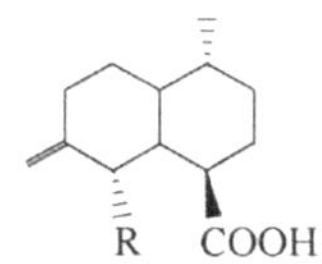

Annuic acid, nor R=OH

Annuic acid formyl ester, nor R=OOCH

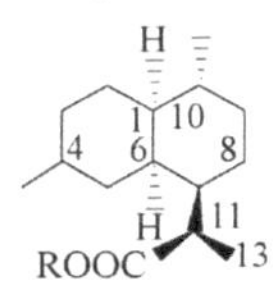

Arteannuic acid(artemisinic acid)

4(5),11(13)-diene R=H

Arteannuic acid,6,7-dehydro

4(5),6(7),11(13)-triene R=H

Arteannuic acid,11(R)-dihydro

4(5)-ene R=H

Arteannuic acid,epoxy

4,5α-epoxy,11(13)-ene R=H

Arteannuic acid,epoxy-dihydro

4,5α-epoxy R=H

Arteannuic acidmethylester

4(5),11(13)-diene R=CH_3

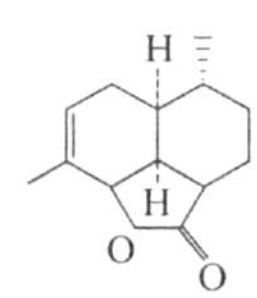

Arteannuin A

Figure 4.1: Structure Diagrams of Sesquiterpenoids Isolated From *Artemisia annua* L.

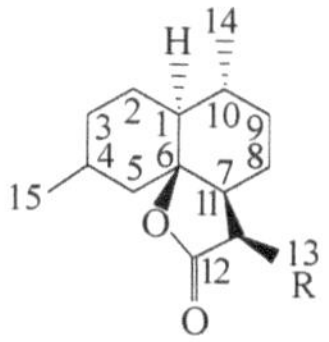

Arteannuin B,epi-deoxy

4(5),11(13)-diene

Arteannuin B,dihydro-epi-deoxy

4(5)-ene

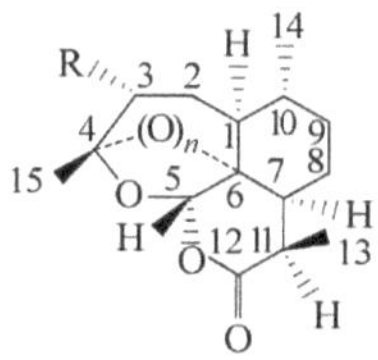

Ateannuin D

R=OH,*n*=1

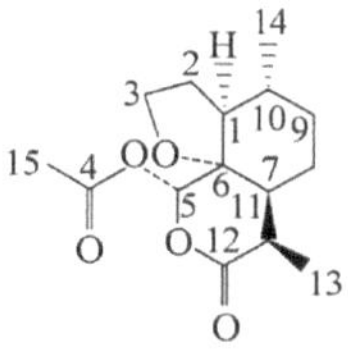

Arteannuin G

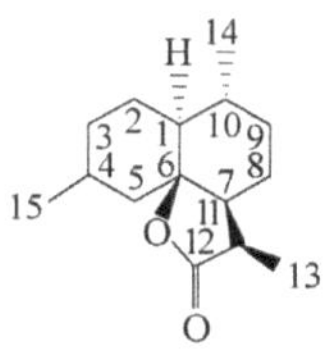

Arteannuin K

5α-OH,3(4)-ene

Arteannuin L

5α-OH,4(15)-ene

Arteannuin M

5α-OH,4α-OH

Arteannuin O

5α-OH,5β-OH

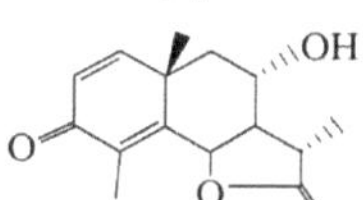

R=OOCH

Artemisin

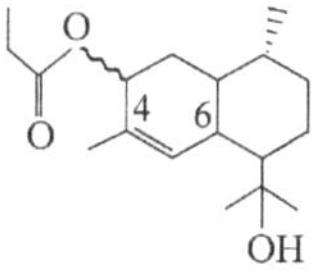

Cadia-4-en-11-ol,3-iso-butyryl

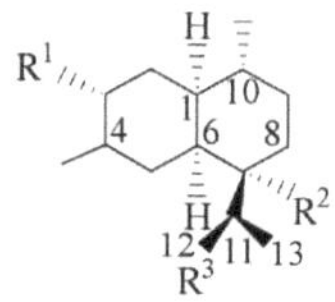

Cadina-4-ene 3α-7-α-dihydroxy

4(5)-ene R^1=OH,R^2=OH, R^3=CH_3

Cadina-4-ene,3α,7α-dihydroxy,3-acetate

(amorpha-4-ene,3α,7α-dihydroxy,

3-acetate)

4(5)-ene

R^1=CH_3COO,R^2=OH,R^3=CH_3

Cadina-4(5),7(11)-dien-12-al

4(5),7(11)-diene

R^1,R^2=H,R^3=CHO

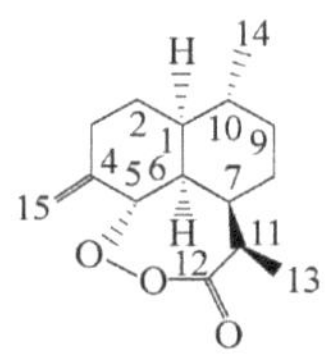

Arteannuin H

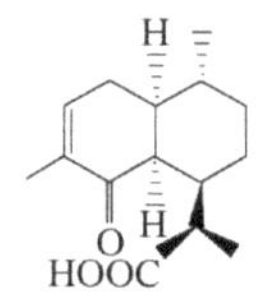

Arteannuin N

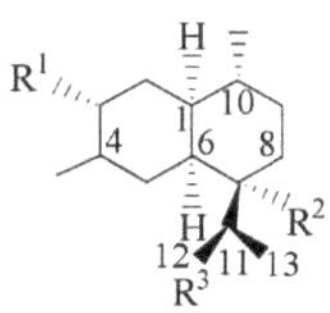

Artemisinol

4(5)-diene R^1,R^2=H,

R^3=CH_2OH

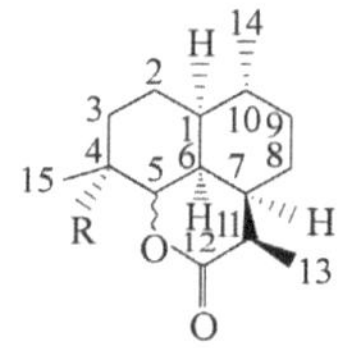

Arteannuin E

5β-O,11(13)-ene R=OH

Arteannuin F (artemisilactone)

5α-O,11(13)-ene R=OH

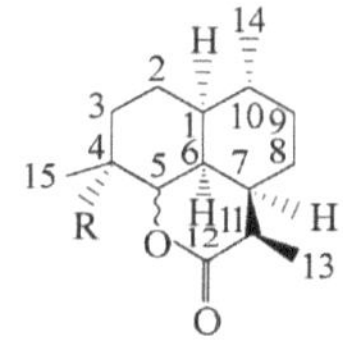

Arteannuin I

5α-O,4(15)-ene R=H

Arteannuin J

5α-O,3(4)-ene R=H

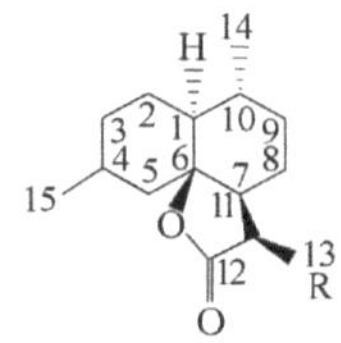

Artemisia dihydroxycadinolide

4-OH,5-OH R=CH_2

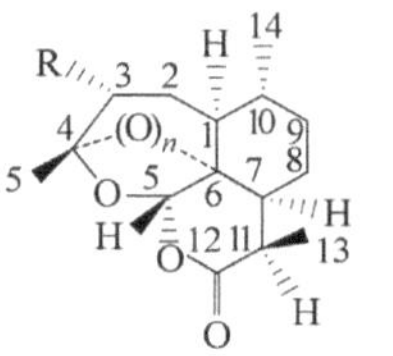

Artemisinin(Qinghaosu)

R=H,*n*=2

Artemisinin,dehydro(artemisitene)

R=H,*n*=2,11(13)-ene

Artemisinin,deoxy

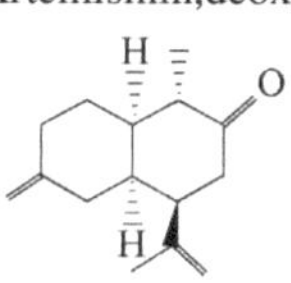

Cadina-4(15),11-dien-9-one

Figure 4.1: (*Continued*)

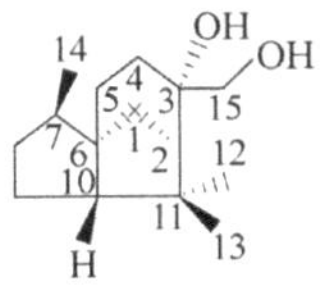

Cedrane,3α,15-dihydroxy

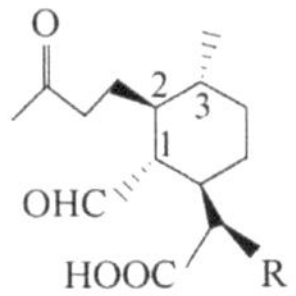

Cyclohexane,1α-aldehyde-2β-[3-butanone]-3α-methyl-6β-[2-propen oicacid](artemisiasecocadinane) R=CH_2

Cyclohexane,1α-aldehyde-2β-[3-butanone]-3α-methyl-6β-[2-propan oicacid] R=CH_3

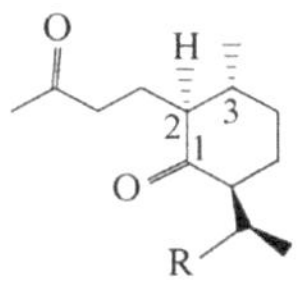

Cyclohexane,1-oxo-2β-[3-butanone]-3α-methyl-6β-[2-propanolform ylester] R=CH_2OOCH

Cyclohexane,1-oxo-2β-[3-butanone]-3α-methyl-6β-[2-propanoicaicd] R=COOH

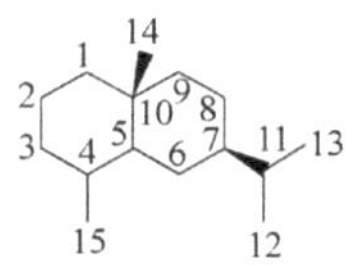

Eudesma-4(15),11-diene,5α-hydroxy
4(15),11(12)-diene,5α-OH
Eudesma-4(15),5(6)-diene,1β-hydroxy
4(15),5(6)-diene,1β-OH

Eudesma-4(15),7(8)-diene,1β-hydroxy
4(15),7(8)-diene,1β-OH
Eudesma-4(15)-ene,1β,6α-dihydroxy
4(15)-ene,1β-OH,61α-OH
Eudesma-4(15),11-diene
4(15),11(12)-diene,5α-H
Eudesma-4(15),11-diene,5α-hydroperoxy
4(15),11(12)-diene,5α-OOH

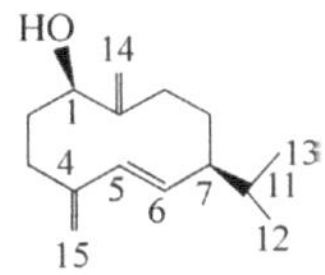

Germacrane-4(15),5E,10(14)-triene,1β-hydroxy

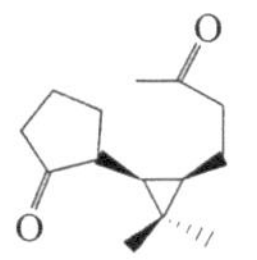

(+)-Nortaylorione

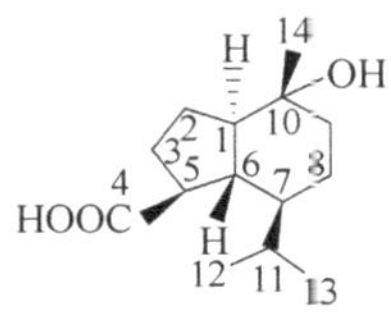

Oplopan-4-oic acid,15-nor-10-hydroxy

Figure 4.1: (*Continued*)

1.2 Other Terpenoids

The yield of volatile oils in traditional Chinese medicine-*A. annua* L. is 0.2%–0.8%, which are mainly monoterpene compounds. To date, there are about 13 monoterpene and diterpene which have been successfully extracted and separated from *A. annua* L. (Table 4.2 and Fig. 4.2).

Triterpenoids and their glycosides are widely distributed across a variety of domains, such as fungi, ferns, monocots, dicots, animals and marine organisms, especially significant in dicotyledons. Traditional Chinese medicine, such as Ginseng, astragalus, licorice, Panax, Campanulaceae, Polygala, and bupleurum, are rich in triterpenes. These compounds exhibit a

Table 4.2: Monoterpenes and diterpenoids isolated from *Artemisia annua* L.

No.	Compound	Part of the Plant	References
1	Fenchone	Aerial part	[22]
2	α-Myrcene hydroperoxide	Aerial part	[31]
3	1,10-Oxy-α-myrcene hydroxide	Seed	[9]
4	β-Mycerene hydroperoxide	Herb	[31]
5	1,10-Oxy-β-myrcene hydroxide	Seed	[9]
6	4-Hydroxy-2-isopropenyl-5-methylene-hexan-l-ol	Seed	[9]
7	5-Hydroxy-2-isopropenyl-5-methylhex-3-en-l-ol,*trans*	Herb	[18]
8	Tricyclene	Aerial part	[22]
9	Phytene,1,2-dihydroxy	Aerial part	[32]
10	Phytene,l hydroxy-2-hydroperoxy	Seed	[9]
11	*trans*-Phytol	Aerial part	[32]
12	Hentriacontan-1-ol-triacontanoate	Herb	[32]
13	*trans*-5-Hydroxy-2-isopropenyl-5-methylhex-3-en-1-ol	Herb	[18]

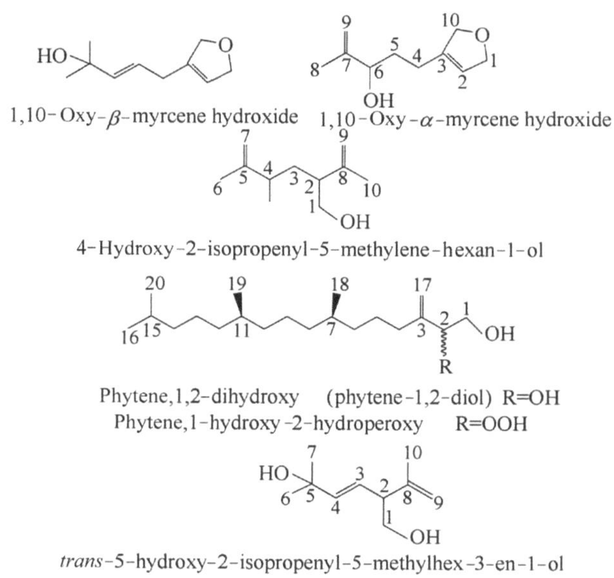

Figure 4.2: Structures of Monoterpenes and Diterpenoids Isolated From *Artemisia annua* L.

Table 4.3: Triterpenoids isolated from *Artemisia annua* L.

No.	Compound	Part of the Plant	References
1	α-Amyrenone	Aerial part	[28]
2	α-Amyrin	Aerial part	[28]
3	β-Amyrin	Aerial part	[28]
4	β-Amyrin,acetate	Aerial part	[33]
5	Baurenol	Aerial part	[28]
6	β-Friedelan-3-ol	Leaf	[34]
7	Friedelin	Leaf	[34]
8	Oleanolic acid	Aerial part	[28]
9	Taraxasterone	Aerial part	[28]
10	Taraxerol acetate	Root	[10]

variety of physiological activities. There are few triterpenes in traditional Chinese medicine-*A. annua* L. So far, about 10 compounds have been isolated (Table 4.3).

As the unique structure of terpenoids in Chinese medicine-*A. annua* L, it is necessary to further investigate the biogenesis pathway of artemisinin; the investigation on terpenoids in traditional Chinese medicine-*A. annua* L is still the objective for many research groups, and some of them have extracted terpenoids from *A. annua* L. by means of tissue culture, microbial metabolism, and so on.

2 Other Compounds

There are a variety of chemical constituents existing in traditional Chinese medicine-*A. annua* L. in addition to terpenoids as represented by artemisinin. Compounds that have been isolated and identified include coumarin, flavonoids, dipeptide, purines, steroids, enyne compounds, and the like. Following is the brief summary of our findings.

2.1 Coumarins

Coumarins isolated from *A. annua* L. are shown in Table 4.4.

2.2 Flavonoids

Flavonoids isolated from *A. annua* L. are shown in Table 4.5.

Table 4.4: Coumarins isolated from *Artemisia annua* L.

No.	Compound	Part of the Plant	References
1	Coumarin	Leaf	[7]
2	6,7-Dihydroxycoumarin(esculetin)	Aerial part	[10]
3	6,7-Dimethoxycoumarin(scoparone)	Herb	[35]
4	7-Hydroxy-6-methoxycoumarin(scopoletin)	Leaf	[3,36]
5	5-Hydroxy-6,7-dimethoxycoumarin(tomentin)	Leaf	[37]
6	7-Hydroxy-5,6-dimethoxycoumarin	Leaf	[38]
7	7-Hydroxy-6,8-dimethoxycoumarin(isofraxidin)	Aerial part	[36]
8	7-Hydroxy-6,8-dimethoxycoumarin,7-O-α-D-glucopyranoside(eleutheroside B_1)	Leaf	[10]

Table 4.5: Flavonoids isolated from *Artemisia annua* L.

No.	Compound	Part of the Plant	References
1	4′,5,7-Trihydroxyflavone(apigenin)	Leaf	[37]
2	4′,5-Dihydroxy-6,7-dimethoxyflavone(cirsimaritin)	Leaf	[38]
3	3,4′,5-Trihydroxy-7-methoxyflavone(rhamnocitrin)	Leaf	[38]
4	3′,4′,5-Trihydroxy-7-methoxyflavone	Leaf	[37]
5	4′,5,7-Trihydroxy-3-methoxyflavone(isokaempferide)	Leaf	[37]
6	4′,5,7-Trihydroxy-3′-methoxyflavone(chrysoeriol)	Leaf	[38]
7	3,3′,4′,7-Tetrahydroxyflavone(fisetin)	Aerial part	[10]
8	3,4′,5,7-Tetrahydroxyflavone(kaempferol)	Aerial part	[39]
9	3′,4′,5,7-Tetrahydroxyflavone(luteolin)	Aerial part	[39]
10	5-Hydroxy-3,3′,4′,7-tetramethoxyflavone(retusin)	Leaf	[29]
11	5-Hydroxy-3,4′,6,7-tetramethoxyflavone	Leaf	[7]
12	3′,5-Dihydroxy-4′,6,7-trimethoxynavone(eupatorin)	Leaf	[38]
13	4′,5-Dihydroxy-3,3′,7-trimethoxyflavone(pachypodol)	Leaf	[29]
14	4′,5-Dihydroxy-3,6,7-trimethoxyflavone(penduletin)	Leaf	[38]
15	4′,5-Dihydroxy-3′,6,7-trimethoxyflavone(cirsilineol)	Leaf	[38]
16	3′,4′,5-Trihydroxy-6,7-dimethoxyflavone(cirsiliol)	Leaf	[38]
17	3,3′,4′,5-Tetrahydroxy-7-methoxyflavone(rhamnentin)	Leaf	[38]
18	3,3′,5,7-Tetrahydroxy-4′-methoxyflavone(tamarixetin)	Leaf	[38]
19	3′,4′,5,7-Tetrahydroxy-3-methoxyflavone	Leaf	[38]
20	3,3′,4′,5,7-Pentahydroxyflavone(quercetin)	Leaf	[37]
21	5-Hydroxy-3,3′,4′,6,7-pentamethoxyflavone(artemetin)	Leaf	[7]
22	3,5-Dihydroxy-3′,4′,6,7-tetramethoxyflavone	Leaf	[3]
23	3′,5-Dihydroxy-3,4′,6,7-tetramethoxyflavone(casticin)	Leaf	[38]
24	4′,5-Dihydroxy-3,3′,6,7-tetramethoxyflavone(chrysosplenetin)	Aerial part	[21]
25	5,7-Dihydroxy-3,3′,4′,6-tetramethoxyflavone(bonanzin)	Aerial part	[40]
26	2′,4′,5-Trihydroxy-5′,6,7-trimethoxyflavone	Leaf	[38]
27	3,3′,5-Trihydroxy-4′,6,7-trimethoxyflavone	Leaf	[36]
28	3′,4′,5-Trihydroxy-3,6,7-trimethoxyflavone(chrysosplenol D)	Leaf	[6]
29	3′,4′,5,7-Tetrahydroxy-3,6-dimethoxyflavone(axillarin)	Leaf	[38]
30	3,5,6,7-Tetrahydroxy-3′,4′-dimethoxyflavone	Leaf	[38]
31	3′,5,7,8-Tetraliydroxy-3,4′-dimethoxyflavone	Leaf	[38]
32	3,3′,4′,5,7-Pentahydroxy-6-methoxyflavone(patuletin)	Aerial part	[39]
33	3,3′,5,6,7-Pentahydroxy-4′-methoxyflavone	Leaf	[37]
34	3,4′,5,7-Tetrahydroxyflavone,3-O-β-D-glucopyranoside(astragalin)	Aerial part	[39]
35	3′,4′,5,7-Tetrahydroxyflavone,7-O-β-D-glucopyranoside(cynaroside)	Leaf	[37]
36	3,3′,4′,5,7-Pentahydroxyflavone,3-O-β-D-glucopyranoside(isoquercitrin)	Aerial part	[39]
37	3,3′,4′,5,7-Pentabydroxyflavone,3′-O-β-D-glucopyranoside	Leaf	[37]
38	3,3′,4′,5,7-Pentahydroxyflavone,7-O-β-D-glucopyranoside(ouercimeritrin)	Leaf	[37]
39	3,3′,4′,5,7-Pentabhydroxy-6-methoxyflavone,3-O-β-D-glucopyranoside	Aerial part	[39]
40	3,3′,4′,5,7-Pentahydroxyflavone,3-O-β-D-rutinoside(rutin)	Aerial part	[39]
41	3,3′,4′,5,7-Pentahydroxy-6-methoxyflavone,3,7-di-O-α-L-rhamnopyranoside(patuletin-3,7-dirhamnoside)	Aerial part	[10]

Table 4.6: Aliphatic and enyne compounds isolated from *Artemisia annua* L.

No.	Compound	Part of the Plant	References
1	Annuadiepoxide	Leaf	[41]
2	Docosan-2-one	Aerial part	[33]
3	Hentriaconatayl-triacontanolate	Aerial part	[42]
4	Hexacosan-1-ol	Aerial part	[33]
5	Hexadecanoic acid(palmitic acid)	Herb	[16]
6	Nonacosan-1-ol	Leaf	[42]
7	*n*-Nonacosane	Aerial part	[33]
8	Ocatacosan-1-ol	Herb	[8]
9	*n*-Pentacosane	Aerial part	[33]
10	Pontiac oxide	Leaf	[41]
11	Triacontane,2,29-dimethyl	Leaf	[42]
12	Tracosan-8-one-23-ol,2-methyl	Leaf	[42]
13	*n*-Tetratriacontane	Aerial part	[43]

2.3 Aliphatic and Enyne Compounds

Aliphatic and enyne compounds isolated from *A. annua* L. are shown in Table 4.6, and the structure of enyne compounds is shown in Fig. 4.3.

2.4 Others

Other compounds isolated from *A. annua* L. are shown in Table 4.7 and Fig. 4.4.

Moreover, β-galactosidase, β-glucosidase enzymes, and other proteins compounds, as well as naphthalene and furanone derivatives isolated from *A. annua* L. were also reported.

To date, some researchers still continue with the in-depth studies on chemical composition of traditional Chinese medicine-*A. annua* L., which is undoubtedly contributing to reveal the traditional efficacy of *A. annua* L. and development of new drugs.

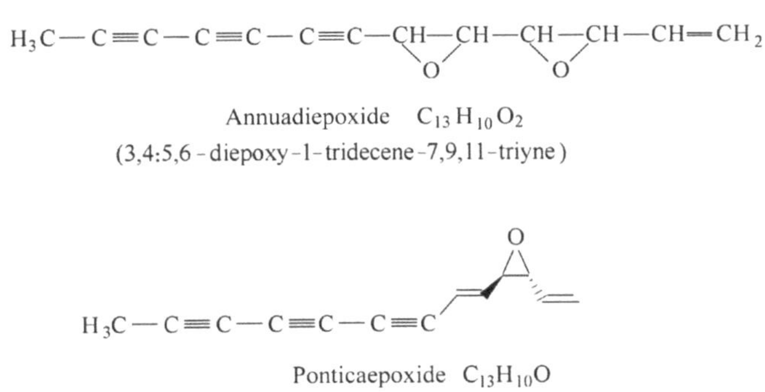

Figure 4.3: Enyne Compounds Isolated From *Artemisia annua* L.

Table 4.7: Other compounds isolated from *Artemisia annua* L.

No.	Compound	Part of the Plant	References
1	Acetophenone,2,4-dihydroxy-6-methoxy	Aerial part	[21]
2	Annephenone	Aerial part	[44]
3	Aurantinamide acetate	Leaf	[7]
4	Balanophonin	Tissue culture	[45]
5	Chromene,2,2,6-trihydroxy	Leaf	[37]
6	Chromene,2,2-dihydroxy-6-methoxy	Leaf	[37]
7	Cinnamaldehyde,4-methoxy	Tissue culture	[45]
8	Coniferaldehyde	Tissue culture	[45]
9	Coniferaldehyde, (2-glyceryl)-O-	Tissue culture	[45]
10	Coniferaldehyde, (2-propcnal)-O-	Tissue culture	[45]
11	Coumaric acid	Aerial part	[21]
12	Phthalate,bis-(hydroxy-2-methyl-propyl)	Aerial part	[10]
13	Purine	Aerial part	[46]
14	Resorcinol	Aerial part	[21]
15	β-Sitosterol	Leaf	[7]
16	Stigmasterol	Herb	[47]
17	Vanillin	Tissue culture	[45]
18	Xanthoxylin	Aerial part	[21]
19	Zeatin	Aerial part	[46]
20	Zenatin	Aerial part	[46]

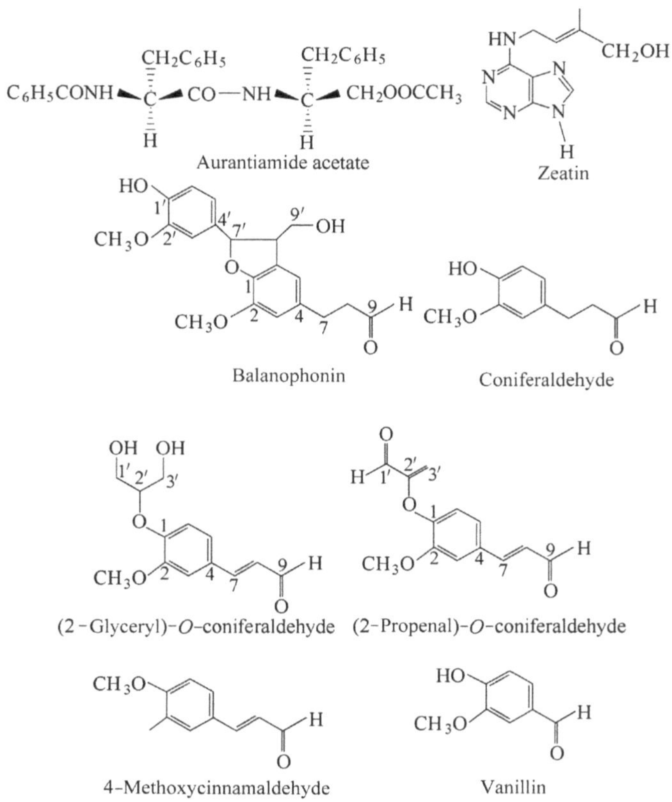

Figure 4.4: Other Eight Compounds Isolated From *Artemisia annua* L.

3 Gas Chromatography Analysis of Volatile Oils

Volatile oils of Chinese medicine-*A. annua* L. with some medicinal value can be used to cure skin diseases, chronic airway diseases, etc. Gas chromatography is widely used in the studies on volatile oils of *A. annua* L. from different areas and sources for the years. The following is a brief overview.

Zhong Yurong et al. [48] reported that the yield of volatile oils extracted from *A. annua* L. produced in Fujian and Xiamen by steam distillation was 0.2%–0.25%. Among five chemical constitutes that were isolated from the volatile oils, camphene, β-pinene, isoartemisia ketone, β-caryophyllene and L-camphor, 20 components, such as α-pinene, myrcene, limonene, 1, 8-cineole, γ-terpinene, artemisia ketone, α-terpineol, *trans*-caryophyllene, *trans*-β-farnesene, bornyl isovalerate, γ-cadinene, δ-cadinene, α-elemene, β-elemene, γ-elemene, β-maaliene, γ-muurolene, *cis*-carveol, copaene, bornyl acetate, and the like were identified after GC-MS analysis.

In the study on the antifungal effect of the volatile oils extracted from *A. annua* L. carried out by Liang Huaqing et al. [49] using GC-MS method, 13 kinds of chemical constituents were identified: L-camphor, 1,8-cineole, α-pinene, bornyl acetate, borneol, salicylic acid, β-terpinene, α-thujene, 2-carene, isobornyl acetate linalyl acetate, etc.

Peng Hong et al. [50] identified 26 kinds of chemical constitutes, such as camphor, borneol, *para*-cymene, 4-terpineol, α-terpineol, myrtenol, piperitone, *cis*-jasmone, α-humulene, γ-selinene, β-selinene, spathulenol, caryophyllene oxide, cadinol, qinghaosu β, and so on from volatile oils that were derived from *A. annua* L. produced in Sichuan via water distillation, with the yield of 0.29%.

Chen Feilong et al. [51] identified 114 kinds of monoterpenes, sesquiterpenes, and long-chain aliphatic chemical constitutes from commercially available volatile oils of *A. annua* L. through steam distillation and supercritical CO_2 extraction, such as β-caryophyllene, caryophyllene oxide, camphor, *trans*-β-farnesene, qinghao acid, hexadecanoic acid, linoleic acid, linoleic acid ethyl ester, deoxyqinghaosu, and so on, of which diphenylene-methane and phenyl-1-naphthylamine were firstly discovered in this plant.

Dong Yan et al. [52] identified 65 kinds of chemical constituents in volatile oils of *A. annua* L. produced in outskirts of Dezhou, Shandong using capillary gas chromatography and mass spectrometry techniques, such as artemisia ketone, camphor, isocaryophyllene, germacrene D, eudesma-4 (14), 11-diene, and so on.

A total of 55 peaks were separated by Qiu Qin et al. [53] using capillary gas chromatography from volatile oils of *A. annua* L. produced in Jinan, Shandong, which were extracted by means of steam distillation (yield of 0.8%). After the relative content of each peak was calculated with normalization method, 49 chemical constituents were identified by gas chromatography mass spectrometry, accounting for more than 89% of volatile oils. The main compounds were cadinene, camphor, α-selinene, α-guaiene [1 (5), 11-guaiadiene], and other sesquiterpenes, terpene alcohols, terpene ketones, esters, as well as a small amount of aldehydes, acids, alkanes, and benzenes.

Liu Liding et al. [54] extracted essential oils through steam distillation from stem, leaves, and flowers above the middle of *A. annua* L. that was produced in Qinling North Slope, Zhouzhi County, Shaanxi in September, mainly consisting of 57 chemical constituents, such as β-pinene, camphene, β-myrcene, 1,8-cinoole, L-camphor, longifolene, phenyl acetate, and the like.

Volatile oils of *A. annua* L. produced in the Netherlands were mainly composed of artemisia ketone (63.9%), artemisia alcohol (7.5%), and α-guaiene [1 (5), 11-guaiadiene] (4.7%) et al. [55].

Twenty kinds of chemical constituents extracted from volatile oils of *A. annua* L. produced in Vietnam have been reported by H.J. Woerdenbag et al. [56], of which camphor, germacrene D, β-caryophyllene, 1,8-cineole, *para*-cymene, *trans*-β-farnesene, etc. accounted for a larger proportion, who also studied the changes in the content of 20 components in volatile oils extracted from *A. annua* L. from June to November in 1 year.

A. Ahmad and L.N. Misra [28] reported that the chemical constituents of volatile oils in *A. annua* L. were mainly artemisia alcohol, camphene hydrate, caryophyllene oxide, chrysanthenone (2-pinen-6-one), α-copaene, β-cubebene, fenchol, α-humulene, myrtenal, myrtenol, α-pinene, pinocarvone, sabinene, α-terpinene, γ-terpinene, α-thujene, etc. (Fig. 4.5).

In the study carried out by F.F. Perazzo et al. [57], the activity of volatile oils derived from *A. annua* L. produced in Brazil on the central nervous system was analyzed, involving 15 chemical constituents, such as α-pinene, β-pinene, sabinene, *p*-cymene, 15 kinds of ingredients 1,8-cineole, α-terpinene, linalool, camphor, α-terpineol, *trans*-caryophyllene, farnesene, germacrene D, and the like.

The antimicrobial and antioxidant activity of volatile oils derived from *A. annua* L. originated from France has been demonstrated in the study conducted by F. Juteau et al. [58]. The chemical constituents of volatile oils that have been analyzed mainly included camphor (44%), germacrene D (16%), *trans*-pinocarveol (11%), β-selinene (9%), β-caryophyllene, and artemisia ketone (3%).

Based on the literature review, 72 kinds of components in volatile oils extracted from *A. annua* L. have been published by Indian scholars, R.S. Bhakuni et al. [10], of which sesquiterpenes and monoterpene were predominant component.

The chemical components of volatile oils derived from *A. annua* L. vary with origin, harvesting and extraction and analysis methods, of which β-camphene, isoartemisia ketone, camphor, borneol, artemisia ketone, caryophyllene, 1,8-cineole, and so on were predominant and commonly seen components. The content of artemisia ketone varies greatly,

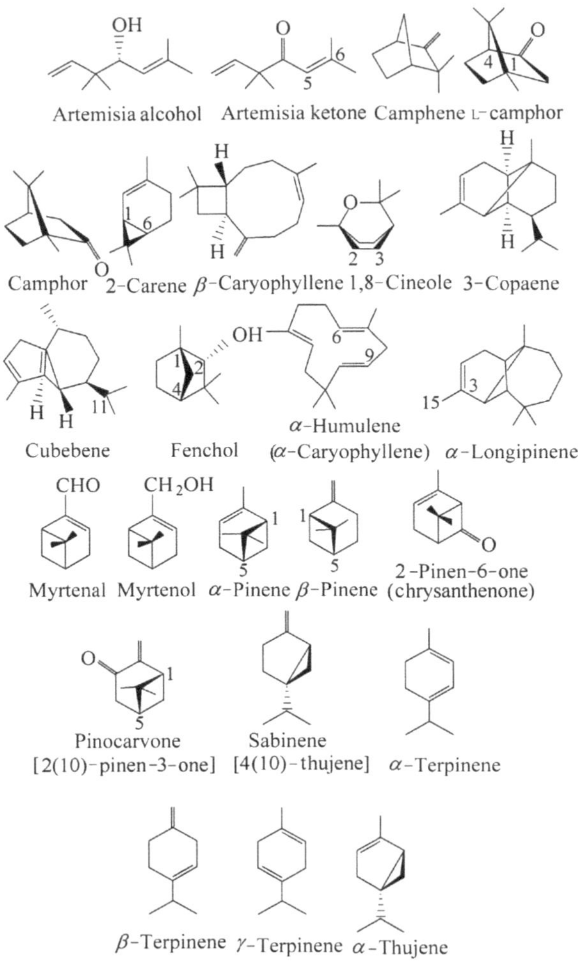

Figure 4.5: Structures of Compounds in Volatile Oils From *Artemisia annua* L.

which is relatively higher in those originated from Bulgaria (80.9%), Netherlands (63.9%) and the United States (63.1%) [10]. The yield of volatile oils was 0.2%–0.8%, in which sesquiterpenes and monoterpene were predominant components.

References

[1] Traditional Chinese medicine research institution. Artemisinin-based antimalarial album (1971~1978). Traditional Chinese medicine research data, 1978;3.

[2] Research Group of Artemisinin Structure (China academy of traditional Chinese medicine and The Chinese Academy of Sciences Institute of Biophysics). A new kind of sesquiterpene lactone—artemisinin. Chin. Sci. Bull. 1977;22(3):142.

[3] Tu YY, Ni MY, Zhong YR, Li LN, Cui SL, Zhang MY, Wang XZ, Liang XT. Study on the constituents of *Artemisia annua L.*. Acta Pharm. Sinica 1981;16(5):366–70.

[4] Tu YY, Ni MY, Zhong YR, et al. Studies on the constituents of *Artemisia annua* and derivatives of artemisinin. Bull. Chin. Mater. Med. 1981;6(2):31.
[5] Tu YY. Chemical studies on Qinghaosu. J. Tradit. Chin. Med. 1982;2(1):3–8.
[6] Tu YY, Ni MY, Zhong YR, et al. Studies on the constituents of *Artemisia annua*: part II. Plant. Med. 1982;44(3):143–5.
[7] Tu YY, Yin JP, Ji L, et al. Studies on the constituents of *Artemisia annua*: part III. Chin. Herbal Med. 1985;16(5):200–1.
[8] Tu YY, Zhu QC, Shen X. Studies on the constituents of *Artemisia annua* young plants. Bull. Chin. Mater. Med. 1985;10(9):419–20.
[9] Brown GD, Liang GY, Sy LK. Terpenoids from the seeds of *Artemisia annua*. Phytochemistry 2003;64(1):303–23.
[10] Bhakuni RS, Jain DC, Sharma RP, et al. Secondary metabolites of *Artemisia annua* and their biological activity. Curr. Sci. 2001;80(1):35–48.
[11] Misra LN, Ahmad A, Thakur RS, et al. Bisnor-cadinanes form *Artemisia annua* and definitive 13C NMR assignments of β-arteether. Phytochemistry 1993;33(6):1461–4.
[12] Brown GD. Annulide, a sesquiterpene lactone from *Artemisia annua*. Phytochemistry 1993;32(2):391–3.
[13] El-Feraly FS, Al-Meshal IA, Khalifa SI. Epi-deoxyarteannuin B and 6,7- dehydroartemisinic acid from *Artemisia annua*. J. Nat. Prod. 1989;52(1):196–8.
[14] Huang JJ, Xia ZQ, Wu LF. Studies on the constituents of *Artemisia annua* L. I. The isolation and structural identification of 11R-(-)-dihydroarmisinic acid. Acta Chim. Sinica 1987;45(6):609–12.
[15] Wu ZH, Wang YY. Studies on structure and synthesis of arteannuin and its related compound XI. The identification of epoxyartemisinic acid. Acta Chim. Sinica 1984;42(6):596–8.
[16] Zhu DY, Zhang SG, Liu BN, et al. Studies on *Artemisia annua* antibacterial active ingredients. Chin. Herbal Med. 1982;13(2):54.
[17] Jeremi D, Jol A, Behbud A, et al. A new type of sesquiterpene lactone isolated from *Artemisia annua* L. arteannuin B. Tetrahedron Lett. 1973;14(32):3039–42.
[18] Sy LK, Brown GD. Deoxyarteannuin B, dihydro-deoxyarteannuin B and *trans*-5-hydroxy-2-isopropenyl-5-methylhex-3-en1ol from *Artemisia annua*. Phytochemistry 2001;58(8):1159–66.
[19] Roth RJ, Acton N. Isolation of epi-deoxyarteannuin B from *Artemisia annua*. Plant. Med. 1987;53(6):576.
[20] Sy LK, Brown GD, Haynes R. A novel endoperoxide and related sesquiterpenes from *Artemisia annua* which are possibly derived from allylic hydroperoxides. Tetrahedron 1998;54(17):4345–6.
[21] Brown GD. Two new compounds from *Artemisia annua*. J. Nat. Prod. 1992;55(12):1756–60.
[22] Misra LN. Arteannuic-C, a sesquiterpene lactone from *Artemisia annua*. Phytochemistry 1986;25(12):2892–3.
[23] Zhu DY, Deng AD, Zhang SG, Xu RS. Structure of artemisilactone. Acta Chim. Sinica 1984;42(9):937–9.
[24] Wei ZX, Pan JP, Li Y. Artemisia G: a sesquiterpene form *Artemisia annua*. Plant. Med. 1992;58(3):300.
[25] Sy LK, Cheung KK, Zhu NY, et al. Structure elucidation of arteannuin O, a novel cadinane diol from *Artemisia annua*, and the synthesis of arteannuins K, L, M and O. Tetrahedron 2001;57(40):8481–93.
[26] Brown GD. Cadinanes from *Artemisia annua* that may be intermediates in the biosynthesis of artemisinin. Phytochemistry 1994;36(3):637–41.
[27] Acton N, Klayman DL. Artemisitene, a new sesquiterpene lactone endoperoxide from *Artemisia annua*. Plant. Med. 1985;51(5):441–2.
[28] Ahmad A, Misra LN. Terpenoids from *Artemisia annua* and constituents of its essential oil. Phytochemistry 1994;37(1):183–6.
[29] Sy LK, Brown GD. Three sesquiterpenes from *Artemisia annua*. Phytochemistry 1998;48(7):1207–11.
[30] de Oliveira CM, Ferracini VL, Foglio MA, et al. Detection, synthesis and absolute configuration of (+)-nortaylorione, a new terpene from *Artemisia annua*. Tetrahedron 1997;8(11):1833–9.
[31] Rücker G, Mayer R, Manns D. α-and β-Myrcene hydroperoxide from *Artemisia annua*. J. Nat. Prod. 1987;50(2):287–9.
[32] Brown GD. Phytene-1,2-diol from *Artemisia annua*. Phytochemistry 1994;36(6):1553–4.
[33] Ulubelen A, Halfon B. Phytochemical investigation of the herba of *Artemisia annua*. Plant. Med. 1976;29(3):258–60.

[34] Zheng GQ. Cytotoxic terpenoids and flavonoids from *Artemisia annua*. Plant. Med. 1994;60(1):54–7.
[35] Chen YG, Yu BY, Dong L, Zhou L. Isolation and identification of *Artemisia annua* and its recursor compounds. Chin. Tradit. Herbal Drugs 2001;32(4):302–3.
[36] Liu HM, Li GL, Wu HZ. Chemical research on artemisinin component. Chin. Pharm. Bull. 1980;15(10):471.
[37] Yang SL, Roberts MF, O'Neill MJ, et al. Flavonoids and chromenes from *Artemisia annua*. Phytochemistry 1995;38(1):255–7.
[38] Yang SL, Roberts MF, Phillipson JD. Methoxylated flavones and coumarins from *Artemisia annua*. Phytochemistry 1989;28(5):1509–11.
[39] Marco JA, Sanz JF, Bea JF, et al. Phenolic constituents from *Artemisia annua*. Die Pharmazie 1990;45(5):382–3.
[40] Tang HQ, Hu J, Yang L, et al. Terpenoids and flavonoids from *Artemisia* species. Plant. Med. 2000;66(4):391–3.
[41] Manns D, Hartmann R. Annuadiepoxide, a new polyacetylene from the aerial parts of *Artemisia annua*. J. Nat. Prod. 1992;55(1):29–32.
[42] Bhakuni RS, Jain DC, Shukla YN, et al. Lipid constituents from *Artemisia annua*. J. Indian Chem. Soc. 1990;67(12):1004–5.
[43] Khan M, Jain DC, Bhakuni RS, et al. Occurrence of some antiviral sterols in *Artemisia annua*. Plant Sci. 1991;75(2):161–5.
[44] Sing AK, Pathak V, Agrawal PK. Annphenone, a phenolic acetophenone from *Artemisia annua*. Phytochemistry 1997;44(3):555–7.
[45] Sy LK, Brown GD. Coniferaldehyde derivatives from tissue culture of *Artemisia annua* and *Tanacetum parthenium*. Phytochemistry 1990;50(5):781–5.
[46] Shukla A, Abad Farooqi AH, Shukla YN. A new adenine derivatives from *Artemisia annua*. J. Indian Chem. Soc. 1997;74(1):59.
[47] Tian Y, Wei ZX, Wu ZH. Studies on the constituents of *Artemisia annua*. Chin. Herbal Med. 1982;13(6):249–51.
[48] Zhong YR, Cui SL. Studies on chemical constituents of volatile oil from *Artemisia annua*. Bull. Chin. Mater. Med. 1983;8(6):31–2.
[49] Liang HQ, Liao SX, Lou DJ, Zhang CS, Chen M. Studies on the antifungal constituents of volatile oil from *Artemisia annua* L.. China J. Chin. Mater. Med. 1990;15(12):734–6.
[50] Peng H, Guo ZD, Zhang JC, Liu LM. Studies on constituents of volatile oil from *Artemisia annua* L. J. Chin. Med. Mater. 1996;19(9):458–9.
[51] Chen FL, He F, Li JL, Luo JB, Wu Z. GC-MS analysis of essential oil from *Artemisia annua* L. by extraction of different methods. J. Chin. Med. Mater. 2001;24(3):176–8.
[52] Dong Y, Liu HL. Comparative study of the chemical constituents of volatile oils from *Artemisia annua* and *Artemisia carvifolia*. J. Chin. Med. Mater. 2004;27(8):568–71.
[53] Qiu Q, Cui ZJ, Liu TL, Tian S. Study on chemical constituents of volatile oil from *Artemisia annua* L. by GC-MS. Chin. Tradit. Patent Med. 2001;23(4):278–80.
[54] Liu LD, Gu JW, Chen JD, Wang YN. Chemical composition of the essential oils of *Artemisia annua* L. and *Artemisia apiacea* H. *Artemisia carvifolia*. Jiangxi Sci. 1996;14(4):234–8.
[55] Zhao B, Wang YC, Ouyang F. The present situation of the extraction and separation of artemisinin medicinal ingredients. Chin. Herbal Med. 1998;29(11):784–6.
[56] Woerdenbag HJ, Pras N, Chan NG, et al. Artemisinin, related sesquiterpenes, and essential oil in *Artemisia annua* during a vegetation period in Vietnam. Plant. Med. 1994;60(3):272–5.
[57] Perazzo FF, Carvalho JCT, Carvalho JE, et al. Central properties of the essential oil and the crude ethanol extract from aerial parts of *Artemisia annua* L.. Pharm. Res. 2003;48(5):497–502.
[58] Juteau F, Masotti V, Bessiere JM, et al. Antibacterial and antioxidant activities of *Artemisia annua* essential oil. Fitoterapia 2002;73(6):532–5.

CHAPTER 5

Studies on Quality Control of Qinghao

Chapter Outline

Qinghao, the dried aerial part of *Artemisia annua*, has been used as a TCM herbal drug in China for about 2000 years. In Chinese Pharmacopeia, Qinghao is described as heat clearing herbal drug bitter in taste, cold in nature, and capable of clearing deficiency heat, relieving steaming bone, releasing summer heat, interrupting malaria, and abating jaundice. To ensure the quality control of Qinghao, the methods of chromatographic identification and quantitative assay with artemisinin and scopoletin as reference substances were developed and 10 batches of Qinghao from different locations of China were tested. The results may contribute significantly to the standardization of Qinghao recorded in Chinese Pharmacopeia.

1 Artemisinin

1.1 Identification of Artemisinin in Qinghao by TLC

To 2 g of the powder add 20 mL of petroleum ether (60–90°C), ultrasonicate for 30 min, filter, evaporate the filter to dryness, dissolve the residue in 1 mL of ethyl acetate as the test solution. Dissolve artemisinin CRS in ethyl acetate to produce a solution containing 1 mg/mL as the reference solution. Carry out the method for thin-layer chromatography, using silica gel as the coating substance and a mixture *n*-hexane and ethyl acetate (4:1) as the mobile phase. Apply separately to the plate 5 μL of each of the aforementioned two solutions. After developing and removal of the plate, dry in air. Spray with a 2% solution of vanillin in 10% sulfuric acid in ethanol, heat at 105°C to the spots clear. The spot in the chromatogram obtained with the test solution corresponds in position and color to the spot in the chromatogram obtained with the reference solution. The TLC chromatograms of **artemisinin** and 10 batches test solution of the samples are shown in Fig. 5.1.

From *Artemisia annua* L. to Artemisinins
http://dx.doi.org/10.1016/B978-0-12-811655-5.00005-2

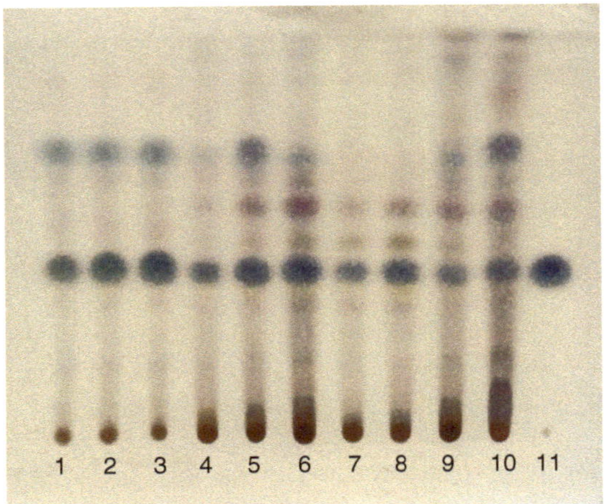

Figure 5.1: TLC Chromatograms of Qinghao and Artemisinin.
1, Zhangjiajie, Hunan; 2, Jishou, Hunan; 3, Youyang, Chongqing; 4, Taiyuan, Shanxi; 5, Nanning, Guangxi; 6, Changxing, Zhejiang; 7, Xiangshan, Beijing; 8, Shenyang, Liaoning; 9, Hefei, Anhui; 10, Wujin, Jiangsu; 11, Artemisinin.

1.2 *Determination of Artemisinin in Qinghao [1] (Aerial Part of* Artemisia annua *L.)*

1.2.1 Chemicals, reagents, and materials

Methanol (HPLC grade) was purchased from Tianjin Shield (Tianjin, China); petroleum, ethanol, sodium hydroxide, and acetic acid (analytical grade) were from Beijing Chemical Reagents Company. Water was purified using a Milli-Q_{plus} system (Milford, Massachusetts, USA). Pure reference standard of artemisinin was purchased from the Chinese National Institute for Control of Pharmaceutical and Biological Products (Beijing, China). Herb samples were collected in different locations of China, and were authenticated as the aerial part of *A. annua* L. by Prof. Youyou Tu (Institute of Chinese Materia Medica, China Academy of Chinese Medical Sciences, Beijing, China.)

1.2.2 Apparatus and chromatographic conditions

Chromatographic analysis was carried out on a ZORBAX RX-C18 column (4.6 × 250 mm, 5 μm) using an Agilent 1100 liquid chromatography system, equipped with the G1311A Quatpump solvent delivery system, autoinjector G1313A, column oven G1316A, photodiode array detector G1315A and managed by a HPCHEM workstation. Reference and test solutions were filtered through a 0.45 mm syringe filter before injection. The mobile phase used was methanol –0.01 M phosphate buffer (45:55, v/v), at a flow rate of 1.0 mL/min. The column temperature was maintained at 30°C and the detection wavelength at 260 nm.

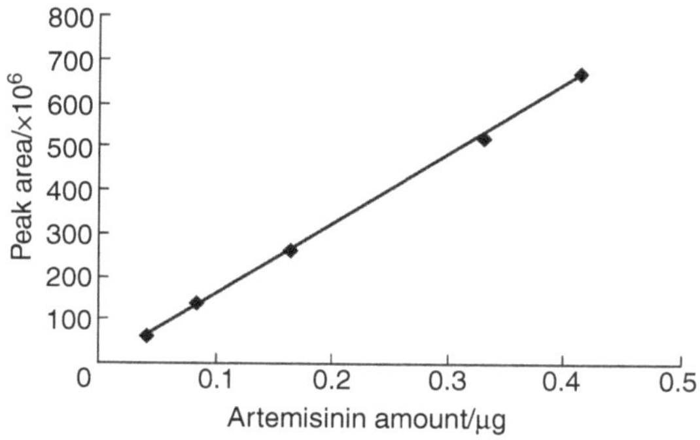

Figure 5.2: Standard Curve of Artemisinin.

1.2.3 Reference solution and calibration curve

A stock solution of **artemisinin** with a known concentration of **0.824** mg/mL was prepared in ethanol. **0.1, 0.2, 0.4, 0.8, and 1.0 mL** of the stock solution were accurately transferred into 10 mL volumetric flask separately and each was diluted to 1 mL with ethanol, then 4 mL of a 0.2% (w/v) sodium hydroxide solution were added. The mixture was mixed well and heated for 30 min in a water bath at 45°C. After cooling to room temperature, 0.08 M acetic acid was added to volume to give five different reference solutions. Each reference solution measuring 10 μL was injected into the HPLC system and the calibration curves were established by plotting peak areas against the corresponding amount of artemisinin and shown in Fig. 5.2. Good response linearity was obtained with excellent regression coefficient value $r^2 = 0.9993$ in the linearity range of 0.0412–0.412 μg and the regression equation was $Y = 1632.6x - 3.9289$.

1.2.4 Test solution

Weigh accurately 1 g of the powder to a stopper conical flask, accurately add 50 mL of petroleum ether (60–90°C) and weigh. Ultrasonicate for 60 min, cool and weigh again, replenish the loss of weight with petroleum ether (60–90°C), mix well, filter. Measure accurately 25 mL of the successive filtrate and evaporate to dryness, dissolve the residue in 5 mL of ethanol. Measure accurately 1 mL of the ethanol solution to a 10 mL volumetric flask, add 4 mL of a 0.2% (w/v) sodium hydroxide solution, mix well and heat for 30 min in a water bath at 45°C, then cool and diluted with 0.08 M acetic acid to volume, shake well and filter, use the successive filtrate as the test solution.

1.2.5 Precision

Precision of the proposed method was obtained by repeatability. Five different injections of the same reference solution were injected and the peak area was calculated. The result was expressed with relative standard deviations (RSD) and the value was 0.91%.

Table 5.1: Results of the recovery test

Original Amount (mg)	Amount Spiked (mg)	Amount Found (mg)	Recovery (%)	Mean (%)	RSD (%)
1.41024	1.56	2.9200	96.78	98.01	1.96
1.33008	1.49	2.8043	98.94		
1.28136	1.46	2.6721	95.25		
1.30368	1.52	2.8124	99.26		
1.30128	1.45	2.7483	99.80		

1.2.6 Stability

Stability was tested at room temperature and samples were analyzed in 0, 4, 8, 12, and 24 h within 1 day. Stability was expressed as the RSD and the value was 1.48%.

1.2.7 Repeatability

To test the repeatability of the assay, five independently prepared samples of the aerial parts of *A. annua* were analyzed in parallel. Variations were expressed as RSD and the value was 2.74%.

1.2.8 Recovery

The recovery test was determined by a standard addition method. The reference compound was spiked into the sample, and then treated as detailed in Section "Test Solution." The mean recovery was calculated according to the following formula: recovery (%) = (amount found − original amount)/amount spiked × 100%, and RSD (%) = (SD/mean) × 100%. The average recoveries of artemisinin were 98.01% and their RSD value was 1.96%. The results are listed in Table 5.1.

1.2.9 Quantitative determination of artemisinin in Qinghao

Accurately inject 5 μL of the reference solution and 5–50 μL of the test solutions, respectively, into the column, and calculated. The contents of **artemisinin** in 10 batches of **Qinghao** were determined and shown in Table 5.2. The representative HPLC chromatograms of **artemisinin** and test solutions of the samples were shown in Fig. 5.3.

Table 5.2: The content of artemisinin in Qinghao

Plant Origin	Content (%)	Plant Origin	Content (%)
Zhangjiajie, Hunan	0.35	Changxing, Zhejiang	<0.0[illegible]
Jishou, Hunan	0.43	Xiangshan, Beijing	0.15
Youyang, Chongqing	0.42	Shenyang, Liaoning	0.02
Taiyuan, Shanxi	0.02	Hefei, Anhui	0.025
Nanning, Guangxi	0.24	Wujin, Jiangsu	<0.0[illegible]

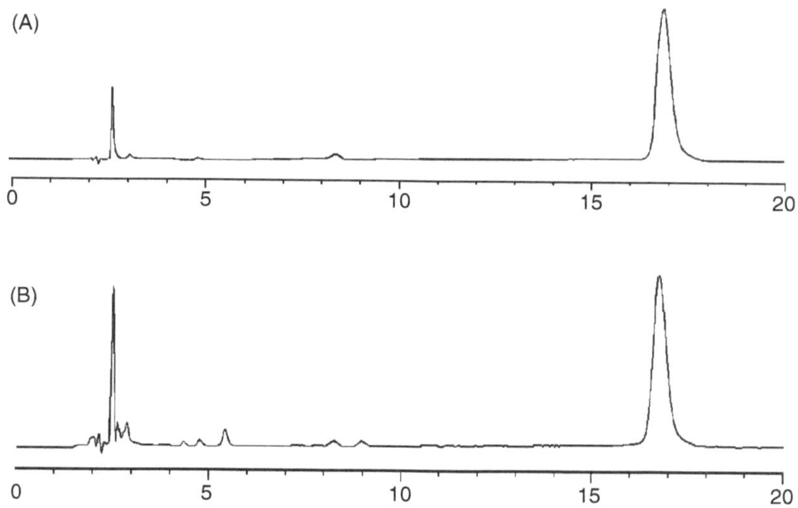

Figure 5.3: HPLC Chromatograms of **Artemisinin** (A) and Test Solution (B).

2 Scopoletin

2.1 Identification of Scopoletin in Qinghao by TLC

To 0.5 g of the powder add 10 mL of methanol, ultrasonicate for 30 min, filter, evaporate the filter to dryness, dissolve the residue in 1 mL of methanol as the test solution. Dissolve scopoletin CRS in methanol to produce a solution containing 0.25 mg/mL as the reference solution. Carry out the method for thin-layer chromatography, using silica gel as the coating substance and a mixture *n*-hexane–ethyl acetate–formic acid (4:6:0.2) as the mobile phase. Apply separately to the plate 3 μL of each of the aforementioned two solutions. After developing and removal of the plate, dry in air. It is observed under UV lamp (365 nm). There were same color of fluorescent spots at the corresponding position between the chromatogram of the sample test and the chromatogram of the reference solution. The TLC chromatograms of scopoletin and 10 batches test solutions of the samples were shown in Fig. 5.4.

2.2 Determination of Scopoletin in Qinghao (Aerial Part of Artemisia annua L.)

2.2.1 Chemicals, reagents, and materials

Methanol (HPLC grade) was purchased from Tianjin Shield (Tianjin, China); water was purified using a Milli-Q_{plus} system (Milford, Massachusetts, USA). **Scopoletin** was separated and purified from ***A. annua*** and its purity was determined to be more than 99% by HPLC-UV peak area normalization method. Herb samples were collected in different locations of China, and were authenticated as the aerial part of *A. annua* L. by Prof. Youyou Tu (Institute of Chinese Materia Medica, China Academy of Chinese Medical Sciences, Beijing, China).

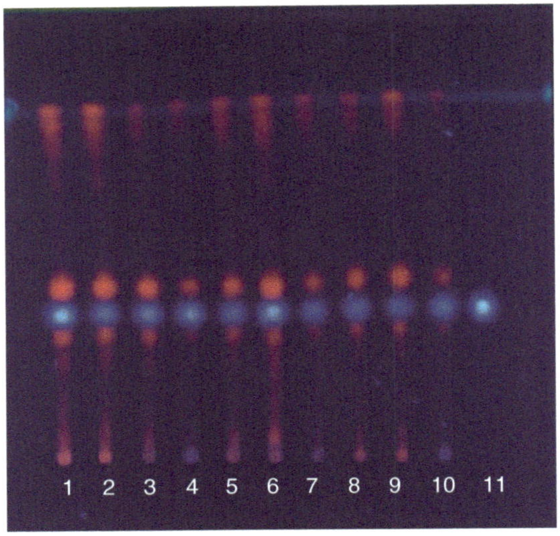

Figure 5.4: TLC chromatograms of Qinghao and Scopoletin.
1, Zhangjiajie, Hunan; 2, Jishou, Hunan; 3, Youyang, Chongqing; 4, Taiyuan, Shanxi; 5, Nanning, Guangxi; 6, Changxing, Zhejiang; 7, Xiangshan, Beijing; 8, Shenyang, Liaoning; 9, Hefei, Anhui; 10, Wujin, Jiangsu; 11, Scopoletin.

2.2.2 Apparatus and chromatographic conditions

Chromatographic analysis was carried out on a YMC-Pack ODS-A column (4.6 × 250 mm, 5 μm) at room temperature using a Waters 201 liquid chromatography system, equipped with photodiode array detector Waters 481 and managed by a Anstar workstation (USA). Reference and test solutions were filtered through a 0.45-μm syringe filter before injection. The mobile phase used was methanol–0.08 M phosphoric acid (40:60, v/v), at a flow rate of 0.8 mL/min. The detection wavelength was at 349 nm.

2.2.3 Reference solution and calibration curve

A reference solution of **scopoletin** with a known concentration of **0.021** mg/mL was prepared in ethanol. 1, 3, 5, 10, 15, 20 μL of the reference solution was injected separately into the HPLC system and the calibration curve was established by plotting peak areas (Y) against the corresponding amount of artemisinin (X) and shown in Fig. 5.5. Good response linearity was obtained with excellent regression coefficient value $r^2 = 0.9999$ in the linearity range of 0.021–0.42 μg and the regression equation was $Y = 4660505X - 5429$.

2.2.4 Test solution

Weigh accurately 0.25 g of the powder to a stopper conical flask, accurately add 10 mL of methanol and weigh. Ultrasonicate for 30 min and weigh again, replenish the loss of weight with methanol, mix well, and filter, use the successive filtrate as the test solution.

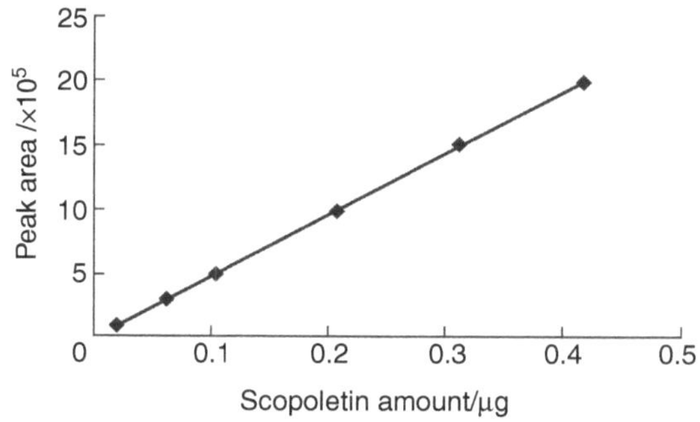

Figure 5.5: Standard Curve of Scopoletin.

2.2.5 Precision

Precision of the proposed method was obtained by repeatability. Five different injections of the same reference solution were injected and calculated the peak area. The result was expressed with relative standard deviations (RSD) and the value was 0.73% for scopoletin.

2.2.6 Repeatability and stability

To test the repeatability of our assay, five independently prepared samples of the aerial parts of *A. annua* were analyzed in parallel. Variations were expressed as RSD and the value was 1.54%. Stability was tested at room temperature and samples were analyzed in 0, 4, 8, 12, and 24 h within 1 day. Stability was expressed as the RSD and the value was 1.65% for scopoletin.

2.2.7 Recovery

The recovery test was determined by a standard addition method. The reference compound was spiked into the sample, and then treated as detailed in Section "Test Solution." The average recovery of **scopoletin** was 99.34% and their RSD value was 2.0%. The results are listed in Table 5.3.

2.2.8 Quantitative determination of scopoletin in Qinghao

Accurately inject 5 μL of the reference solution and 5 μL of the test solution, respectively, into the column, and the content was calculated on the basis of the peak area. The contents

Table 5.3: Results of the recovery test

Original Amount (mg)	Amount Spiked (mg)	Amount Found (mg)	Recovery (%)	Mean (%)	RSD (%)
0.1586	0.1593	0.3117	96.11	99.34	2.0
0.1570	0.1593	0.3183	101.2		
0.1574	0.1593	0.3175	100.5		
0.1562	0.1593	0.3137	98.90		
0.1572	0.1593	0.3166	100.0		

Table 5.4: The content of scopoletin in Qinghao

Plant Origin	Content (%)	Plant Origin	Content (%)
Zhangjiajie, Hunan	0.15	Changxing, Zhejiang	0.097
Jishou, Hunan	0.12	Xiangshan, Beijing	0.14
Youyang, Chongqing	0.13	Shenyang, Liaoning	0.052
Taiyuan, Shanxi	0.092	Hefei, Anhui	0.037
Nanning, Guangxi	0.066	Wujin, Jiangsu	0.03

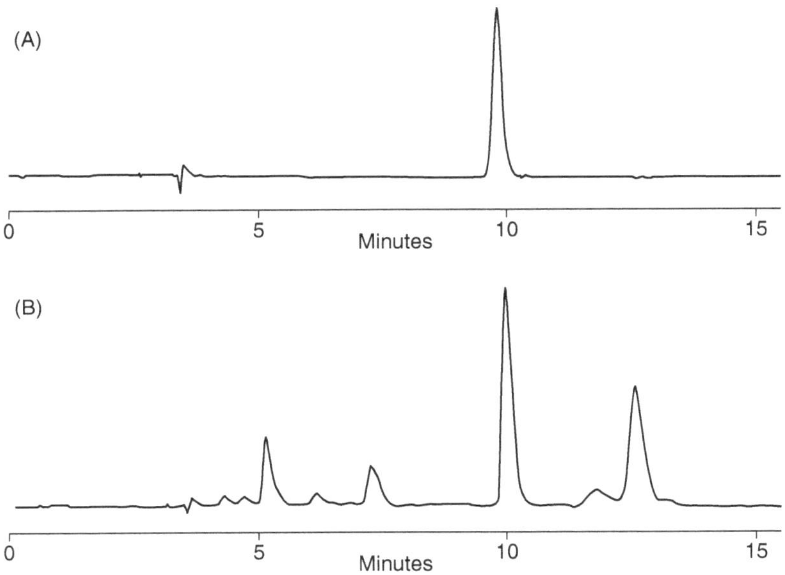

Figure 5.6: HPLC Chromatograms of **Scopoletin** (A) and Test Solution (B).

of **scopoletin** in 10 batches of Qinghao were determined and shown in Table 5.4. The representative HPLC chromatograms of scopoletin and extracted solution of the samples were shown in Fig. 5.6.

Reference

[1] Zhao S, Zeng M. Determination of Qinghaosu in Artemisia annua L. By high performance liquid chromatography. Chin. J. Pharm. Anal. 1986;6(1):3–5.

CHAPTER 6

Studies on Pharmacological Actions of Artemisia annua

Chapter Outline

Artemisia annua is a commonly used Chinese herbal medicine, which is bitter in taste and cold in nature. Assigned to hepatic, biliary, and renal meridians, it has the traditional functions of clearing summer heat, removing hectic heat and antimalaria. Clinically, it is used for the treatment of symptoms such as fever due to *Yin* deficiency, affection due to exogenous summer heat, and malaria and jaundice due to dampness heat. Its functions and uses have been documented in historical works about Chinese materia medica. *Pictorial Classic of Herbal* described that "*Artemisia annua* is the best medicine for treating bone-heat syndrome and fever in chronic consumptive diseases and was used alone in ancient prescriptions." *A. annua* can also be used for treating "ulcer," as documented in *Shennong's Herbal Classic*. According to

From *Artemisia annua* L. to Artemisinins
http://dx.doi.org/10.1016/B978-0-12-811655-5.00006-4

the *Newly Compiled Materia Medica of the Tang Dynasty*, "crude stalks of *Artemisia annua* can be applied to incised (metal-inflicted) wounds, and effectively stop bleeding, promote tissue regeneration and alleviate pain." In addition, *Peaceful Holy Benevolent Prescriptions and Miraculous Emergency Prescriptions* also mentioned that *A. annua* had been used for treating "unstoppable purulence in ears" and "dental swelling and pain." In modern times, a great many explorative studies have been conducted on the pharmacological actions of *A. annua*. In this chapter, studies on pharmacological actions of *A. annua* beyond antimalaria are summarized.

1 Studies on Antipyretic, Heat-Resistant, Antiinflammatory, Analgesic, and Antimicrobial Actions of Artemisia annua [1]

In recent tests and clinical studies, *A. annua*, used alone [2–6] or in compound prescriptions [7–13], has demonstrated favorable efficacy in treating patients with deficiency heat, hyperpyrexia, summer heat, and fever of undetermined origin. It can be used for treating multiple pyretic diseases, including cold, acute tonsillitis, rheumatoid arthritis, acute cholecystitis, infection in urinary system, and functional low-grade fever. Clinical evidences indicate that the heat-clearing action of *A. annua* is widely used, which correlates not only to its function of relieving exterior syndrome and expelling evils, but also to other functions such as reducing fire, cooling blood to remove toxic materials, dissipating dampness to promote bile flow, and dredging vessels to reduce swelling. In view of this, experimental studies on its antipyretic, heat-resistant, antiinflammatory, analgesic, and antimicrobial actions were conducted; different medicinal parts (leaves, stalks) and batches obtained in different harvesting seasons were compared.

Experimental materials

A. annua L. was collected from Beijing. Some drugs were purchased from Beijing Crude Drugs Co., Ltd.

Water extract of *A. annua* was the concentrate of water decoction of *A. annua*. An aqueous solution of 300% (1 mL of water extract of *A. annua* was equivalent to 3 g of crude drug) was used in drug efficacy tests, and an aqueous solution of 500% (1 mL of water extract of *A. annua* was equivalent to 5 g of crude drug) was used in the acute toxicity test. Ethyl acetate extract was the portion extracted with ethyl acetate from the water extract of *A. annua*; *n*-butanol extract was the portion extracted with *n*-butanol from the portion remaining after extraction with ethyl acetate; the residual aqueous portion was the residual solution obtained following extraction of the water extract with ethyl acetate and *n*-butanol.

Samples provided for antimicrobial tests included aqueous decoction of crude *A. annua*, ethyl ether extract, and ethanol extract of crude *A. annua*.

Animals: Body weights of Kunming mice were 17–26 g, and the body weight differences of test mice in each batch were <4 g. Body weights of rats were 160–230 g, and the body weight differences of test rats in each batch were ≤60 g. All rats were male; for pain relief test by hot plate method mice were female, while in other tests both male and female mice were used.

Test microorganisms: Fifty clinical isolates of *Staphylococcus aureus*, *Escherichia coli*, and *β-hemolytic streptococcus* each, obtained from Peking University First Hospital, were used. Ten isolates of *Streptococcus pneumoniae* were obtained from several hospitals including China-Japan Friendship Hospital. Each isolate was subcultured and preserved by our laboratory on tryptic soy agar (TSA) slants. TSA containing 5% rabbit blood was used for culture of *β-hemolytic streptococcus* and *S. pneumoniae*.

Culture media: TSA was purchased from Oxoid (United Kingdom); TSA blood plates were prepared with TSA containing 5% defibered rabbit blood.

Methods and results

In the fever reduction test, artificial fever in rats induced by fresh yeast and normal body temperature of rats were measured; in the summer heat clearance test, resistance to high temperatures was detected; in the antiinflammation test, effects were observed on yeast-induced arthritis in rats and mice, and on egg white-induced joint swelling in mice and xylene-induced ear inflammation in mice; in the pain relief test, the pain threshold of tail-flick response of mice and acetic acid-induced writhing response of mice were detected.

1.1 Water-Extracted Portion

1.1.1 Fever reduction test

1. Effect on artificial fever in rats induced by fresh yeast: Before the fresh yeast suspension was injected, the anal temperature of each rat was measured twice (with an interval of 1 h between the two measurements), and the mean value was recorded as the normal anal temperature. A fresh yeast suspension of 15% was prepared using sterile physiological saline and injected subcutaneously to the neck or back of rats at a dose of 2.0 mL/100 g body weight. The extract was intragastrically administered 4 h after injection of yeast, and anal temperatures at 1, 2, 3, and 4 h after drug administration were measured. Results were expressed as differences of anal temperatures at different time points. According to results in Table 6.1, single-dose intragastric administration of water extract of *A. annua* (1/3–1/2 LD_{50} dose group) demonstrated very significant antipyretic effects in rats with artificial fever induced by fresh yeast, and the action could last for about 4 h.
   a. Comparison of crude drug samples obtained in different harvesting seasons: The extract was intragastrically administered to rats at 5 h after injection of fresh yeast [see paragraph (1) under "Fever reduction test" for the procedure], and anal temperatures at 1, 2, 3, and 4 h after drug administration were measured. According to results in Table 6.2, the antipyretic action of water extract of the first batch of *A. annua* (preflowering period) was significantly stronger than that of water extract of the second batch (bud period). Comparison of 1/3 LD_{50} equivalent doses of the two batches (*t* test) showed significant differences at 3 and 4 h after drug administration (*t* values were 2.15 and 2.29, respectively; $P < .05$).
   b. Comparison of antipyretic actions of different medicinal parts (leaves and stalks): Commercial *A. annua* (preflowering period) was obtained from Beijing. Leaves

Table 6.1: Antipyretic effects of water extract of *Artemisia annua* on fresh yeast-induced fever in rats

Drug	Dose (g/kg)	Number/ Rats	Normal Body Temperature (°C)	Body Temperature 4 h After Fever Induction (°C)	Body Temperatures at Different Time Points After Drug Administration (°C) ($\overline{X} \pm SE$)			
					1 h	2 h	3 h	4 h
Control (distilled		6	37.50 ± 0.24	38.20 ± 0.27	1.10 ± 0.39	1.40 ± 0.28	1.20 ± 0.82	1.10 ± 0.68
water)	27.0 × 1	6	37.10 ± 0.22	38.10 ± 0.24	1.20 ± 0.63	1.50 ± 0.64	1.60 ± 0.74	1.40 ± 0.79
Water extract of	54.0 × 1	6	37.60 ± 0.09	38.40 ± 0.15*	0.40 ± 0.52**	0.20 ± 0.95	0.40 ± 0.79	0.30 ± 0.62*
Artemisia annua	81.0 × 1	5	37.60 ± 0.17	38.60 ± 0.17		−1.20 ± 0.31**	−0.60 ± 0.62**	−0.30 ± 0.60**

*Compared with the control group, $P < .05$.

**Compared with the control group, $P < .01$.

Table 6.2: Comparison of antipyretic effects of water extract of *Artemisia annua* obtained in different harvesting seasons on artificial fever in rats induced by fresh yeast

Sample	Dose (g/kg)	Number/ Rats	Normal Body Temperature (°C)	Body Temperature 5 h After Yeast Injection (°C)	Body Temperatures at Different Time Points After Drug Administration (°C) ($\overline{X} \pm SE$)			
					1 h	2 h	3 h	4 h
Control (distilled water)		7	37.90 ± 0.10	39.00 ± 0.09	0.20 ± 0.07	0.00 ± 0.16	0.10 ± 0.14	0.10 ± 0.10
The first batch of samples (preflowering period)	1/6 LD_{50} 19.6	7	37.60 ± 0.14	39.40 ± 0.24	−0.50 ± 0.16†	−1.10 ± 0.29**	−0.70 ± 0.27¶	−0.50 ± 0.52
	1/3 LD_{50} 39.2	7	38.00 ± 0.15	39.50 ± 0.23	−0.30 ± 0.46**	−1.10 ± 0.26**	−1.20 ± 0.32**	−1.00 ± 0.36*
	1/2 LD_{50} 55.8	7	38.00 ± 0.10	39.20 ± 0.13	−1.90 ± 0.48**	−1.20 ± 0.47**	−0.80 ± 0.19**	−0.40 ± 0.15
	1/6 LD_{50} 20.2	7	38.00 ± 0.22	39.00 ± 0.28	−0.15 ± 0.23	−0.20 ± 0.34	−0.40 ± 0.10	−0.30 ± 0.29
The second batch of samples (bud period)	1/3 LD_{50} 40.4	7	37.60 ± 0.13	39.00 ± 0.20	−0.50 ± 0.36*	−0.50 ± 0.18	−0.10 ± 0.35	−0.14 ± 0.18
	1/2 LD_{50} 60.7	6‡	38.20 ± 0.11	39.30 ± 0.13	−0.90 ± 0.35**	−0.60 ± 0.31¶	−0.60 ± 0.31	−0.40 ± 0.41

*Compared with the control group, $P < .05$.

**Compared with the control group, $P < .01$.

†$P < .001$.

‡Animal death occurred.

¶$P \approx .05$.

and stalks were separated, and water extract samples of needed concentrations were prepared. The procedure in the fever reduction test was carried out 1 h after single-dose intragastric administration of 1/3 LD_{50} dose to rats. According to results in Table 6.3, both leaves and stalks of *A. annua* demonstrated significant antipyretic effects in rats with artificial fever induced by fresh yeast; the action duration and intensity were similar.

2. Effect on normal body temperature of rats: Anal temperature of rats was measured twice, and the mean value was recorded as the normal body temperature before drug administration. The extract was intragastrically administered q.a.m. and q.p.m. for 2 days. Anal temperature was measured at 1, 2, and 3 h after administration of the last dose. Results were expressed as differences between anal temperatures before and after drug administration. According to results in Table 6.4, water extract of *A. annua* (1/6 LD_{50}, 27.0 g/kg group) significantly lowered normal body temperature of rats.

1.1.2 High temperature resistance test

After mice were randomized, normal axillary skin temperature was measured twice before drug administration, and the mean value was the body temperature before drug administration. The drug was intragastrically administered b.i.d. for 3 days. One hour after administration of the last dose, body temperature was measured before the mice were put into the incubator and heated. The mice were placed in cascade cages (one mouse per room) and heated by batch in an incubator (45 ± 0.5°C) for 20 min. Then, they were immediately taken out, and the body temperature of each mouse was measured. Differences between skin temperatures before and after heating and between skin temperatures after heating and before drug administration were calculated, and *t* test was carried out. According to results in Table 6.5, water extract of *A. annua* (doses of 30.0 and 60.0 g/kg) demonstrated very significant effect on enhancing resistance to high temperature (skin temperature after heating in the drug administrated group was significantly lower than that of mice in the control group.)

1.1.3 Antiinflammation test

1. Effect on yeast-induced joint swelling in mice and rats: The drug was intragastrically administered to healthy mice or rats by batch for 4 times. One hour after intragastric administration of the last dose, 0.05 or 0.1 mL of 10% medicinal yeast powder supernatant was subcutaneously injected to the right ankle joints of mice or rats. Volume differences of voix pedis and ankle joints at 1, 2, 3, 4, and 5 h following inflammation induction were measured using a voix pedis volume tester, as an indicator of joint swelling. The results were analyzed by *t* test. According to results in Tables 6.6 and 6.7, water extract of *A. annua* significantly inhibited yeast-induced arthritis in mice and rats.
2. Effect on egg white-induced joint swelling in mice: The drug was intragastrically administered to mice for 6 times. One hour after administration of the last dose,

Table 6.3: Comparison of antipyretic effects of different medicinal parts on artificial fever in rats induced by fresh yeast

Medicinal Part	Dose (g/kg)	Number/Rats	Normal Body Temperature (°C)	Body Temperature 5 h After Fever Induction (°C)	Body Temperatures at Different Time Points After Drug Administration (°C) ($\bar{X} \pm SE$)			
					1 h	2 h	3 h	4 h
Control (distilled water)		10	37.10 ± 0.22	38.90 ± 0.16	0.60 ± 0.28	0.20 ± 0.14	−0.50 ± 0.14	0.90 ± 0.17
Water extract of leaves	1/3 LD_{50} 39.2	9	37.20 ± 0.09	39.40 ± 0.13	−0.40 ± 0.23**	−0.70 ± 0.26*	−1.10 ± 0.18*	−1.40 ± 0.24
Water extract of stalks	1/3 LD_{50} 39.2	9	37.20 ± 0.11	39.30 ± 0.08	−0.20 ± 0.17**	−0.30 ± 0.18*	−0.60 ± 0.23	−0.90 ± 0.19

*Compared with the control group, $P < .05$.
**Compared with the control group, $P < .01$.

Table 6.4: Effect of water extract of *Artemisia annua* on normal body temperature of rats

Medicinal Part	Dose (g/kg)	Number/ Rats	Normal Body Temperature (°C)	Body Temperatures at Different Time Points After Drug Administration (°C) ($\bar{X} \pm SE$)		
				1 h	2 h	3 h
Control (distilled water)		10	38.80 ± 0.13	−0.40 ± 0.45	−0.30 ± 0.42	−0.60 ± 0.49
Water extract of *Artemisia annua*	27.0 × 4	10	38.90 ± 0.15	−1.20 ± 0.70*	−1.00 ± 0.58*	−0.90 ± 0.51
	54.0 × 4	10	38.70 ± 0.14	−0.60 ± 0.65	−0.40 ± 0.45	−0.40 ± 0.63

*Compared with the control group, $P < .01$.

Table 6.5: Effect of water extract of *Artemisia annua* on body temperature of mice in a high-temperature environment

Drug	Dose (g/kg)	Number/ Mice	Body Temperature Changes (°C) ($\bar{X} \pm SE$)			
			Body Temperature Before Drug Administration	Body Temperature After Heating for 20 min	Difference Between Body Temperatures Before and After Heating	Difference Between Body Temperatures After Heating and Before Drug Administration
Control (distilled water)		11	36.30 ± 0.23	37.70 ± 0.07	1.80 ± 0.21	1.30 ± 0.22
	15.0 × 6	10	36.10 ± 0.16	37.50 ± 0.21	2.10 ± 0.25	1.40 ± 0.36
Water extract of *Artemisia annua*	30.0 × 6	10	36.50 ± 0.17	36.80 ± 0.08	1.10 ± 0.82	0.26 ± 0.16**
	60.0 × 6	10	36.50 ± 0.10	36.80 ± 0.13	1.22 ± 0.45*	0.40 ± 0.19**

*Compared with the control group, $P < .05$.
**Compared with the control group, $P < .01$.

0.05 mL of fresh egg white was subcutaneously injected to ankle joints of mice. The test was carried out following the aforementioned method for yeast-induced joint swelling. According to results in Table 6.8, water extract of *A. annua* significantly inhibited egg white-induced arthritis in mice. Test results of the two batches were similar.

3. Effect on xylene-induced ear inflammation in mice: One hour after intragastric administration of a single dose in mice, 0.05 mL of xylene was applied to left ears of mice. After 15 min, left and right ear auricles of mice were separately isolated using a round stainless steel punch with a diameter of 0.8 cm and weighed. The difference between weights of left and right auricles was used as an indicator of swelling. According to results in Table 6.9, water extract of *A. annua* in the large-dose group (100 g/kg) significantly inhibited xylene-induced ear inflammation of mice.

Table 6.6: Effect of water extract of *Artemisia annua* on yeast-induced foot swelling in mice

Test Batch	Drug	Dose (g/kg)	Number/ Mice	Degree of Foot Swelling at Different Time Points After Injection of the Inflammatory Agent (mL) ($\bar{X} \pm SE$)				
				1 h	2 h	3 h	4 h	5 h
The first batch	Control (distilled water)		10	0.16 ± 0.01	0.15 ± 0.01	0.12 ± 0.01	0.12 ± 0.01	0.13 ± 0.01
	Water extract of	30.0 × 4	12	0.05 ± 0.01†	0.06 ± 0.01†	0.06 ± 0.01†	0.05 ± 0.01†	0.07 ± 0.01**
	Artemisia annua	60.0 × 4	11	0.09 ± 0.01†	0.10 ± 001*	0.11 ± 0.01	0.08 ± 0.01	0.10 ± 0.01
The second batch	Control (distilled		10	0.02 ± 0.01	0.049 ± 0.000	0.071 ± 0.010	0.054 ± 0.020	0.02 ± 0.01
	water)	15.0 × 4	10	0.07 ± 0.01	0.07 ± 0.01†	0.07 ± 0.01*	0.066 ± 0.010	0.078 ± 0.010
	Water extract of	30.0 × 4	10	0.04 ± 0.01	0.015 ± 0.010**	0.042 ± 0.010*	0.027 ± 0.010	0.028 ± 0.013
	Artemisia annua	60.0 × 4	10	0.04 ± 0.01	0.052 ± 0.010	0.057 ± 0.010	0.051 ± 0.010	0.036 ± 0.010

*Compared with the control group, $P < .05$.
**Compared with the control group, $P < .01$.
†$P < .001$.

Table 6.7: Effect of water extract of *Artemisia annua* on yeast-induced foot swelling in rats

Drug	Dose (g/kg)	Number/Rats	Degree of Foot Swelling at Different Time Points After Injection of the Inflammatory Agent (mL) ($\bar{X} \pm SE$)				
			1 h	2 h	3 h	4 h	5 h
Control (distilled water)		6	0.86 ± 0.09	0.71 ± 0.09	0.54 ± 0.08	0.37 ± 0.07	0.34 ± 0.09
Water extract of *Artemisia annua*	30.0 × 4	6	0.38 ± 0.06**	0.36 ± 0.08*	0.21 ± 0.06**	0.26 ± 0.09	0.21 ± 0.05
	60.0 × 4	6	0.61 ± 0.06	0.69 ± 0.05	0.48 ± 0.03	0.33 ± 0.04	0.29 ± 0.03

*Compared with the control group, $P < .05$.
**Compared with the control group, $P < .01$.

Table 6.8: Effect of water extract of *Artemisia annua* on egg white-induced foot swelling in mice

Test Batch	Drug	Dose (g/kg)	Number/Mice	Effect on Foot Swelling at Different Time Points After Injection of the Inflammatory Agent (mL) ($\bar{X} \pm SE$)				
				1 h	2 h	3 h	4 h	5 h
The first batch	Control (distilled water)		10	0.06 ± 0.01	0.04 ± 0.01	0.03 ± 0.00	0.01 ± 0.01	−0.01 ± 0.00
	Water extract of *Artemisia annua*	15.0 × 6	9	0.05 ± 0.01	0.07 ± 0.00	0.03 ± 0.00	0.01 ± 0.00	0.00 ± 0.00
		30.0 × 6	11	0.06 ± 0.01	0.03 ± 0.00	0.02 ± 0.00	0.00 ± 0.00	0.00 ± 0.00
		60.0 × 6	11	0.04 ± 0.00	0.01 ± 0.00	0.00 ± 0.00	0.00 ± 0.00	0.00 ± 0.00
	Prednisone	0.1 × 1	10	0.03 ± 0.00	0.00 ± 0.01	0.00 ± 0.00	0.00 ± 0.01	0.01 ± 0.00
The second batch	Control (distilled water)		8	0.15 ± 0.02	0.06 ± 0.01	0.08 ± 0.01	0.05 ± 0.01	0.03 ± 0.00
	Water extract of *Artemisia annua*	15.0 × 6	7	0.10 ± 0.01	0.11 ± 0.01	0.05 ± 0.01	0.02 ± 0.01	0.03 ± 0.01
		30.0 × 6	8	0.07 ± 0.02	0.12 ± 0.02	0.07 ± 0.01	0.03 ± 0.00	0.01 ± 0.00
		60.0 × 6	8	0.05 ± 0.00	0.05 ± 0.00	0.03 ± 0.01	0.01 ± 0.01	0.00 ± 0.00
	Prednisone	0.1 × 1	8	0.09 ± 0.00	0.04 ± 0.00	0.00 ± 0.00	0.00 ± 0.00	0.00 ± 0.01

Table 6.9: Effect of water extract of *Artemisia annua* on xylene-induced ear inflammation in mice

Test Batch	Drug	Dose (g/kg)	Number/Mice	Weight of Ear Auricle (mg) ($\bar{X} \pm SE$)	Inhibition Rate (%)
The first batch	Control (distilled water)		12	25.80 ± 3.56	
	Water extract of *Artemisia annua*	25.0	12	20.00 ± 2.22	22.5
		50.0	12	24.50 ± 3.49	5.0
		100.0	12	16.80 ± 2.01*	53.5
The second batch	Control (distilled water)		10	20.20 ± 1.63	
	Water extract of *Artemisia annua*	100.0	8	14.00 ± 2.53	29.7

*Compared with the control group, $P < .05$.

1.1.4 Pain relief test

1. Effect on the pain threshold of tail-flick response in mice: Mouse prescreening was conducted by the routine method (water bath temperature: 49 ± 0.5°C). Those with tail flick within 3 s and without tail flick within 30 s were excluded, and eligible mice were used for the test by batch. Pain threshold changed at different points 1 h following intragastric administration of the last dose (the time between entry of the tail into water bath and occurrence of tail flick was the pain threshold) was measured. Table 6.10 showed the results. Test results of the two batches indicated that water extract of *A. annua* significantly prolonged the time to pain threshold response of mice and demonstrated certain pain-relieving action.
2. Effect on acetic acid-induced writhing response of mice: One hour following intragastric administration of the last dose, freshly prepared 1% acetic acid in physiological saline was intraperitoneally injected to mice (0.1 mL/mouse). The number of writhing responses of mice within 30 min following injection of acetic acid was observed. According to results in Table 6.11, water extract of *A. annua* in the low-dose group (13.1 g/kg) significantly reduced the number of writhing responses of mice (the pain relief rate was 58.3%, $P < .02$).

1.2 Solvent-Extracted Portion

1.2.1 Fever reduction test

Considering that water extract of *A. annua* demonstrated significant antipyretic effects, ethyl acetate extract, *n*-butanol extract, and residual aqueous portion were separately intragastrically administered, and 1 h following administration, the effects on body temperature of rats with artificial fever induced by fresh yeast were observed. According to results in Table 6.12, water-extracted portion, ethyl acetate extract, and *n*-butanol extract of *A. annua* all demonstrated significant antipyretic effects, while the residual aqueous portion had no effect.

1.2.2 High temperature resistance test

Considering that water extract of *A. annua* demonstrated effects in enhancing resistance to high temperature, the test was carried out on the ethyl acetate extract, *n*-butanol extract, and residual aqueous portion separately. According to results in Table 6.13, both ethyl acetate extract and *n*-butanol extract demonstrated effects on enhancing resistance to high temperature, and the action of *n*-butanol extract was particularly strong (differences between body temperatures after and before heating were significant); the residual aqueous portion had no effect.

1.2.3 Antiinflammation test

Considering that water extract of *A. annua* demonstrated very significant antiinflammatory effects, segmental extraction was performed and the test was conducted with the extracts

Table 6.10: Effect of water extract of *Artemisia annua* on tail-flick response of mice

Test Batch	Drug	Dose (mg/kg)	Number/Mice	Pain Threshold Value (s) ($\bar{X} \pm SE$)				
				Before drug Administration	1 h After drug administration	2 h After drug administration	3 h After drug administration	4 h After drug administration
The first	Control (distilled		15	9.90 ± 0.96	1.40 ± 1.73	−0.40 ± 2.01	−3.00 ± 1.04	−2.81 ± 1.31
batch	water)	19.6 × 3	14	10.70 ± 0.87	4.70 ± 2.41	7.60 ± 1.33**	1.80 ± 2.34	1.60 ± 1.94*
	Water extract of	39.2 × 3	12	11.80 ± 1.15	9.10 ± 2.58*	1.20 ± 1.86	2.70 ± 2.14*	−4.80 ± 1.59
	Artemisia annua	58.8 × 3	13	9.90 ± 1.04	6.20 ± 2.71	4.10 ± 2.11	3.50 ± 1.54**	5.10 ± 1.43
	Bucinperazine	0.1 × 1*	13	11.20 ± 1.78	15.20 ± 4.08**	4.40 ± 3.04	−0.10 ± 1.52	0.80 ± 3.09
The	Control (distilled		14	12.80 ± 1.13	0.00 ± 1.76	−2.80 ± 0.89	7.30 ± 2.58	2.10 ± 1.24
second	water)	13.1 × 3	14	13.50 ± 1.38	9.10 ± 3.03**	3.30 ± 1.46**	1.00 ± 2.51	−0.26 ± 2.47
batch	Water extract of	19.6 × 3	15	12.30 ± 0.96	15.00 ± 3.45**	0.10 ± 1.78	0.20 ± 2.17	5.90 ± 3.36
	Artemisia annua	39.2 × 3	12	10.90 ± 1.26	4.70 ± 3.22	−3.60 ± 1.23	1.90 ± 4.73	12.80 ± 3.61**

*Compared with the control group, $P < .05$.

**Compared with the control group, $P < .01$.

Bucinperazine was subcutaneously administered.

Table 6.11: Effect of water extract of *Artemisia annua* on writhing response of mice

Test Batch	Drug	Dose (g/kg)	Number/ Mice	Number of Writhing Responses (times) ($\bar{X} \pm SE$)	Inhibition rate (%)
The first batch	Control (distilled water)		11	30.70 ± 5.46	
	Water extract of *Artemisia annua*	13.1 × 3	10	12.80 ± 2.66*	58.3
		19.6 × 3	11	22.10 ± 2.14	28.0
		39.2 × 3	11	17.90 ± 4.16	41.7
	Pethidine	0.05 × 1	11	0.00 ± 0.00**	100.0
The second batch	Control (distilled water)		9	47.50 ± 6.99	
	Water extract of *Artemisia annua*	19.6 × 3	10	30.00 ± 4.90†	36.8
		39.2 × 3	6	33.80 ± 8.01	28.8

*Compared with the control group, $P < .05$.
**Compared with the control group, $P < .01$.
†$t = 2.09$ ($P < .05$, $t = 2.12$).

Table 6.12: Effect of segmental extracts of *Artemisia annua* on body temperature of rats with artificial fever induced by fresh yeast

Drug	Dose (g/kg)	Number/ Rats	Normal Body Temperature (°C)	Body Temperature 5 h After Fever Induction (°C)	Body Temperatures at Different Time Points After Drug Administration (°C) ($\bar{X} \pm SE$)			
					1 h	2 h	3 h	4 h
Control (distilled water)		10	37.10 ± 0.22	38.90 ± 0.16	0.60 ± 0.28	0.20 ± 0.14	0.50 ± 0.14	0.90 ± 0.17
Ethyl acetate extract	2.0	10	37.20 ± 0.19	39.20 ± 0.19	−0.80 ± 0.43**	−0.40 ± 0.23*	−0.70 ± 0.26	−1.40 ± 0.16†
n-Butanol extract	4.0	10	37.30 ± 0.07	39.20 ± 0.21	−0.20 ± 0.28*	−0.20 ± 0.21	−0.50 ± 0.22	−0.90 ± 0.27

*Compared with the control group, $P < .05$.

**Compared with the control group, $P < .01$.

†Approximately potent $t = 1.94$, $t = 2.10$, $P < .05$.

Table 6.13: Effect of segmental extracts of *Artemisia annua* on body temperature of mice in a high-temperature environment

Drug	Dose (g/kg)	Number/ Mice	Body Temperature Changes (°C) ($\bar{X} \pm$ SE)			
			Body Temperature Before Drug Administration	Body Temperature After Heating for 20 min	Difference Between Body Temperatures Before and After Heating	Difference Between Body Temperatures After Heating and Before Drug Administration
Control (distilled water)		8	35.60 ± 0.68	41.60 ± 0.31	5.90 ± 0.27	6.00 ± 0.24
Water extract of whole plant	100.0 × 4	8	35.80 ± 0.12	39.50 ± 0.24	3.90 ± 0.43*	3.70 ± 0.30**
Ethyl acetate extract	1.0 × 4	8	36.10 ± 0.08	40.30 ± 0.29	5.50 ± 0.28**	4.10 ± 0.27**
n-Butanol extract	2.0 × 4	7	35.80 ± 0.13	38.80 ± 0.34	2.80 ± 0.88	3.00 ± 0.46**
Aqueous solution after extraction	100.0 × 4	8	35.60 ± 0.08	41.70 ± 0.19	6.20 ± 0.27	6.00 ± 0.25

*Compared with the control group, $P < .01$.
**$P < .001$.

separately. The ethyl acetate extract, *n*-butanol extract, and residual aqueous portion all demonstrated significant antiinflammatory effects (Table 6.14).

1.3 Antiinflammation Test of Scopoletin

Considering that extracts of *A. annua* demonstrated very significant antiinflammatory effects, the antiinflammation test against yeast-induced foot swelling of mice was further conducted using scopoletin isolated from *A. annua*. According to results in Table 6.15, scopoletin demonstrated significant antiinflammatory effects, indicating it is an active antiinflammatory ingredient of *A. annua*.

1.4 Antimicrobial Tests

1.4.1 Methods

Agar dilution method was used: *A. annua* preparations of different concentrations were added to melted TSA (or TSA plates) medium to prepare plates containing different concentrations of *A. annua*. The plates were then placed in a 37°C incubator overnight for sterility test. A small quantity of colony was scraped off from the 24-h slant culture of the test microorganism and dispersed in 2 mL of 0.05% yeast powder solution to prepare a suspension. After the turbidity was calibrated to 1/2 of that of 1# McFarland standard, the suspension was further diluted in the yeast powder solution for use. The dilution factors for different

Table 6.14: Effect of segmental extracts of *Artemisia annua* on yeast-induced foot swelling in mice

Drug	Dose (g/kg)	Number/Mice	Degree of Foot Swelling at Different Time Points After Injection of the Inflammatory Agent (mL) ($\bar{X} \pm SE$)				
			1 h	2 h	3 h	4 h	5 h
Control (distilled water)		10	0.10 ± 0.01	0.08 ± 0.01	0.05 ± 0.00	0.03 ± 0.01	0.03 ± 0.01
Water extract of whole plant	100.0 × 3	10	0.02 ± 0.01**	0.02 ± 0.01**	0.02 ± 0.01	0.03 ± 0.01	0.00 ± 0.01*
Ethyl acetate extract	2.0 × 3	10	0.06 ± 0.01*	0.03 ± 0.01**	0.02 ± 0.01**	0.01 ± 0.01*	0.01 ± 0.01*
n-Butanol extract	4.0 × 3	10	0.06 ± 0.01*	0.04 ± 0.01*	0.03 ± 0.01	0.02 ± 0.01	0.02 ± 0.01
Aqueous solution after extraction	200.0 × 3	10	0.05 ± 0.01*	0.03 ± 0.01**	0.02 ± 0.01**	0.02 ± 0.01	0.03 ± 0.01

*Compared with the control group, $P < .05$.
**Compared with the control group, $P < .01$.

Table 6.15: Effect of scopoletin on yeast-induced foot swelling in mice

Test Batch	Drug	Dose (g/kg)	Number/Mice	Degree of Foot Swelling at Different Time Points After Injection of the Inflammatory Agent (mL) ($\bar{X} \pm SE$)				
				1 h	2 h	3 h	4 h	5 h
The first batch	Control (distilled water)		15	0.04 ± 0.01	0.03 ± 0.01	0.04 ± 0.01	0.02 ± 0.00	0.01 ± 0.00
	Scopoletin	120.0	15	0.03 ± 0.01	0.01 ± 0.01**	0.02 ± 0.00**	0.02 ± 0.01	0.00 ± 0.00
		60.0	15	0.04 ± 0.01	0.02 ± 0.01	0.02 ± 0.01**	0.02 ± 0.01**	0.00 ± 0.00
		30.0	15	0.03 ± 0.01	0.01 ± 0.00**	0.01 ± 0.01†	0.01 ± 0.01*	0.00 ± 0.00**
The second batch	Control (distilled water)		10	0.10 ± 0.01	0.07 ± 0.01	0.05 ± 0.00	0.03 ± 0.01	0.03 ± 0.01
	Scopoletin	120.0	10	0.05 ± 0.01†	0.03 ± 0.01**	0.01 ± 0.01†	0.02 ± 0.01	0.00 ± 0.01**

*Compared with the control group, $P < .05$.
**Compared with the control group, $P < .01$.
†$P < .001$.

microorganisms were different, being 1:50 for *S. aureus*, 1:40 for *E. coli* and 1:20 for *β-hemolytic streptococcus*; no dilution was performed for *S. pneumoniae*. After dilution, every 10 μL of the suspension contained 6×10^4 to 6×10^5 PFU microorganisms. Ten microliter of each diluted suspension was added, using a micropipet, to plates containing different concentrations of drug and also to a drug-free plate as control. Following inoculation, the plates were incubated at 37°C for 24 h. Inhibition of microbial growth at different drug concentrations was observed and recorded. The inhibitory percentage of different concentrations and 50% inhibitory concentrations (IC_{50}) were calculated.

1.4.2 Results

1. Segmental solvent extract (A and B denote to ethyl ether extract and ethanol extract of *A. annua*, respectively). See Tables 6.16–6.19 for in vitro inhibition of four bacteria including *S. aureus* by crude extracts of *A. annua*. It can be known from the test results that extracts of *A. annua* have the strongest inhibitory action against *S. aureus* among the four bacteria.
   Comparison of IC_{50} (Table 6.20) showed that the inhibitory action of crude extracts of *A. annua* against *S. aureus* was the strongest; the same result was obtained by comparing different preparations. Comparison of the crude extracts showed that the inhibitory action of *A. annua* B was stronger, but responses of different microorganisms varied, and the mechanisms remain unknown.
2. Antimicrobial action of artemisic acid: Agar dilution method was used for this test. Inhibitory action of artemisic acid (lotion) against more than 10 bacteria including

Table 6.16: Inhibitory action of crude extracts of *Artemisia annua* against growth of *Staphylococcus aureus*

Crude Extract of *Artemisia annua*	Number of Strains Inhibited by Different Concentrations					
	200 mg/mL	100 mg/mL	50 mg/mL	25 mg/mL	12.5 mg/mL	0 (Control)
Water decoction of *Artemisia annua*	50/50	50/50	50/50	50/50	0/50	0/50
Artemisia annua A	50/50	50/50	38/50	0/50	0/50	
Artemisia annua B	50/50	50/50	50/50	7/50	0/50	

Table 6.17: Inhibitory action of crude extracts of *Artemisia annua* against growth of *E. coli*

Crude Extract of *Artemisia annua*	Number of Strains Inhibited by Different Concentrations				
	200 mg/mL	100 mg/mL	50 mg/mL	25 mg/mL	0 (Control)
Water decoction of *Artemisia annua*	50/50	5/50	0/50	0/50	0/50
Artemisia annua A	5/50	0/50	0/50	0/50	
Artemisia annua B	50/50	25/50	5/50	0/50	

Table 6.18: Inhibitory action of crude extracts of *Artemisia annua* against growth of *β-hemolytic streptococcus*

Crude Extract of *Artemisia annua*	Number of Strains Inhibited by Different Concentrations			
	200 mg/mL	100 mg/mL	50 mg/mL	0 (Control)
Water decoction of *Artemisia annua*	5/50	0/50	0/50	0/50
Artemisia annua A	50/50	0/50	0/50	
Artemisia annua B	50/50	50/50	0/50	

Table 6.19: Inhibitory action of crude extracts of *Artemisia annua* against growth of *Streptococcus pneumoniae*

Crude Extract of *Artemisia annua*	Number of Strains Inhibited by Different Concentrations			
	200 mg/mL	100 mg/mL	50 mg/mL	0 (Control)
Water decoction of *Artemisia annua*	0/10	0/10	0/10	0/10
Artemisia annua A	10/10	6/10	0/10	
Artemisia annua B	10/10	4/10	0/10	

Table 6.20: Comparison of inhibitory action of crude extracts of *Artemisia annua* against bacterial growth

Test Bacteria	IC_{50} (mg/mL)		
	Water decoction of *Artemisia annua*	*Artemisia annua* A	*Artemisia annua* B
Staphylococcus aureus	17.6	41.7	32.1
E. coli	131.9	263.9	93.3
β-Hemolytic streptococcus	263.9	141.4	70.7
Streptococcus pneumoniae	>200.0	93.3	107.1

S. aureus (the results were expressed as the minimum inhibitory concentrations) was measured. Table 6.21 showed the results.

Artemisic acid showed certain inhibitory action against some Gram-positive bacteria (e.g., *Bacillus subtilis*, *S. aureus*, *Staphylococcus albus*, etc.); the minimum inhibitory concentrations were 0.5–1.0 mg/mL.

1.5 Acute Toxicity Test

Using mice (male:female = 1:1) intragastrically administrated with water extract of *A. annua* (self-collected) for single dose, the median lethal dose (LD_{50}) with its confidence interval

Table 6.21: Inhibitory action of artemisic acid against growth of more than ten bacteria including *Staphylococcus aureus*

	Test Bacteria	Control (Without Drug)	Artemisic Acid Concentration (mg/mL) 2.0	1.0	0.5	0.25	Minimum Inhibitory Concentration (mg/mL)
Gram-positive	*Staphylococcus aureus*	+	–	–	+	+	1.0
	Staphylococcus albus	+	–	–	+	+	1.0
	Enterococcus	+	+	+	+	+	>2.0
	Bacillus subtilis	+	–	–	–	+	0.5
Gram-negative	*E. coli*	+	+	+	+	+	>2.0
	Proteobacteria	+	+	+	+	+	>2.0
	Pseudomonas aeruginosa	+	+	+	+	+	>2.0
	Salmonella typhi	+	+	+	+	+	>2.0
	Shigella flexneri	+	+	+	+	+	>2.0
	Shigella sonnei	+	+	+	+	+	>2.0

+, Bacterial growth; –, bacterial growth inhibited.

(P = .95) was calculated as 161.3 ± 4.8 g crude drug/kg by simple probit method. The acute toxicity dose of water extract of the first batch (preflowering period) purchased from Beijing was 117.2 ± 26.5 g crude drug/kg. The acute toxicity dose of water extract of the second batch (bud period) purchased from Beijing was 121.3 ± 12.5 g crude drug/kg.

Discussion

1. Animal tests indicate that extracts of *A. annua* have significant antipyretic, high temperature-resistant, antiinflammatory, and analgesic actions. In vitro antimicrobial tests indicate that extracts of *A. annua* have certain antimicrobial action, providing a scientific basis for traditional functions of *A. annua*. Active antiinflammatory and antimicrobial substances have been confirmed through tests, including scopoletin and artemisic acid; other active ingredients await further investigation.
2. As to medicinal parts, in antimalarial studies, leaves are the principal active part, while stalks and straws are inactive. Referring to antipyretic indicators, both stalks and leaves are active, and the efficacy of plants in the preflowering leafy period is higher than in the flowering period. Therefore, the medicinal part is still defined as the overground part, while the harvesting season is preferentially the preflowering leafy period.

2 Studies on Pharmacological Actions of Volatile Oils of Artemisia annua

Studies on volatile oils of *A. annua* over multiple years indicate that volatile oils of *A. annua* have medicinal values and may be used for treating many diseases, including skin diseases and chronic bronchitis [14–16]. Essential oils contained in *A. annua* have significant antifungal action against *Trichophyton gypseum*, *Trichophyton rubrum*, *Acrothesium floccosum*, *Trichophyton tonsurans*, and so on [17]. Studies on pharmacological actions of volatile oils of *A. annua* are briefly discussed in the subsequent sections.

2.1 Studies on Expectorant, Antitussive, and Antiasthmatic Actions

- **Expectorant action:** Expectorant test was carried out by phenol red method in mice. The results indicated that volatiles have significant expectorant action (Table 6.22).
- **Antitussive action:** The test was carried out by fumigation with strong ammonia in mice. The results indicate that volatiles have favorable antitussive action (Table 6.23).
- **Antiasthmatic action:** Antiasthmatic test was carried out by constant-pressure histamine-acetylcholine spraying method in whole guinea pigs. Single-dose intragastric administration of volatile oils of *A. annua* 0.24 mL/kg (mice 1/10 LD_{50}) could significantly reduce the number of animals giving grade-IV responses to histamine, indicating volatile oils have certain antiasthmatic action (Table 6.24).

Table 6.22: Expectorant action of volatile oils of *Artemisia annua*

Group	Dose (mL/kg)	Number of Animals/ Animals	Phenol Red Concentration (mg/mL) ($\bar{X} \pm SE$)	Action Intensity (%)
Control group		10	0.37 ± 0.02	100
Artemisia annua group	0.24	12	0.75 ± 0.09	202.7
Artemisia annua group	0.48	13	0.84 ± 0.11	227.6
Artemisia annua group	0.96	14	0.97 ± 0.09	262.2

Table 6.23: Antitussive action of volatile oils of *Artemisia annua*

Group	Dose (mL/kg)	Number of Animals/ Animals	Number of Coughs $\bar{X}$	Antitussive (%)
Control group		32	6.50	0
Codeine group	50*	30	4.13	36.5
Artemisia annua group	0.24	18	2.94	54.8
Artemisia annua group	0.12	30	4.80	26.2

*50 mg/kg intraperitoneal injection.

Table 6.24: Antiasthmatic action of volatile oils of *Artemisia annua*

Group	Dose (mL/kg)	Number of Animals/ Animals	Number of Animals Giving Grade-IV Response to Histamine/Animals	Response Rate (%)	Significance (*P* Value)
Control group		21	18	85.71	
Artemisia annua group	0.24	22	12	54.54	<0.05

2.2 *Toxicity Tests of Volatile Oils of* Artemisia annua

2.2.1 Acute toxicity test

The median lethal dose (LD_{50}) for mice, calculated by probit plotting, was 2.34 ± 0.09 mL/kg.

2.2.2 Subacute toxicity test

Subacute toxicity test was carried out in mice by intragastrically administering the lotion of volatile oils of *A. annua* at 0.3 mL/kg (1/8 LD_{50}) and 0.15 mL/kg (1/16 LD_{50}) for 30 consecutive days. All animals in each group survived during the administration period. After drug administration for 30 days, histological morphology of main organs was observed. No significant toxic changes were found.

2.2.3 Effect of volatile oils of Artemisia annua on serum alanine transaminase level

The effect of volatile oils of *A. annua* on the serum alanine transaminase levels of mice and rabbits was investigated.

1. Mice: The study period was 1 month. The drug was intragastrically administered. Two-dose groups were set.

   Thirty healthy mice (6–8 mice per litter) weighing 14–19 g were randomized by body weight and litter into three groups. Each mouse was given water or the drug at a dose of 0.2 mL/10 g. Mice in the control group were given tap water. The low administration dose of volatile oils of *A. annua* was equivalent to 1/16 of LD_{50}, and the high dose was equivalent to 1/8 of LD_{50}. The drug was administered continually for 1 month. Blood was collected using a capillary through the posterior venous plexus before drug administration and on days 14 and 30 after drug administration. Then, using the method for ultramicroanalysis of serum alanine transaminase, 10 μL of each serum sample was transferred using a microsyringe to a small test tube and the content of alanine transaminase in serum was determined.
2. Rabbits: The study process was as same as that in mice, but only one-dose group was set.
   Ten gray rabbits weighing 1.9–2.9 kg were randomized by body weight into two groups. Each rabbit was given water or the drug at a dose of 30 mL/kg. Rabbits in the control group were given water. Rabbits in the drug administration group were given volatile oils of *A. annua* at a dose equivalent to 1/20 of LD_{50}. The drug was administered continually for 1 month. Blood was collected through the aural median artery of rabbits before, during and after drug administration using capillaries. Serum alanine transaminase was determined by the same method as that for mice.

2.2.4 Conclusions

Statistically, serum alanine transaminase levels on day 14 after drug administration and before drug administration in the high-dose group of mice were significantly different ($P < .05$), while the levels on day 30 after drug administration were not significantly different from those before drug administration ($P > .05$). The same results were obtained in the drug administration group of rabbits; however, limited by the animal number, statistical processing was not allowed. Therefore, according to the above results, large doses of volatile oils of *A. annua* (1/8 LD_{50} in mice, 1/20 LD_{50} in rabbits) seem to be able to transiently elevate the serum alanine transaminase level.

2.3 Formulation Studies

The above studies indicate that volatile oils of *A. annua* can be used for treating chronic trachitis. Therefore, oil gelatin pills were prepared for use. The manufacturing process was as follows.

2.3.1 Formula

Artemisia annua oil	20 g
Edible vegetable oil	60 g
To make	1000 pills

2.3.2 Preparation

1. Dissolve *A. annua* oil in refined edible vegetable oil (solid fatty acids were removed at around 0°C), and adjust to a suitable concentration.
2. Dissolve medicinal gelatin and q.s. medicinal glycerin in 70°C distilled water, and filter to produce a gelatin solution of a suitable concentration.
3. Transfer the *A. annua* oil and the gelatin solution of suitable concentrations in a pill machine to make gelatin pills.
4. After molding, drying, washing, disinfection, and selection of the obtained gelatin pills, package the product.
   The process flow of *A. annua* oil gelatin pills is shown in Fig. 6.1.
5. Format: Each pill contains 20 mg of *A. annua* oil.
6. Dosage and administration: Take 1–2 pills orally once, t.i.d.
7. Storage: Preserve in tightly closed containers, store in a cool and shady place, and protect from light.

2.4 Clinical Studies

According to pooled data, in 1357 patients with trachitis continually treated for two or more courses, the average total response rate was 92.56%; the excellent response rate was 57.26%.

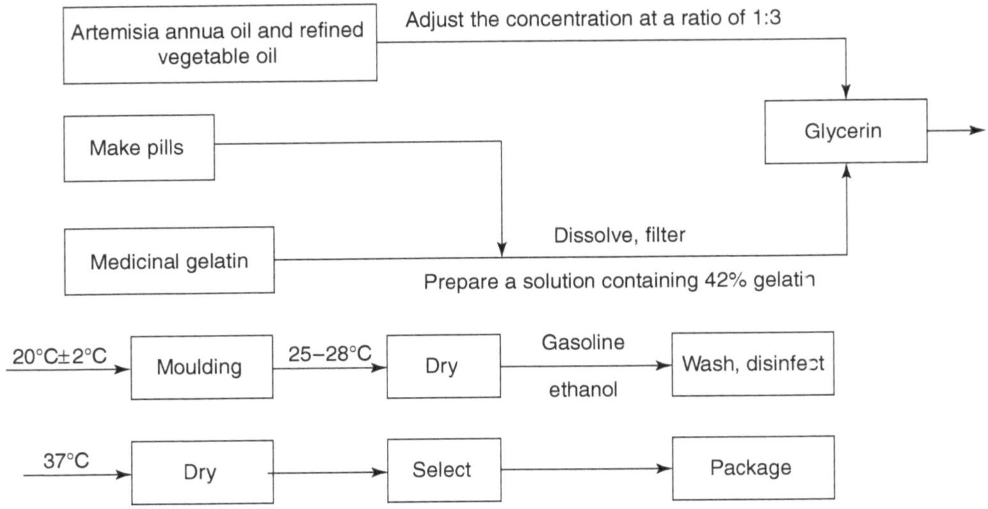

Figure 6.1: Process Flow of *Artemisia annua* Oil Gelatin Pills.

Satisfactory results were obtained in treating both cough and asthma. According to follow-up results of part patients, the effectiveness stability rate was 76–78% after drug withdrawal for 10 months, indicating the drug has certain long-term efficacy.

Results of clinical observation and laboratory tests indicate volatile oils of *A. annua* act against chronic trachitis mainly by the following mechanisms:

1. Antiinflammatory action: After treatment, expectoration of viscous and purulent phlegm was significantly reduced; the quantity of DNA in phlegm, an indicator of the severity of inflammation, significantly decreased, while the quantity of lysozyme, an indicator of humoral immunity, significantly increased. Patients with a total WBC count >10,000 mm^{-3} and a neutrophil percent (>5%) showed a tendency to return to normal after treatment; after treatment, slimy yellow tongue fur was changed into thin and white tongue fur in most patients, red or nearly red tongue texture was changed into pale red tongue texture, and rapid or slippery pulse was changed into slow pulse. Changes of tongue fur and pulse suggest that this product has functions of clearing heat, eliminating phlegm, dissolving dampness, and removing water retention.
2. Regulating autonomic nerve functions: After treatment, patients' sympathetic nerve and parasympathetic nerve functions were gradually equilibrated, suggesting *A. annua* oil can regulate hyperfunction of parasympathetic nerves and tension inadequacy of sympathetic nerves.
3. Inhibiting abnormal secretion of bronchial mucosal glands: The quantity of phlegm was significantly reduced. This is possibly because *A. annua* oil can control infective inflammation or allergies so that glandular secretion is reduced or because the drug can directly or indirectly inhibit glandular secretion by regulating activity of sympathetic nerves.

4. Regulating endocrinal functions: The content of 17-hydroxycorticosteroids in urine was increased, indicating the function of the hypothalamic–pituitary–adrenal cortex system has been improved.
5. Improving ventilation: Through pulmonary function examination, the pulmonary ventilation function of part patients was improved in the Affiliated Hospital of Anhui Medical College, indicating *A.* annua oil gelatin pills can reduce airway obstruction.
6. Improving right heart load: In part patients, pulmonary P wave subsided and ST-T changes recovered in ECG. The phenomena were also observed in the Affiliated Hospital of Anhui Medical College. This is possibly because patients' right heart loads have been improved.

Side effects of this product are mild. At the normal dose, only a few patients presented with dry mouth, belching, and gastric discomfort, which generally subsided spontaneously within a few days without need of particular treatment or continued treatment. No significant influence on heart, liver, kidney, and hematopoietic functions has been found in ECG, liver, and kidney function tests or blood and urine routine tests, indicating the dose currently used is within the safety margin.

3 Progress in Pharmacological Action Studies and Clinical Application of Artemisia annua

In addition to antimalarial action, *A. annua* has also functions of modulating immunity, inhibiting schistosomiasis, expelling phlegm, reducing cough, alleviating asthma, and resisting viruses and bacteria; clinical use for treating discoid lupus erythematosus, trachitis, schistosomiasis, diarrhea, and fever has also been reported, as discussed in the subsequent sections.

3.1 Immunomodulation Function [18–20]

Following sequential extraction of dry whole plant of *A. annua* with petroleum ether, ethyl ether, ethyl acetate, ethanol and water, petroleum ether extract, ethyl ether extract, ethyl acetate extract, ethanol extract, and water extract were obtained, respectively. The effects of these extracts on human complements, T-lymphocyte proliferation, and yeast polysaccharide-stimulated neutrophil chemiluminescence were investigated. All extracted exhibited significant inhibitory action in a dose-dependent manner in immunoassays. Ethyl acetate extract and ethanol extract demonstrated the strongest activity by the classical pathway of complements. In chemiluminescent assay, both ethyl ether extract and ethyl acetate extract demonstrated significant inhibitory activity. In T-lymphocyte proliferation test, ethyl ether extract showed the strongest inhibitory activity.

The effect of artemisinin on immunity was observed by mouse foot pad test, lymphocyte transformation test, specific immunity rosette test, and plaque test. It was found that artemisinin significantly inhibit humoral immunity and promote cellular immunity, indicating the compound has immunomodulatory action. Noori et al. [20] found that artemisinin has inhibitory action against delayed-type hypersensitivity.

3.2 Antischistosoma Activity [21–23]

Artemisinin has antischistosoma activity and can kill schistosomal larvae throughout the entire administration period. The antischistosoma active groups are possibly the peroxide bridges in its structure, acting possibly by affecting glucose metabolism of the parasites.

3.3 Antiviral Action

Qian et al. [18] found in chick embryo infection test that artemisinin could inhibit influenza virus but not kill the virus. In 2005, Romero et al. [24] reported the action of artemisinin against hepatitis B virus.

In laboratories, it has been found that water extract of *A. annua* has activity against herpes simplex virus and hepatitis B virus in vitro; using activity tracing techniques, it is found that the antiviral ingredient is condensed tanin. A study was carried out in vitro to investigate the anti-HSV-2 and anti-HBV activity of a condensed tanin (CTA) isolated from the water extract of *A. annua*. The results showed that CTA had significant anti-HSV-2 activity; the CC_{50} of the positive comparator, acyclovir (ACV), and CTA were 6.84 and 3.69 mg/mL, respectively; the IC_{50} were 0.162 and 0.138 mg/mL, respectively, indicating CTA is less cytotoxic than ACV, while the two compounds have equivalent antiviral activity. According to culture results of HepG2.2.1.5 cells in CTA-containing media, CTA demonstrated mild cytotoxicity at 2.5 mg/mL and significant inhibitory action against HepG2.2.1.5 cell-expressed HBeAg at 0.156–2.5 mg/mL and yielded an HBeAg inhibition rate of 90.45% on day 12 of culture at 0.625 mg/mL, indicating CTA has potential anti-HBV activity [25].

3.4 Antimicrobial Action

Juteau et al. [26] carried out a study on volatile oils of *A. annua* growing in France, and found that the substance showed antimicrobial and antioxidative activity. Analytical results indicate the ingredients mainly include camphor (44%), germacrene D (16%), *trans*-pinocarveol (11%), β-selinene (9%), β-caryophyllene, and artemisia ketone (3%).

Experimental studies indicate that both artemisic acid and artemisinin B have inhibitory activity against multiple bacteria and fungi [27,28].

In a study [29], a solution containing 1250 mg/L *A. annua* ointment acting on *S. aureus* and *E. coli* for 1 min and *Candida albicans* for 3 min produced a mean killing rate up to 99.90%.

The antimicrobial effect decreased following decreases of the concentration of *A. annua* ointment and increased with temperature; when the bacterial suspension contains 10% calf serum, the antimicrobial effect was also significantly affected. No significant decrease was found in the antimicrobial effect after the *A. annua* solution was sealed and placed in a 37°C constant-temperature water bath for 3 months. *A. annua* ointment has favorable performance in killing bacterial propagules, but the antimicrobial effect can be significantly influenced by organics.

3.5 *Antiendotoxin Action*

A. annua extract has antiendotoxin action. Tests indicate that *A. annua* extract and artemisinin can lower concentrations of lipid peroxides, lysosomal acid phosphatase, endotoxin, tumor necrosis factor-alpha in hepatic mitochondria, and P450 in hepatic microsomes of rats; enhance the activity of superoxide dismutase in hepatic mitochondria; decrease the death rate of mice with endotoxin shock; prolong the mean survival of mice; and also have certain protective effects on histological morphology of liver and lungs [30].

Serious traumas, hemorrhagic shock, large area burns, pancreatitis, and ischemic reperfusion injury can all impair functions of the intestinal mucosal barrier and reduce immunity and lead to displacement of bacteria and endotoxin in the intestinal tract, resulting in intestinal endotoxemia. Injection by Gram-negative bacteria may also cause endotoxemia. Endotoxin may induce excessive inflammatory responses, and is closely associated with multiple organ dysfunction syndrome and multiple organ failure. Studies in recent years indicate NO is one of the final common media of cascade chain reaction in the inflammatory response and also a key medium responsible for septic shock. *A. annua* can significantly inhibit elevation of serum NO concentrations of mice with endotoxemia, and is valuable for prevention and treatment of multiple organ dysfunction syndrome and multiple organ failure secondary to endotoxemia [31].

3.6 *Pharmacological Action of* Artemisia annua *Prescriptions*

A. annua and Turtle Carapace Decoction is a representative prescription for treating fever due to *Yin* deficiency in the later stage of seasonal prevalent diseases. Chen, referring to related literatures, considers that *A. annua* and Turtle Carapace Decoction can be well responded to by all clinical symptoms of "*Yin*-related fever" [32].

Zhang clinically treated 200 patients using self-designed *A. annua* and Gypsum Decoction, and found that the prescription was rapidly effective in reducing fever in patients affected by warm heat, excessive evil heat, and exuberant vital *Qi* [33].

Nasal bleeding is a common and frequent clinical disease, which often occurs in summer, particularly in children. Treatment of the disease is oriented for clearing summer heat, benefiting vital energy, and promoting the generation of body fluid, cooling blood, and stopping bleeding. Favorable results were obtained using the water decoction of 15 g of *A. annua* and 50 g of rock candy, one dose per day, with 7 days as a course of treatment [34].

Li et al. [35] obtained favorable results in treating long-term fever in patients with malignant cancers with *A. annua* and *Anemarrhena asphodeloides* Bunge Decoction.

3.7 Antifeedant Activity Against Pests

According to a report [36], bioactivity test results of water extract of *A. annua* indicate that the crude extract of *A. annua* has antifeedant activity against the six test pests. It has very strong antifeedant activity against *Odontotermes formosanus* (Shiraki), *Tribolium castaneum* (Herbst), and *Rhizopertha dominica* (Fabricius), and also strong antifeedant activity against *Aphis gossypii* Glover, *Tetranychus urticae* Koch, and *Etiella zinckenella* Treitschke. When the crude extract of *A. annua* is used, it is preferentially diluted in 1:500 or 1:800; significant differences were found between the treatment and control.

The research group of Maggi in Argentina [37] reported the study on using extract of *A. annua* and artemisinin for control of pests in pumpkin seedlings, which indicates that both extract of *A. annua* and artemisinin have pest-resistant activity.

3.8 Other Biological Activities and Clinical Application

Acetone extract of *A. annua* applied to skin can expel mosquitoes for more than 2 h. Perazzo et al. [38] found in a study on volatile oils of *A. annua* growing in Brazil that volatile oils of *A. annua* could activate the central nervous system. *A. annua* extract and artemisinin compounds have the function of killing pests and regulating autonomic growth [39,40]. Allen et al. [41] investigated the efficacy of *A. annua* in treating chicken coccidiosis.

Liu [42] obtained favorable results in using water decoction of fresh *A. annua* to treat viral and epidemic conjunctivitis. Most cases of yellow fluid ulcer are caused by summer dampness and heat obstruction in skin. *A. annua* is cold in nature and bitter in taste, and has the function of clearing summer heat. External washout with the water decoction can dredge follicular orifices and sweat glands, and regulate *Qi* so as to expel evils. Pharmacological tests indicate that the water decoction of *A. annua* has strong inhibitory action against *Staphylococcus epidermidis*, *Micrococcus catarrhalis*, *S. aureus*, and *Pseudomonas aeruginosa*. Wang and Xu obtained satisfactory effects in treating yellow fluid ulcer with the water decoction of *A. annua* [43].

In recent years, more and more research groups have conducted antineoplastic studies of artemisinin compounds, and some new progress has been made [44–46].

References

[1] Huang L, Liu JF, Liu LX, et al. A study on antipyretic and anti-inflammatory effects of Chinese medicine *Artemisia annua*. China J Chin Mater Med 1993;18(1):44–8.

[2] Li KG, Qian RS, Li ZL. Observation of the efficacy of *Artemisia annua* decoction in treating dengue fever. Chin Tradit Herb Drugs 1985;16(6):256–8.

[3] Lin L, Cai SJ, Lin HM. Experience in clinical use of *Artemisia annua* for treating fever. Shanghai J Tradit Chin Med 2003;37(2):17–8.

[4] Tang R, Zhang WC. Use of *Artemisia annua* in treatment of seasonal febrile disease. Sichuan J Tradit Chin Med 2004;22(7):27.

[5] Jiang MQ. Clinical experience in using *Artemisia annua* for treating fever. Sichuan J Tradit Chin Med 2002;20(4):25.

[6] Li CH, Zhao Y, Lin QY. An introduction to antipyretic ingredients of *Artemisia annua*. Guangxi J Tradit Chin Med 1997;20(5):269–70.

[7] Hou L. Treatment of high fever with Jinghao Chaibai Decoction. Hebei J Tradit Chin Med 1985;7(4):28.

[8] Zhou XC. Treatment of high fever in colds with *Artemisia annua* and Starwort Root Radix Stellariae-based preparations. J Pract Tradit Chin Intern Med 1988;2(3):131.

[9] Peng QC, Xiao BR, Hong XQ, et al. Observation of the efficacy of Qingyin Injection in treating acute upper respiratory tract infection. J Luzhou Med Coll 1989;12(2):109–12.

[10] Feng WY, Peng QC, Qian YL, et al. A study on preparation and clinical use of Qingyin Injection. J Luzhou Med Coll 1989;12(3):161–5.

[11] Qian YL, Xiao SH, Dong Z. A study on antipyretic and toxicological tests of Qingyin Injection. J Luzhou Med Coll 1989;12(3):171–5.

[12] Peng QC, Lu WS, Wang HC, et al. Observation of the antipyretic efficacy of Qingyin Injection against acute high fever. J Luzhou Med Coll 1989;12(4):259–62.

[13] Yang DY, Yao GY, Chen WH. Combination of *Artemisia annua* for treating symptoms caused by summer dampness and heat. Henan J Tradit Chin Med 1999;14(1):48–9.

[14] Xiamen Medicine Research Institute. A study on treatment of chronic bronchitis with volatile oils of *Artemisia annua*. Xiamen Med J 1980;(3):1–4.

[15] Feng WY, Qiu HR, Zheng JA, et al. Development and clinical use of *Artemisia annua* Oil Liniment. J Luzhou Med Coll 1989;12(1):1–3.

[16] Chen FL, He F, Li JL, et al. GC-MS analysis of ingredients of volatile oils of *Artemisia annua* extracted by different methods. China J Chin Mater Med 2001;24(3):176–8.

[17] Lu HS. Development and utilization values and cultivation techniques of *Artemisia annua*. Forest By-Product Spec. China 2002;(1):6–7.

[18] Qian RS, Li ZL, Yu JL, et al. Immunological and anti-viral actions of artemisinin. J Tradit Chin Med 1981;22(5):462–6.

[19] Zang QZ, Zheng ZY, Qi SB, et al. Immunomodulatory action of artemisinin. Nucl Tech 1984;(2):55–7.

[20] Noori S, Naderi GA, Hassan ZM, et al. Immunosuppressive activity of a molecule isolated from *Artemisia annua* on DTH responses compared with cyclosporin A. Int Immunopharmacol 2004;4(10–11):1301–6.

[21] Le WJ, Wang GF, You JQ, et al. An experimental study on treatment of animal schistosomiasis with artemisinin derivatives. Pharm Bull 1980;15(4):38.

[22] Institute of Chinese Materia Medica, China Academy of Chinese Medical Sciences. Collection and Quality Research of Commonly Used Chinese Materia Medica—*Artemisia annua*, 1990, 6–7.

[23] Xiao SH. Development of antischistosomal drugs in China, with particular consideration to praziquantel and the artemisinins. Acta Trop 2005;96(2–3):153–67.

[24] Romero MR, Eferth T, Serrano MA, et al. Efect of artemisinin/artesunate as inhibitors of hepatitis B virus pro-duction in an "in vitro" replicative system. Antiviral Res 2005;68(2):75–83.

[25] Zhang JF, Tan J, Pu Q, et al. A study on antiviral activity of Condensed Tannin of *Artemisia annua*. Res Dev Nat Prod 2004;16(4):307–11.

[26] Juteau F, Masoti V, Bessiere JM, et al. Antibacterial and antioxidant activities of *Artemisia annua* essential oil. Fitoterapia 2002;73(6):532–5.
[27] Bhakuni RS, Jain DC, Sharma RP, et al. Secondary metabolites of *Artemisia annua* and their biological activity. Curr Sci 2001;80(1):35–48.
[28] Dhingra V, Pakki SR, Narasu ML. Antimicrobial activity of artemisinin and its precursors. Curr Sci 2000;78(6):709–13.
[29] Wang HQ, Tian GQ, Liu QZ, et al. A study on antimicrobial efficacy of *Artemisia annua* solution and influencing factors. Chin J Disinfect 2004;21(2):112–3.
[30] Tan YQ, Zhao Y, Lin QY. An experimental study on anti-endotoxin action of *Artemisia annua* extract. China J Chin Mater Med 1999;24(3):166–71.
[31] Wang JC, Zhao D. Effects of *Artemisia annua* on serum NO concentration of mice with endotoxemia. Natl Med J China 2000;35(7):469.
[32] Chen YJ. Progress in clinical studies on *Artemisia annua* and Turtle Carapace Decoction. J Chin Med Res 2002;18(5):51–3.
[33] Zhang SC. Treatment of 200 patients with exogenous high fever with *Artemisia annua* and Gypsum Decoction. Beijing J Tradit Chin Med 1999;(2):27.
[34] Duan JP. Treatment of nasal bleeding with *Artemisia annua* and Rock Candy Decoction. J New Chin Med 2005;37(2):31.
[35] Li XD, Sun J, Luan ZP. Treatment of 34 patients with long-term fever in malignant cancers with *Artemisia annua* and Anemarrhena asphodeloides bunge Decoction. Tradit Chin Med Res 2005;18(6):46–7.
[36] Zhu F, Lei CL, Wang J. A preliminary study on antifeedant activity of crude extract of *Artemisia annua* against several pests. Natl Enemies Insects 2003;25(1):16–9.
[37] Maggi ME, Mangeaud A, Carpinela MC, et al. Laboratory evaluation of *Artemisiaannua* L. extracts and artemisinin activity against Epilachna paenulata and Spodoptera eridania. J Chem Ecol 2005;31(7):1527–36.
[38] Perazzo FF, Carvalho JCT, Carvalho JE, et al. Central properties of the essential oil and the crude ethanol ex-tract from aerial parts of *Aremisiaannua* L. Pharamacol Res 2003;48(5):497–502.
[39] Bagchi GD, Jain DC, Kumar S. Arteether: a potent plant growth inhibitor from *Artemisiaannua*. Phytochemistry 1997;45(6):1131–3.
[40] Dayan FE, Hemandez A, Alen SN, et al. Comparative phytotoxicity of artemisinin and several sesquiterpene analogues. Phytochemistry 1999;50(4):607–14.
[41] Allen PC, Lydon J, Danforth HD. Effects of components of *Artemisia annua* on coccidia infections in chickens. Poult Sci 1997;76(8):1156–63.
[42] Liu ZG. Treatment of epidemic conjunctivitis with washout of *Artemisia annua*. J New Chin Med 2003;35(1):8.
[43] Wang ST, Xu ZX. Treatment of 28 patients with yellow fluid ulcer with *Artemisia annua*. China Naturopathy 2000;8(10):7.
[44] Zheng GQ. Cytotoxic terpenoids and flavonoids from *Artemisia annua*. Plant Med 1994;60(1):54–7.
[45] Singh NP, Lai H. Selective toxicity of dihydroartemisinin and holotransferrin toward human breast cancer cels. Life Sci 2001;70(1):49–56.
[46] Lai H, Singh NP. Oral artemisinin prevents and delays the development of 7,12-dimethylbenz [α] anthracene (DMBA)-i nduced breast cancer in the rat. Cancer Lett 2006;231(1):43–8.

CHAPTER 7

Studies on the Cultivation and Breeding of Artemisia annua *L.*

Chapter Outline

In this chapter, we first review the biological characteristics of *Artemisia annua*, and then focus on major works in our laboratory on the biotechnology of *A. annua*, including tissue culture, hairy roots culture, and shoot clusters induction of *A. annua*. In addition to the modification of *A. annua* via biotechnology, the space breeding for a high-yield strain SP-18 is also reported.

From *Artemisia annua* L. to Artemisinins
http://dx.doi.org/10.1016/B978-0-12-811655-5.00007-6

1 Biological Characteristics of Artemisia annua and Its Progress

1.1 Cultivation of Artemisia annua

As chemical synthesis of artemisinin is very expensive, field cultivating of *A. annua* remains the only viable source of artemisinin for commercial use. In the past years, the main source of artemisinin came from Chinese wild *A. annua*, and China provided enough materials for clinical research of antimalarial drugs in Southeast Asia [1]. Along with the increase of interests and wide applications of artemisinin, more and more researchers attempted to domesticate medical *A. annua* as a cultivated crop [2–4]. As a short-day plant, *A. annua* plant flowers before it accumulates enough biomass for the production of artemisinin in tropical region; it is not suitable for growing in tropical region. Many countries including Australia [3], Belgium [5], India [6], Switzerland [7], America [2,8–10], and Vietnam [11] had tried to culture *A. annua* as a crop and obtained good results.

Although asexual propagation of *A. annua* is simple, breeding germplasm of *A. annua* must be grown under long-day conditions to avoid early flowering. The most practical and economical method for commercially used artemisinin is sowing with seeds. *A. annua* seeds can be stored at a low temperature for at least 3 years. Chen et al. [12] investigated the various influential factors on artemisinin yield of *A. annua*, including seeds sources, sowing time, ecotypes, and samples from different parts of the plant. The results of seeds sources showed that the germplasm from Guangdong (south of China) had the highest artemisinin content (0.714%, DW). The results of sowing time showed that sowing on February 14 and March 14 could achieve high artemisinin than other times; the artemisinin content reached 0.46 and 0.681% (DW), respectively. They also found that light had significant effects on the content of artemisinin and artemisinin content could reach 0.947% in plants without shading. The comparison of artemisinin contents in different parts of plant showed that artemisinin content of leaves (0.787%) was higher than those of flowers (0.347%) and inflorescences (0.122%). Ferreira et al. [13] reported that sowing in late spring or early summer could obtain high yield of biomass in the north temperate zone, and the artemisinin content reached the highest before flowering with the photoperiod of about 13.5 h at this time. If the seeds were sowed too late, *A. annua* plant would flower when the plant was still short and the production of artemisinin would be reduced. In India, *A. annua* was sowed and grown successfully in the middle of December, while in Tasmania of Australia, the yield of the plants sowed in October was twice than that of in November [4]. In Indiana of USA, seedlings of *A. annua* were transplanted every month from April to July; the results showed that aromatic oil content reached its highest at the full flowering stage and the highest aromatic oil content was found in seedlings transplanted in May and June [2].

Nitrogen fertilizer can promote the growth of *A. annua*; however, reports concerning its effects on the accumulation of artemisinin are few. Srivastava and Sharma [14] reported that microelement such as boron could increase the content of artemisinin, but whether the tested soil was boron deficient and how *A. annua* plant used boron remained unclear. Greenhouse experiments showed that when the soil pH changed from 5.5 to 7.7, both the yields of leaves and the content of artemisinin kept relatively unchanged [3]. Charles et al. [15] found that the content of artemisinin would significantly reduce when *A. annua* was subjected to water stress 2 weeks before harvesting. Shukla et al. [16] reported that plant growth regulators promoted not only the growth of *A. annua*, but also the content increase of artemisinin; however, how to effectively use such growth regulators is still largely unknown.

In field, when *A. annua* planting density changed from 1 to 20 plants/m^2, the dry weight of raw material varied from 2 to 6.8 tons/ha, and planting density did not affect the contents of artemisinin and artemisinic acid [4]. In Indiana, USA, under three levels of nitrogen fertilizer (0, 67, 134 kg/ha) conditions, the biomass of *A. annua* increased gradually with planting density from 2.7 to 11.1 plants/m^2 in each nitrogen fertilizer level, and the maximum biomass was obtained with density of planting 2.7 plants/m^2 in nitrogen fertilizer level of 67 kg/ha [2]. The hybrid *A. annua* from Switzerland cultivated in Campinas, Brazil, with the plant spacing 0.3 × 0.5 m, 2.15 tons/ha dry leaves were harvested from September to November, and 6 kg artemisinin and 14.6 kg artemisinic acid were extracted from these leaves [5].

1.1.1 Harvest

As the content of artemisinin is very low, *A. annua* cultivated in field should be harvested during the time when the artemisinin content reached its highest, and both the biomass and the content of artemisinin should be taken into account for the optimum harvest time selection.

1.1.2 Postharvest handling

Before extraction of artemisinin from *A. annua*, lots of plant materials need to be dried. Ferreira et al. [17] compared freeze drying, oven drying (40°C), and open air drying on the yield of artemisinin by HPLC-CE detection; their results showed that open air drying of samples gave significantly higher artemisinin yield (0.13%) than oven drying (0.10%) and freeze drying (0.02%). Open air drying for 2, 4, 6, and 8 days did not affect artemisinin content, but microwave drying greatly reduced artemisinin (0.02% air drying vs 0.002% microwave drying). However, when the artemisinin standard solution was drying by microwave for 5 min at 50% power, no artemisinin loss was detected by HPLC-EC [13].

1.2 Reproductive Physiology of Artemisia annua

The nodding capitulum of *A. annua* is greenish-yellow and its diameter is about 2–3 mm, with calyces composed of imbricated bracts. Glandular trichomes distribute on the whole surface of floral organs [18]. *A. annua* is naturally cross-pollinated by insects and wind action, which is unusual in the Compositae family [19]. Although Ferreira et al. [10] found that the structure of *A. annua* capitulum was suitable for self-pollination, Peter-Blanc [20], however, showed that self-pollination of *A. annua* was rare and difficult to produce seeds, indicating that self-incompatibility is present in *A. annua* like other species of the Compositae family [21], which makes it difficult to preserve the high-yield strains by sexual reproduction. This might be the reason for the differences of artemisinin content in different plants.

Ferreira et al. [10] reported that *A. annua* was a determinate short-day plant, nonjuvenile plants were sensitive to photoperiod signal, and flowered after 2-week induction of short-day photoperiod. Vegetative growth *A. annua* flowered after 2-week induction by 8-, 10-, or 12-h short-day photoperiod in greenhouse, while by 16-, 20-, or 24-h long-day photoperiod, the plants remained vegetative. In field studies in the northern temperate zone, the first flower buds were visible on 4 September when the day length was 12:57 h at Lafayette, IN, USA.

1.3 Physiology and Biochemistry of Artemisia annua

One of the remarkable biochemical characteristics of *A. annua* is its containment of both artemisinin and its derivatives. There are many members of *Artemisia* genus; do other *Artemisia* plants also contain artemisinin? To address this problem, Hu et al. [22] investigated various *A. annua*, collected from different regions of China and five species of *Artemisia* genus to see whether they could produce artemisinin. Results showed that among these plants, only *A. annua* L. contained artemisinin. Klayman [23] analyzed several plants of *Artemisia* genus such as *Artemisia ludoviciana*, *Artemisia vulgaris*, *Artemisia schidtiana*, *Artemisia pontica*, *Artemisia arbuscula*, and *Artemisia dracunculus*, but no artemisinin was detected in these plants. Now at least 40 components of essential oil were separated from *A. annua* [13], in addition to many nonvolatile sesquiterpenoids, such as artemisinin, artemisinic acid, and artemisinin B. Artemisinin was mainly found in leaves, young stems, flower buds, and flowers of *A. annua* [10,24–29].

Chinese scientists [30] studied the content of *A. annua*, and found that the highest content of artemisinin was obtained at the flowering stage. The flowering time is significantly affected by the terrain, temperature, soil, and fertilizer, so the suitable time of harvesting *A. annua* should be at the exuberant leaf stage before flowering. The plantlets of *A. annua* cultivated with fertilizer were higher and stronger than the wild type, and the contents of artemisinin were also higher. Chen and Zhang [31] found that if *A. annua* were grown in manual control environment and the basic nutrients could be met, the nutrient contents and the substrate in growing environment had little impact on the artemisinin contents. However, a high temperature (30°C) and strong illumination could increase the yield of artemisinin by hundreds of times. Ferreria et al. [10] found that the content of artemisinin in inflorescence was higher than in leaf, and the highest content was present at the full flowering phase of *A. annua*. This result was consistent with studies performed in field or in greenhouse [9,10,28,32]. However, some researchers found that the artemisinin content was the maximum in early flowering period [3,11,24,26,33,34]. The production of artemisinin and the leaves of *A. annua* reached the highest after growth of 5 months, with the maximum artemisinin content of 0.86% (DW) [34]. Then, the artemisinin content began to decrease gradually. Singh et al. [28] also found that the highest content and yield of artemisinin occurred at 50% flowering and full flowering phase. Elhag et al. [35] observed that the high artemisinin production strains were characterized as tall robust plants with long internodes, open branching, dense leaves, and thick stems. Liersch et al. [26] tested two kinds of hormone growth regulators, daminozide and chlormequat, on *A. annua* strain 811, and found that daminozide was able to increase the artemisinin content by 30% over untreated plants. The aforementioned results indicate that the suitable harvest time of *A. annua* should be from late vegetative growth phase to reproductive growth phase.

Another biochemical characterization of *A. annua* is that plants grown in different regions could produce specific secondary metabolites. For example, the major secondary metabolite of *A. annua* grown in China was artemisia ketone [36], but it was not detected in *A. annua* grown in Vietnam [11]. Another example is that the artemisinin content in *A. annua* grown in Hainan (0.87%) (south of China) was higher than *A. annua* grown in the North of China, such as Beijing (0.18%). However, the artemisinin B content of *A. annua* grown in Beijing was higher than grown in Hainan. The aforementioned results showed that the environmental factors played important roles in *A. annua* secondary metabolism.

1.4 Structural, Biochemical, and Physiological Functions of the Glandular Trichomes of Artemisia annua

Trichomes and glands are common to the Compositae and contain volatile monoterpenes and sesquiterpenes [37]. Duke and Paul [38] described the origin of the 10-celled biseriate structure glands in leaves based on scanning and transmission electron microscopy. They found that in the youngest leaf primordia, the differentiation of foliar cells into gland cells had already began. The earliest stage of gland formation was the one-cell stage, and a single epidermal cell enlarged and produced above the leaf surface at this stage. After considerable expansion, this cell divided anticlinally, and then both of the resulting cells divided periclinally and transformed into two basal cells and two apical cells. Six cells were formed after another periclinal division of the two apical cells. At last, a 10-celled biseriate structure was formed of two stalk cells, two basal cells, and three pairs of secretory cells. At the 1- to 4-cell stage, vacuoles were relatively small and plastids were proplastids with only a few unstacked thylakoids. At the 6-cell stage, all the cells contained chloroplasts with few stacked thylakoids and no starch grains. The only distinguishing feature between chloroplasts of glands and mesophyll tissue was the lack of starch grains. At the 10-cell stage, the 6 secretory cells began to separate from cuticular surface and formed a two-lobe capsule and were used for preserving the cytoplasm. The secretory cells contained extensive endoplasmic reticulum, and the plastids of apical cell pairs looked like amoeboid and did not contain thylakoids, whereas the plastids of the four apical secretory cells were amoeboid with thylakoids.

The basal cell pair contained thylakoids, and the chloroplasts of stalk cells contained thylakoids and starch grains. With the development and differentiation of the 10 cells, the detachment of the cuticle from the external cell walls of the gland cells continued down to the 2 stalk cells above the basal cells and formed a subcuticular space bordered by the 6 most apical cells of the gland. At this stage, the two basal cells contained chloroplasts and relatively large vacuoles.

As the gland matured, the two cells generally did not contain chloroplast and plastids were polymerized to larger proplastids. The two apical cells looked like metastatic cell and the cell wall was growing inward, thereby expanding the surface area to facilitate the secretion of metabolites. The ratio of chloroplasts stroma to thylakoid was higher than that in mesophyll cells. Near the plasma membrane adjacent to cell walls bordering the subcuticular space, the cytoplasm was enriched in smooth endoplasmic reticulum. In addition, Duke et al. [39] reported that artemisinin and artemisitene in foliar tissue were localized entirely in the subcuticular space of capitate glands of *A. annua*. As virtually all the detectable artemisinin and artemisitene in glanded could be extracted by organic solvent without visible damage to the leaf epidermal cells, while in *A. annua* biotype without glands contained neither artemisinin nor artemisitene, the variation of artemisinin content in different tissue, different grown stage, and environmental factors of natural *A. annua* could be explained by these aforementioned results. In addition, immunohistochemical localization of artemisinin seems necessary to confirm these glands as the sites of artemisinin synthesis. Weather et al. [40] and Liu et al. [41] reported high artemisinin levels in *A. annua* hairy root cultures which did not contain glands. These results might provide some basis for the artemisinin biosynthesis in tissues outside the glands. As glands localize on surface of plants, it may provide a place to locate artemisinin and other terpenoids as phytoalexin to protect plants against pathogens.

1.5 Genetic Improvement

For a long time, the genetic improvement of *A. annua* witnessed a slow pace due to the complicated metabolism of *A. annua* and many unknown genetic mechanisms, and many efforts were made on the hereditary aspects to improve the content of artemisinin. A high-yield Chinese clone of *A. annua* (the content was over 1%) was induced to flower under controlled greenhouse conditions with pollen from Italian, Yugoslavian, and Spanish origins [7]. These hybrids, cultivated in Switzerland in 1990 and 1991, produced in average 0.64, 0.73, and 0.95% artemisinin, respectively, for a yield of dry leaves of about 2000 kg/ha, indicating that the artemisinin content was controlled by additive genetic factors. Ferreira and Janick [42] analyzed the genetic features of *A. annua* grown in field and greenhouse, and found that artemisinin content was determined by hereditary features. The difference of artemisinin content was found very small for the same nonflowering plants grown in the field and in the greenhouse under long days. Hence, growth in greenhouse under long days could be an effective means for screening high artemisinin-yielding strains. The selected superior strains could be induced to flower under short days and intercrossed to obtain higher yield artemisinin clones. Wallaart et al. [43] found that the averaged artemisinin level in tetraploid *A. annua* plants induced with colchicine was higher than that of the wild type, and the leaves were bigger compared with the diploid wild type. However, the tetraploid plants were smaller than the wild-type plants, and the averaged biomass of the tetraploid plants was lower compared with the wild-type plants. Therefore, the artemisinin yield of tetraploids was decreased by 25%. Generally speaking, tetraploid *A. annua* could be a useful starting material for a breeding program to obtain larger and faster growing plants than higher levels of artemisinin.

The artemisinin content was highly variable in different countries, sometimes as low as 0.01% was reported [44]. In the United States, the mean artemisinin content in *A. annua* was reported ranging from 0.05% to 0.21%, but some individual plants could produce up to 0.42% at the full flowering stage [10]. Swiss researchers reported that a strain from China could contain up to 1.1 % artemisinin [7]. RAPD analyses of selected chemotypes from a decade old introduced population in India were carried out by Indian scientists [45]. They found that the wide phytochemical diversity existed among these plants; this diversity was largely from the genetic diversity. These results once again showed the high prospects of selecting and breeding of superior artemisinin containing lines by genetic means.

2 *Tissue Culture of* Artemisia annua *L.*

The establishment of *A. annua* tissue culture system is an indispensible step for improving *A. annua* by biotechnology. He et al. [46] carried out callus induction, regeneration of plantlets from *A. annua*, and the changes of artemisinin contents in early 1980s. The following are some related works.

2.1 Callus Induction and Plant Regeneration

Tender leaf and stem were used as explants to induce callus. The results showed that both of them could be induced to form callus and leaf was easier to form callus than stem. The callus-inducing rate of tender leaves was higher than that of old leaves, and old stems could hardly be induced into callus. Explants began to grow white or flaxen callus after culturing for 5–10 days and it reached vigorous growth after a period of 15–20 days.

Improved MS, Miller, N6, B5, and White medium were used to induce callus, respectively. The comparison results indicated that *A. annua* from Beijing and Hainan province could produce callus on the other four mediums except for White medium; the callus induced in

N6 medium was loose and grew faster than the other media. Improved B5, Miller, and MS medium are better for subculturing of the callus. Callus of *A. annua* from Hainan province grew faster under the condition of light than dark culture.

Calli subcultured for 3–4 times were transferred to improved MS medium containing BA, after being subcultured for several times under light condition; the calli were differentiated into plantlets. The regenerated plantlets had accessed to flowering period after transplanting outside for 140 days.

2.2 Determination of Artemisinin Content

Extracts of callus, callus with differentiated buds, and regenerated plants of *A. annua* were analyzed by thin-layer chromatography and chromogenic reaction, and identified by ultraviolet absorption spectrum after alkali treatment. The results indicated that the Rf value of callus with differentiated buds was in accordance with that of artemisinin standard. Then the chromatographic band of callus with differentiated buds was eluted for ultraviolet absorption detection, and the results showed that its ultraviolet absorption peak was the same as that of artemisinin standard, while callus of *A. annua* did not show chromogenic reaction. The results also showed that the biosynthesis of artemisinin was accompanied by the regeneration of buds in callus of *A. annua* from Hainan. The artemisinin content was 0.008% of the dry weight, while the artemisinin content of the flowering plant regenerated from callus reached 0.92% of the dry weight, which was much higher than that of the natural plant (0.56%). These results indicated that the biosynthesis of artemisinin was closely related to the highly differentiated organ in *A. annua*.

3 Hairy Root Cultures of Artemisia annua L.

Given that some secondary metabolites exist only in highly differentiated tissues of plant, the contents of these secondary metabolites are very low in vitro cultured cells. Hairy root cultures induced by *Agrobacterium rhizogenes* are in a highly differentiated state and have attracted great interests. Advantages of hairy root cultures over cell, tissue, and organ cultures for secondary metabolites production include high growth rate, genetic stability, and unlimited branch. Our laboratory conducted out hairy root cultures of *A. annua* for artemisinin production since the early 1990s. The following is a brief review of our results [47].

3.1 Hairy Root, Untransformed Root, and Callus Cultures of Artemisia annua

Transformed hairy root of *A. annua* line 025 was obtained when infected with *A. rhizogenes* ATCC15834. Hairy root, untransformed root, and callus of *A. annua* L. were cultured in the same medium to compare their biomass. The results showed that the biomass of hairy roots

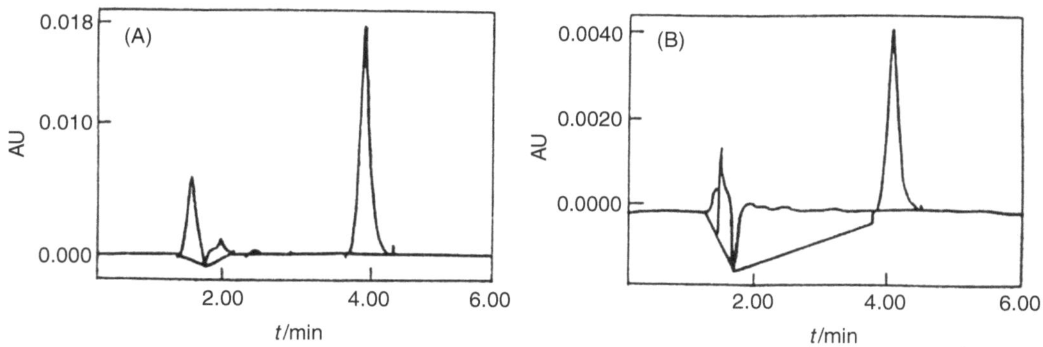

Figure 7.1: HPLC Chromatograms of Artemisinin Standard Solution and a Hairy Root Extract.
(A) Artemisinin standard; (B) hairy roots extract.

was much higher than those of untransformed roots and callus, while there was no significant difference between nontransformed callus and untransformed roots.

No artemisinin was detected in callus of *A. annua*, and a trace amount of artemisinin was detected in the nontransformed roots, while some relatively high content of artemisinin was detected in the hairy roots (Fig. 7.1).

3.2 Selection of Artemisinin High-Yielding Hairy Root Lines

A total of 747 hairy roots were clipped and cultured on carbenillin-containing medium to eliminate surplus agrobacteria, and 53 fast-growing hairy root lines were selected. After subculturing for 3 times, seven hairy root lines were selected. The biomass and artemisinin content exhibited significant differences among the seven selected hairy root lines (Fig. 7.2), and HR-9 line gave the highest artemisinin production, reaching 33.25 mg/month·L.

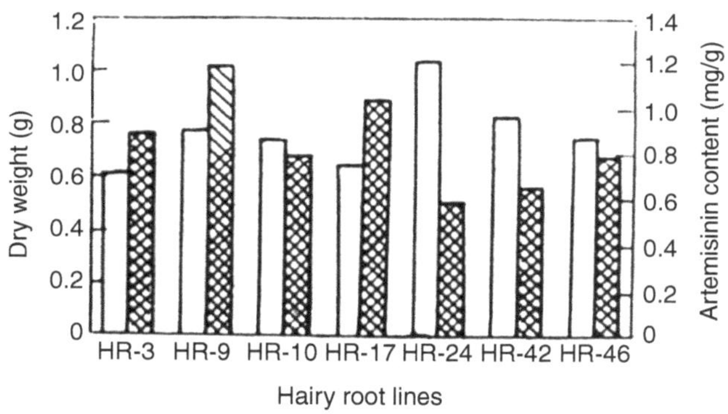

Figure 7.2: Comparison of Growth and Artemisinin Contents Among Different Hairy Root Lines.

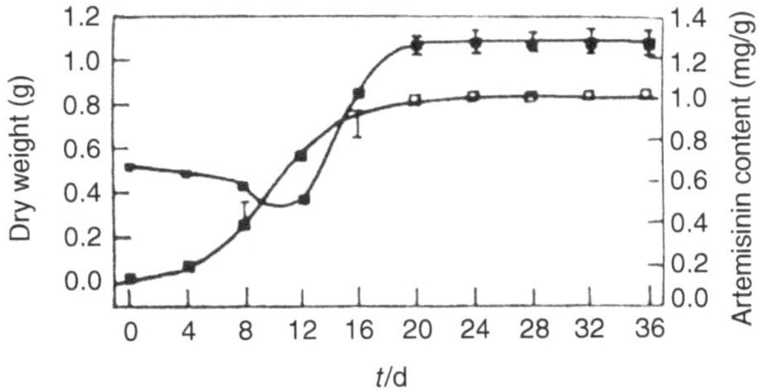

Figure 7.3: Curves of Growth and Artemisinin Contents of Hairy Roots in Liquid Medium.

3.3 *Dynamics of Growth and Artemisinin Biosynthesis of Hairy Root of* Artemisia annua

The dynamics of growth and artemisinin biosynthesis of HR-9 were measured to determine the suitable culture conditions. The characteristics of artemisinin content of hairy roots were evidently "related to growth." The artemisinin content decreased slowly during the exponential stage, increased while the growth rate slowed down, and kept constant after the growth stopped (Fig. 7.3). The highest artemisinin production reached 33.25 mg/L at 21 days, and the optimum culture interval for hairy roots of *A. annua* was 21 days.

4 *Induction of Shoot Clusters of* Artemisia annua *L.*

The biosynthesis of artemisinin was closely related to the differentiation of the tissues, and the artemisinin content was extremely low in calli, hairy roots, and suspension cells. It was reported that *A. annua* adventitious buds without roots could produce artemisinin and the content was stable, so shoot cultures for artemisinin production has attracted great attention. Our group had established an *A. annua* shoot clusters-inducing system [48], and the main physical and chemical factors affecting the induction rate of *A. annua* shoot clusters are reviewed as follows.

4.1 *Factors Influencing the Induction of* Artemisia annua *Shoot Clusters*

4.1.1 *Effects of medium components on the induction rate of shoot clusters*

MS medium was used as the basic medium in our study. Different combinations of hormones such as 6-BA, KT, ZT, 2,4-D, NAA, and IBA were used to induce shoot clusters with different explants such as leaf disks, leaves, stem segments, and leaf with stalks. The results showed that shoot clusters could be highly induced using MS medium supplemented with a combination of 2 mg/L 6-BA with NAA under three different concentrations (0.05, 0.1, and 0.15 mg/L), and the induction rate was 98.7, 97, and 96.8%, respectively. The effects of different concentrations of NAA (0.05, 0.1, and 0.15 mg/L) on the vitrification rate of

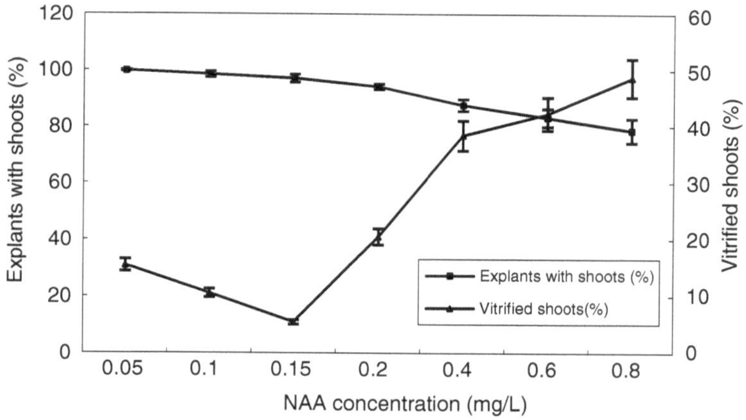

Figure 7.4: Effects of NAA Concentration in Combination With 2.0 mg/L 6-BA on Shoot Clusters Induction Ratio and Vitrified Shoots Ratio of *A. annua*.

shoot clusters were significantly different, and reaching 15.4, 10.5, and 5.4%, respectively (Fig. 7.4). The appropriate recombination of hormones for the induction of *A. annua* shoot clusters was 2.0 mg/L 6-BA and 0.15 mg/L NAA, and this medium was called induction medium. Most of the shoot clusters obtained by other combinations of hormones became vitrified, brown, and prone to callus formation, and the induction rate was also affected.

Different basic media had significantly different influences on the cell growth and the biosynthesis of secondary metabolites, which could be used to search for influential chemical factors on the biosynthesis of secondary metabolites of plants.

Six different basic media (MS, N6, DCR, B5, WS, and White) were used in our study to investigate the effects on the induction of shoot clusters. The results showed that MS, WS, and White media inhibited the induction of shoot clusters differentially. HPLC analysis showed that the highest yield of artemisinin was in the DCR medium, followed by the MS medium. Compared with the DCR medium, the content of K^+ and N source of the MS medium was 4–5 times higher than that of the DCR medium, while the content of Ca^{2+} of the DCR medium was higher by nearly one-third than that of the MS medium. These results indicated that K^+ and N source were favorable to the induction of shoot clusters, while Ca^{2+} could enhance the biosynthesis of artemisinin.

Agar is used not only as a common hardening agent in medium, but also as a regulator on the moisture of the culture vessel, hence it could affect the vitrification of shoot clusters. We found 0.8% agar was suitable both for the induction and the growth of shoot clusters, and the vitrification rate of shoot clusters was very low, about 95% of the shoot clusters were normal at this agar concentration. When the concentration of agar was too high, shoot clusters showed obvious water shortage; although the vitrification rate was low, the induced shoot clusters soon turned to yellow. When the concentration of agar was too low, the ratio of FW/DW was higher than 20, the induced shoot clusters showed obvious water excess, and vitrification was acute. In addition, artemisinin content decreased significantly at too high or too low agar concentrations, artemisinin content was the highest at 0.8% agar, and the ratio of FW/DW was about 10. These results suggested that the biosynthesis of artemisinin is closely related to the growing state of shoot clusters.

The leaf disk explants of *A. annua* were inoculated in the induction medium with different pH values, the induction rate was calculated, and artemisinin contents were determined by HPLC. The results indicated that the suitable pH value for the induction of shoot clusters was from 5.5 to 5.8, and that for the biosynthesis of artemisinin was 5.8.

4.1.2 Differences of shoot clusters induction rate in explants of different culture time and different genotypes

Leaf disks from different parts of 025 seedlings culturing for 1–8 weeks were used as explants to induce shoot clusters under the same culture condition. The results showed that culture time of explants had no obvious influence on the induction rate of shoot clusters, and the induction rates all reached more than 95% (Fig. 7.5).

Leaf disks from different *A. annua* genotypes (001, 014, 021, 025, and 032) of the same culture time were used as explants to induce shoot clusters under the same condition. The shoot cluster induction rate of 025 was the highest, about 98% (Fig. 7.6). This result indicates that different genotypes of the same species respond differently to the same exogenous hormones. Therefore, different plant genotypes should use different hormone combinations to obtain more shoot clusters in terms of shoot cluster induction rate.

4.1.3 Effects of light, temperature, and photoperiod on the induction of shoot clusters

The leaf disk explants of *A. annua* were cultured under light intensity of 0–6000 Lux to investigate the influence of light intensity on the induction rate of shoot clusters. The results

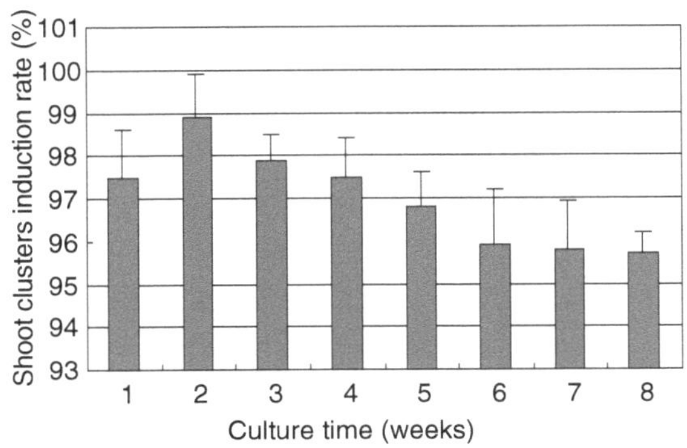

Figure 7.5: Effects of Explants Culture Time on Shoot Clusters Induction Rate of *A. annua*.

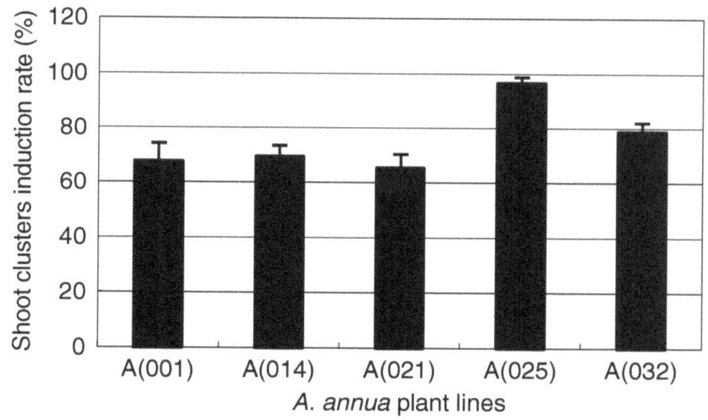

Figure 7.6: Comparison of Shoot Clusters Induction Rate in Different Genotypes of *A. annua*.

indicated that the induction rate of shoot clusters was not affected by light conditions. Even in complete darkness, the induction rate of shoot clusters could be as high as 80% of that of light conditions. Compared with the light-induced shoot clusters, the dark-induced shoot clusters grew faster, but appeared thin and weak, and each cluster contained only a small number of shoots.

The results under different temperatures (15–40°C) showed that too high or too low temperature would significantly affect the induction rate of shoot clusters [28]. The induction rate could reach more than 97% at the optimum temperature (25°C). A change of temperature with ±3°C had little effects on the induction rate of shoot clusters.

The germ-free seedlings cultured at the same temperature were treated under different photoperiods (8, 12, 16, 20, and 24 h) for 4 weeks before the induction of shoot clusters. Then the leaf disks were used for shoot clusters induction and cultured under the aforementioned five photoperiods (8, 12, 16, 20, and 24 h) to assess the induction effect of shoot clusters. The results showed that shoot clusters were nearly impossible to be induced under continuous illumination even if the concentration of cytokinin was increased to 4.0 or 8.0 mg/L. If the seedlings treated with different photoperiods were changed to 16 h photoperiod for 2 weeks, then the leaf disks could restore the ability to induce shoot clusters. These results showed that both the seedlings and the leaf disks could reach the highest induction rate of shoot clusters (97.6%) under 16 h photoperiod.

4.2 Regulation of Physicochemical Factors on the Growth of Shoot Clusters and Artemisinin Biosynthesis

4.2.1 Effect of constituents of culture medium on the growth of shoot clusters and artemisinin biosynthesis

4.2.1.1 Inorganic constituents

A single-factor trail was conducted with different concentrations of NH_4^+ and NO_3^- and different ratios of these two ions; the results showed that NH_4^+ and NO_3^- could regulate the growth of shoot clusters and artemisinin biosynthesis independently, and high concentrations of NH_4^+ and NO_3^- were not good for the accumulation of artemisinin. NO_3^- had greater influence on the growth of shoot clusters, and its concentration had obviously positive correlation with the growth rate (dry weight) of shoot clusters within a certain range of concentration. NH_4^+ and NO_3^- had coordinated regulating effect on the growth of shoot clusters and artemisinin biosynthesis. The ratio of NH_4^+/NO_3^- within 0.27–0.35 L^{-1} was beneficial to the growth of shoot clusters, and the ratio within 0.35–0.52 L^{-1} was conducive to the biosynthesis of artemisinin (Fig. 7.7). Results of different concentrations of Mg^{2+} showed that the concentration of Mg^{2+} within 0.750×10^{-3} to 6.004×10^{-3} M had little effect on the growth of shoot clusters, and the concentration of Mg^{2+} at 0.188×10^{-3} M was most conducive to the biosynthesis of artemisinin. However, the medium without Mg^{2+} severely inhibited the growth of shoot clusters and the biosynthesis of artemisinin (Fig. 7.8). When different concentration of Mn^{2+} was added to the medium, it had little impact on the growth of shoot clusters, while it significantly affected the biosynthesis of artemisinin. The detection of artemisinin content showed that artemisinin content reached its maximum when the concentration of Mn^{2+} was 1.5×10^{-4} M.

4.2.1.2 Other constituents

When MS medium with the addition of 2.0 mg/L 6-BA and 0.15 mg/L NAA (ie, shoot clusters-inducing medium) was used as shoot clusters subculture medium, the shoot clusters grew into plantlets because of the apical dominance after subculturing for three generations (Fig. 7.9). After comparing different hormone combinations, the most suitable culture medium for shoot clusters was determined as the MS medium with the addition of 4.0 mg/L 6-BA and 0.05 mg/L NAA.

Results of different concentrations of sucrose showed that 2–3% sucrose was the most favorable condition for the growth of shoot clusters, and the ratio of FW/DW was about

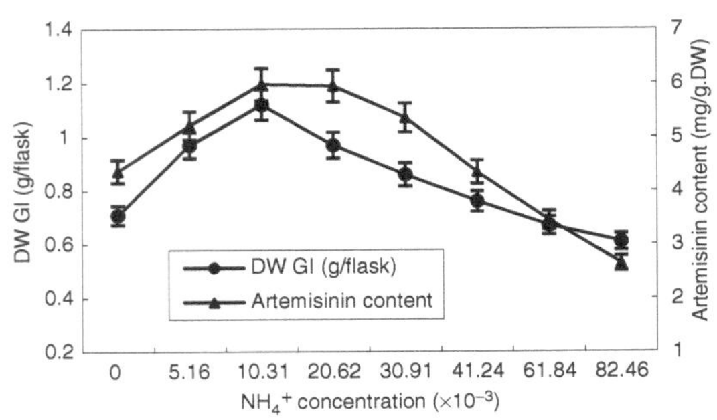

Figure 7.7: Effect of NH_4^+ Concentration on the Growth and Artemisinin Biosynthesis.

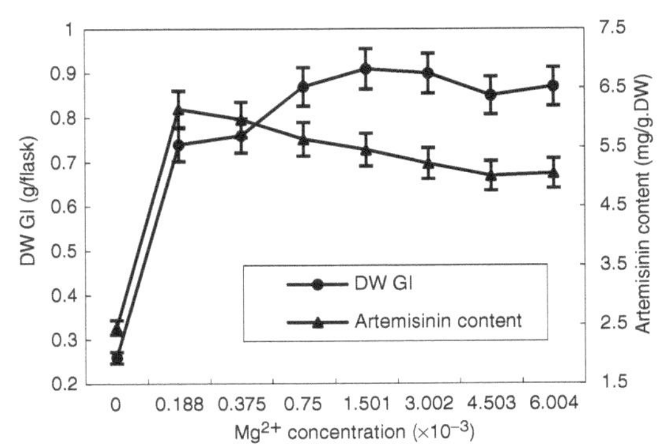

Figure 7.8: Effect of Mg^{2+} Concentration on the Growth and Artemisinin Biosynthesis.

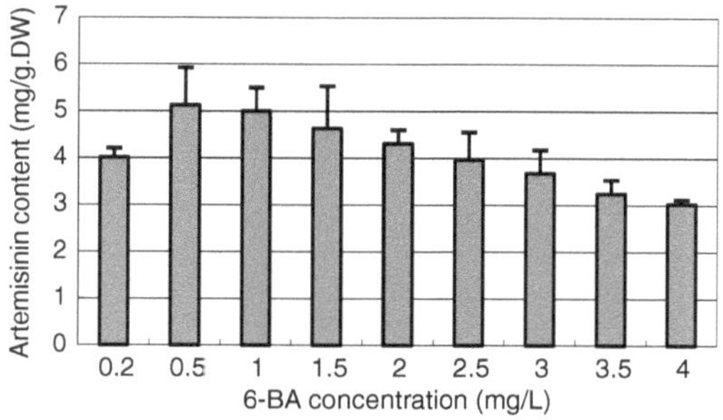

Figure 7.9: Effect of 6-BA Concentration in Subculture Medium on Artemisinin Content of *A. annua* Shoot Clusters.

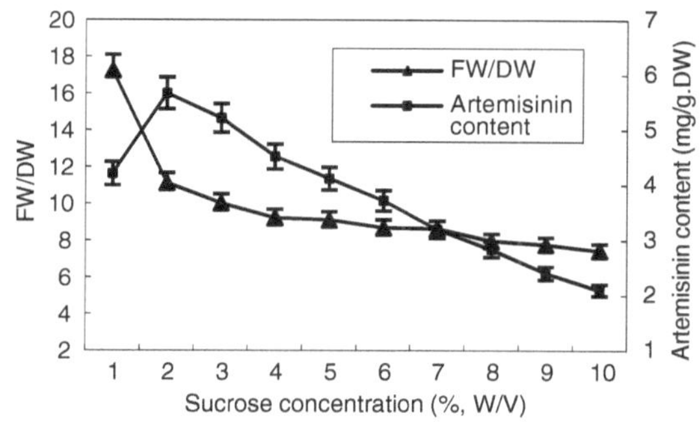

Figure 7.10: Effect of Sucrose Concentration on the Growth and Artemisinin Biosynthesis of *A. annua* Shoot Clusters.

10–11 (Fig. 7.10). The detection of artemisinin content showed that 2% sucrose was the most conducive one to the biosynthesis of artemisinin. Too high (>3%) or too low (<1%) concentration of sucrose might cause the vitrification of shoot clusters. The possible reason could be that sucrose in the medium not only provided carbon source, but also adjusted the osmotic potential between the medium and the plant tissues. So too low or too high concentration of sucrose in the medium would affect the osmotic potential of plant tissues and resulting in either excessive absorption or loss of water. Hence, the cultured shoot clusters showed growth characteristic of stress resistance, producing harmful substances, such as ethylene, and accelerating the vitrification of shoot clusters.

When the concentration of sucrose was fallen to 2% and the concentrations of other important nutrients in the medium, such as inositol, VB1, VB6, and NA, were adjusted appropriately, the leaves of shoot clusters became greener and grew faster, and artemisinin content increased by 5%. This indicated that these organic ingredients had influences on the growth of shoot clusters and artemisinin biosynthesis, and 10 mg/L GA_3 and 0.6 g/L casein hydrolysate in the medium were conducive to the growth of shoot clusters and artemisinin biosynthesis.

4.2.2 Effects of light, temperature, and growth cycle on the growth of shoot clusters and artemisinin biosynthesis

It could be seen from Fig. 7.11 that high light intensity promoted obviously the growth of shoot clusters and artemisinin biosynthesis. There was no significant difference in the growth of shoot clusters and artemisinin content under the light intensity of 5000 and 6000 Lux, suggesting that the growth of shoot clusters had certain limited demand for light intensity, weak light intensity inhibited the growth of shoot clusters and artemisinin biosynthesis, and the growth of shoot clusters and artemisinin biosynthesis were strongly suppressed under the condition of complete darkness.

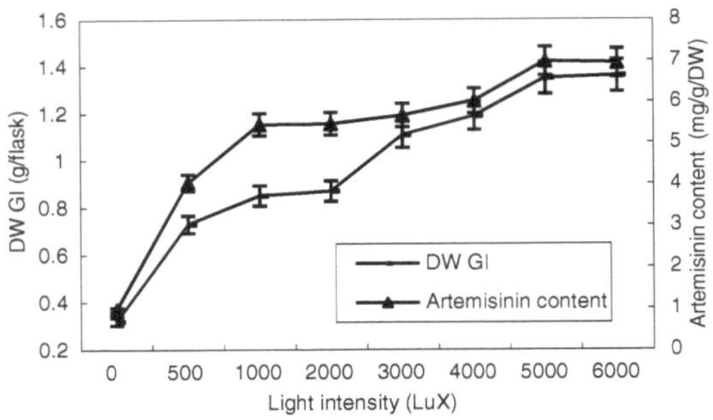

Figure 7.11: Effect of Light Intensity on the Growth and Artemisinin Biosynthesis of *A. annua* Shoot Clusters.

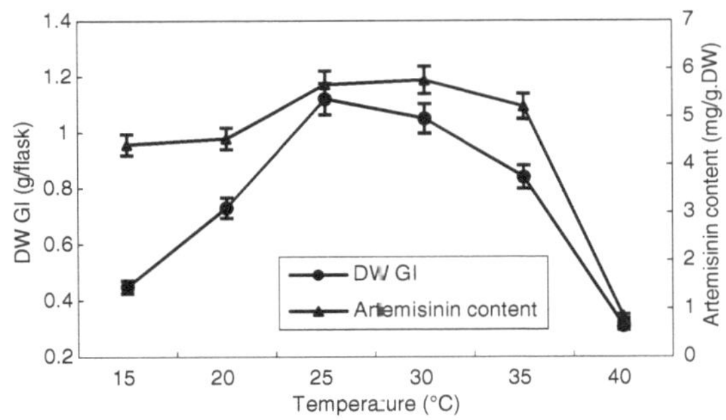

Figure 7.12: Effect of Temperature on the Growth and Artemisinin Biosynthesis of *A. annua* Shoot Clusters.

As shown in Fig. 7.12, the most suitable temperature for the growth of *A. annua* was 25°C and artemisinin content changed little between 25 and 30°C; artemisinin content was slightly higher when the temperature was 30°C. When cultured at 35 or 40°C, the growth of shoot clusters and artemisinin biosynthesis were significantly inhibited and shoot clusters became water-lost, withered, and yellow, even black to death, so high temperature had obvious lethal effect on shoot clusters. From 15 to 25°C, the culture temperature had stronger promotion effects on the growth of shoot clusters than that of artemisinin biosynthesis.

The growth curve of shoot clusters showed that cultivating for 20 days was the turning point of shoot clusters growth (Fig. 7.13). The growth of shoot clusters changed from increased rapidly to slowly, and then to more slowly after this point. Therefore, the subculture of shoot clusters was carried out at every 3 weeks, and the biomass of shoot clusters was about 60% of the total cultures at this time. Artemisinin content in the culturing shoot clusters varied with culture time, reaching its maximum of about 6.35 mg/L (DW) in the late growth phase, while its minimum of about 3.15 mg/L (DW) before the most vigorous growth phase (Fig. 7.14).

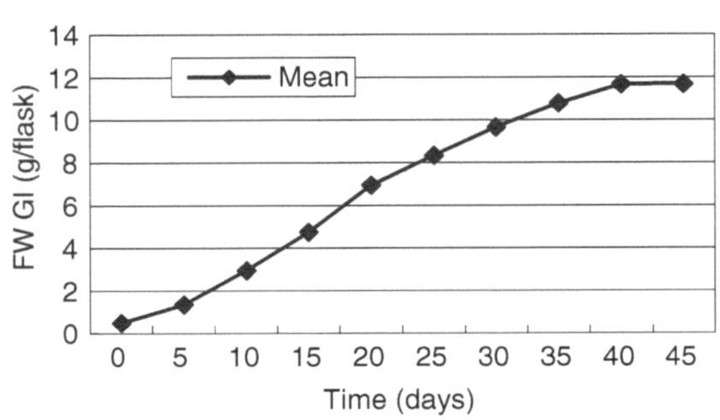

Figure 7.13: Growth Curve of Shoot Clusters During One Subculture Cycle in Solid Medium.

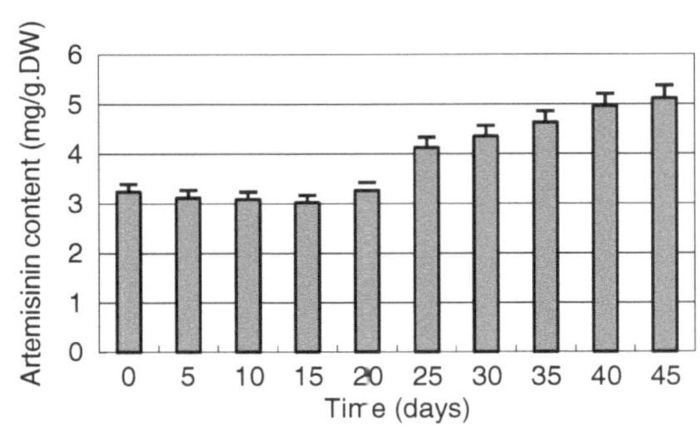

Figure 7.14: Changes of Artemisinin Content of Shoot Clusters During One Subculture Cycle.

These results indicated that artemisinin content of the shoot clusters was closely related to their growth stages.

5 Artemisinin Production in Bioreactors

The most successful example for the production of artemisinin in bioreactor was performed by the scientists from the Institute of Chemical Metallurgy, now the Institute of Process Engineering of the Chinese Academy of Sciences (CAS) and the Institute of Botany, CAS [49,50]. A novel mist bioreactor was modified to produce artemisinin by *A. annua* hairy roots and adventitious shoot cultures. The original ultrasonic mist bioreactor was improved to make full use of the bioreactor space, to reduce the loss of nutrient solution, and to prevent bacterial contamination. A novel internal ultrasonic mist bioreactor was used for the production of artemisinin by *A. annua* adventitious shoot and hairy root cultures. This bioreactor worked well, and the growth speed of adventitious shoots in the bioreactor was significantly higher than that in flask cultures. The growth speed of adventitious shoots in the bioreactor was 2.4 times of solid cultures and 2.1 times of cultures in flasks. The artemisinin content cultured in the bioreactor was 1.5 times of solid cultures and 1.8 times of cultures in flasks. The yield of artemisinin cultured in the bioreactor was 2.9 times of solid cultures and 3.2 times of cultures in flasks. These results showed that the mist bioreactor provided a suitable condition for the rapid growth and the biosynthesis of a large amount of artemisinin of *A. annua* adventitious shoots.

Nair et al. [51] found that the suspended cell cultures produced artemisinin, while the suspended callus cultures did not. Tawfiq et al. [52] did not detect artemisinin in the suspended cell and tissue cultures. Based on their experiments in culturing *Rauvolfia serpentina* adventitious shoots, Fulzele et al. [53] thought that bud cultures could be an ideal system for the production of secondary metabolites. They successfully performed bud cultures of *A. annua* in a bioreactor, but the content of artemisinin did not increase.

All the aforementioned researches and applications of biotechnology have provided strong theoretical basis for the industrialization of artemisinin. However, relatively low artemisinin contents in both the originally used *A. annua* plant and in vitro cultured *A. annua* via biotechnology constitute still the biggest obstacle for the cost reduction of artemisinin production.

6 Space Breeding of Artemisia annua *L.*

Space breeding is a technology of breeding new plant strains by sending plant seeds, tissues, organs, or living individuals to the space with recoverable spacecraft, such as satellites and aircrafts, to induce beneficial mutations through space mutagenic factors. Beginning at the middle of 1987, China has carried out dozens of breeding tests, including grain, cotton, rapeseed, fruit, medicinal plants, and other plant varieties, and has bred a group of new plant strains with high yield, high quality, early maturity, and multiresistance. Space mutagenesis has smaller damage, larger variation range, and more beneficial mutations compared with conventional radiation mutagenesis, and is an important means in germplasm development of medicinal plants. Space breeding of *A. annua* is a product of space high-tech combined with genetic breeding technology of Chinese medicinal plant *A. annua*. To obtain high-yield *A. annua* strain, the seeds of *A. annua* were sent to recoverable satellite Jianbing No. 1 in 1996, and high-yield strain SP-18 was screened out by asexual reproduction. The differences in growth characteristics, yield, and artemisinin content of high-yield strain SP-18 planted in the north region and the south region of China are reviewed in the following sections.

6.1 Screening of Space-High-Yield Strain

A total of 27 aseptic seedlings were germinated from *A. annua* seeds returned from Jianbing No. 1 satellite. The artemisinin content of the seedling No. 18 reached 1.26%, which is 0.5% higher than the average level of wild type No. 001 strain, and kept stable after subculturing for several times. The No. 18 strain was transplanted in the greenhouse after shoots cluster inducing, seedling strengthening, rooting, and seedling hardening. The shoots cluster induction rate of No. 18 strain was more than 90% and the average time of seedling strengthening and rooting was about 5 days less than that of No. 001 strain. Then this strain was named as SP-18 (space plant breeding No. 18).

6.2 Field Cultivation of Space-High-Yield Strain SP-18

SP-18 strain was cultivated in the greenhouse, at the suburb of Beijing and Youyang of Sichuan province, respectively. The plants all had long internode, stout stem, outstretched branches, and broad leaves. The plant of SP-18 strain looked tower-shaped; it was different from that of No. 001 strain which was irregular shaped, and the blade of SP-18 was yellow-green while No. 001 strain was green (Fig. 7.15). The branch number and plant height of SP-18 strain in the fast-growing stage and budding stage were once higher than that of No. 001 strain. The two strains planted in different regions and different planting time had largely the same average leaf water content. The planting time of *A. annua* grown in the north and the south had no obvious impacts on later growth characteristics, growth speed, and biomass.

6.3 Comparison of Artemisinin Content of SP-18 Planted in the South and in the North of China

When the planting density was 100 × 100 cm, the growth potential, plant height, branch number, and maximum diameter of *A. annua* were better than other planting density. *A. annua* was a heliophyte and it needed enough light in its growing process. *A. annua* plants grew well at the planting density of 100 × 100 cm for sharing sufficient sunlight and absorbing

Figure 7.15: SP-18 Strain and 001 Strain Planted in the Suburb of Beijing.

large range of soil nutrients. Especially for SP-18 strain, the plant is tower-shaped and has a large diameter; its leaves would turn yellow or dead at late growth due to little light stage if planted overcrowded. *A. annua* had highest biomass at the planting density of 60 × 60 cm; the biomass of SP-18 strain reached 3085.95 kg/ha (DW), about 1 time higher than that of No. 001 strain which was only 1675.8 kg/ha (DW). Taking the artemisinin content of SP-18 strain grew in Youyang of Sichuan province at fast-growing stage, for example, the artemisinin production of SP-18 strain reached 28.35 kg/ha, which was 20.25 kg/ha more than that of No. 001 strain, so it was regarded as a high-yield strain. When the planting density was 100 × 100 cm, the biomass of SP-18 strain was 2460 kg/ha (DW), while the artemisinin production was only 22.65 kg/ha. Thus, moderate dense planting could be beneficial to increasing the artemisinin production.

When planted in the suburb of Beijing, the biomass of SP-18 strain was 1 time higher than that of No. 001 strain, but their artemisinin contents manifested no significant difference. When planted in Youyang of Sichuan province, the artemisinin content of SP-18 strain was about 0.4% higher than that of No. 001 strain. The growth of *A. annua* was affected not only by climate conditions, such as temperature, humidity, illumination, and so on, but also by the soil conditions. Soil composition analysis showed that the N/P/K ratio in the soil of Youyang of Sichuan province was close to 4:1:4, while in the soil of Beijing suburb it was about 1:1:4. The N/P/K ratio in MS medium was close to 10:1:10, and the artemisinin content of SP-18 strain seedlings was 0.5% higher than that of No. 001 strain under the condition of tissue culture. Hence, the content of hydrolysable nitrogen in the soil played an important role in the biosynthesis of artemisinin, and low proportion of hydrolysable nitrogen might decrease the artemisinin content.

Table 7.1: Comparison of artemisinin content in different strains which were planted in different regions and harvest in different periods

Strains	Beijing Suburb (%)	Youyang of Sichuan Province (%)
SP-18	Flourishing growth period	
	0.65 (±0.2)	0.92 (±0.12)
	Initial bud period	
	0.53 (±0.14)	0.78 (±0.11)
001	Flourishing growth period	
	0.63 (±0.07)	0.49 (±0.08)
	Initial bud period	
	0.55 (±0.08)	0.43 (±0.01)

Artemisinin content varies significantly in the different development periods of *A. annua*. Artemisinin content is lower at the early vegetative stage and full-flowering stage, and higher at the fast-growing stage before flowering and budding stage. Flowering of *A. annua* is influenced by many factors, such as climate, soil, and fertilizer, so the harvest time has obvious influences on artemisinin production and artemisinin content. Table 7.1 shows that the artemisinin content of the two strains planted in the south and the north region both reached the highest in the fast-growing stage, so it is the best stage for harvesting *A. annua*. The artemisinin content of SP-18 strain in the fast-growing stage was about 0.13% higher than that in the budding stage, while the artemisinin content of No. 001 strain was about 0.07% higher in the former stage than that in the later stage.

After space mutagenesis, plants usually showed production effects, quality effects, morphological effects, and stimulating effects, and all these effects also occurred in the space breeding of SP-18 strain. In terms of production effects, the production of SP-18 strain was 1 time higher than that of 001 strain. In terms of quality effects, vegetative-propagated SP-18 strain had stable artemisinin content, which was 0.5% higher than that of 001 strain, and the artemisinin content of SP-18 strain planted in the field was 0.4% higher than that of 001strain at the fast-growing stage. There were also obvious variations on the morphology of SP-18 strain, such as long internode, stout stem, stretched branches, broad leaves, and so on. Owing to the induced mutation by space flight, the induction rate of shoot clusters increased and the time for seedling-strengthening and rooting shortened. Thus, SP-18 strain had the characteristics of rapid vegetative propagation, genetic stability, high artemisinin content, and high yield, which makes it a new high-yield strain suitable for planting in large scale.

Acknowledgments

The authors would like to thank Dr. Sa Geng, Long Zhang, Guoqin Cai, Mingbo Qin et al. for their contributions to this work.

References

[1] WHO. UNDP/World Bank/WHO Special Programme for Research and Training in Tropical Diseases (1994–1995), The role of artemisinin and derivatives in the current treatment of malaria. Report of an Informal Consultation convened by WHO in Geneva, 1993, pp. 27–9.
[2] Simon JE, Charles D, Cebert E, Grant L, Janick J, Whipkey A. *Artemisia annua* L.: a promising aromatic and medicinal. In: Janick J, Simon JE, editors. Advances in New Crops. Portland, OR: Timber Press; 1990. p. 522–6.
[3] Laughlin JC. Effect of agronomic practices on plant yield and antimalarial constituents of *Artemisia annua* L. Acta Hort. 1993;331:53–61.
[4] Laughlin JC. Agricultural production of artemisinin: a review. Trans R Soc Trop Med Hyg 1994;88(Suppl. 1): 21–2.
[5] Magalhaes PM. A experimentacao agricola com plantas medicinais e aromaticas. Atualidades Cientific\as 1994;3:31–56.
[6] Singh A, Kaul CC K, Mahajan VP, Singh A, Misra LN, Thakur RS, Husain A. Introduction of *Artemisia annua* in India and isolation of artemisinin, a promising antimalarial drug. Indian J Pharm Sci 1986;48:137–8.
[7] Delabays N, Benakis A, Collet G. Selection and breeding for high artemisinin (qinghaosu) yielding strains of *Artemisia annua*. Acta Hort 1993;330:203–7.
[8] Simon JE, Cebert E, *Artemisia annua*: a production guide. In: Simon JE, Clavio LZ, editor. Third National Herb Growing and Marketing Conference, 1988, pp. 78–83.
[9] Morales MR, Charles DJ, Simon JE. Seasonal accumulation of artemisinin in *Artemisia annua* L. Acta Hort 1993;334:416–20.
[10] Ferreira JFS, Simon JE, Janick J. Developmental studies of *Artemisia annua*: flowering and artemisinin production under greenhouse and field conditions. Plant Med 1995;61:167–70.
[11] Woerdenbag HJ, Pras N, Chan NG, Bang BT, Bos R, van Uden W, Van YP, Van Boi N, Batterman S, Lugt CB. Artemisinin, related sesquiterpenes, and essential oil in *Artemisia annua* during a vegetation period in Vietnam. Plant Med 1994;60:272–5.
[12] Chen HR, Chen M, Zhong FL, Chen FT, Huang JS, Zhang MX, Huang JR. Factors that influenced the effects of *Artemisia annua*. Bull Chin Mater Med 1986;11:9–11. (in Chinese)
[13] Ferreira JFS, Simon JE, Janick J. *Artemisia annua*: botany, horticulture, pharmacology. Hortic Rev 1997;19: 319–71.
[14] Srivastava NK, Sharma S. Influence of micronutrient imbalance on growth and artemisinin content in *Artemisia annua*. Indian J Pharm Sci 1990;52:225–7.
[15] Charles DJ, Simon JE, Shock CC, Feibert EBG, Smith RM. Effect of water stress and post-harvest handling on artemisinin content in the leaves of *Artemisia annua* L. In: Janick J, Simon JE, editors. New Crops. New York, NY: Wiley; 1993. p. 640–2.
[16] Shukla A, Farooqi-Abad AH, Shukla YN, Sharma S. Effect of triacontanol and chlormequat on growth, plant hormones and artemisinin yield in *Artemisia annua* L. Plant Grow Regul 1992;11:165–71.
[17] Ferreira JFS, Charles D, Simon JE, Janick J. Effect of drying methods on the recovery and yield of artemisinin form *Artemisia annua* L. HortScience 1992;27:650.
[18] Ferreira JFS, Janick J. Floral morphology of *Artemisia annua* with special reference to trichomes. Int J Plant Sci 1995;156:807–15.
[19] McVaugh R. Compositae. In: Anderson WR, editor. Flora Novo-Galiciana: a descriptive account of the vascular plants of Western Mexicog, vol. 12. Ann Arbor, MI, USA: University of Michigan Press; 1984.
[20] Peter-Blanc C. Developpement et biologie de la reproduction de l'*Artemisia annua* L. Travail de diplome, Univ. Switzerland: De Lausanne; 1992.
[21] North C. Plant Breeding and Genetics in Horticulture. London: Macmillan; 1979.
[22] Hu SL, Xu QC, Liu JF, Gu YX. The plant resource of *Artemisia annua*. Bull Chin Mater Med 1981;9:13–6 (in Chinese).
[23] Klayman DL. Qinghaosu (artemisinin): an antimalarial drug from China. Science 1985;228:1049–55.

[24] Acton N, Klayman DL. Artemisitene, a new sesquiterpene lactone endoperoxide form *Artemisia annua*. Plant. Med. 1985;51:445–6.
[25] Zhao SS, Zeng MY. Spektrometrische hochdruck-lussigkeitschromatographische (HPLC) undersuchungen zur analytic von qinghaosu. Plant Med 1985;51:233–7.
[26] Liersch R, Soicke H, Stehr C, Tuller HV. Formation of artemisinin in *Artemisia annua* during on vegetation period. Plant Med 1986;52:387–90.
[27] Martinez BC, Staba J. The production of artemisinin in *Artemisia annua* L. tissue cultures. Adv Cell Cult 1988;6:69–87.
[28] Singh A, Vishwakarma RA, Husain A. Evaluation of *Artemisia annua* strains for higher artemisinin production. Plant Med 1988;54:475–6.
[29] Madhusudanan KP. Mass spectral studies on artemisinin, dihydroartemisinin and arteether. Indian J Chem 1989;28B:751–4.
[30] Anti-malaria group of department of pharmaceutical chemistry, Sichuan Institute of Traditional Chinese Medicine. Chin Tradit Herbal Drugs 1979;1:5–12. (in Chinese)
[31] Chen FT, Zhang GH. Studies of several physiological factors on artemisinin synthesis in *Artemisia annua*. Plant Physiol Commun 1987;5:26–30. (in Chinese)
[32] Pras N, Visser JF, Batterman S, Woerdenbag HJ, Maingre TM, Lugt CB. Laboratory selection of *Artemisia annua* L. for high artemisinin yielding types. Phytochem Anal 1991;2:80–3.
[33] El-Sohly HN. A large scale extraction technique of artemisinin from *Artemisia annua*. J Nat Prod 1990;53:1560–4.
[34] Woerdenbag HJ, Pras N, Bos R, Visser JF, Hendriks H, Malingré TM. Analysis of artemisinin and related sesquiterpenoids from *Artemisia annua* L. by combined gas chromatography/mass spectrometry. Phytochem Anal 1991;2:215–9.
[35] Elhag HM, El-Domiaty MM, El-Feraly FS, Mossam JS, El-Olemi MM. Selection and micropropagation of high artemisinin producing clones of *Artemisia annua* L. Phytother Res 1992;6:20–4.
[36] Charles DJ, Cebert E, Simon JE. Characterization of the essential oil of *Artemisia annua* L. J Ess Oil Res 1991;3:33–9.
[37] Mehrotra S, Mehrotra BN, Aswal BS, Sharma HP. Leaf surface studies of some medicinal artemisias. Int J Crude Drug Res 1990;28:103–19.
[38] Duke SO, Paul RN. Development and fine structure of glandular trichomes of *Artemisia annua* L. Int J Plant Sci 1993;154:107–18.
[39] Duke MV, Paul RN, Elsohly HN, Sturtz G, Duke SO. Localization of artemisinin and artemisitene in foliar tissues of glanded and glandless biotypes of *Artemisia annua* L. Int J Plant Sci 1994;155:365–72.
[40] Weathers PJ, Cheetham RD, Follansbee E, Theoharides K. Artemisinin production by transformed roots of *Artemisia annua*. Biotechnol Lett 1994;16:1281–86.
[41] Liu CZ, Wang YC, Ye HC, Li GF. Production of artemisinin by hairy root cultures of *Artemisia annua* L. Biotechnol Lett 1997;19:927–9.
[42] Ferreira JFS, Janick J. Relationship of artemisinin content of tissue-cultured, greenhouse-grown, and field-grown plants of *Artemisia annua*. Plant Med 1995;61:351–5.
[43] Wallaart TE, Pras N, Quax W. Seasonal variations of artemisinin and its biosynthetic precursors in tetraploid *Artemisia annua* plants compared with the diploid wild-type. Plant Med 1999;65:723–8.
[44] Trigg PI. Qinghaosu (artemisinin) as an antimalarial drug. Econ Med Plant Res 1990;3:20–5.
[45] Sangwan RS, Sangwan NS, Jain DC, Kumar S, Ranade SA. RAPD profile based genitic characterization of chemotypic variants of *Artemisia annua* L. Biochem Mol Biol Int 1999;47:935–44.
[46] He XC, Zeng MY, Li GF, Liang Z. Callus induction and regeneration of plantlets from *Artemisia annua* and changes of qinghaosu contents. Acta Bot Sin 1983;25:87–90. (in Chinese)
[47] Liu BY, Ye HC, Li GF, Chen DH, Geng S, Zhang YP, Chen JL. Studies on dynamics of growth and biosynthesis of artemisinin in hairy roots of *Artemisia annua* L. Chin J Biotichnol 1999;14:249–54.
[48] Geng S, Ji SD, Yuan JY, Lu LD, Ye C, Li GF. Factors affecting induction of shoot cluster of *Artemisia annua*. Chin Tradit Herbal Drugs 2004;35:566–72.

[49] Liu CZ, Wang YC, Guo C, Ouyang F, Ye HC, Li GF. Production of artemisinin by shoot cultures of *Artemisia annua* L. in a modified inner-loop mist bioreactor. Plant Sci 1998;135:211–7.
[50] Liu CZ, Wang YC, Ouyang F, Ye HC, Li GF. Production of artemisinin by *Artemisia annua* hairy root culture in an internal loop airlift bioreactor. Acta Bot Sin 1999;41:181–3.
[51] Nair MSR, Acton N, Klayman DL, Kendrick K, Basile DV, Mante S. Production of artemisinin in tissue cultures of *Artemisia annua*. J Nat Prod 1986;49:504–7.
[52] Tawfiq NK, Anderson LA, Roberts MF, Phillipson JD, Bray DH, Warhurst DC. Antiplasmodial activity of *Artemisia annua* plant cell cultures. Plant Cell Rep 1989;8:425–8.
[53] Fulzele DP, Sipahimalani AT, Heble MR. Tissue cultures of *Artemisia annua*: organogenesis and artemisinin production. Phytother Res 1991;5:149–53.

CHAPTER 8

Metabolic Engineering of Artemisia annua *L.*

Chapter Outline

From *Artemisia annua* L. to Artemisinins
http://dx.doi.org/10.1016/B978-0-12-811655-5.00008-8

In this chapter, we first give a review on the recent progress on artemisinin biosynthetic pathway and its molecular regulation in the literature, and then report our recent works in the field in our laboratory, including. The establishment of genetic transformation system of *A. annua*. The isolation and function identification of the genes involved in artemisinin biosynthesis. The overexpression of the genes involved in artemisinin biosynthesis (such as *ADS, FPS*) and their effects on artemisinin biosynthesis. The inhibition of the genes involved in the competing pathways of artemisinin biosynthesis (such as *SQS, CPS*) and their effects on artemisinin biosynthesis. The effects of *ipt* gene expression on the physiological and biochemical characteristics of *A. annua*; and, finally, the effects of the flower-promoting genes (*FPF1* and *CO*) on *A. annua* flowering time and the link between flowering and artemisinin biosynthesis.

1 *Progress in the Molecular Regulation of Artemisinin Biosynthesis*

Artemisinin is an effective antimalarial monomer first isolated from *A. annua* by Chinese scientists. It is a new sesquiterpene lactone with an endoperoxide bridge; its needle-like crystals are colorless and odorless with melting point 156–157°C, with the molecular formula $C_{15}H_{22}O_5$, and molecular weight of 282.1 [1]. Artemisinin is a new structure type of drug which has a completely different mode of action unlike earlier antimalarial drugs, and it is also the fastest onset, the best efficacy, and the least toxic compound of all the antimalarial drugs, particularly on cerebral malaria and chloroquine-resistant malaria [2]. As artemisinin is effective to chloroquine-resistant *Plasmodium falciparum* malaria, it has been treasured by various parties since its discovery, and recommended as the most effective antimalarial drug by the World Health Organization (WHO).

A. annua L. belongs to genus *Artemisia* of the Compositae family and is an annual herb. *A. annua* is native in China; it is now widely distributed throughout the world. Currently, medicinal artemisinin is mainly extracted directly from the leaves of *A. annua* plant. Considering the complex, time-consuming, and laborious extraction process, as well as large differences in quality of *A. annua* from different sources, artemisinin comes at high cost of production, and it is difficult to meet the market demand [3]. Artemisinin can be chemically synthesized, but the high cost, toxicity and low yields make the commercial production unsatisfactory [4]. Rather, scientists around the world turn to studying the regulation of artemisinin biosynthesis and exploring ways to increase the yield of artemisinin. In recent years, with the rapid development of molecular biology techniques and the accumulation of knowledge on artemisinin biosynthesis, some of the key genes of artemisinin biosynthetic pathway have been cloned, with which high-yield artemisinin production by genetic engineering has become the new hotspot. In the following, we will give a short review on the latest progresses in this field.

1.1 The Biosynthetic Pathway of Artemisinin

1.1.1 From acetyl-CoA to farnesyl diphosphate (FPP)

Artemisinin is a new sesquiterpene lactone; the biosynthetic pathway of artemisinin belongs to plant isoprenoid pathway. Recent studies have shown that at least two pathways exist for plant isoprenoid biosynthesis: the mevalonic acid (MVA) pathway and 1-deoxy-D-xylulose-5-phosphate (DXP) pathway. Artemisinin and other sesquiterpenes are synthesized through the MVA pathway which occurs in the cytosol. First, three molecules of acetyl-CoA generate 3-hydroxy-3-methylglutaryl-CoA (HMG-CoA) by intermolecular condensation under the catalysis of acetyl-CoA thiolase and HMG-CoA synthase [5]. Next, HMG-CoA generates MVA by the reduction of 3-hydroxy-3-methylglutary-CoA reductase (HMGR). Then, through the catalysis of MVA kinase, MVA is phosphorylated to generate MVA-5-phosphate which is later phosphorylated by MVA-5-phosphate kinase to generate MVA-5-pyrophosphate, and by the decarboxylation of MVA-5-pyrophosphate decarboxylase, MVA-5-pyrophosphate is converted into isopentenyl diphosphate (IPP). During these processes, the formation of MVA is irreversible, so HMGR is considered as the first rate-limiting enzyme of MVA pathway [6–8]. Finally, through the isomerization of isopentenyl diphosphate isomerase, IPP becomes DMAPP, and these two activated isoprene units form geranyl diphosphate (GPP) by head-to-tail condensation according to electrophilic reaction mechanism catalyzed by farnesyl diphosphate synthase (FPS). Later the product GPP reacts with another IPP unit and generates fanesyl diphosphate (FPP). FPP is the common precursor of various sesquiterpenes, and different sesquiterpene synthases catalyze the formation of different sesquiterpenes. The schematic diagram from acetyl-CoA to FPP is shown in Fig. 8.1. Schramek et al. [9] reported that the artemisinin biosynthetic precursor IPP is from both the MVA pathway and the DXP pathway, and GPP is likely to be synthesized in plastids and transported to cytoplasm, forming FPP with another IPP unit.

1.1.2 From farnesyl diphosphate to artemisinin

In recent years, with the isolation of enzyme genes involved in artemisinin biosynthesis and the separation of some important intermediates in artemisinin biosynthetic pathway, the pathway from FPP to dihydroartemisinic acid has been elucidated.

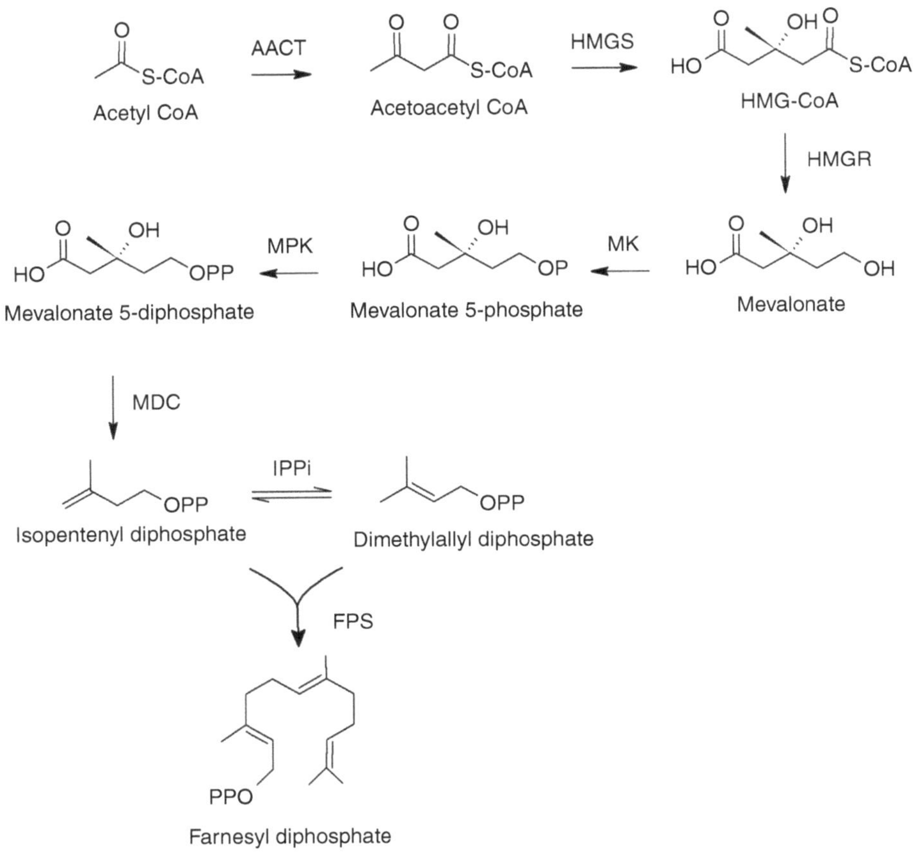

Figure 8.1: Mevalonic Acid (MVA) Pathway.
AACT, Acetoacetyl-coenzyme A thiolase; *FPS*, farnesyl diphosphate synthase; *HMGR*, 3-hydroxy-3-methylglutaryl-CoA reductase; *HMGS*, 3-hydroxy-3-methylglutaryl-CoA synthase; *IPPi*, isopentenyl diphosphate isomerase; *MDC*, mevalonate 5-diphosphate decarboxylase; *MK*, mevalonate kinase; *MPK*, mevalonate 5-phosphate kinase.

Akhila et al. [10] studied the biosynthetic pathway of artemisinin by radioisotope tracer method; they proposed the following process: farnesyl diphosphate (FPP) → artemisinic acid → dihydroartemisinic acid → artemisinin. In this process, FPP formed an unknown sesquiterpenoids intermediate after enzymatic reaction, and this step is considered to be an important rate-limiting step in the formation of artemisinin. The work of isolation and identification of sesquiterpene intermediate from *A. annua* have been carried out for many years, and have made important progresses in the past 2 decades. In 1999, Bouwmeester et al. [11] isolated amorpha-4,11-diene, an important sesquiterpenoids intermediate of artemisinin biosynthetic pathway, from *A. annua* leaves for the first time, and further isolated the enzyme which can catalyze the formation of amorpha-4,11-diene, amorpha-4,11-diene synthase, the enzyme catalyzes cyclization of FPP to amorpha-4,11-diene and also is one of the key enzymes of artemisinin biosynthesis. In 1999, Wallaart et al. [12,13] isolated dihydroartemisinic acid (0.17%, DW) and dihydroartemisinic acid hydroperoxide (0.14%, DW) from *A. annua* for the first time. In addition, dihydroartemisinic acid was converted to artemisinin in mimic in vivo photochemical conditions; the intermediate is dihydroartemisinic acid hydroperoxide. The separation of dihydroartemisinic acid and dihydroartemisinic acid hydroperoxide and their in vitro transformation provide a powerful evidence for the nonenzymatic photochemical reaction from dihydroartemisinic acid to artemisinin in *A. annua*.

In 2005, Bertea et al. [14] revealed the presence of a number of compounds, such as artemisinic alcohol, dihydroartemisinic alcohol, artemisinic aldehyde, and dihydroartemisinic aldehyde for the first time by analyzing the terpenoids in *A. annua* leaves and glandular trichomes. In addition, they also demonstrated the presence of some biosynthetic enzymes, such as amorpha-4,11-diene synthase, artemisinic alcohol dehydrogenase, and dihydroartemisinic aldehyde dehydrogenase in the glandular trichomes and leaves. From these results, they speculated that artemisinin biosynthetic pathway includes amorpha-4,11-diene hydroxylation to artemisinic alcohol, followed by oxidation to artemisinic aldehyde, reduction of the C11—C13 double bond to dihydroartemisinic aldehyde and oxidation to dihydroartemisinic acid.

In 2006, Keat et al. [15] isolated the gene of amorpha-4,11-diene monooxygenase (*CYP71AV1*) for the first time, specifically expressed in glandular trichomes. It encodes 488 amino acids, catalyzes amorpha-4,11-diene to form artemisinic aldehyde, artemisinic aldehyde, and artemisinic acid.

In 2009, Teoh et al. [16] isolated an aldehyde dehydrogenase (*ALDH1*) cDNA from *A. annua*; the cDNA encoded a 499 amino acid protein with a molecular weight of 53.8 kDa. Expression of the cDNA in *Escherichia coli* and characterization of the purified recombinant enzyme revealed that *ALDH1* catalyses the NAD(P)-dependent oxidation of both artemisinic aldehyde and dihydroartemisinic aldehyde into the corresponding artemisinic acid and dihydroartemisinic acid.

To sum up, the downstream of artemisinin biosynthetic pathway (from FPP to artemisinin) can be summarized in Fig. 8.2.

1.2 Related Enzymes of Artemisinin Biosynthesis

The enzymes related to artemisinin biosynthesis mainly include 3-hydroxy-3-methylglutaryl CoA reductase (HMGR), farnesyl diphosphate synthase (FPS), sesquiterpene synthase (cyclase), amorphadiene hydroxylase (cytochrome P450), artemisinic aldehyde Δ11(13) double bond reductase (DBR2), aldehyde dehydrogenase (ALDH1), and alcohol dehydrogenase (ADH1); they will be reviewed separately as follows.

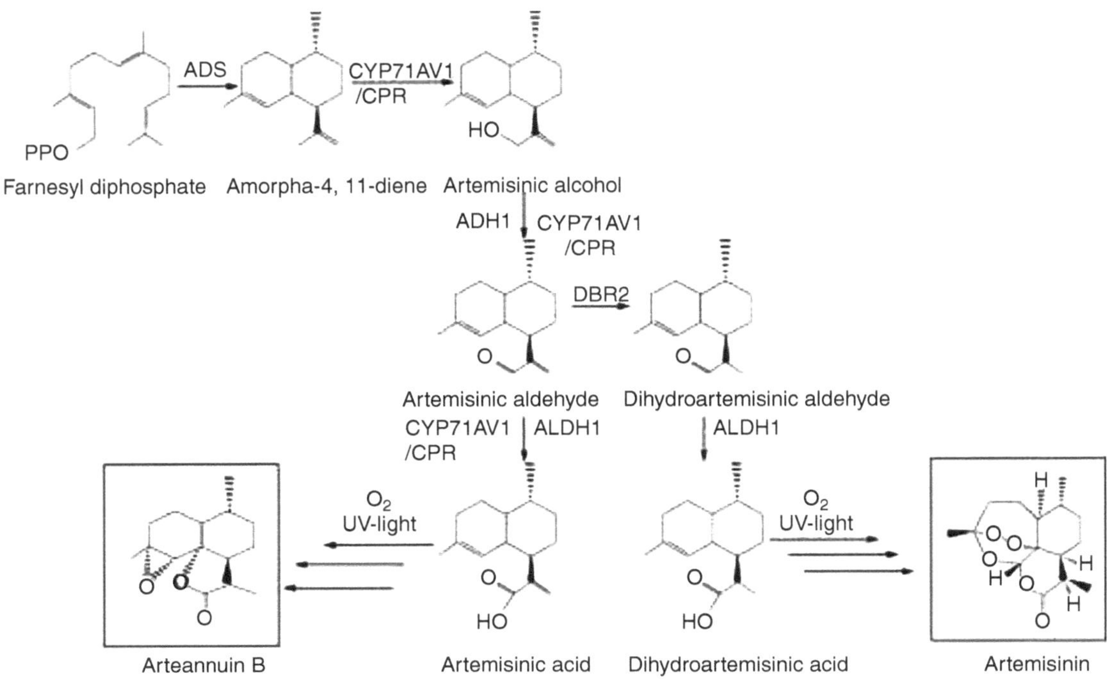

Figure 8.2: Biosynthetic Pathway From FPP to Artemisinin.

1.2.1 3-Hydroxy-3-methylglutaryl CoA reductase (HMGR)

HMGR catalyzes HMG-CoA to form mevalonic acid (MVA). As the formation of MVA is an irreversible process, HMGR is regarded as a rate-limiting enzyme of isoprenoid pathway in animals, plants, fungi, and even insects [17]. Many studies have shown that the expression level of *HMGR* gene (including transcription and translation level) or enzyme activity is positively associated with the content of isoprenoids product [8,18,19]. Chappell and Nable [20] reported that tobacco suspension cell cultures treated with fungal elicitor could lead to the accumulation of sesquiterpenoids (capsidiol) in medium; meanwhile, instantaneous peak of HMGR was also detected. In 1995, Chappell et al. [17] reported that the HMGR activity of the transgenic tobacco, by the introduction of HMGR gene of hamster into tobacco under the control CaMV35S promoter, increased 3–6 times, and the accumulation of total steroids increased 3–10 times. However, the content of sterolal end products, such as sitosterol, campesterol, and stigmasterol only increased 2 times, but the content of sterolal intermediate increased more than 100 times. These results further indicated that the total content of sterols is controlled by HMGR activity, and one or more enzymes in the late stages of sterols biosynthesis are involved in the regulation of the relative content of sterolal end products.

Further researches showed that there exist a gene family associated with HMGR [8,18]; in this family, the expression of different homologous genes may control the flow of "carbon flow" of mevalonate pathway in the cytoplasm. Results from some model plants (such as potato) showed that HMGR is a key enzyme with "macro" regulation role in mevalonate pathway, the branches of this pathway are respectively regulated by specific members of *HMGR* gene family, and the expression of the members of HMGR II subfamily is closely related to the synthesis of sesquiterpenoids phytoalexin [18]. HMGR, as an early stage enzyme of mevalonate pathway, decides the distribution of "carbon flow," but the synthesis of end products is controlled by the key enzymes in each branch, such as squalene synthase, sesquiterpene synthase, and so on.

HMGR genes of *A. annua* have been cloned by Kang et al. (GenBank accession No. U14624, U14625) and Chen et al. of our laboratory (GenBank accession No. AF 142473), but function analysis and genetic transformation of *A. annua HMGR* are still not available.

1.2.2 Farnesyl diphosphate synthase (FPS)

Farnesyl diphosphate synthase is a kind of 1,4-isoprene transferase; it catalyzes the condensation of IPP and DMAPP to form GPP, and also the condensation of GPP and IPP to form FPP. Its reaction mechanism is: DMAPP (GPP) loses a small amount of electrons to form positive carbon ions (namely the ion pairs of C1 and pyrophosphoric acid group) due to their head diphosphate groups. And C4 of IPP due to the existence of the conjugated double bond forms a carbon atom with rich electrons, under the action of farnesyl diphosphate synthase, and electrophilic C1 of DAMPP (GPP) attacks C4 of IPP, which causes electrophilic polymerization reaction. In recent years, a variety of isoprene transferase cDNA of plants have been cloned, including *Arabidopsis thaliana* [21], *Lupinus albus* [22], *Parthenium hysterophorus* [23], *A. annua* [24], *Oriza sativa* [25], *Gossypium hirsutum* [26], *Lycopersicon esculentum* [27], and *Zea mays* (GenBank accession No. L39789).

FPP locates in the multibranch site of isoprenoid pathway; it is generally believed that the synthesis of FPP is strictly regulated, so it is possible that FPS plays a role in the regulation of isoprenoid biosynthetic pathways. It was shown that after being treated with elicitor, the accumulation of sesquiterpenoids phytoalexin in pepper fruit [28] and cotton suspension cell cultures [29] were positively correlated with RNA level, protein expression, and enzyme activity of FPS.

In 1996, Matsuchita et al. [24] cloned an *FPS* cDNA from *A. annua*. The sequence analysis showed that the cDNA encoded a protein of 343 amino acid residues with a calculated molecular weight of 39.42 kDa. The amino acid sequence of FPS showed 76, 84, and 72% homology with those from plants, *A. thaliana*, *L. albus*, and *Z. mays*, respectively. The FPS protein also showed 46 and 45% homology with those of rat and human, respectively. The two conservative domains found in all the polyisoprene transferases also existed in *A. annua* FPS [30,31]. After expressed in *E. coli*, FPS activity could be detected in vitro [24]. In 2003, Hemmerlin et al. [32] cloned two *FPS* genes, *FDS-1* and *FDS-2*, from *Artemisia tridentata*; the homology of these two cDNAs was 83%, and the homology of *A. tridentata FDS-1* and *A. annua FPS* was 97%. Their results suggested that the *FDS-1* was likely to be involved in biosynthesis of sesquiterpenoids phytoalexin.

In 1998, two cDNAs of farnesyl diphosphate synthase (*FPS1* and *FPS2*) were cloned from *A. annua* in our laboratory (Chen et al., GenBank accession No. AF136602 and AF112881). The cDNA sequences of *FPS1* and *FPS2* showed the homology of more than 98% to that of *FPS* cloned by Matsushita et al. [24]. The results of function analysis showed that the FPS1 has farnesyl diphosphate synthase activity [33]. In addition, the effects of overexpression endogenous *FPS1* gene on the regulation of artemisinin biosynthesis were studied [34], and the results showed that FPS enzyme activity of transgenic *A. annua* was 2–3 times higher than that of the untransgenic control, and artemisinin content of the transgenic *A. annua* was about 30% higher than that of the control, indicating that *FPS* gene has certain regulatory effect on artemisinin biosynthesis.

1.2.3 Sesquiterpene synthase (cyclase)

In the enzymes involved in artemisinin biosynthetic pathway, the most studied is sesquiterpene synthase (cyclase). Sesquiterpene synthase (cyclase) catalyzes the intramolecular cyclization and various oxidative and reductive modifications of farnesyl diphosphate, and finally forms sesquiterpenoids. In addition to the artemisinin, *A. annua* also contains other sesquiterpenoids, such as germacrene D, β-farnesene, β-caryophyllene, α-copaene, artemisinic acid, and so on. Besides, there also exist a series of sesquiterpene synthases (sesquiterpene synthase gene family) in *A. annua*. Since the high demand of sesquiterpene synthase for the stereospecificity of substrate, different stereo structures of sesquiterpenoids are catalyzed by different sesquiterpene synthases.

A number of sesquiterpene synthases have been cloned from *A. annua*, such as *epi*-cedrol synthase [35,36], amorpha-4,11-diene synthase [37–39], β-caryophyllene synthase [40], (*E*)-β-farnesene synthase [41], germacrene A synthase [42], aside from a new unknown function sesquiterpene synthase (sesquiterpene synthase, GenBank accession No. AF304444) [43]. In the aforementioned sesquiterpene synthases, only amorpha-4,11-diene, catalyzed by amorpha-4,11-diene synthase from FPP acting as a precursor of artemisinin, is directly involved in the biosynthesis of artemisinin; the other sesquiterpene synthases compete the common substrate FPP with amorpha-4,11-diene synthase. In 2003, Liu et al. [44] of our group cloned squalene synthase from *A. annua*; it catalyzes two molecules of farnesyl diphosphate to generate squalene, and competes the same substrate FPP with amorpha-4,11-diene synthase.

1.2.3.1 Separation and purification of amorpha-4,11-diene synthase

In 1999, Bouwmeester et al. [11] separated amorpha-4,11-diene synthase from *A. annua* for the first time; the enzyme catalyzed FPP to form amorpha-4,11-diene which is an intermediate sesquiterpenoid of artemisinin biosynthesis. Partial purified enzyme had the typical characteristics of sesquiterpene synthase, such as a wide pH range (the optimal pH value is 6.5–7.0), the molecular weight was 56 kDa, and K_m was 0.6 μM.

1.2.3.2 Gene cloning, function analysis, and heterologous expression of amorpha-4,11-diene synthase

From 1999 to 2000, Mercke et al. [37], Chang et al. [38], and Wallaart et al. [39] reported successively works on gene cloning, expression, and function analysis of amorpha-4,11-diene synthase (ADS) of *A. annua*. Full-length cDNA of amorpha-4,11-diene synthase of *A. annua* was about 2100 bp with a 1641 bp open reading frame coding for 546 amino acids; the calculated molecular weight of the coding protein was 63.9 kDa, slightly higher than that of the amorpha-4,11-diene synthase purified from *A. annua* (56 kDa). Speculated pI value was about 5.6. pI value and molecular weight of *A. annua* amorpha-4,11-diene synthase were close to that of the sesquiterpene synthases reported in many other plants [45–48]. After expressed in *E. coli*, amorpha-4,11-diene synthase could catalyze FPP to form amorpha-4,11-diene. The enzyme was located in cytoplasm, due to the lack of plastid targeting sequence, which is consistent with the biosynthesis of sesquiterpenoids carried out in cytoplasm [49,50].

Li et al. [51] of our group also cloned an amorpha-4,11-diene synthase gene from artemisinin high-yielding *A. annua* line 001. The results of function analysis showed that the enzyme could cyclize FPP to form amorpha-4,11-diene. As amorpha-4,11-diene synthase is considered as a key enzyme of artemisinin biosynthesis, amorpha-4,11-diene synthase was overexpressed in *A. annua* to investigate its role in artemisinin biosynthesis [52]. Their results showed that the highest artemisinin content of the transgenic *A. annua* was about 41% higher than that of nontransgenic plants, and the content of artemisinic acid and dihydroartemisinic acid in transgenic plants also increased 47 and 79%, respectively. Their results demonstrate the regulatory role of ADS in artemisinin biosynthesis.

Amorpha-4,11-diene synthase gene of *A. annua* was also transferred into tobacco by Wallaart et al. [39] (tobacco does not contain endogenous amorpha-4,11-diene synthase). Their results showed that the enzyme activity of amorpha-4,11-diene synthase could be detected in transgenic tobacco, and the level of amorpha-4,11-diene in transgenic tobacco leaves was 0.2–1.7 ng/g FW.

1.2.4 Amorpha-4,11-diene hydroxylase (cytochrome P450 monooxygenase)

Cytochrome P450 is an ancient gene superfamily widely distributed in various organisms; its protein structures, catalyzed substrates, and reaction types are polymorphic [53]. Cytochrome P450 plays an important role in the biosynthesis of plant secondary metabolites that are mainly associated with plant resistance to insect and stress. The molecular weight of P450 protein is about 45–62 kDa. There are many P450 proteins in plants; according to the homology of amino acid sequences, amino acid sequences are classified as the same family if their homology is greater than 40%, and classified as the same subfamily if their homology is greater than 55%. In 2005, Bertea et al. [14] reported the presence of a number of compounds in *A. annua*, such as artemisinic alcohol, artemisinic aldehyde, dihydroartemisinic alcohol, and dihydroartemisinic aldehyde, and also demonstrated the existence of some enzyme activities, such as amorpha-4,11-diene synthase, artemisinic alcohol dehydrogenase, and dihydroartemisinic aldehyde dehydrogenase in the glandular trichomes and leaves of *A. annua*. In 2006, Keat et al. [15] cloned the gene of amorpha-4,11-diene monooxygenase (*CYP71AV1*) for the first time. The gene encoded a protein of 488 amino acids, and catalyzed amorpha-4,11-diene to form artemisinic aldehyde, and then artemisinic aldehyde and artemisinic acid. Expression analysis showed that *CYP71AV1* was expressed specifically in the glandular trichomes. The cloning and function identification of amorpha-4,11-diene hydroxylase has a great significance for the elucidation of artemisinin biosynthetic pathway. At the same time, Ro et al. [54] also cloned the *CYP71AV1* gene from *A. annua*. By introducing *CYP71AV1* and other genes into engineering yeast, they successfully made engineering yeast to heterologously synthesize the precursor of artemisinin–artemisinic acid, and the content of artemisinic acid in engineering yeast reached 100 mg/L.

1.2.5 Artemisinic aldehyde Δ11(13) double bond reductase (DBR2)

In 2008, Zhang et al. [55] found that the extract of *A. annua* buds had artemisinic aldehyde C11—C13 double bond reductive activity. They purified aldehyde Δ11(13) reductase from *A. annua* buds, and successfully isolated aldehyde Δ11(13) reductase encoding gene—*DBR2*—from *A. annua* secretory glandular trichomes cDNA library based on the sequence results of enzyme protein. The coding region of *DBR2* cDNA was 1331 bp, encoding a 415 amino acid protein, and the relative molecular weight of DBR2 was 45.6 kDa.

1.2.6 Aldehyde dehydrogenase (ALDH1)

In 2009, Teoh et al. [16] cloned an aldehyde dehydrogenase (*ALDH1*) gene from *A. annua*; the cDNA encoded a 499 amino acid protein with a molecular weight of 53.8 kDa. *ALDH1* was only expressed in the aerial part of *A. annua*, and the highest expression was in the secretory glandular trichomes, but no expression in roots. The expression pattern of *ALDH1* was similar to the other artemisinin biosynthetic enzyme genes, suggesting that it might be a gene involved in artemisinin biosynthesis. Prokaryotic expression and in vitro functional identification found that ALDH1 had the functions of oxidizing both artemisinic aldehyde and dihydroartemisinic aldehyde into the corresponding artemisinic acid and dihydroartemisinic acid.

1.2.7 Alcohol dehydrogenase (ADH1)

In *A. annua*, the glandular trichomes are thought to be the site of biosynthesis and storage of artemisinin. In the course of investigating the biosynthesis of artemisinin in *A. annua* glandular trichomes, a gene encoding a putative alcohol dehydrogenase (ADH1) was examined [56]. Sequence analysis and in vitro characterization revealed that ADH1 is an NAD-dependent alcohol dehydrogenase of the medium chain dehydrogenase/reductase superfamily, with specificity toward artemisinic alcohol. This specificity and the evidence for strong glandular trichome expression indicate a role for ADH1 in the formation of artemisinic aldehyde in the artemisinin pathway of *A. annua*.

1.2.8 Dihydroartemisinic aldehyde reductase (Red1)

In 2010, Rydén et al. [57] cloned a dihydroartemisinic aldehyde reductase gene *Red1* from *A. annua* flowers. Red1 is a kind of oxidoreductases which can use a variety of substrates, such as ketones, menthol, and neomenthol. In addition, Red1 also has a higher affinity to dihydroartemisinic aldehydes and monoterpenes. In the artemisinin biosynthetic pathway, Red1 might play a negative role by reducing dihydroartemisinic aldehyde to dihydroartemisinic alcohol.

1.2.9 Cinnamyl alcohol dehydrogenase (CAD)

According to the EST sequence of the alcohol dehydrogenase from *A. annua* secretory glandular trichomes cDNA library, Li et al. [58] of our laboratory isolated the full-length cDNA of *A. annua* cinnamyl alcohol dehydrogenase gene *AaCAD* by RACE strategy (GenBank accession No. EU417964) in 2012. The cDNA of *AaCAD* was 1364 bp, ORF encoded a 361 amino acid protein (ACB54931), and the predicted theoretical pI was 5.54. The results of in vitro functional identification showed that AaCAD had strong bona fide CAD activity. When cinnamic aldehyde, sinapyl aldehyde, and coniferyl aldehyde were used as substrates, AaCAD could effectively reduce them to their corresponding alcohols with the NADPH as the cofactor at pH 7.5. Similarly, the reverse reaction could also occur efficiently at pH 9.5 with $NADP^+$ as the cofactor. Further analyses showed that AaCAD could also reduce artemisinin biosynthetic intermediate artemisinic aldehyde to artemisinic alcohol at pH 7.5, and at the same time, its reverse reaction could also occur albeit less efficiently under alkaline conditions (pH 9.5), but its catalytic activity was much lower than the former.

1.3 Gene Regulation of Artemisinin Biosynthesis

The regulation of artemisinin biosynthesis refers to the regulation of artemisinin biosynthesis in vitro cultures, the regulation of artemisinin biosynthesis in the whole plant, and the gene regulation of artemisinin biosynthesis. The gene regulation of artemisinin biosynthesis has been the focus in recent years. Since 1980s, scientists began to explore the ways of artemisinin industrialization by plant cell and tissue culture technology. Various attempts were made by culturing *A. annua* callus, adventitious buds, and hairy roots, but all failed to significantly improve the artemisinin content of the cultures. With the rise of genetic engineering of plant secondary metabolism, two ideal vectors of plant genetic engineering (Ri plasmid from the *Agrobacterium rhizogenes* and Ti plasmid from *Agrobacterium tumefaciens*) were applied to the regulation of artemisinin biosynthesis, and then *A. annua* hairy roots in vitro culture system mediated by Ri plasmid and *A. annua* adventitious buds in vitro culture system mediated by Ti plasmid were established, which provide the foundation for further studies on the regulation of artemisinin biosynthesis by genetic engineering.

Our laboratory has focused on the molecular regulation of artemisinin by the genetic transformation since 1980s, including the regulation of artemisinin biosynthesis in *A. annua* hairy roots by cotton *cadinene synthase* (*CAD*) gene [59]; the regulation of artemisinin biosynthesis in *A. annua* hairy roots and plants by cotton *FPS* gene [60,61]; the effects of *A. tumefaciens* isopentenyl transferase (*ipt*) gene expression on the physiological and chemical characteristics of *A. annua* [62]; the effects of *FPF1*, *CO* gene on *A. annua* flowering time and on the linkage between flowering and artemisinin biosynthesis [63,64]; the regulation of artemisinin biosynthesis by overexpressing *A. annua* endogenous *FPS* gene [34]; the regulation of artemisinin biosynthesis by overexpressing *A. annua amorpha-4,11-diene synthase* (*ADS*) gene [52]; effect of antisense *squalene synthase* (*SQS*) gene expression on the increase of artemisinin content in *A. annua* [65]; and artemisinin biosynthesis enhancement by downregulation of the *β-caryophyllene synthase* (*CPS*) gene [65]. More details of the aforementioned works will be described in later sections.

2 Establishment of Genetic Transformation System of Artemisia annua

2.1 Ri Plasmid-Mediated Genetic Transformation of Artemisia annua

Root-inducing plasmid (Ri plasmid) is a double-stranded circular DNA; it is physically separated from chromosomal DNA of *A. rhizogenes* and can replicate independently. The size of Ri plasmid varies from 180 to 250 kb. If the injured plant tissues are infected by *A. rhizogenes*, the T-DNA of Ri plasmid will be transformed into plant cells and integrated into their genome. Therefore, the Ri plasmid is an ideal vector for plant genetic engineering.

As Ri plasmid can infect plants on the wound and cause them to produce hairy roots, the Ri plasmid-transformed hairy root cultures have been used to produce high level of secondary metabolites since the 1980s. Compared with the large-scale plant cell culture, the cultivation of the hair root has many advantages. First, the hairy root phenotype is characterized by fast growth, lateral branching, and lack of geotropism; Second, its secondary metabolites synthesis and genetic background are relatively stable because hairy root itself is under the organ differentiation level. Third, there are no chimeras in the hairy root initiated from single cell, which is advantageous to screen high-yield hairy root lines. In some plants, the transformants transformed by Ri plasmid are easier to regenerate.

Next, we introduce the main factors influencing Ri plasmid-mediated genetic transformation of *A. annua*.

2.1.1 Effects of different Agrobacterium rhizogenes on the morphology and induction rate of hairy roots

A. rhizogenes strains ATCC15834, R1000, and A4 were used to infect *A. annua*, and pBI121 was used as the plant expression vector. The leaves of artemisinin high-yielding line 001were used as explants. The results indicated that the induction rate of hairy roots varied significantly when infected with different *A. rhizogenes* strains. The highest induction rate (71.8%) was ATCC15834, followed by A4 and the lowest was R1000. In addition, the morphology of *A. annua* hairy roots induced by different *A. rhizogenes* strains also varied significantly. The ATCC15834-induced hairy roots were very stout, multibranched, and with extremely thick hairs; they were markedly different from normal roots and exhibited typical characteristics of hairy roots in morphology. As such hairy roots grow fast and their morphological characteristics remain stable between generations, the corresponding hairy roots can be easily selected. Most of the hairy roots induced by A4 and R1000 are similar to normal roots, and they often have fewer hairs and grow slowly.

2.1.2 Effects of Artemisia annua genotype on the induction rate and morphological characteristics of hairy roots

The Sichuan artemisinin high-yielding lines 001 and 025 were used as receptor plants and were induced by *A. rhizogenes* ATCC15834. The results showed that the induction rate varied with different *A. annua* genotypes. The hairy root induction rate of 001 was about twice that of 025. The hairy roots derived from 001 were stockier than that of 025 and with more dense branches. Hence, both the type of *A. rhizogenes* and the genotype of *A. annua* affected the induction rate. The induction rate of hairy roots was determined not only by the pathogenecity of *A. rhizogenes*, but also by the control of plant genes.

2.1.3 Effects of seedling age and explants types on the induction rate of hairy roots

The germ-free seedlings of *A. annua* line 001 were cultured for 5–50 days and were induced by *A. rhizogenes* ATCC15834; no obvious differences were observed in the induction rate (60.02–67.5%) during 5–20 days subculture period, while cultured after 30 days, the induction rate dropped significantly (43.9–55.3%). Given that the biomass of the seedlings was higher after subculture for 15–20 days, the seedlings subcultured for 15–20 days were chosen for plant genetic transformation.

In addition, different explants, such as intact leaves, leaf disks, stem segments, and main stems were used to compare the induction rate. The results showed that the induction rate of hairy roots was different among different explants. The highest induction rates were obtained in intact leaves and leaf disks which were 57.1 and 57.4%, respectively. The induction rate of stem segments was 5.1%, which was the lowest of all the tested explants. The induction rate of the main stems used as explants could reach 80%; however, the main stems were limited by the source of explants. The morphology of hairy roots induced from intact leaves and that from leaf disks were significantly different, intact leaf-induced hairy roots were stout and grew fast, while the growth potential of hairy roots induced from leaf disks was weak. Considering the induction rate, the morphological characteristics of hairy roots and the source of materials, the intact leaves could act as suitable explants for hairy roots induction.

2.1.4 Effect of preculture period of Artemisia annua, culture period, and concentration of Agrobacterium rhizogenes on the induction rate of hairy roots

Leaf explants were used to investigate the effects of preculture on the induction rate of hairy roots. The results showed that preculture had no effect on the induction rate of hairy roots. Whether the leaf explants were precultured or not, the induction rate was about 70%. However, the time of starting hairy roots was reduced from 12–14 days to 7 days if the explants were precultured for 2 days.

The culture period of *A. rhizogenes* had a significant impact on the induction rate of hairy roots, and the highest induction rate (94.2%) was observed when the OD_{660nm} was 0.75. When the concentration (OD_{660nm} = 0.75) of *A. rhizogenes* ATCC15834 was diluted to 5–20 times, little effects on the induction rate of hairy roots were observed, while too high or too low dilution could indeed decrease the induction rate of hairy roots.

2.1.5 Effect of phenols on the induction rate of hairy roots

Phenolic compounds, such as acetosyringone, potato extracts, dinitrophenol, and dichlorophenol were added to the *A. rhizogenes* culture medium to investigate the effects on the induction rate of hairy roots. The hairy root induction rates after the addition of the aforementioned phenolic compounds were 86.4, 90.5, 87.1, and 86.3% respectively; the results suggested that these phenolic compounds were not suitable for the hairy root induction of *A. annua*.

2.2 Ti Plasmid-Mediated Genetic Transformation of **Artemisia annua**

The frequency of exogenous gene transferred to plant cells and the frequency of plant regeneration in vitro culture conditions are two critical limiting factors to the establishment of an efficient plant genetic transformation system. For the plant for which genetic transformation is difficult, scientists are, on the one hand, seeking to establish an efficient tissue culture system to increase the frequency of plant regeneration, and, on the other hand, devoting to improve the transgenic methods and invent new ones. So far, *Agrobacterium*-mediated transformation is one of the most widely used and most efficient method in plant genetic engineering. The transferring ability of exogenous gene from *Agrobacterium* to plant cells and efficient bud differentiation frequency of plant tissues are two key factors to affect the establishment of *Agrobacterium*-mediated genetic transformation system. The ability of explants to accept exogenous genes is related to both the pathogenicity of *Agrobacterium* on specific plant materials and the competent state of plant tissue itself. In 2005, Han et al. [66] of our laboratory studied the factors influencing Ti plasmid-mediated genetic transformation of *A. annua* to optimize the transformation system; their main results are as follows.

2.2.1 Influence of Agrobacterium tumefaciens strain and genotype of Artemisia annua on transformation efficiency

A. annua seeds of artemisinin high-yielding lines, 001 and NJ, were collected from Sichuan Province and Hunan Province, China, respectively. The *A. tumefaciens* strains LBA4404, EHA105, harboring the binary vector pBIGFP, were used for genetic transformation. The results indicated that *A. tumefaciens* strains and plant genotypes played some very important roles in the transformation efficiency. EHA105 was superior to LBA4404, and 001 was superior to NJ when infected by *A. tumefaciens* EHA105. Moreover, both the content of artemisinin and the shoot clusters induction rate of 001 were higher than those of NJ. Hence, 001 was used as the transgenic receptors.

Of leaves, stem segments and root segments used as explants, leaves were the optimum explants for *A. annua* transformation because the shoot induction rate of leaves was the highest. In addition, leaves were more sensitive to be infected by *A. tumefaciens*. When the cotyledons or hypocotyls from 8-day-old seedlings were used as explants, the transformation efficiency was very low, and when the seeds were used as explants, the transformed seeds were merely able to germinate, unable to root on the selected medium, and transgenic plants could not be obtained.

2.2.2 Influence of preculture period on transformation efficiency

The preculture process was not necessary for the transformation of *A. annua*, and on the contrary, longtime preculture could decrease the induction rate of Kan-resistant shoots. The possible reason could be that shoot induction happened at early stage, hence the cells in wounded leaf stalks could miss the optimum period for shoot induction after being precultured for a long period. After 4-day preculturing and 3-day coculture, some leaves began to root at the petiole on selection medium.

2.2.3 Influence of the composition of infection medium on transformation efficiency

A. tumefaciens grew better in LB medium than in MS medium. However, the results indicated that LB medium was not suitable for the transformation of *A. annua*. The reason might be that some components in LB medium, such as tryptone, yeast extract, were likely to inhibit shoot induction of *A. annua*.

2.2.4 Influence of cocultivation on transformation efficiency

Coculture is one of the most important steps for genetic transformation of plants, and T-DNA is incorporated into plant genomic DNA during this process. In our study, after liquid cocultivation, the rate of Kan-resistant shoots was 24.4% for the infected leaves, while the rate was 32.4% for the control. As the leaves of the control were not infected by *Agrobacterium*, but only cultured in liquid MS medium for 2 days, we thought many false positive Kan-resistant shoots were produced after liquid cocultivation. Another interesting phenomenon is that the sensitivity of shoot induction to Kan was decreased after the treatment of liquid MS medium.

It was reported that the transformation efficiency could be improved by adding hormones to coculture medium because hormones could accelerate cell division of explants and maintain cell activity, consequently favoring the cell growth after transformation. However, in our study, the addition of hormones to coculture medium resulted in false positive Kan-resistant shoots, the possible reason could be that the shoot induction of *A. annua* happened at early stage and could start in a short period. Therefore, the addition of hormones to coculture medium might make the shoot induction occur and the shoots continue developing even on the selection medium.

It is reported that acetosyringone (AS), a phenolic compound produced by wounded plant cells, could induce the transcription of the virulence genes of *Agrobacterium*, so AS has been used to improve transformation efficiency for many plants, especially for those which failed to produce AS by themselves, such as most of the monocotyledons and some dicotyledons. However, it was not so effective to the transformation of *A. annua* when 10 mg/L AS was added to the medium (Table 8.1) in our case.

Coculture period has some great influence on the transformation efficiency. The *Agrobacterium* which only survived in the wound of plant over 16 h could induce tumor formation. Therefore, too short coculture period is not favorable for plant transformation. However, too long coculture period could result in overgrowth of *Agrobacterium*, and is harmful to plant cells. For the transformation of *A. annua*, the optimum coculture period is 2–3 days (Table 8.2).

Table 8.1: Effects of AS on the Kan-resistant shoot induction frequency of 001

Cocultivation Medium	+AS	−AS
001	19.5% (n = 113)	17.5% (n = 137)

Table 8.2: Effects of cocultivation periods on the Kan-resistant shoot induction efficiency of 001

Cocultivation Period (day)	1	2	3	4
001	0% (*n* = 66)	12.8% (*n* = 78)	19.4% (*n* = 67)	2.8% (*n* = 71)

Based on the aforementioned results, the optimized conditions for genetic transformation of *A. annua* are the leaves with stalk as the explants with the coculture period of 2–3 days in MS solid medium.

3 *Genetic Transformation of* Artemisia annua *by Heterologous Related Genes and Their Influences on Artemisinin Biosynthesis*

3.1 *Hairy Root Cultures and Analysis of Artemisinin Content of* Artemisia annua *Transformed by Cotton* Cadinene Synthase *(CAD) Gene*

The biosyntheses of artemisinin and cotton toxin all belong to isoprenoid metabolic pathway; artemisinic acid, the main predecessor of artemisinin, has a typical cadinene skeleton. It was reported that cadinene and its analogues were presented in extracts taken from *A. annua* plants. Hence, artemisinin and cotton toxin might share similar biosynthetic pathway.

To further increase the contents of artemisinin and its precursors of *A. annua*, Chen et al. [59] by our *A. rhizogenes*-mediated transformation system, the cDNA encoding cotton cadinene synthase (provided by Prof. Xiaoya Chen, Shanghai Institutes for Biological Sciences, the Chinese Academy of Sciences) driven by CaMV 35S promoter was transferred into *A. annua* by Ri plasmid-mediated transformation. Hairy root was successfully obtained from leaves infected by *A. rhizogenes* strain ATCC15834. The hairy roots cultured on MS with carbenicillin showed robust main root and more branches with thick root hair (Fig. 8.3).

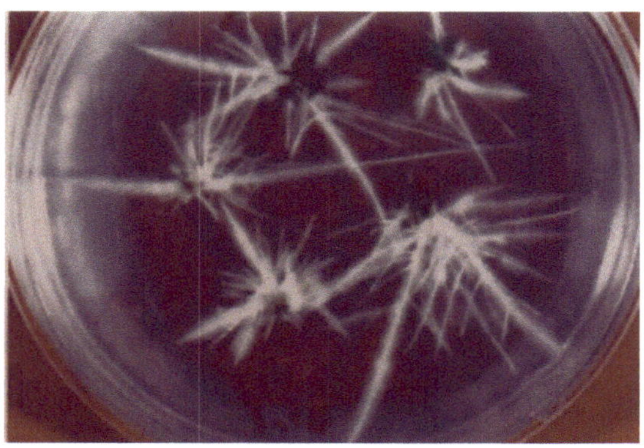

Figure 8.3: Hairy Roots Induced by *A. rhizogenes* 15834 Containing the *CAD* Gene.

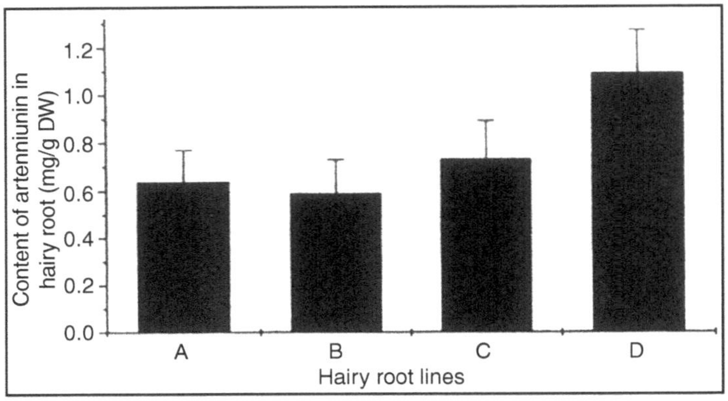

Figure 8.4: The Content of Artemisinin Detected by HPLC in the Transgenic Hairy Root Compared With That of the Nontransgenic Control.

PCR and Southern blotting analysis showed that exogenous gene was integrated into the genome of *A. annua*. RT-PCR analysis indicated that the exogenous gene was expressed at transcriptional level. The artemisinin content of C-37 line with the expressed exogenous gene at transcriptional level was detected by HPLC, and the artemisinin content of the transgenic hairy root was shown increased at a certain degree compared with the nontransgenic control (Fig. 8.4).

3.2 Culturing of Cotton Farnesyl Diphosphate Synthase *(FPS) Gene Transformed Hairy Roots and Artemisinin Content Analysis*

Farnesyl diphosphate (FPP) is located in multiple branch sites of the isoprenoid metabolic pathway; it is generally believed that the biosynthesis of FPP is under strict regulation. Therefore, FPS may play a regulatory role in isoprenoid metabolic pathway. FPP is an important intermediate product of artemisinin biosynthesis. It is known that FPP forms the first cyclic compound amorpha-1,4-diene by the cyclization of amorpha-4,11-diene synthase (ADS), so the yield of artemisinin is affected directly by the content of FPP in *A. annua*. To increase the biosynthesis of artemisinin, the cDNA encoding a cotton farnesyl diphosphate synthase (FPS) was inserted into the plant expression vector under a CaMV 35S promoter, and then transferred into *A. annua* via *A. rhizogenes* strain ATCC15834 in our laboratory [60]. The results of PCR and Southern blotting indicated that the exogenous gene was integrated into the genome of *A. annua*. Northern blotting analysis showed that the exogenous gene was expressed at the transcriptional level at least in three transgenetic hairy roots lines. Growth yield analysis of *A. annua* showed that the growth yield of the transgenic hairy roots lines was significantly higher than that of the control (Fig. 8.5). HPLC detection of artemisinin content showed that artemisinin content of the transgenetic hairy roots was about 2.0–3.0 mg/g (DW), an increase of 3–4 times more than that of the control (Fig. 8.6).

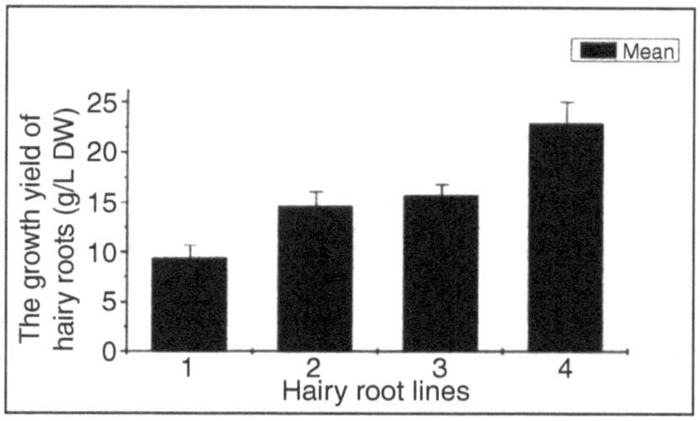

Figure 8.5: The Growth Yield of the Transgenic Hairy Roots Lines and the Control.
1, Control hairy roots; 2, hairy roots of F-l line; 3, hairy roots of F-24 line; 4, hairy roots of F-26 line.

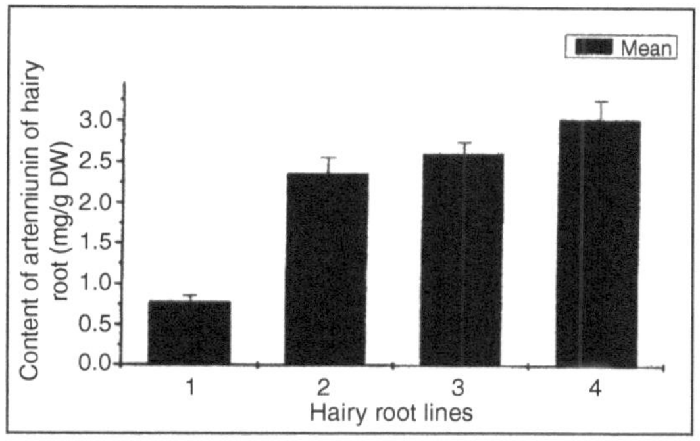

Figure 8.6: Contents of Artemisinin in Hairy Roots Detected by HPLC.
1, Control hairy roots; 2, hairy roots of F-l line; 3, hairy roots of F-24 line; 4, hairy roots of F-26 line.

3.3 Ti-Mediated Transformation of Artemisia annua *With a Recombinant* Farnesyl Diphosphate Synthase *(FPS) Gene and Artemisinin Content Analysis*

Chen et al. [61] of our laboratory has successfully transferred the cDNA encoding cotton *FPS* into *A. annua* by *A. tumefaciens* LBA4404-mediated transformation. The content of artemisinin in the transgenetic plants was about 10 mg/g DW, an increase of 2–3 times more than that of the control (Fig. 8.7).

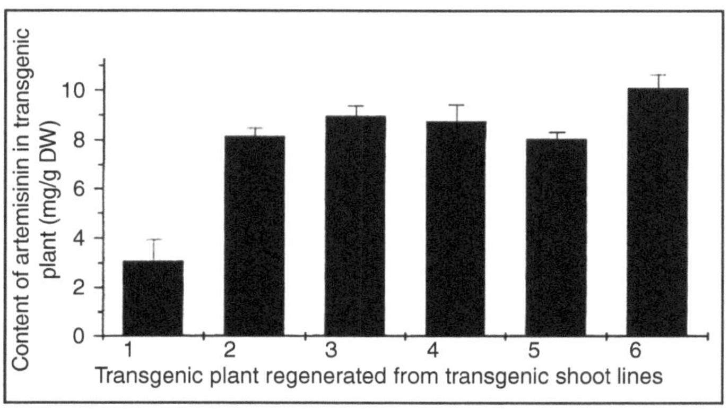

Figure 8.7: Contents of Artemisinin in *A. annua* Transgenic Plants Detected by HPLC.
1, Control, untransformed plant; 2, transgenic plant F-1 line; 3, transgenic plant F-4 line; 4, transgenic plant F-61 line; 5, transgenic plant F-62 line; 6, transgenicplant F-73 line.

4 Cloning and Function Analysis of the Genes Involved in Artemisinin Biosynthesis

4.1 Cloning and Enzymatic Analysis of Farnesyl Diphosphate Synthase (FPS) Gene From Artemisia annua

In the proceeding section, we introduced our studies on the genetic transformation of *A. annua* with heterologous *FPS* gene. Although the content of artemisinin in *FPS* transgenetic plants was considerably higher than that of the wild-type control, the extent of increase did not match our expectations. To investigate the regulations of endogenous *FPS* gene on artemisinin biosynthesis, Zhao et al. [33] of our laboratory cloned endogenous *FPS* gene from the cDNA library of *A. annua*, constructed plant expression vectors, and carried out genetic transformation of *A. annua*. The following is a review of their work.

4.1.1 Cloning and sequence analyzing of AaFPS1 cDNA

Two DNA fragments, 850 and 800 bp, were cloned using subsection cloning strategy from a high-yield *A. annua* strain 025 cDNA library. Sequencing analysis showed that the two fragments were partly overlapped *AaFPS* genes. Then the whole *AaFPS1* cDNA was amplified using a pair of specific primers designed according to the 3′ and 5′ ends.

Sequencing of the *AaFPS1* cDNA revealed a 1333 bp fragment. The nucleotide sequence presented 96% identity with a reported native *AaFPS* cDNA [24], encoding a protein of 343 aa, with molecular weight of 39 kDa (Fig. 8.8). The protein sequence of AaFPS1 presented 93, 45, and 98% identity with those from *Helianthus annuus*, human, and the reported native *A. annua*, respectively. The amino acid sequence of AaFPS1 contained also five distinctly conserved domains among all known FPSs. The five conserved domains were 46–52, 88–105, 61–65, 91–92, and 226–256 (Fig. 8.8, boxed). Compared with the reported native *A. annua FPS* cDNA sequence, *AaFPS1* cDNA presented several amino acid residues differing in the coding region (Fig. 8.8, shadow) because of single nucleotide mutation: 4-T(ACC)→I(ATC),113-P(CCC)→L(CTC),212-V(GTG)→M(ATG),268-P(CCC)→A(GCC),275-T(ACC)→V(GTC),322-H(CAC)→R(CGC).

These changes might suggest the polymorphisms of DNA associated with different genotypes of *A. annua* in different studies.

4.1.2 Expression and function identification of the AaFPS1

In vitro enzymatic catalysis was conducted using transformed and induced bacteria cell extracts. The results showed that extracts from *AaFPS1* transformed *E. coli* BL21 (BL21-*AaFPS1*) afforded easily measurable levels of FPS (farnesyl diphosphate synthase) activity, while *E. coli* BL21 (no insert) and *E. coli* BL21-*AaFPS1* (uninduced) gave rise to little detectable FPS catalytic activity. The observed activity was dependent on the addition of GPP (gerenyl diphosphate) or DMAPP (dimethylallyl diphosphate) to the reaction mixture because

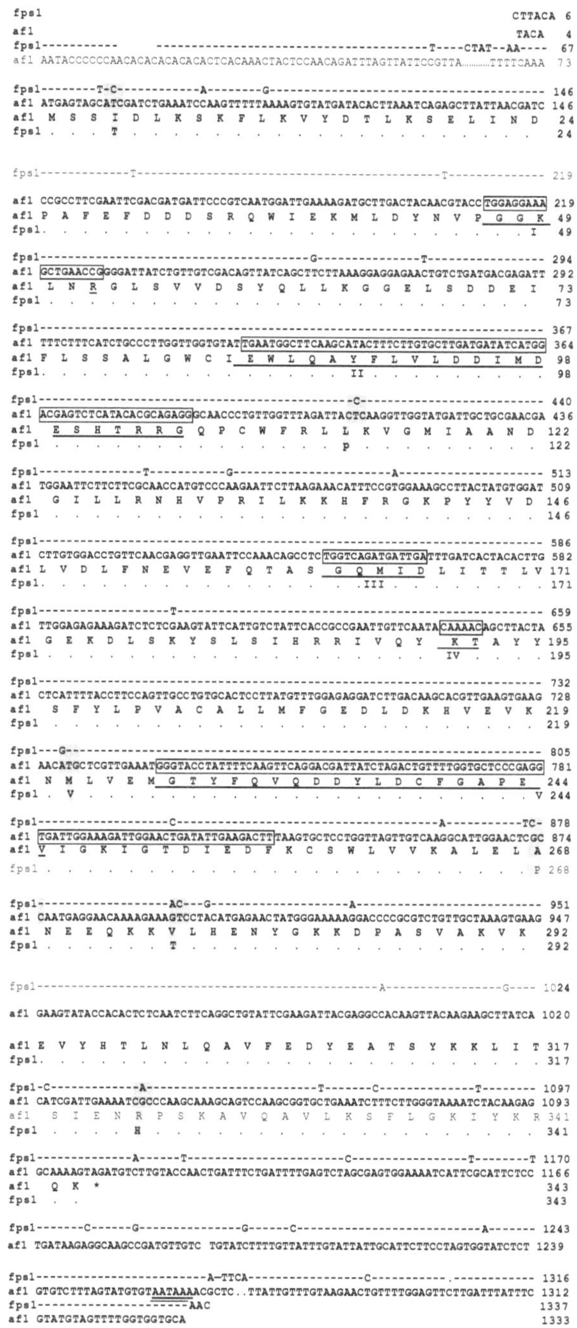

Figure 8.8: The Complete Nucleotide Sequence of *AaFPS1* cDNA (af1) and Alignment With a Reported Native *AaFPS* cDNA (fps1).
Stop codon is denoted by an *asterisk*; identical residues are represented by *dots*; different residues are represented by *shaded boxes*; the putative polyadenylation signal is denoted by *double underlines*; the five conserved domains presented in many prenyltransferases are *boxed*.

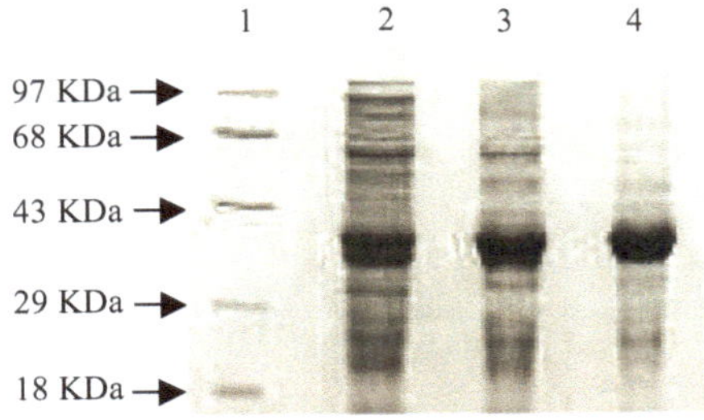

Figure 8.9: SDS–PAGE of AaFPS1 Purified by Different Methods.
1, Molecular markers; 2, cell-free extract from *E. coli* BL21-*AaFPS1*; 3, ammonium sulfate fraction; 4, recovered active fractions from Q-sepharose FF.

assays carried out in the absence of allylic diphosphates showed little activity. Further analysis of the reaction products confirmed that the cell extracts of *E. coli* BL21-*AaFPS1* had no activity to convert FPP with IPP (isopentanyl diphosphate) into GGPP (geranyl geranyl diphosphate) in the presence of Mn^{2+}. Therefore, the protein encoded by the cDNA had the activity of FPS but not GGPPS (GGPP synthase), indicating the gene encoded FPS of *A. annua*.

4.1.3 Kinetic studies

The results of a typical purification were shown in Fig. 8.9. One microgram purified protein was used for 100 μL reaction buffer.

Kinetic studies were conducted using the acid-lability assay to measure the formation of 4-^{14}C-FPP from 4-^{14}C-IPP and unlabeled GPP. Some basic kinetic parameters of the enzyme were evaluated. The time-initial velocity curve for GPP and IPP showed that the reaction speed kept constantly only in the initial stages, and velocity decreased with time lapsing. The optimal fitting function of the curve was the exponential decay function with correlation coefficient $r = 0.99$.

Initial speed of different concentrations of GPP and FPP was calculated by solving the slope of the time–velocity curve function. The substrate–velocity curve for GPP and IPP fitted for the Michaelis–Menton equation ($r = 0.99$). The apparent K_m 29.4 μmol for IPP and 17.2 μmol for GPP were calculated from the curve function.

4.2 Molecular Cloning, Escherichia coli *Expression, and Molecular Analysis of* Amorpha-4,11-diene Synthase *(ADS) Gene From* Artemisia annua

Bouwmeester et al. [11] reported that amorpha-4,11-diene was the likely olefinic sesquiterpene intermediate of biosynthesis of artemisinin, and at the same time, they isolated and purified the protein of amorpha-4,11-diene synthase (ADS). The low abundance of amorpha-4,11-diene and the high activity of amorpha-4,11-diene synthase in *A. annua* plant suggested that the cyclization of FPP into amorpha-4,11-diene was a rate-limiting step in artemisinin biosynthesis. Therefore, increasing the expression level of *ADS* gene in *A. annua* would be a promising direction to increase the content of artemisinin. In 2000–01, several groups [37–39] had reported the cloning and function identification of the *ADS* gene from different genotypes of *A. annua*. Our laboratory collected a high artemisinin yielding *A. annua* strain 001 from Sichuan Province of China. To investigate whether the activity of ADS from high artemisinin-yielding *A. annua* strain was higher than that of wild type (low artemisinin-yielding strain), Li et al. [51] in our group cloned the *ADS* gene from the high artemisinin yielding *A. annua* strain 001, overexpressed it in *E. coli*, identified its function, analyzed its gene structure and its expression pattern in *A. annua*, and the following is a review on their work.

4.2.1 Cloning of *Artemisia annua ADS1* cDNA

A 2067 bp cDNA (*ADS1*) was isolated from high artemisinin-yielding *A. annua* strain 001 (0.65% DW) using RT-PCR strategy. The deduced amino acid sequence of ADS1 was aligned with three other published sequences of ADS (Fig. 8.10), of which a share of 99% identity was found. Two amino acid discrepancies were found in the ADS1 compared with others.

4.2.2 *Escherichia coli* expression of *ADS1* and purification of ADS1

To facilitate the purification of ADS1, His-tag strategy was used. The entire coding region of *ADS1* was inserted into prokaryotic expression vector pET30a(+), including a hexa-histidyl leader sequence under the control of the T7 RNA polymerase promoter. The recombinant ADS1 was extended with 17 amino acids after fusion to the His-tag and its deduced molecular weight was 69.3 kDa. The recombinant ADS1 was purified by immobilized metal affinity chromatography on a column loaded with Ni^{2+}. SDS–PAGE showed that an efficient purification of the recombinant enzyme was achieved (Fig. 8.11).

4.2.3 Characterization of recombinant ADS1

Purified recombinant ADS1 was used for in vitro enzyme assays. The detection of gas chromatography–mass spectrometry (GC–MS) showed that the main product was amorpha-4,11-diene (Fig. 8.12). In the pH-optimum assay, the recombinant ADS1 exhibited maximum activity at around pH 6.5–7.0, at which most of the sesquiterpene syntheses had their maximum activities. The K_m of the recombinant ADS1 was determined at pH 7.0, 37°C in the presence of Mg^{2+}. The apparent K_m value was 3.4 mmol/L, which had no significant difference with prior reported enzymes (1.6–3.3 mmol/L) [67].

4.2.4 Cloning of *ADS1* genomic DNA

PCR amplification product using *A. annua* genomic DNA as a template and *ADS1* as gene specific primers showed a 4392 bp band ranging from the 5′-untranslated region (5′-UTR) to the polyadenylation site. Comparisons of *ADS1* genomic DNA (*gADS1*) and cDNA

```
WALLT-ADS    msl teekpirpianfppsiwgdqfliyqkqveqgveqivndlkkevrqllkealdipmkhanllklidei   70
MERCKE-ADS   msl teekpirpianfppsiwgdqfliyekqveqgveqivndlkkevrqllkealdipmkhanllklidei   70
CHANG-ADS    msl teekpirpianfppsiwgdqfliyekqveqgveqivndlkkevrqllkealdipmkhanllklidei   70
LIU-ADS      msl teekpirpianfppsiwgdqfliyekqveqgveqivndlkkevrqllkealdipmkhanllklidei   70

WALLT-ADS    qrlgipyhfereidhalqciyetygdnwngdrsslwfrlmrkqgyyvtcdvfnnykdkngafkqslandv   140
MERCKE-ADS   qrlgipyhfereidhalqciyetygdnwngdrsslwfrlmrkqgyyvtcdvfnnykdkngafkqslandv   140
CHANG-ADS    qrlgipyhfereidhalqciyetygdnwngdrsslwfrlmrkqgyyvtcdvfnnykdkngafkqslandv   140
LIU-ADS      qrlgipyhfereidhalqciyetygdnwdgdcsslwfrlmrkqgyyvtcdvfnnykdkngafkqslandv   140

WALLT-ADS    egllelyeatsmrvpgeiiledalgftrsrlsimtkdafstnpalfteiqralkqplwkrlprieaaqyi   210
MERCKE-ADS   egllelyeatsmrvpgeiiledalgftrsrlsimtkdafstnpalfteiqralkqplwkrlprieaaqyi   210
CHANG-ADS    egllelyeatsmrvpgeiiledalgftrsrlsmmtkdafstnpalfteiqralkqplwkrlprieaaqyi   210
LIU-ADS      egllelyeatsmrvpgeiiledalgftrsrlsimtkdafstnpalfteiqralkqplwkrlprieaaqyi   210

WALLT-ADS    pfyqqqdshnktllklaklefnllqslhkeelshvckwwkafdikknapclrdrivecyfwglgsgyepq   280
MERCKE-ADS   pfyqqqdshnktllklaklefnllqslhkeelshvckwwkafdikknapclrdrivecyfwglgsgyepq   280
CHANG-ADS    pfyqqqdshnktllklaklefnllqslhkeelshvckwwkafdikknapclrdrivecyfwglgsgfepq   280
LIU-ADS      pfyqqqdshnktllklaklefnllqslhkeelshvckwwkafdikknapclrdrivecyfwglgsgyepq   280

WALLT-ADS    ysrarvfftkavavitliddtydaygtyeelkifteaverwsitcldtlpeymkpiyklfmdtytemeef   350
MERCKE-ADS   ysrarvfftkavavitliddtydaygtyeelkifteaverwsitcldtlpeymkpiyklfmdtytemeef   350
CHANG-ADS    ysrarvfftkavavitliddtydaygtyeelkifteaverwsitcldtlpeymkpiyklfmdtytemeef   350
LIU-ADS      ysrarvfftkavavitliddtydaygtyeelkifteaverwsitcldtlpeymkpiyklfmdtytemeef   350

WALLT-ADS    lakegrtdlfncgkefvkefvrnlmveakwaneghipttеehdpvviitgganlltttcylgmsdiftke   420
MERCKE-ADS   lakegrtdlfncgkefvkefvrnlmveakwaneghipttеehdpvviitgganlltttcylgmsdiftke   420
CHANG-ADS    lakegrtdlfncgkefvkefvrnlmveakwaneghipttеehdpvviitgganlltttcylgmsdiftke   420
LIU-ADS      lakegrtdlfncgkefvkefvrnlmveakwaneghipttеehdpvviitgganlltttcylgmsdiftke   420

WALLT-ADS    svewavsapplfrysgilgrrlndlmthkaeqerkhssssle symkeynvneeyaqtliykevedvwkdi   490
MERCKE-ADS   svewavsapplfrysgilgrrlndlmthkaeqerkhssssle symkeynvneeyaqtliykevedvwkdi   490
CHANG-ADS    svewavsapplfrysgilgrrlndlmthkaeqerkhssssle symkeynvneeyaqtliykevedvwkdi   490
LIU-ADS      svewavsapplfrysgilgrrlndlmthkaeqerkhssssle symkeynvneeyaqtliykevedvwkdi   490

WALLT-ADS    nreylttkniprpllmaviylcqflevqyagkdnftrmgdeykhliksllvypmsi   546
MERCKE-ADS   nreylttkniprpllmaviylcqflevqyagkdnftrmgdeykhliksllvypmsi   546
CHANG-ADS    nreylttkniprpllmaviylcqflevqyagkdnftrmgdeykhliksllvypmsi   546
LIU-ADS      nreylttkniprpllmaviylcqflevqyagkdnftrmgdeykhliksllvypmsi   546
```

Figure 8.10: Alignment of the Deduced Amino Acids of Amorpha-4,11-diene Synthase From High-Yield *A. annua* Strain 001 With Other Reported Sequences.
WALLT-ADS, GenBank accession number: AY006482; MERCKE-ADS, GenBank accession number: AF138959; CHANG-ADS, GenBank accession number: AJ251751; LIU-ADS, amorpha-4,11-diene synthase from high yield *A. annua* strain 001. *Arrows* indicate the mutation amino acids.

suggested that the *gADS1* contained seven exons ranging in size from 87 to 382 nucleotides (corresponding to 29–127 amino acids), and six introns separating these exons ranging in size from 97 to 1 698 nucleotides (Fig. 8.13). The 3'-untranslated region (3'-UTR) of *gADS*1 was identical to the 3'-UTR of the *ADS1* cDNA, indicating that the isolated *ADS1*genomic gene corresponded to the expressed gene.

4.2.5 Southern blotting analysis of ADS1 gene

A. annua strain 001 genomic DNA (20 μg) digested with *Nco* I; *EcoR* I was used for Southern blotting analysis of *ADS1* gene. Southern blotting was performed using [α-^{32}P]-radioactively

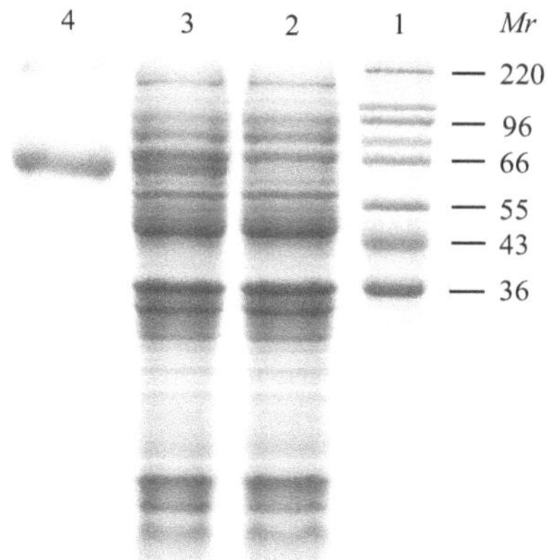

Figure 8.11: SDS–PAGE of Proteins Produced in *E. coli*.
1, Protein molecular mass standard; 2, soluble fractions expressed by *E. coli* harboring pET-*ADS1* without isopropylthio-β-galactoside (IPTG) induction; 3, soluble fractions expressed by *E. coli* harboring pET-*ADS1* induced by IPTG; 4, purified recombinant amorpha-4,11-diene synthase.

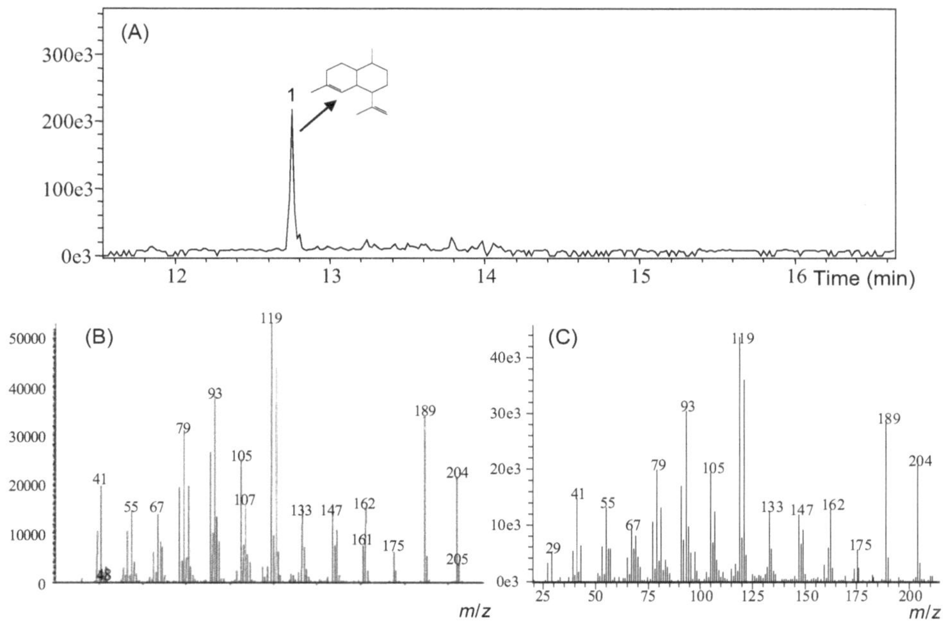

Figure 8.12: Gas Chromatography–Mass Spectrometry (GC–MS) Analysis of the Reaction Product of Amorpha-4,11-diene Synthase From *A. annua* Strain 001.
(A) Total ion chromatogram of the reaction product using FPP as substrate; (B) mass spectrum of the authentic amorpha-4,11-diene [39]; (C) mass spectrum of the peak 1.

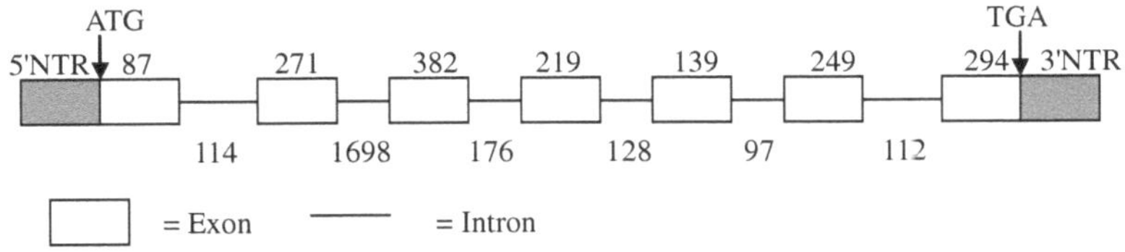

Figure 8.13: Genomic Organization of *ADS1* Gene.
The figure indicates the position of intron, exon, 5′-UTR, and 3′-UTR. Numbers above intron are the number of base pairs of the exon. Numbers below intron are the number of base pairs of the intron.

labeled full-length *ADS1* cDNA as a probe. The results showed four hybridizing bands for *Nco* I-digested and *Eco*R I-digested genomic DNA (Fig. 8.14). As there was neither *Nco* I nor *Eco*R I site in *gADS1*, it was postulated that there are at least four gene copies per genome.

4.2.6 Organ-specific expression pattern of ADS1

RNA samples obtained from roots, stems, leaves of the vegetation period, and flowers were analyzed by RT-PCR for tissue-specific expression analysis of *ADS1*. RT-PCR results showed that *ADS1* was actively transcribed in leaf and stem of the vegetation period and inflorescences, but not expressed in root (Fig. 8.15).

The *ADS1* cDNA isolated from a high artemisinin-yielding *A. annua* strain 001 showed 97 and 99% identity in nucleotide and amino acid sequences, respectively, to a published *A. annua ADS* gene. The results showed that the two mutations in *ADS1* had no influence on its catalytic properties. Therefore, the catalytic properties of ADS1 were not the cause for the

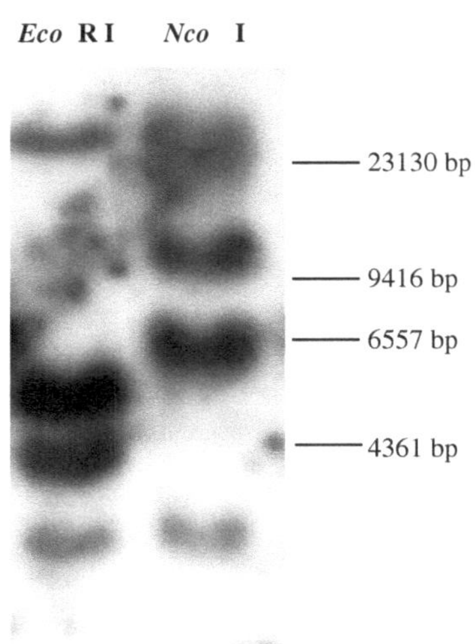

Figure 8.14: Southern Bloting Analysis of *gADS1* in the *A. annua* Genome.

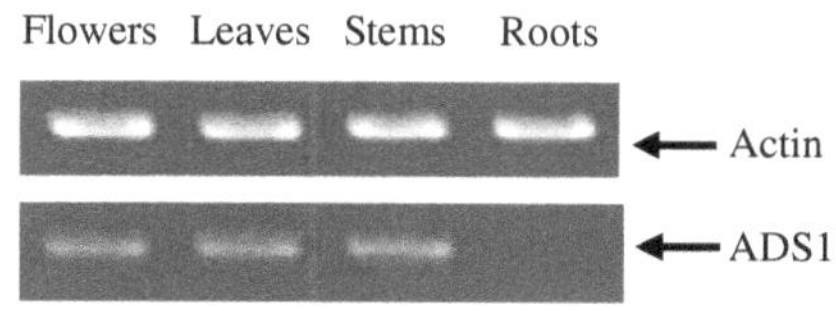

Figure 8.15: RNA Analysis for *ADS1* From *A. annua* Strain 001.

high-yield of this strain. Chang et al. [38] reported that there was one copy of *ADS* in the *A. annua* genome. Our current results [51] showed that there were at least four copies of *ADS1* in *A. annua* strain 001 genome. The more copies of *ADS* might be one of the reasons for the high artemisinin-yielding property of *A. annua* strain 001.

4.3 Molecular Cloning, Escherichia coli *Expression and Genomic Organization of* Squalene Synthase *(SQS) Gene From* Artemisia annua

Sesquiterpene synthases and squalene synthase are key branch point enzymes in isoprenoid biosynthetic pathway for the synthesis of sesquiterpenoid phytoalexins and steroids/steroidal alkaloids, respectively. Vogeli and Chappell [68] reported that squalene synthase activity was suppressed, while the biosynthesis of sesquiterpenoid phytoalexin was induced when fungal elicitors were added to tobacco suspension cultures. Kudakasseril et al. [69] reported that the steroid inhibitors miconazole, AMO 1618, CCC, and MER29 increased both the incorporation of ^{14}C-IPP into artemisinin and the production of artemisinin in cell-free extracts and shoot cultures. Thus, it is in theory desirable to clone the squalene synthase gene of *A. annua* and block the steroids biosynthetic pathway to shunt carbon flow toward the artemisinin biosynthesis by antisense technology. To this end, Liu et al. [44] of our group isolated a cDNA encoding *squalene synthase* (*AaSQS*) from a high-yield *A. annua* strain 001 and overexpressed it in *E. coli*. The following is a review of their work.

4.3.1 cDNA cloning of AaSQS

RACE strategies were used to isolate the full-length squalene synthase of *A. annua* (*AaSQS*) cDNA. First-strand cDNA derived from total RNA was used as the template with two internal nested degenerate primers and an oligo (dT)-adapter primer to obtain the 3′ end of the cDNA. A similar strategy was used to obtain the 5′ end sequences, except that a specific internal primer was used for the initial reverse transcription and a poly (A) tail was added to the 3′ terminus of the resulting first-strand cDNA. Sequences derived from the 5′- and 3′-RACE were used to construct the full-length 1539 bp AaSQS cDNA (Fig. 8.16). The cDNA contained an open reading frame (ORF) of 1257 bp, preceded by 23 bp of 5′-nontranslated region (NTR) and followed by 229 bp of 3′-NTR and 30 bp of poly(A)$^+$ tail. Classical polyadenylation signal (AATAAA) was not found in the 3′-NTR of cDNA except for one

```
-23                                                                   GGATTTGGATCTTGAAGAAGAAA
1     ATGAGTAGTTTGAAAGCAGTATTGAAACACCCAGATGACTTTTATCCATTATTGAAGTTGAAAAGGCTGCAAAGAAAGCCGAAAAA
1      M  S  S  L  K  A  V  L  K  H  P  D  D  F  Y  P  L  L  K  L  K  M  A  A  K  K  A  E  K
88    CAGATCCCATCACAACCTCACTGGGCTTTCTCTTATTCCATGCTTCATAAAGTTTCTAGAAGTTCGCTCTTGTTATTCAACAACTT
30     Q  I  P  S  Q  P  H  W  A  F  S  Y  S  M  L  H  K  V  S  R  S  F  A  L  V  I  Q  Q  L
175   AATCCCCAGCTTCGTGATGCTGTCTGCATCTTTTATTTGGTTCTTCGCGCTCTTGATACTGTTGAGGATGATACAAGCATAGCTGCT
59     N  P  Q  L  R  D  A  V  C  I  F  Y  L  V  L  R  A  L  D  T  V  E  D  D  T  S  I  A  A
262   GATATCAAAGTACCCATTCTGATCGCCTTTCATAAGCATATCTACAATCGTGATTGGCACTTTGCATGTGGTACAAAGGAATACAAA
88     D  I  K  V  P  I  L  I  A  F  H  K  H  I  Y  N  R  D  W  H  F  A  C  G  T  K  E  Y  K
349   GTTCTCATGGACCAGTTCCACCATGTTTCTACTGCCTTTCTGGAACTTAAGAGAGGTTATCAGGAGGCAATTGGGGATATAACCATG
117    V  L  M  D  Q  F  H  H  V  S  T  A  F  L  E  L  K  R  G  Y  Q  E  A  I  G  D  I  T  M
436   AGAATGAGCGCTGGGATGGCAAAATTTATATGTAAAGAGGTTGAGACAGTTGATGATTATGATGAGTTTTGTCATTATGTTGCGGGA
146    R  M  S  A  G  M  A  K  F  I  C  K  E  V  E  T  V  D  D  Y  D  E  F  C  H  Y  V  A  G
523   CTTGTTGGAATAGGGTTGTCAAAGCTCTTCCATTCTTCAGGCACGGAAATTTTGTTTTCTGATTCTATCTCCAATTCGATGGGTTTA
175    L  V  G  I  G  L  S  K  L  F  H  S  S  G  T  E  I  L  F  S  D  S  I  S  N  S  M  G  L
610   TTTCTTCAGAAGACAAATATCATTAGAGATTATCTCGAGGATATTAATGAGATACCTAAGTCACGCATGTTTTGGCCTCGTGAGATC
204    F  L  Q  K  T  N  I  I  R  D  Y  L  E  D  I  N  E  I  P  K  S  R  M  F  W  P  R  E  I
697   CGGAGTAAATATGTTAATAAGCTTGAGGACCTGAAATATGAAGAGGACTCTGAGAAAGCCGTTCAGTGCTTAAATGATATGGTGACA
233    R  S  K  Y  V  N  K  L  E  D  L  K  Y  E  E  D  S  E  K  A  V  C  C  L  N  D  M  V  T
784   AATGCTTTGATACACATTGAAGACTGCTTAAAGTATATGTCTCAGTTGAAAGATCCAGCCATCTTCAGGTTTTGTGCAATACCACAG
262    N  A  L  I  H  I  E  D  C  L  K  Y  M  S  Q  L  K  D  P  A  I  F  R  F  C  A  I  P  Q
871   ATCATGGCAATTGGAACACTTGCTTTATGCTACAATAACATTGAGGTTTTCAGGGGTGTAGTCAAATTGAGACGTGGTCTAACTGCC
291    I  M  A  I  G  T  L  A  L  C  Y  N  N  I  E  V  F  R  G  V  V  K  L  R  R  G  L  T  A
958   AAAGTAATTGATCGGACTAAAACAATGGCTGATGTGTACCAGGCTTTTTCTGATTTTTCTGACATGCTTAAGTCCAAGGTTGACATG
320    K  V  I  D  R  T  K  T  M  A  D  V  Y  Q  A  F  S  D  F  S  D  M  L  K  S  K  V  D  M
1045  CATGATCCCAATGCCCAAACGACCATCACTAGGTTAGAAGCAGCTCAGAAAATTTGCAAAGACTCTGGCACCCTTAGCAACAGGAAA
349    H  D  P  N  A  Q  T  T  I  T  R  L  E  A  A  Q  K  I  C  K  D  S  G  T  L  S  N  R  K
1132  TCTTACATAGTTAAGAGAGAGTCAAGCTACAGTGCAGCTTTGCTTGCTCTGCTTTTCACTATACTGGCTATTCTTTATGCTTATCTC
378    S  Y  I  V  K  R  E  S  S  Y  S  A  A  L  L  A  L  L  F  T  I  L  A  I  L  Y  A  Y  L
1219  TCTGCTAATCGACCCAATAAAATCAAATTCACTTTGTAAGAATCTCGCGGTGATACGTAAGCAAAGAAGATCCAGTCTTGAAATGTA
407    S  A  N  R  P  N  K  I  K  F  T  L  *
```

Figure 8.16: ***AaSQS*** **cDNA Sequence and Its Deduced Amino Acid Sequence.**

AATAAA sequence in 1257 bp. The cDNA encoded a speculated protein of 418 amino acid residues, with a predicted molecular weight of 47.7 kD and a calculated pI of 8.45.

4.3.2 Comparison of the amino acid sequence AaSQS with other squalene synthases

Alignment of the amino acid sequence for AaSQS and some other cloned squalene synthases by Clustal W program revealed high identities and similarities of AaSQS to those squalene

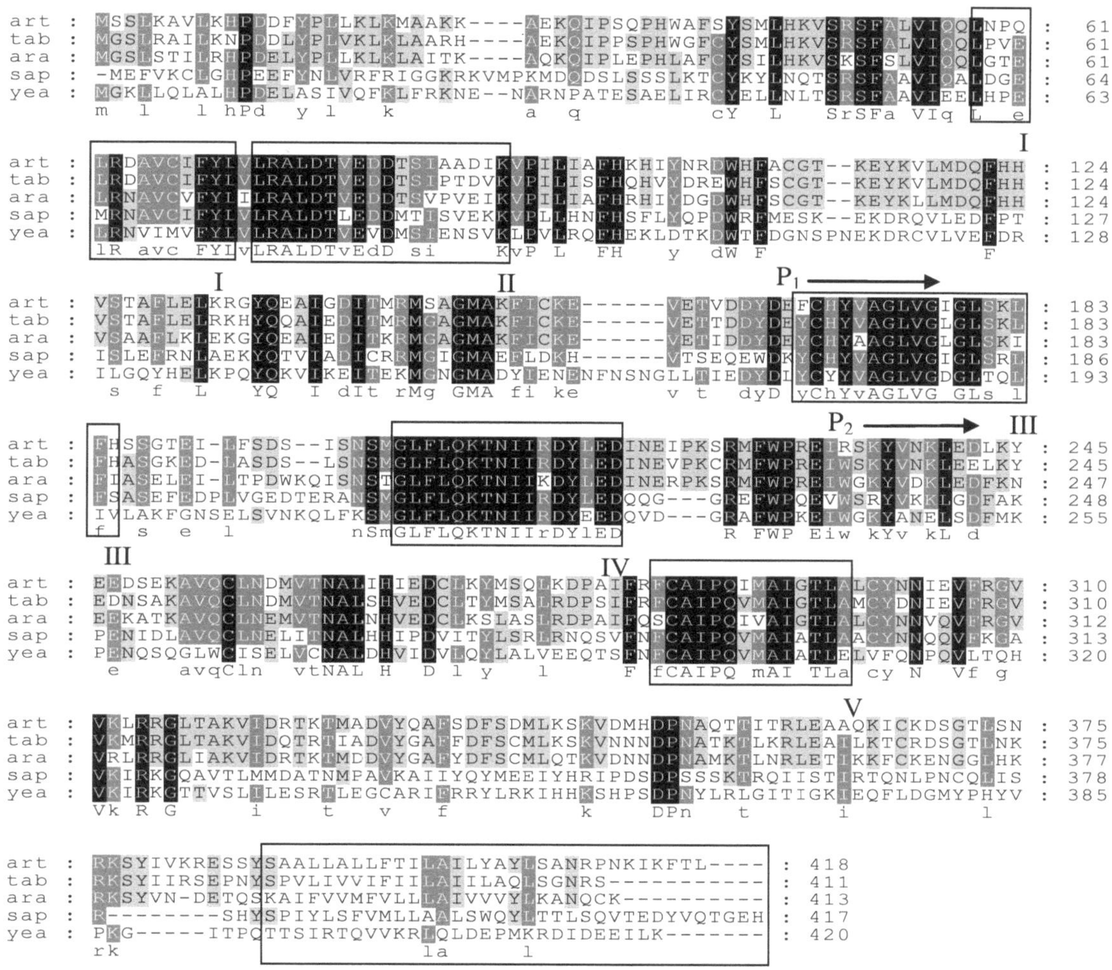

Figure 8.17: Alignment of the Deduced Amino Acid Sequences of Five Squalene Synthases. art, Squalene synthase from *A. annua*; tab, squalene synthase from tobacco; ara, squalene synthase from *A. thaliana*; sap, squalene synthase from human; yea, squalene synthase from yeast. Two *arrows* mark the regions used to design two nested degeneration PCR primers for 3′-RACE. The six conserved regions identified by Robinson et al. [70] are *boxed* and *numbered* below. The *shading colors* of the residues correspond to the level of similarity: *black*, 100% similarity; *dark gray*, ≥80% similarity; *light gray*, ≥60% similarity.

synthases (Fig. 8.17). The entire coding sequence of AaSQS was 70, 77, 44, and 39% identical, respectively, to squalene synthases from *A. thaliana*, tobacco, human, and yeast. Six highly conserved regions identified in SQS by Robinson et al. [70] were also found within AaSQS amino acid sequence. Three of these regions (III, IV, and V) were highly conserved, while regions I and II were less conserved. Region VI of all the SQSs, including AaSQS, showed a low identity, but their amino acids were all very hydrophobic. Region VI was postulated to be a membrane targeting signal for SQS.

4.3.3 Genomic organization of the AaSQS

PCR was performed to isolate the genomic DNA of *AaSQS* (g*AaSQS*). The sequence of *gAaSQS* revealed a gene of 3852 bp in length from the mRNA start site to the polyadenylation site. The 3′-NTR of *gAaSQS* was identical to the 3′-NTR of the *AaSQS* cDNA, indicating that the isolated *gAaSQS* corresponds to an expressed gene. Comparison of the full *gAaSQS* sequence with the coding region of the *AaSQS* cDNA showed a complex organization of exons and introns. The *gAaSQS* contained 14 exons ranging in size from 11 to 195 nucleotides (coreesponding to 4–65 amino acids). Thirteen introns separated these exons and ranged in size from 71 to 540 nucleotides (Fig. 8.18). The exon–intron organization of the *gAaSQS* gene was quite similar to that of tobacco.

4.3.4 Escherichia coli expression and purification of AaSQS

The carboxy-terminal truncated (99 bp) AaSQS cDNA was subcloned into a bacterial expression vector pET30a(+) in frame with N-terminal and C-terminal HiS_6-tag, and then the recombinant vector pETSSA was introduced into *E. coli* strain BL21(DE3). Expression was induced at 28°C by the addition of 0.5 mmol/L IPTG (isopropyl-beta-D-thiogalactoside) to exponentially growing cell (OD_{600} = 0.5) and cultured at 37°C for another 4, 8, or 12 h, and then a small amount of bacterial cultures were harvested and detected by SDS–PAGE. The results showed that the soluble fractions of AaSQS reached its highest level induced by 0.5 mmol/L IPTG for 8 h. The recombinant AaSQS was purified by immobilized metal affinity chromatography on Ni^{2+} columns and eluted by 300 mmol/L imidazole. SDS–PAGE detection showed a protein of predicted molecular mass of 51 kD, which is in consistency with the predicted relative molecular mass of recombinant AaSQS (Fig. 8.19).

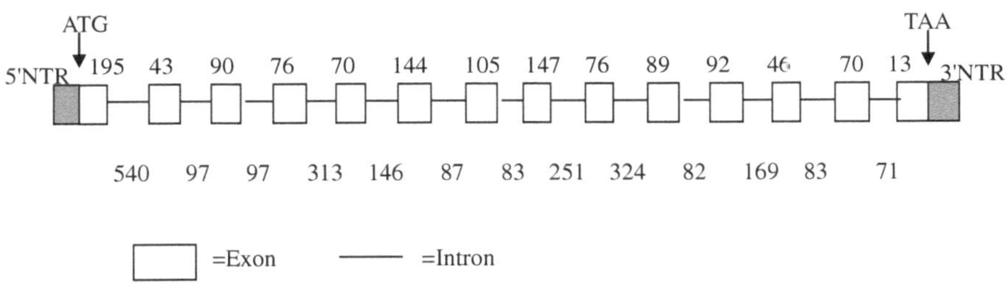

Figure 8.18: Organization of *AaSQS* Gene Showing Exon and Intron Localization as well as the 5′- and 3′-Nontranslated Regions (NTR).
Numbers above exons indicate the number of base pairs in the exon. Numbers below introns indicate the number of base pairs in the intron.

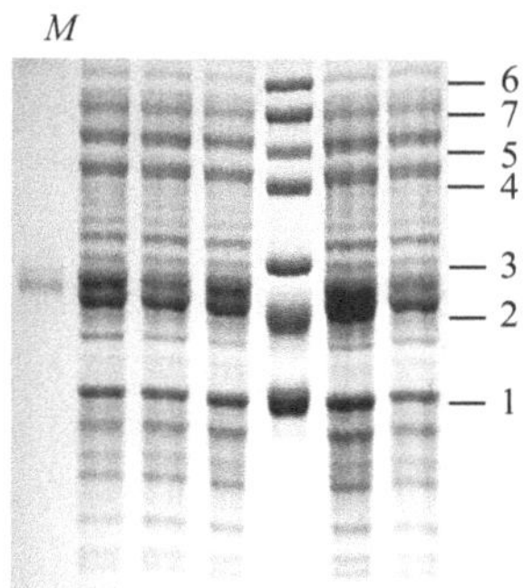

Figure 8.19: SDS-PAGE of Carboxy-Terminal Truncated AaSQS Produced in *E. coli* Strain BL21 (DE3).
1, Total protein expressed by *E. coli* strain BL21 (DE3) harboring pET30a(+); 2, total protein expressed by *E. coli* strain BL21 (DE3) harboring pETSSA; 3, protein relative molecular marker; 4–6, soluble fractions expressed by *E. coli* harboring pETSSA induced by 0.5 mmol/L IPTG for 8, 4, 12 h; 7, elution of 300 mmol/L imidazole.

4.3.5 Function identification of AaSQS

Purified recombinant AaSQS was used for in vitro catalyzing reaction. The results of GC–MS detection showed that the single product was squalene (Fig. 8.20), which indicates that the protein encoded by the *AaSQS* had the activity of SQS.

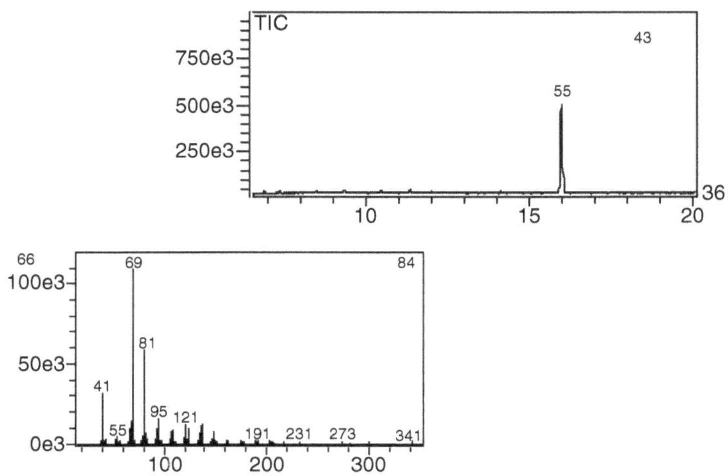

Figure 8.20: Gas Chromatography–Mass Spectrometry (GC–MS) Analysis of Recombinant AaSQS Catalyzing Reaction Product.
(A) Total ion chromatograms of the reaction products; (B) mass chromatograms of peak 1; (C) mass chromatograms of authentic squalene.

5 *Effects of Overexpression of* Artemisia annua *Farnesyl Diphosphate Synthase on the Artemisinin Content in* Artemisia annua

Farnesyl diphosphate synthase (FPS) is a prenyltransferase located at the multiple branch points in the isoprenoid biosynthetic pathway (Fig. 8.21); hence, FPS is thought to play a regulatory role in the isoprenoid biosynthetic pathway. When *Gossypium arboreum FPS* was overexpressed in low artemisinin-yielding *A. annua* strain 025 (artemisinin content was about 0.3% DW), the content of artemisinin in the transgenic *A. annua* was two- to threefold greater than that of the nontransgenic control [61]. In this section, a review is given on *A. annua* endogenous *FPS* overexpression in high artemisinin-yielding *A. annua* strain 001 (artemisinin content was about 0.6% DW) to increase the content of artemisinin in transgenic *A. annua* [34].

5.1 Recombinant Plant Expression Vector and Molecular Detection of the Transgenic Plants

The seeds of high artemisinin-yielding *A. annua* strain 001 were preserved in our laboratory. The recombinant plasmid pBIFPS1 contained the endogenous *FPS* gene of *A. annua*. DNA gel blotting analysis of the transgenic *A. annua* indicated that the foreign *FPS* gene had been integrated into the genome of *A. annua*. The results of RNA gel blotting and protein gel

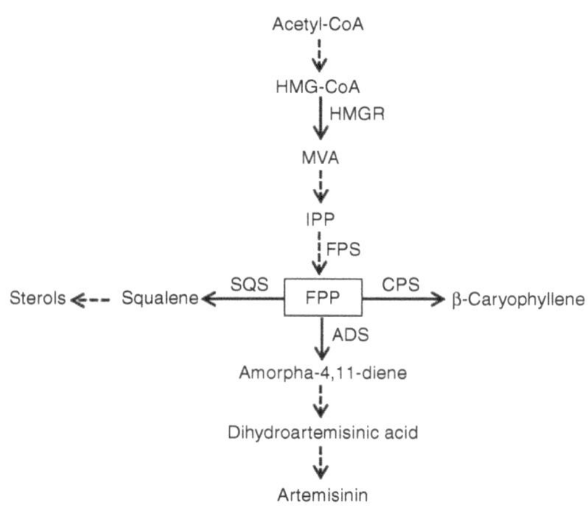

Figure 8.21: Diagram of the Mevalonic Acid Pathway Leading to the Synthesis of Artemisinin and Other Isoprenoid End Products.
ADS, Amorpha-4,11-diene synthase; *CPS*, β-caryophyllene synthase; *FPP*, fanesyl diphosphate; *FPS*, fanesyl diphosphate synthase; *HMG-CoA*, 3-hydroxy-3-methylglutary-CoA; *HMGR*, 3-hydroxy-3-methylglutary-CoA reductase; *IPP*, isopentenyl diphosphate; *MVA*, mevalonic acid; *SQS*, squalene synthase.

blotting indicated that the *FPS* gene was overexpressed at transcriptional and translational levels in the transgenic *A. annua*.

5.2 Analysis of FPS Enzyme Activity and Artemisinin Content of the Transgenic Artemisia annua

In vitro enzyme activity assay showed that FPS activity in the transgenic *A. annua* was approximately one- to twofold greater than that of the nontransgenic plants. The enzyme activity in transgenic line F6 was the highest (Fig. 8.22). The artemisinin content of the transgenic *A. annua* was 24.2–34.4% higher than that of the nontransgenic plants. The artemisinin content in transgenic line F6 was also the highest, and reached approximately 0.9% DW (Fig. 8.23). It can be seen from Figs. 8.22 and 8.23 that the increased enzyme activity in line F6 is consistent with the enhancement of artemisinin content.

5.3 Effects of Overexpressing the Endogenous FPS *on the Regulation of Artemisinin Biosynthesis*

As FPP is located at multiple branch points in the isoprenoid biosynthetic pathway, the biosynthesis of FPP is thought to be tightly regulated and FPS is thought to play a regulatory role in the isoprenoid biosynthetic pathway. After elicitor treatment, the accumulation of the sesquiterpenoid phytoalexin in pepper fruit or cotton suspension cell cultures was positively correlated with increased FPS mRNA levels, protein levels, and enzyme activity. Overexpressing a yeast FPS in tobacco resulted in an increase of the FPS activity

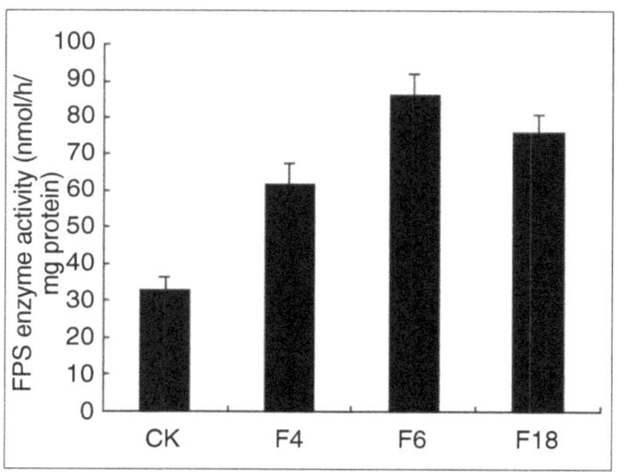

Figure 8.22: Farnesyl Diphosphate Synthase (FPS) Activity Analysis of the Transgenic *A. annua* (Lines F4, F6, and F18) and the Nontransgenic *A. annua* (CK).

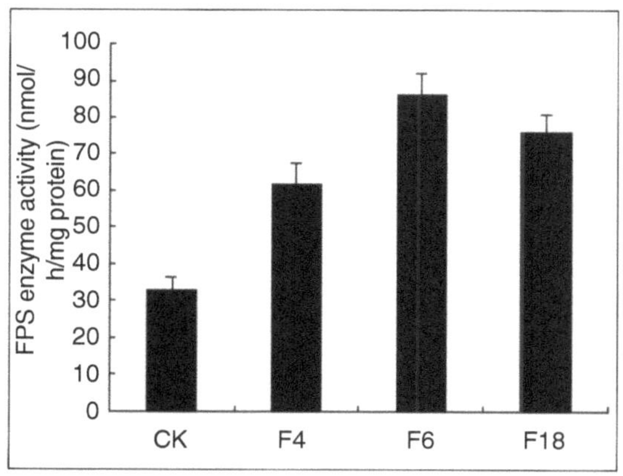

Figure 8.23: Artemisinin Content of the Transgenic *A. annua* (Lines F4, F6, and F18) and the Nontransgenic *A. annua* (CK). Data are the mean ± SD of three experiments.

of the transgenic tobacco by 12 times, and an increase of sterol content by 4 times. We overexpressed *A. annua* FPS in high artemisinin-yielding *A. annua*. The highest artemisinin content in the transgenic *A. annua* was approximately 0.9% DW, which was 34.4% higher than that in the nontransgenic *A. annua*. These results show directly that FPS plays a regulatory role in artemisinin biosynthesis.

It is interesting to compare the artemisinin content of overexpressing *G. arboreum* FPS *A. annua* [61] with that of overexpressing endogenesis FPS plants [34]. *G. arboreum* FPS was overexpressed in low artemisinin-yielding *A. annua* strain 025 (artemisinin content approximately 0.3% DW). The artemisinin content in the transgenic *A. annua* was two- to threefold greater than that in the nontransgenic *A. annua* and the highest content was approximately 1.0% DW [61]. For the overexpressed *A. annua* FPS in high artemisinin-yielding *A. annua* strain 001 (artemisinin content approximately 0.6% DW), the artemisinin content in the transgenic *A. annua* was 24.2–34.4% higher than that in the nontransgenic *A. annua* and the highest artemisinin content was approximately 0.9% DW [34]. Although the increases of artemisinin content in overexpressed *G. arboreum* FPS were higher than those in overexpressed *A. annua* FPS, the highest artemisinin contents in the transgenic *A. annua* were similar for these two cases (1.0% vs 0.9%). This indicates that although overexpression of FPS can increase the artemisinin content in *A. annua*, the function of FPS is weakened when the background content of artemisinin is high in *A. annua*. One possible reason for the modest increase of artemisinin content in overexpressed endogenesis FPS is that artemisinin has phytotoxic properties and it inhibits root growth and cell division. Some reports showed that if the toxic secondary metabolites were greatly accumulated in cells by genetic engineering, the toxicity of these secondary metabolites would become the limiting factor of increasing production. Therefore, an upper limitation might exist to the artemisinin content in *A. annua*, although artemisinin is stored in glandular trichomes. Another possible reason is that artemisinin biosynthesis is a complex enzymatic process; there are other regulatory enzymes in addition to FPS in the pathway. Hence, increasing the activity of one of the enzymes could only give a limited effect on the promotion of artemisinin content.

6 Effects of Overexpressing of ADS Gene on the Regulation of Artemisinin Biosynthesis

In artemisinin biosynthetic pathway, amorpha-4,11-diene synthase (ADS) catalyzes the first step of the sesquiterpene branch pathway to artemisinin biosynthesis (Fig. 8.21), and this reaction is speculated as the rate-limiting step of artemisinin biosynthesis according to the general rule of metabolic regulation. Hence, over expression of this enzyme in *A. annua* may increase the carbon flow of artemisinin biosynthesis, thereby increasing the content of artemisinin. At the same time, the intermediate products of downstream can also be detected to analyze the role of each step in the biosynthetic pathway of artemisinin.

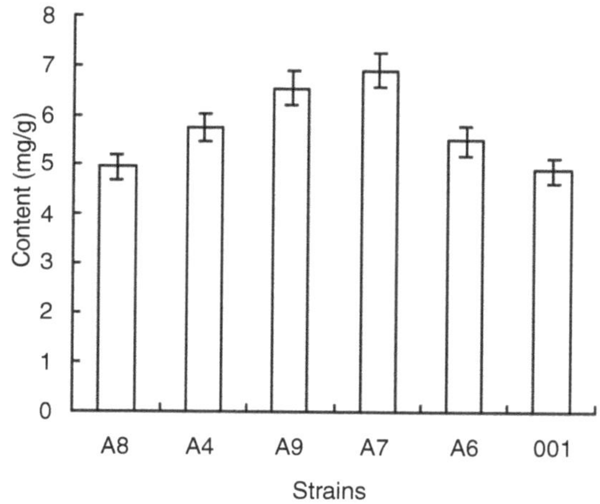

Figure 8.24: Artemisinin Content in Transgenic Plants.
A4, A6, A7, A8, A9, *ADS* transgenic lines; 001, CK.

In 2007, Li [52] of our group overexpressed *ADS* gene in *A. annua*; the artemisinin content of some of the transgenic *A. annua* increased significantly and the highest lines increased by 41% compared with the control (Fig. 8.24). In addition, the content of artemisinic acid and dihydroartemisinic acid of the transgenic *A. annua* also increased over the control, and the two highest lines increased by 47 and 79%, respectively (Fig. 8.25). These results indicate that the synthesis of amorpha-4,11-diene is a rate-limiting step in artemisinin biosynthetic pathway. In addition, they also suggest that the conversion of artemisinic acid or dihydroartemisinic acid to artemisinin may be a rate-limiting step in downstream of artemisinin biosynthesis.

7 Effect of Antisense Squalene Synthase Gene Expression on the Increase of Artemisinin Content in Artemisia annua

Squalene synthase (SQS) is located in the endoplasmic reticulum membrane, a crucial branch point enzyme of the isoprenoid pathway for synthesizing sterols or triquiterpenes from FPP (Fig. 8.21). The substrate FPP catalyzed by SQS is also the precursor for sesquiterpenes biosynthesis. In elicitor-treated tobacco cell cultures, the induction of sesquiterpene phytoalexins biosynthesis was correlated with the suppression of sterol biosynthesis, and the enzyme activity of SQS decreased rapidly. In *A. annua* cell cultures, the artemisinin content was increased after adding inhibitor of squalene epoxidase (natifine) in the culture medium. In this section, our work on the effects of inhibiting SQS gene expression on the biosynthesis of squalene and artemisinin by antisense RNA technology is reported [65].

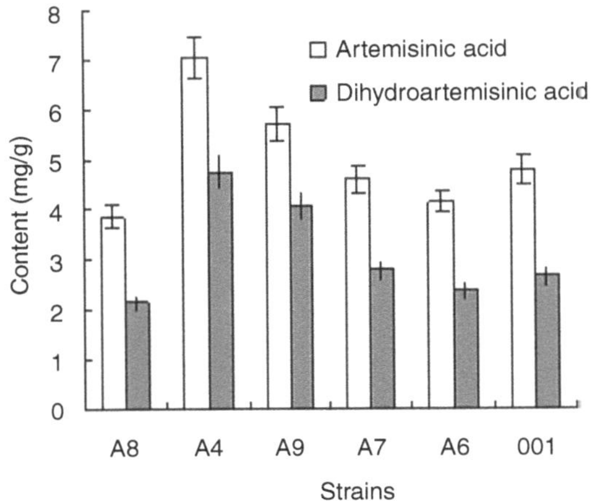

Figure 8.25: Contents of Artemisinic Acid and Dihydroartemisinic Acid in Transgenic Plants. A4, A6, A7, A8, A9, *ADS* transgenic lines; 001, CK.

7.1 Detection of Squalene Content in Transgenic Plants

A high artemisinin-yielding *A. annua* strain 001 was preserved in our laboratory. Recombinant plasmid pBISQS contained antisense squalene synthase gene. RT-PCR verified the suppressed expression of the endogenous *A. annua* SQS gene in some transgenic lines (ASQ_3, ASQ_5). To determine the effects of inhibiting squalene synthase gene expression on sterols biosynthesis, the leaves of transgenic lines (ASQ_3, ASQ_5) and the control 001 were selected to detect squalene content. The results of GC–MS showed that in ASQ_3 and ASQ_5 transgenic lines, squalene content was decreased by 19.4 and 21.6%, respectively, in comparison with the control (Fig. 8.26).

7.2 Detection of Artemisinin Content in Transgenic Plants

The content of artemisinin was detected by HPLC. The results indicated that artemisinin content of ASQ_3 and ASQ_5 transgenic lines was increased by 23.2 and 21.5%, respectively, compared with that of the control (Fig. 8.27). In contrast to ASQ_2 transgenic line, the artemisinin content manifested no obvious variation compared with the control.

7.3 Regulation of Artemisinin Biosynthesis by Expressing Antisense Squalene Synthase Gene

Wentzinger et al. [71] treated tobacco cell suspensions with 50 nM squalestatin, specific inhibitor of squalene synthase. Their results showed that SQS activity decreased by 80% after being treated with 50 nM squalestatin for 3 days, the biosynthesis of sterol was inhibited, and the content of FPP was increased. In addition, they also found that the transcription level of HMGR increased by fivefold in the tobacco cells treated with 500 nM squalestatin, indicating

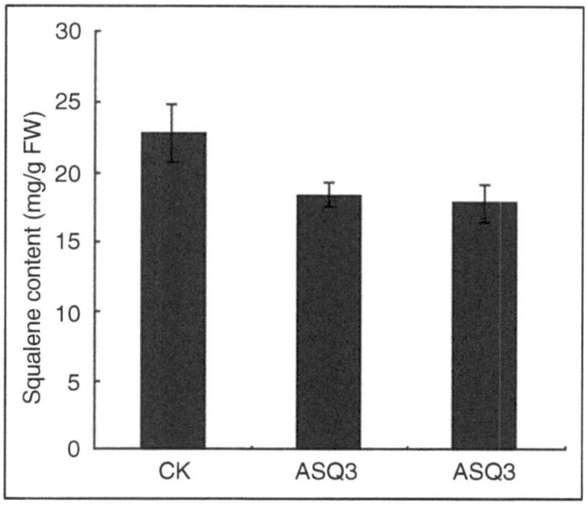

Figure 8.26: Analysis of Squalene Content of the Transgenic Plants and the Control.
CK, 001 control; ASQ_3 and ASQ_5, transgenic line.

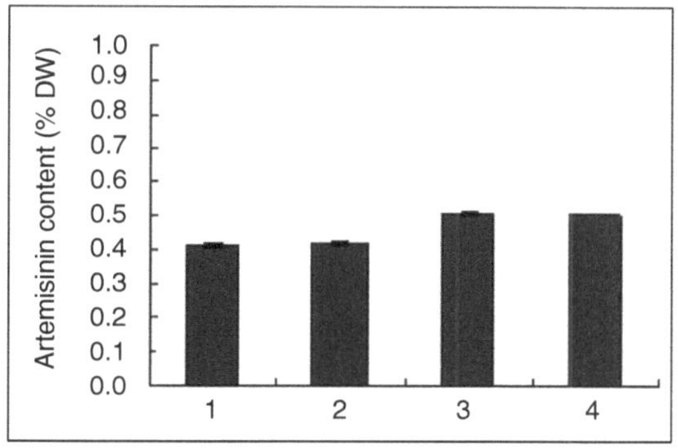

Figure 8.27: Analysis of Artemisinin Content of the Transgenic Plants and the Control.
1, 001 control; 2–4, transgenic lines ASQ_2, ASQ_3, and ASQ_5.

that the inhibition of SQS triggered positively feedback regulation of HMGR activity. As artemisinin is synthesized by the mevalonic acid pathway, and the biosynthesis of artemisinin and squalene used the common precursor FPP, SQS activity inhabitation is considered as an effective way to enhance artemisinin biosynthesis.

To investigate the effects of inhibiting the squalene synthase gene expression on the artemisinin biosynthesis, we introduced antisense squalene synthase gene into *A. annua* via *Agrobacterium*-mediated transformation, and transgenic plants were obtained. RT-PCR detection showed that the expression of antisense squalene synthase gene could partially suppress the endogenous SQS activity. The results of squalene content detection showed that the squalene content of the transgenic lines decreased by nearly 20% in comparison with the control. Although the SQS gene expression of the transgenic plants was inhibited, both the growth and the phenotype of the transgenic plants showed no obvious difference to those of the control, and this was consistent with the results of transgenic tobacco [72]. Artemisinin content of transgenic line ASQ_3 and ASQ_5 was increased by 23.2 and 21.5%, respectively, compared with that of the control.

8 *Effect of Antisense β-Caryophyllene Synthase Gene Expression on the Increase of Artemisinin Biosynthesis in* Artemisia annua

β-Caryophyllene synthase (CPS) catalyzes FPP to produce β-caryophelene. CPS is a sesquiterpene cyclase in the multibranch sites of isoprenoid metabolic pathway and competes with artemisinin biosynthesis for common precursor FPP (Fig. 8.21). Antisense technology is an effective means of inhibiting the expression of target gene. To increase artemisinin content of *A. annua*, the antisense fragment (750 bp) of β-caryophelene synthase gene was inserted into the plant expression vector pBI121, and then introduced into *A. annua* by *Agrobacterium*-mediated transformation [73], and its successful inhibition to the expression of endogenous *CPS* gene in the transgenic plants was confirmed by RT-PCR analysis. Meanwhile, the expression of the key enzyme genes involved in artemisinin biosynthesis, such as *ADS*, *ALDH1*, *DBR2*, and *CYP71AV1*, was enhanced at various degrees compared with that of the control. The underlying reason could be that the depression of *CPS* expression in the transgenic plants led more FPP diverted to the branch of artemisinin synthesis, which in turn induced high expression of the genes involved in artemisinin biosynthesis.

The detection of β-caryophyllene content showed that all the transgenic plants had significantly lower β-caryophyllene content than the untransgenic control, and the transgenic lines T2, T3, and T9 had a β-caryophyllene content reduction of 54.1, 62.7, and 50% than the control, respectively (Fig. 8.28). The RT-PCR analysis results showed that the inhibition of endogenous *CPS* gene in these three lines was stronger than other transgenic lines, which indicated that the inhibition degree of endogenous *CPS* was correlated positively with the content of β-caryophyllene. The determination results of artemisinin content showed that compared with the control, except for transgenic line T4, the artemisinin content of the other transgenic lines all increased at various degrees, and T10 had the highest increase of 54.9% (Fig. 8.29). Such results indicate that the inhibition of biosynthetic pathway competing for precursor of artemisinin biosynthesis is a potentially promising way to increase artemisinin production.

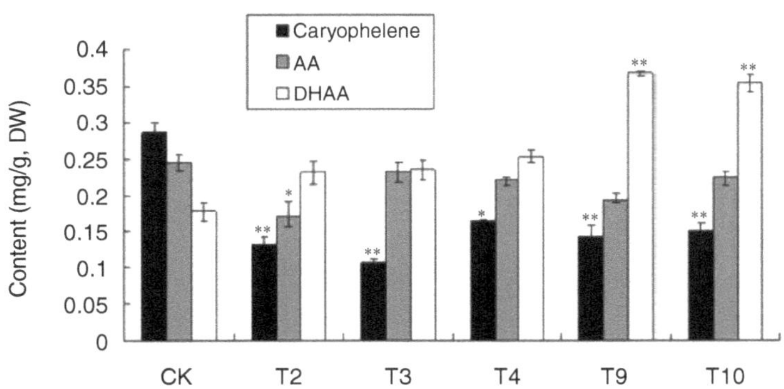

Figure 8.28: The Contents of β-Caryophelene, AA, and DHAA in *asCPS* Transgenic and Control Plants.

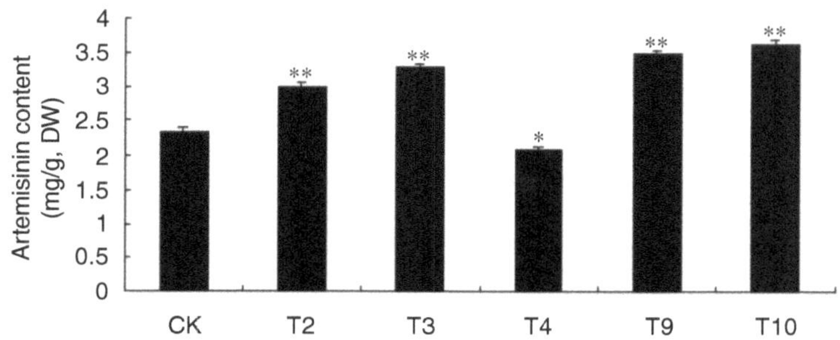

Figure 8.29: Artemisinin Content of *asCPS* Transgenic and Control Plants.

9 *Effects of* ipt *Gene Expression on the Physiological and Chemical Characteristics of* Artemisia annua

Cytokinins are a group of plant hormones that regulate many aspects of growth and development, including cell division, cell enlargement, cell differentiation, apical dominance, chloroplast degradation, senescence, and nutrient transport. The enzyme *isopentenyl transferase* (*ipt*) was isolated from the *A. tumefaciens* by Akiyoshi et al. [74], a rate-limiting step in cytokinin biosynthesis. Studies on the transformation of *ipt* gene are mainly focused on dicotyledonous plants, such as *Solanaceae*, especially tobacco in the literature; *ipt* gene is able to delay leaf senescence, and improve the quality and production of tobacco. Many reports indicated that the overexpression of *ipt* gene was a viable way to elevate cytokinin levels in transgenic plants.

Previous studies indicated that the capitate glands on the leaf surface of *A. annua* might be the main sites for artemisinin synthesizing/storing, and the specialized chloroplasts of the capitate glands seemed to play important roles in artemisinin biosynthesis. To study the function of cytokinins during the growth, development, and production of artemisinin, the *ipt* gene of *A. tumefaciens* under the control of the CaMV 35S promoter was transferred into *A. annua* in our laboratory, and the relationship between chlorophyll, cytokinin, and artemisinin levels was investigated [62]. Here are our main results.

9.1 *Expression of the* ipt *Gene in Transgenic* Artemisia annua *Plants and Their Physiological and Biochemical Changes*

Plant materials used in this study was *A. annua* strain 025. The results PCR and Southern blot analysis showed that the chimeric *ipt* gene was integrated into the genome of transgenic *A. annua* plants. RT-PCR and Northern blot analysis indicated that the *ipt* gene was expressed at the transcriptional level. The transgenic *A. annua* exhibited the loss of apical dominance, and axillary bud growth in the transgenic *A. annua* was significantly greater than that of wild-type plants, common features found in *ipt* transgenic plants. In addition, the elevated cytokinin level in transgenic *A. annua* plants also resulted in obstructing in root growth and decreasing in root mass.

The results of cytokinins content detection showed that the content of cytokinins in transgenic plants was elevated by two- to threefold than that of the control, and elevated endogenous cytokinins in transgenic plants stimulated chlorophyll accumulation by 20–60% at the same time. Artemisinin content detection showed that the content of artemisinin in transgenic plant was increased by 30–70% compared with control. Moreover, the increases of cytokinins, chlorophyll, and artemisinin level showed positive correlations (Fig. 8.30).

9.2 Overexpression of the ipt Gene Increased the Biosynthesis of Cytokinins, Chlorophyll, and Artemisinin

Presently, two different views exist on the effects of the elevated cytokinins content on the chlorophyll content in transgenic plants. One is from Estruch et al. [75] that the content of chlorophyll in the chimeric rolC gene transgenic plants was reduced because the elevated cytokinins level resulted from overexpression of the rolC gene retarded the normal development of chloroplasts. The other is from Jia et al. [76] that the etiolated plastids were turned into chloroplasts when cytokinin was applied to the albino leaves or cotyledon which were maintained in dark, so the biosynthesis of chlorophyll was improved. Our results supported the second view. We found that the elevated endogenous cytokinin level indeed improved the biosynthesis of chlorophyll in transgenic *A. annua* plants. Smigocki et al. [77]

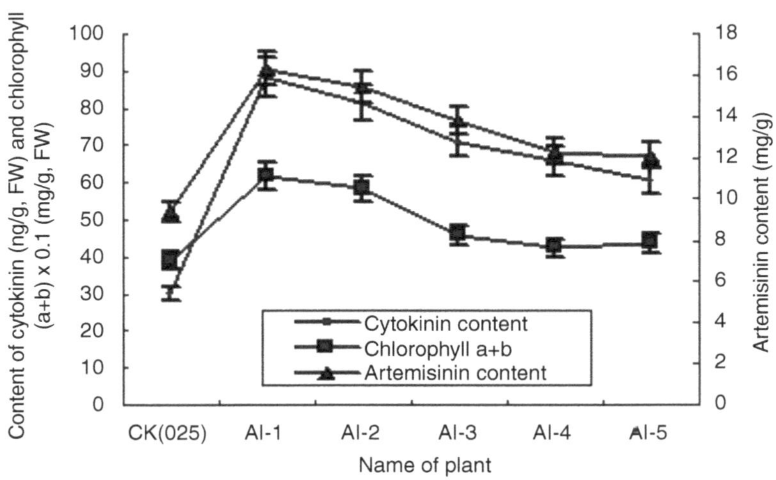

Figure 8.30: Contents Relationship of Cytokinin, Chlorophyll, and Artemisinin in *A. annua* Transgenic Plants.

also reported that the chlorophyll a/b content in the *ipt* gene transgenic plants was double that of the control. Our results showed that the contents of cytokinin, chlorophyll, and artemisinin had positive correlations in the *ipt* gene transgenic *A. annua* plants, which is coincident with the fact that the biosynthesis of artemisinin could only be achieved after a certain degree of tissue differentiation. The increase of chlorophyll content indicated the metabolism improvement in leaves of the transgenic plants; this, in turn, showed the advancement of the growth and development in the whole transgenic plant, by which the biosynthesis of artemisinin was improved.

10 The Effects of FPF1, CO Gene on Artemisia annua Flowering Time and on the Linkage Between Flowering and Artemisinin Biosynthesis

We found that the artemisinin biosynthesis increased with a progressive increase in plant growth, artemisinin content kept relatively low at younger plants, then increased gradually with plant development, and reached a maximum prior to flowering or at full flowering stage. However, it was not known if this is just a coincidence or if some inner linkage exists between flowering and artemisinin biosynthesis. To this end, the flower-promoting gene *FPF1* (*flowering promoting factor 1*) and *CO* (*CONSTANS*) of *A. thaliana* was transferred into *A. annua* via *A. tumefaciens*-mediated transformation, respectively [63,64]. Our main results are given as follows.

10.1 The Effects of Flower-Promoting Genes on the Flowering Time of Transgenic Artemisia annua

By *A. tumefaciens*-mediated transformation, we successfully obtained *FPF1* and *CO* transgenic *A. annua*, respectively. The molecular detections showed that *FPF1* and *CO* genes were integrated into the genome of *A. annua* and were expressed at the transcriptional level, respectively. Under short-day conditions, the flowering time of *FPF1* transgenic plants was about 20 days earlier than the nontransformed plants, and the flowering time of *CO* transgenic plants was about 15 days earlier. These results showed that both *FPF1* and *CO* could promote *A. annua* flowering, which was consistent with the results of *A. thaliana*.

10.2 No Existence of Direct Linkage Between Flowering and Artemisinin Biosynthesis

The changes of artemisinin content were analyzed in the transgenic plants and control before and postflowering. Vegetative-growth transgenic and nontransgenic plants cultured in a greenhouse were treated with short-day to induce flowering, those whose growth was still at long-day were used as controls of flowering. During this time, both short-day treated and

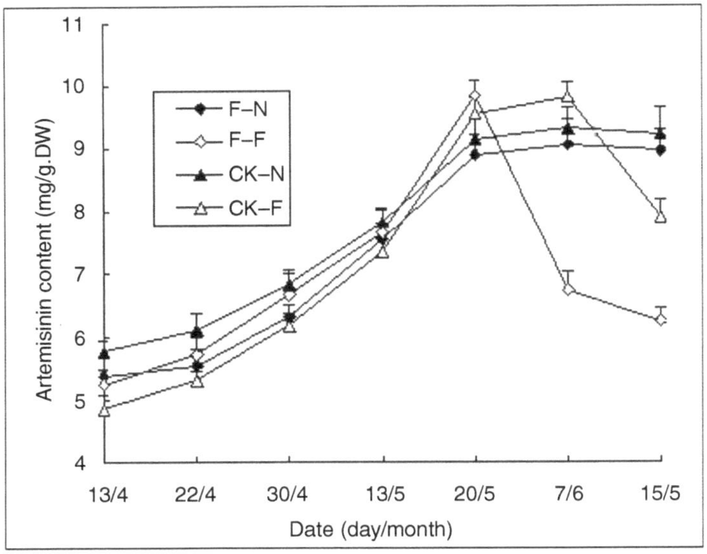

Figure 8.31: Comparison of Artemisinin Content in Flowering and Nonflowering FPF1 Transgenic CK and CK Plants.
F-F, Flowering FPF1 transgenic plants; F-N, nonflowering FPF1 transgenic plants; CK-F, flowering CK (nontransformed) plants; CK-N, nonflowering CK (nontransformed) plants.

nontreated plants were collected every week or every 10 days to determine the artemisinin content in different development stages (vegetative growth period, before flowering, full flowering, and postflowering). It was observed that after short-day treatment for 20 days, the *FPF1* transgenic plants generated flower buds, but the nontransgenic plants were still in vegetative growth at this time. The artemisinin contents of the *FPF1* transgenic plants and the control were shown in Fig. 8.31. From Fig. 8.31, it can be seen that before the flowering, for the short-day-treated *FPF1* transgenic plants, the short-day-treated nontransgenic plants, and nonshort-day-treated ones, their artemisinin content all increased with the plant development. The artemisinin content of the short-day-treated *FPF1* transgenic plants reached its maximum before the emergence of flowering buds; the artemisinin content of the short-day-treated nontransgenic plants is close to this maximum value at the same time. However, for the nonshort-day-treated transgenic and nontransgenic plants, although remaining vegetative, their respective artemisinin content showed no significant difference in comparison with those of the short-day-treated ones. After flowering, the artemisinin content of *FPF1* transgenic plants began to decrease significantly, but for the nonshort-day-treated *FPF1* transgenic plants and control plants, being still in vegetative growth, their artemisinin contents still kept at a high level. The artemisinin content of *CO* transgenic plants was similar to *FPF1* transgenic plants. The aforementioned results indicated that the foreign gene could indeed promote *A. annua* to flower earlier, but as the artemisinin content of the flowering transgenic plants and that of the nontransgenic control plants did not manifest any significant differences, it can be concluded that the increase of artemisinin content before flowering is not induced by flowering itself.

Acknowledgments

The authors thank Dahua Chen, Sa Geng, Yan Liu, Yujun Zhao, Junli Han, Zhenqiu Li, Yugang Song, and Jianlin Chen et al. for their contributions to this chapter.

References

[1] Coordinating research group for the structure of artemisinin. A new type of sesquiterpene lactone—artemisinin. Chin Sci Bull 1977;22:142. (in Chinese)
[2] Klayman DL. Qinghaosu (artemisinin): an antimalarial drug from China. Science 1985;228:1049–55.
[3] Wallaart TE, Pras N, Beekman AC, Quax WJ. Seasonal variation of artemisinin and its biosynthetic precursors in plants of *Artemisia annua* of different geographical origin: proof for existence of chemotypes. Plant Med 1999;66:57–62.
[4] Schmid G, Hofheinz W. Total synthesis of qinghaosu. J Am Chem Soc 1983;105:624–5.
[5] Mann J. Secondary Metabolism (R.Q. Cao, Trans.). China Science Press; Beijing. 1983. p. 88. (in Chinese).
[6] Bach TJ. Synthesis and metabolism of mevanoic acid in plants. Plant Physiol. Biochem. 1987;25:163–78.
[7] Kleining H. The role of plastid in isoprenoid biosynthesis. Annu Rev Plant Physiol Plant Mol Biol 1989;40:39–59.
[8] Choi D, Word BL, Bostock RM. Differential induction and suppression of potato 3-hydroxyl-3-methylglutaryl coenzyme A reductase genes in response to phytophores infestans and to its elicitor arechidonic acid. Plant Cell 1992;4:1333–44.
[9] Schramek N, Wang HH, Römisch-Margl W, Keil B, Radykewicz T, Winzenhörlein B, Beerhues L, Bacher A, Rohdich F, Gershenzon J, Liu BY, Eisenreich W. Artemisinin biosynthesis in growing plants of *Artemisia annua*. A $^{13}CO_2$ study. Phytochemistry 2010;71:179–87.
[10] Akhila A, Thakur RS, Popli SP. Biosynthesis of artemisinin in *Artemisia annua*. Phytochemistry 1987;16:1927–30.
[11] Bouwmeester HJ, Wallaart TE, Janssen MHA, van Loo B, Jansen BJM, Posthumus MA, Schmidt CO, Kraker Jan-willem D, Konig WA, Franssen MCR. Amorpha-4,11-diene synthase catalyses the first probable step in artemisinin biosynthesis. Phytochemistry 1999;52:843–54.
[12] Wallaart TE, van Uden W, Lubberink HGM, Woerdenbag HJ, Pras N, Quax WJ. Isolation and identification of dihydroartemisinic acid from *Artemisia annua* and its role in the biosynthesis of artemisinin. J Nat Prod 1999;62:430–3.
[13] Wallaart TE, Pras N, Quax WJ. Isolation and identification of dihydroartemisinic acid hydroperoxide from *Artemisia annua*: a novel biosynthetic precursor of artemisinin. J Nat Prod 1999;62:1161–2.
[14] Bertea CM, Freije JR, van der Woude H, Verstappen FWA, Perk L, Marquez V, De Kraker J-W, Posthumus MA, Jansen BJM, de Groot A, Franssen MCR, Bouwmeester HJ. Identification of intermediates and enzymes involved in the early steps of artemisinin biosynthesis in *Artemisia annua*. Plant Med 2005;71:40–7.
[15] Teoh KH, Polichuk DR, Reed DW, Nowak G, Covello PS. *Artemisia annua* L. (Asteraceae) trichome-specific cDNAs reveal CYP71AV1, a cytochrome P450 with a key role in the biosynthesis of the antimalarial sesquiterpene lactone artemisinin. FEBS Lett 2006;580:1411–6.
[16] Teoh KH, Polichuk DR, Reed DW, Covello PS. Molecular cloning of an aldehyde dehydrogenase implicated in artemisinin biosynthesis in *Artemisia annua*. Botany 2009;87:635–42.

[17] Chappell J, Wolf F, Proulx J, Cuellar R, Saunders C. Is the reaction catalyzed by 3-hydroxyl-3-methylglutaryl coenzyme A reductase a rate-limiting step for isoprenoid biosynthesis in plants? Plant Physiol 1995;109:1337–43.
[18] Yang ZB, Park HS, Lacy GH, Cramer CL. Differential activation of potato 3-hydroxy-3-methylglutaryl coenzyme A reductase genes by wounding and pathogen challenge. Plant Cell 1991;3:397–405.
[19] Joost O, Bianchini G, Bell AA, Benedict CR, Magill CW. Differential induction of 3-hydroxy-3-methylglutaryl CoA reductase in two cotton species following inoculation with Verticillium. Mol Plant Microbe Interact 1995;8:880–5.
[20] Chappell J, Nable R. Induction of sesquiterpenoid biosynthesis in tobacco cell suspension cultures by fungal elicitors. Plant Physiol 1987;85:469–73.
[21] Delourme D, Lacroute F, Karst F. Cloning of an *Arabidopsis thaliana* cDNA coding for farnesyl diphosphate synthase by functional complementation in yeast. Plant Mol Biol 1994;26:1867–73.
[22] Attucci S, Aitken SM, Ibrahim RK, Gulick PJ. A cDNA encoding farnesyl pyrophosphate synthase in white lupine. Plant Physiol 1995;108:835–6.
[23] Pan Z, Herickhoff L, Backhaus RA. Cloning, characterization, and heterologous expression of cDNAs for farnesyl diphosphate synthase from the guayule rubber plant reveals that this prenyltransferase occurs in rubber particles. Arch Biochem Biophys 1996;332:196–204.
[24] Matsushita Y, Kang WY, Charlwood BV. Cloning and analysis of a cDNA encoding farnesyl diphosphate synthase from *Artemisia annua*. Gene 1996;172:207–9.
[25] Sanmiya K, Iwasaki T, Matsuoka M, Miyao M, Yamamoto N. Cloning of a cDNA that encodes farnesyl diphosphate synthase and the blue-light-induced expression of the corresponding gene in the leaves of rice plants. Biochim Biophys Acta 1997;1350:240–6.
[26] Liu CJ, Meng YL, Hou SS, Chen XY. Cloning and sequencing of a cDNA encoding farnesyl pyrophosphate synthase from *Gossypium arboreum* and its expression pattern in the developing seed of *Gossypium hirsutum* cv. "Sumian-6". Acta Bot Sin 1998;40:703–10. (in Chinese)
[27] Gaffe J, Bru J-P, Causse M, Vidal A, Stamitti-Bert L, Carde J-P, Gallusci P. LEFPS1, a tomato pyrophosphate gene highly expressed during early fruit development. Plant Physiol 2000;123:1351–62.
[28] Hugueney P, Bouvier F, Badillo A, Quennemet J, d'Harlingue A, Camara B. Developmental and stress regulation of gene expression for plastid and cytosolic isoprenoid pathways in pepper fruits. Plant Physiol 1996;111:619–26.
[29] Liu CJ, Heinstein P, Chen XY. Expression pattern of genes encoding farnesyl diphosphate synthase and sesquiterpene cyclase in cotton suspension-cultured cells treated with fungal elicitors. Mol Plant Microbe Interact 1999;12:1095–104.
[30] Ashby MN, Edwards PA. Elucidation of deficiency in two yeast coenzyme Q mutants: characterization of the structural gene encoding hexaprenyl pyrophosphate synthetase. J Biol Chem 1990;265:13157–64.
[31] Math SK, Hearst JE, Poulter CD. The *crtE* gene in *Erwinia herbicola* encodes geranylgeranyl diphosphate synthase. Proc Natl Acad Sci USA 1992;89:6761–4.
[32] Hemmerlin A, Rivera SB, Erickson HK, Poulter CD. Enzymes encoded by the farnesyl diphosphate synthase gene family in the big sagebrush *Artemisia tridentata* ssp. spiciformis. J Biol Chem 2003;278:32132–40.
[33] Zhao YJ, Ye HC, Li GF, Chen DH, Liu Y. Cloning and enzymology analysis of farnesyl pyrophosphate synthase gene from a superior strain of *Artemisia annua* L. Chin Sci Bull 2003;48:63–7.
[34] Han JL, Liu BY, Ye HC, Wang H, Li ZQ, Li GF. Effects of overexpression of *Artemisia annua* farnesyl diphosphate synthase on the artemisinin content in *Artemisia annua*. J Integr Plant Biol 2006;48:482–7.
[35] Hua L, Matsuda SPT. The molecular cloning of 8-*epi*-cedrol synthase from *Artemisia annua*. Arch Biochem Biophys 1999;369:208–12.
[36] Mercke P, Crock J, Croteau R, Brodelius PE. Cloning, expression, and characterization of *epi*-cedrol synthase, a sesquiterpene cyclase from *Artemisia annua* L. Arch Biochem Biophys 1999;369:213–22.
[37] Mercke P, Bengtsson M, Bouwmeester HJ, Posthumus MA, Brodelius PE. Molecular cloning, expression, and characterization of amorpha-4,11-diene synthase, a key enzyme of artemisinin biosynthesis in *Artemisia annua* L. Arch Biochem Biophys 2000;381:173–80.

[38] Chang YJ, Song SH, Park SH, Kim SU. Amorpha-4,11-diene synthase of *Artemisia annua*: cDNA isolation and bacterial expression of a terpene synthase involved in artemisinin biosynthesis. Arch Biochem Biophys 2000;383:178–84.
[39] Wallaart TE, Bouwmeester HJ, Hille J. Amorpha-4,11-diene synthase: cloning and functional expression of a key enzyme in the biosynthetic pathway of the novel antimalarial drug artemisinin. Planta 2001;21:460–5.
[40] Cai Y, Jia JW, Crock J, Lin ZX, Chen XY, Croteau R. A cDNA clone for β-caryophyllene synthase from *Artemisia annua*. Phytochemistry 2002;61:523–9.
[41] Picaud S, Brodelius M, Brodelius PE. Expression, purification and charact erization of recombinant (*E*)-β-farnesene synthase from *Artemisia annua*. Phytochemistry 2005;66:961–7.
[42] Bertea CM, Voster A, Verstappen FW, Maffei M, Beekwilder J, Bouwmeester HJ. Isoprenoid biosynthesis in *Artemisia annua*: cloning and heterologous expression of a germacrene A synthase from a glandular trichome cDNA library. Arch Biochem Biophys 2006;448:3–12.
[43] Liu Y, Ye HC, Li GF. Cloning, heterologous expression and molecular analysis of a novel sesquiterpene synthase gene form *Artemisia annua*. Acta Bot Sin 2002;44:1450–5.
[44] Liu Y, Ye HC, Li GF. Molecular cloning, *E. coli* expression and genomic organization of squalene synthase form *Artemisia annua*. Acta Bot Sin 2003;45:608–13.
[45] Facchini PJ, Chappell J. Gene family for an elicitor-induced sesquiterpene cyclase in tobacco. Proc Natl Acad Sci USA 1992;89:11088–92.
[46] Chen XY, Chen Y, Heinstein P, Davidsson VJ. Cloning, expression and characterization of (+)-δ-cadinene synthase: a catalyst for cotton phytoalexin biosynthesis. Arch Biochem Biophys 1995;324:255–66.
[47] Crock J, Wilding M, Croteau R. Isolation and bacterial expression of a sesquiterpene synthase cDNA clone from peppermint (*Mentha* × *piperita*, L.) that produces the aphid alarm pheromone (*E*)-beta-farnesene. Proc Natl Acad Sci USA 1997;94:12833–8.
[48] Colby SM, Crock J, Dowdle-Rizzo B, Lemaux PG, Croteau R. Germacrene C synthase from *Lycopersicon esculentum* cv. VFNT cherry tomato: cDNA isolation, characterization, and bacterial expression of the multiple product sesquiterpene cyclase. Proc Natl Acad Sci USA 1998;95:2216–21.
[49] Mc-Garvey DJ, Croteau R. Terpenoid metabolism. Plant Cell 1995;7:1015–26.
[50] Bohlmann J, Meyer-Gauen G, Croteau R. Plant terpenoid synthases: molecular biology and phylogenetic analysis. Proc Natl Acad Sci USA 1998;95:4126–33.
[51] Li ZQ, Liu Y, Liu BY, Wang H, Ye HC, Li GF. The cloning, bacterial expression, and molecular analysis of amorpha-4,11-diene synthase from a high-yield strain of *Artemisia annua* L. J Integr Plant Biol 2006;48:1486–92.
[52] Z.Q. Li. Functional analysis and transformation of genes involved in artemisinin biosynthesis. Doctoral Dissertation, Institute of Botany, Chinese Academy of Sciences, Beijing, 2007, p. 81–91. (in Chinese)
[53] Porter TD, Coon MJ. Cytochrome P450. Multiplicity of isoforms, substrates, calalytic and regulatory mechanisms. J. Biol. Chem. 1991;266:13469–72.
[54] Ro DK, Paradise EM, Ouellet M, Fisher KJ, Newman KL, Ndungu JM, Ho KA, Eachus RA, Ham TS, Kirby J, Chang MCY, Withers ST, Shiba Y, Sarpong R, Keasling JD. Production of the antimalarial drug precursor artemisinic acid in engineered yeast. Nature 2006;440:940–3.
[55] Zhang Y, Teoh KH, Reed DW, Maes L, Goossens A, Olson DJ, Ross AR, Covello PS. The molecular cloning of artemisinic aldehyde Delta11(13) reductase and its role in glandular trichome-dependent biosynthesis of artemisinin in *Artemisia annua*. J Biol Chem 2008;283:21501–8.
[56] Paddon CJ, Westfall PJ, Pitera DJ, Benjamin K, Fisher K, McPhee D, Leavell MD, Tai A, Main A, Eng D, Polichuk DR, Teoh KH, Reed DW, Treynor T, Lenihan J, Fleck M, Bajad S, Dang G, Dengrove D, Diola D, Dorin G, Ellens KW, Fickes S, Galazzo J, Gaucher SP, Geistlinger T, Henry R, Hepp M, Horning T, Iqbal T, Jiang H, Kizer L, Lieu B, Melis D, Moss N, Regentin R, Secrest S, Tsuruta H, Vazquez R, Westblade LF, Xu L, Yu M, Zhang Y, Zhao L, Lievense J, Covello PS, Keasling JD, Reiling KK, Renninger NS, Newman JD. High-level semi-synthetic production of the potent antimalarial artemisinin. Nature 2013;496:528–32.
[57] Rydén AM, Ruyter-Spira C, Quax W, Osada H, Muranaka T, Kayser O, Bouwmeester H. The molecular cloning of dihydroartemisinic aldehyde reductase and its implication in artemisinin biosynthesis in *Artemisia annua*. Plant Med 2010;76:1778–83.

[58] Li X, Ma DM, Chen JL, Pu GB, Ji YP, Lei CY, Du ZG, Liu BY, Ye HC, Wang H. Biochemical characterization and identification of a cinnamyl alcohol dehydrogenase from *Artemisia annua*. Plant Sci 2012;193–194:85–95.
[59] Chen DH, Meng YL, Ye HC, Li GF, Chen XY. Culture of transgenic *Artemisia annua* hairy root with cotton cadinene synthase gene. Acta Bot Sin 1998;40:711–4. (in Chinese)
[60] Chen DH, Liu CJ, Ye HC, Li GF, Liu BY, Meng YL, Chen XY. Ri-mediated transformation of *Artemisia annua* with a recombinant farnesyl diphosphate synthase gene for artemisinin production. Plant Cell Tiss Org Cult 1999;57:157–62.
[61] Chen DH, Ye HC, Li GF. Expression of a chimeric farnesyl diphosphate synthase gene in *Artemisia annua* L. transgenic plants via *Agrobacterium tumefaciens*-mediated transformation. Plant Sci 2000;155:179–85.
[62] Geng S, Ma M, Ye HC, Liu BY, Li GF, Chong K. Effects of *ipt* gene expression on the physiological and chemical characteristics of *Artemisia annua* L. Plant Sci 2001;160:691–8.
[63] Wang H, Ge L, Ye HC, Chong K, Liu BY, Ye HC. Studies on the effects of *fpf1* gene on *Artemisia annua* flowering time and on the linkage between flowering and artemisinin biosynthesis. Plant Med 2004;70:347–52.
[64] Wang H, Liu Y, Chong K, Liu BY, Ye HC, Li ZQ, Yan F, Li GF. Earlier flowering induced by over-expression of *CO* gene does not accompany increase of artemisinin biosynthesis in *Artemisia annua*. Plant Biol 2007;9:442–6.
[65] Wang H, Song YG, Shen HY, Liu Y, Li ZQ, Wang HH, Chen JL, Liu BY, Ye HC. Transgenic Plants-Advances and Limitations (Effect of Antisense Squalene Synthase Gene Expression on the Increase of Artemisinin Content in *Artemisia annua*), edited by Yelda Özden Çiftçi. Rijeka: InTech; 2012. p. 397–406.
[66] Han JL, Wang H, Ye HC, Liu Y, Li ZQ, Zhang Y, Zhang YS, Yan F, Li GF. High efficiency of genetic transformation and regeneration of *Artemisia annua* L. via *Agrobacterium tumefaciens*-mediated procedure. Plant Sci 2005;168:73–80.
[67] Picaud S, Olofsson L, Brodelius M, Brodelius PE. Expression, purification, and characterization of recombinant amorpha-4,11-diene synthase from *Artemisia annua* L. Arch Biochem Biophys 2005;436:215–26.
[68] Vogeli U, Chappell J. Induction of sesquiterpene cyclase and suppression of squalene synthetase activities in plant cell cultures treated with fungal elicitor. Plant Physiol 1988;88:1291–6.
[69] Kudakasseril GJ, Lam L, Staba J. Effect of sterol inhibitors on the incorporation of 14C-isopentenyl pyrophosphate into artemisinin by a cell-free system from *Artemisia annua* tissue cultures and plants. Plant Med 1987;53:280–4.
[70] Robinson GW, Tsay YH, Kienzle BK, Smith-Monroy CA, Bishop RW. Conservation between human and fungal squalene synthetase. Similarities in structure, function, and regulation. Mol Cell Biol 1993;13: 2706–17.
[71] Wentzinger L, Bach T, Hartmann M. Inhibition of squalene synthase and squalene epoxidase in tobacco cells triggers an up-regulation of 3-hydroxy-3-methylglutaryl coenzyme a reductase. Plant Physiol 2002;130: 334–46.
[72] Zhang Y, Liu Y, Wang H, Ye HC, Li GF. Regulation of squalene synthase gene expression in tobacco by antisense transformation with an *Artemisia annua* squalene synthase gene. J Agric Biotech 2005;13:416–22. (in Chinese)
[73] Chen JL, Fang HM, Ji YP, Pu GB, Guo YW, Huang LL, Du ZG, Liu BY, Ye HC, Ye HC, Wang H. Artemisinin biosynthesis enhancement in transgenic *Artemisia annua* plants by downregulation of the β-caryophyllene synthase gene. Plant Med 2011;77:1759–65.
[74] Akiyoshi D, Klee H, Amasino R, Nester EW, Gordon MP. T-DNA of *Agrobacterium tumefaciens* encodes an enzyme of cytokinin biosynthesis. Proc Natl Acad Sci USA 1984;81:5994–8.
[75] Estruch JJ, Chriqui D, Grossmann K, Schell J, Spena A. The plant oncogene rolC is responsible for the release of cytokinins from glucoside conjugates. EMBO J 1991;10:2889–95.
[76] Jia YT, Ma M, Qu GP, Qian ZX, Lin ZP. The cloning of *RolC* gene and over expression of cytokinins in *Nicotiana tabacum*. Acta Bot Sin 1998;40:211–5.
[77] Smigocki A, Neal JW Jr, McCanna I, Douglass L. Cytokinin-mediated insect resistance in *Nicotiana* plants transformed with the *ipt* gene. Plant Mol Biol 1993;23:325–35.

PART 2

Artemisinin

CHAPTER 9

Introduction to Artemisinin

In 1820, the French chemist Pierre-Joseph Pelletier and Joseph-Bienaimé Caventou separated quinine from cinchona bark. Subsequently, a series of quinolones compounds were synthesized, among them chloroquine, created in the 1930s, showed effective antimalarial activity in clinical trials during World War II, and therefore it was introduced into clinical practice in 1947 for both the treatment and prevention of malaria. However, in the early 1960s, some strains of malaria parasites developed chloroquine resistance and rapidly spread out, which made a great challenge to treat malaria. So, there was an urgent need for new antimalarials with novel structure in the world.

In the early 1970s, Chinese scientists discovered a new type of antimalarial drugs—artemisinins. All these remarkable scientific discoveries were achieved within about 10 years, making an extraordinary and unique China story that is worth to be told to the world community.

In 1971, a team led by Youyou Tu, who first found neutral ethyl ether extract of *Artemisia annua* L., showed 100% effectiveness in inhibiting malaria parasites, achieving the success of Qing Hao antimalarial research. Later in 1972, from that neutral part they isolated a sesquiterpene lactone, artemisinin, containing a peroxy group with a formula of $C_{15}H_{22}O_5$ and a molecular weight of 282. Owing to its novel structure, artemisinin acts totally different from the previous quinoline drugs in chemical nature and pharmacologic mechanism. Thus, the problem of chloroquine-resistant malaria was able to be solved, which was recognized widely as a new breakthrough of antimalarial agents. Therefore, in representation of the 523 group at ICMM, Youyou Tu was invited to the National Congress of Science and Technology in 1978 and was granted award. And 1 year later, artemisinin project was granted the National Scientific Discovery Award for the Antimalarial Drug-Qinghaosu. In 1982, Youyou Tu attended the National Congress for Awards in Science and Technology. As the first inventor of artemisinin project, she received a medal and a certificate for the antimalarial drug invention. Artemisinin is by far the first original Chinese innovative drugs to be appreciated worldwide.

In October 1981, in the fourth meeting of the scientific working group on the chemotherapy of malaria, an international conference in the name of "special conference on qinghaosu" was held by the United Nations Development Programme/the World Bank/the World Health Organization (WHO) special program for research and training in malaria chemotherapy group of tropical diseases in Beijing. Youyou Tu delivered the chief report "Studies on the

From *Artemisia annua* L. to Artemisinins
http://dx.doi.org/10.1016/B978-0-12-811655-5.00009-X

Chemistry of Qinghaosu" to experts from China, the United States, Britain, and India, which was highly praised. And it was believed that the great significance of the discovery laid not only in a new antimalarial agents, but also, more importantly, in the novel unique chemical structure, which would indicate the direction for further design and synthesis of new drugs. As mentioned earlier, a series of artemisinin derivatives, including dihydroartemisinin, artemether, arteether, and artesunate, were developed successively. In the international conference, many subjects on artemisinin were reported, such as the pharmacology, toxicology, clinical application, and others, laying the foundation of the further internationalization of artemisinins. The China Ministry of Health granted an artemisinin New Drug Certificate to the Institute of Chinese Materia Medica of China Academy of Chinese Medical Sciences in 1986 and a dihydroartemisinin New Drug Certificate in 1992, respectively. In brief, the impact of artemisinin on public health is immediate and tremendous. Millions of lives have been saved by the use of artemisinins.

WHO ever stated "expanding access to artemisinin-based combination therapies (ACTs) in malaria-endemic countries has been integral to the remarkable recent success in reducing the global malaria burden. No alternative antimalarial medicine is currently available offering the same level of efficacy and tolerability as ACTs. The emergence of artemisinin resistance in the Greater Mekong Subregion (GMS) is therefore a matter of great concern. Resistance to other antimalarial medicine was also detected first in GMS, eventually appearing elsewhere. In Africa there is evidence that the spread of resistance coincided with increases in child mortality and morbidity." The public awareness of this serious issue must have been raised.

Artemisinin, with its unique chemical structure and superior efficacy, has attracted many domestic and foreign scholars to in-depth studies. In addition to antiparasite, these studies also open a promising future for artemisinins in the fields of anticancer and immunology. During more than 40 years since artemisinin was discovered, studies on *A. annua* L. resources and artemisinin-related compounds have remained hot spots all the time. A search of PubMed for key word "artemisinin" identified more than 3000 related articles published in the last 3 years.

The study field of artemisinin is very wide. In this part the authors are going to present some artemisinin research data concerning chemical structure and properties, methods of extraction, synthesis, determination, pharmacology, toxicology, clinical research, and so on.

CHAPTER 10

Structure and Properties of Artemisinin

Chapter Outline

1 Overview of Structure of Artemisinin [1–5]

Artemisinin is an innovative compound derived from *Artemisia annua* L., which is first found and named by the Institute of Chinese Materia Medica, China Academy of Chinese Medical Sciences (ICMM-CACMS). The CAS Registry Number of artemisinin (Qinghaosu, QHS) is 63968-64-9, and its molecular formula is $C_{15}H_{22}O_5$, which is a novel sesquiterpene lactone with a structural unit of 1,2,4-trioxane including a peroxy group rare in nature. Its molecule consists of seven chiral centers, and biogenetically falls into amorphane type; it is characterized by *cis*-connection between A and B rings, *trans* relationship between isopropyl and bridgehead hydrogen, and interruption of A-ring carbon skeleton by an oxygen atom in artemisinin.

The name of artemisinin was originally Qinghaosu in Pinyin. Later, it was named as artemisinine due to its relation to the generic name, Artemisia; however, "ine" is generally the postfix of such nitrogen compounds as alkaloids and amino acids, whereas artemisinin is a nonnitrogen compound, so Chemical Abstracts recommend "artemisinin" as its English name. Today, artemisinin and Qinghaosu are frequently used.

From *Artemisia annua* L. to Artemisinins
http://dx.doi.org/10.1016/B978-0-12-811655-5.00010-6

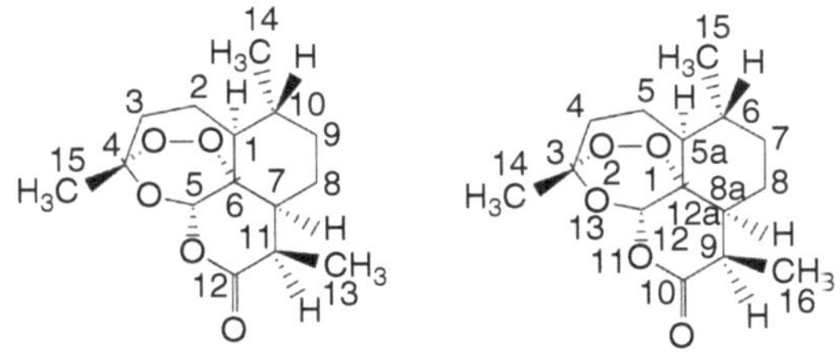

Figure 10.1: The Structure of Artemisinin.

The structure of artemisinin is illustrated in Fig. 10.1, and there are two structure numberings as follows.

In recent years, the latter numbering is accepted in the *Chinese Pharmacopoeia*, and thus the name of artemisinin is (3R,5αS,6R,8αS,9R,12S,12αR)-octahydro-3,6,9-trimethyl-3,12-epoxy-12H-pyrano[4,3-*j*]-1,2-benzodioxepin-10(3H)-one.

Artemisinin is white needle crystal: melting point 156–157°C, $[\alpha]_D^{17}$ = +66.3° (c = 1.64, CCl_4), HRMS *m/e* 282.1472M$^+$, microanalysis (C 63.72%, H 7.86%), without double bond and UV absorption. It is freely soluble in chloroform (CCl_4), acetone, ethyl acetate, and benzene; soluble in ethanol and ether; slightly soluble in cold petroleum ether; and insoluble in water. Owing to its special peroxy group, it is susceptible to reducing substances [5]. Artemisinin is decomposed at more than 190°C.

Its IR spectrum (KBr) has revealed that it contains a δ-lactone (1745 cm^{-1}) and peroxy groups (831, 881 cm and 1115 cm^{-1}).

^{1}H-NMR (CCl_4, 100 MHz $(Me_3Si)_2O$, δ): 0.93 (d, 3H, J = 6 Hz, 15-CH_3); 1.06 (d, 3H, J = 6 Hz, 16-CH_3); 1.36 (s, 3H, 14-CH_3); 3.26 (m, 1H, 9-H, irradiation on this proton causes signal at 1.06 ppm to change from doublet to singlet); 5.68 (s, 1H, 12-H).

^{13}C-NMR ($CHCl_3$, 22.63 MHz): a broadband decoupled spectrum reveals signals of 15 carbon atoms equivalent to sesquiterpenoid skeleton; an off-resonance decoupled spectrum (δ) reveals: 79.5, 105 (s, 2-C); 32.5, 33, 45, 50, 93.5 (d, 5-CH); 25, 25.1, 35.5, 37 (t, 4-CH_2); 12, 19, 23 (q, 3-CH_3); 172 (s, —C=O).

Different research groups reported the NMR data and signal assignments for the artemisinin in literature [6–12].

After measuring crystallographic parameters with X-ray diffractometry, artemisinin is orthorhombic: space group D_2^4-$P2_12_12_1$, unit cell parameters a = 24.077 Å, b = 9.443 Å, c = 6.356 Å; experimental d_0 = 1.30 g/cm^3, calculated d_c = 1.296 g/cm^3, number of molecules in unit cell Z = 4.

2 Chemical Properties and Reactions of Artemisinin [1,4,6,13]

Structurally, artemisinin molecule contains such active groups as peroxy, acetal, ketal, and lactone groups, which can react under different conditions, including reduction reactions of different types, and reactions with acids and bases.

2.1 Reaction of Peroxy Group

Artemisinin is refluxed in triphenylphosphine and xylene solution under nitrogen atmosphere, agitated with formaldehyde and water. The organic layer is rinsed with water. Water layer and acidic solution are then combined. After alkalization, the combined solution is extracted with peroxide-free ether and dried with anhydrous sodium sulfate to remove ether. Triphenylphosphine is weighed. The result shows that the mole number of triphenylphosphine consumed is close to that of artemisinin, demonstrating that artemisinin molecule contains a peroxy fragment.

2.2 Borohydride Reduction [1,2,4,6,13]

Artemisinin is dissolved in methanol. Solid sodium borohydride is slowly added successively under stirring at 0–5°C, and then continues to stir for 0.5 h. The reaction solution is neutralized with glacial acetic acid, and the solvent is evaporated under vacuum to obtain macrocrystalline dihydroartemisinin, which is a hemiacetal compound derived by sodium borohydride reducing artemisinin.

Such mild reductants as lithium borohydride and potassium borohydride may also reduce artemisinin to dihydroartemisinin.

General lactone structures cannot be reduced by borohydrides under reaction conditions of artemisinin (0–5°C, methanol solution). It is neither clear how the peroxy group in the artemisinin molecule promotes the borohydride reduction of δ-lactone, nor clear why it is only reduced to hemiacetal, but not further reacted to alcohols.

2.3 Other Reduction Reactions

In catalytic hydrogenation of artemisinin over Pd/$CaCO_3$ in methanol at room temperature and atmospheric pressure, the peroxide is reduced to the epoxide, that is, deoxoartemisinin. During this reaction process, what is originally derived from the reaction is oily matter. If it is dissolved in a small quantity of acetone in *n*-hexane for 4–5 days, it will spontaneously change into deoxoartemisinin crystal, with a yield of 75%; if it is dissolved in benzene containing a small quantity of *p*-toluenesulfonic acid, it will entirely change into

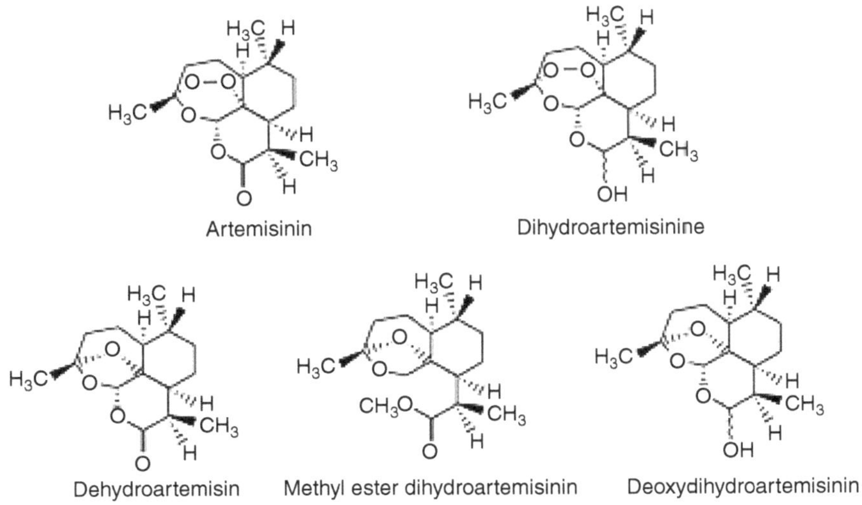

Figure 10.2: Structures of Several Reduction Products of Artemisinin.

deoxoartemisinin crystal rapidly, with a yield up to 96%; when in diazomethane, it will change into methyl ester compounds by methyl esterification.

Catalytic hydrogenation of dihydroartemisinin over $Pd/CaCO_3$ at room temperature and atmospheric pressure will lead to a loss of the peroxy group and obtain an epoxide, that is, deoxydihydroartemisinin.

Zinc/acetic acid reduction may also convert artemisinin into deoxoartemisinin with a high yield.

For deoxoartemisinin, the lactone structure can be reduced to deoxy dihydroartemisinin over diisobutylaluminium hydride at low temperatures. Deoxoartemisinin can be recovered from oxidation of deoxy dihydroartemisinin over chromic anhydride/pyridine (Fig. 10.2).

A restudy of chemistry of artemisinin by Sy et al. [14] in 1997 found four products illustrated in Fig. 10.3 during reduction of artemisinin by lithium aluminum hydride ($LiAlH_4$). Of them, those illustrated in Fig. 10.3A–B are newly found products.

2.4 Reactions With Bases [1,4,15,16]

Artemisinin is dissolved in methanol. In addition, potassium carbonate is dissolved in water. The resulting solution is slowly added to the artemisinin/methanol solution while stirring, mixing well to a clear liquid. At a constant temperature of 20–22°C for 1 h, add water to the solution, extract with ether twice; the ether layer is rinsed with a little water twice, while the water layer is acidized with 10% hydrochloric acid to a PH of 2, then extract with ether thrice; the ether layer is rinsed with water until neutral, dried with anhydrous sodium sulfate

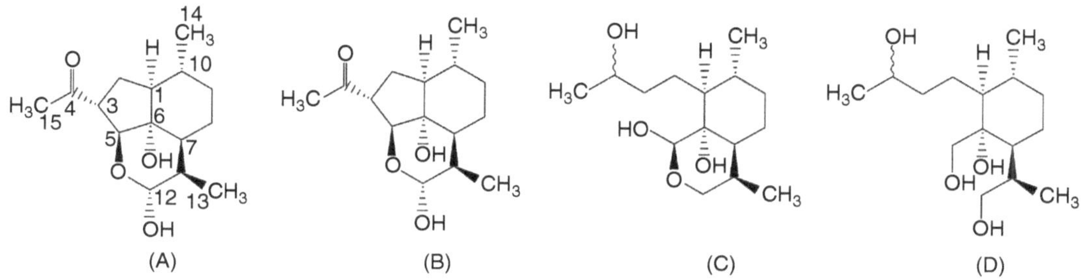

Figure 10.3: Structures of Reduction Products of Artemisinin by $LiAlH_4$.

for 2–3 h, and evaporated under vacuum; the resulting residues are stored in a refrigerator overnight; when semisolids are precipitated, add a little methanol, cool down, precipitate needle crystal, that is, compound 1 in Fig. 10.4, filter, recrystallize once, and obtain more purified crystal.

Dissolve artemisinin in ethanol, add 0.2% sodium hydroxide solution, react in a water bath at a constant temperature of 50°C for 0.5 h, and thus quantitatively produce Q292, with a maximum UV absorbance at 292 nm and an absorptivity of 1.65×10^4 L/(mol·cm). At pH 5.38–6.04, Q292 can be quantitatively converted into the compound Q260, with a maximum UV absorbance at 260 nm and an absorptivity of 1.12×10^4 L/(mol·cm).

2.5 Reactions With Acids [1,4,17]

Shake artemisinin with a mixture of concentrated H_2SO_4 to dissolve, and stand at 25°C for 16–17 h. The resulting solution is sundown, with mild fluorescence. Pour the reaction solution into isometric ice water, mix well, and extract with CCl_4 thrice; the CCl_4 layer is rinsed with water until neutral, dried with anhydrous sodium sulfate, and evaporated under vacuum; recrystallize the resulting coarse crystal twice, and obtain the flaky crystal of the compound 2 in Fig. 10.4, with a melting point of 144–146°C, $[\alpha]_D^{10} = -16.2°$ ($c = 2.1$, CCl_4).

Add hydrochloric acid to artemisinin in ethanol, and react in a water bath at a constant temperature of 55°C for 5 h. The resulting reaction products have maximum absorbance at 254 nm. NMR indicates destruction of the oxygen bridge and into a product with α,β-unsaturated carbonyl structure in acidic solution, but its chemical structure remains unclear.

2.6 Electrochemical Properties

Chen et al. [18,19] used multiple electrochemical methods to investigate the reduction of the peroxy group of artemisinin molecule on the mercury electrode, and found that when the reduction potential was around 0.0 V (vs Ag/AgCl), the electrode process was an irreversible reduction; number of electrons for reduction reaction $n = 2$, half-wave potential $E_{1/2} = 0.012$ V, electron transfer coefficient $\alpha = 0.66$, apparent standard rate constant for the

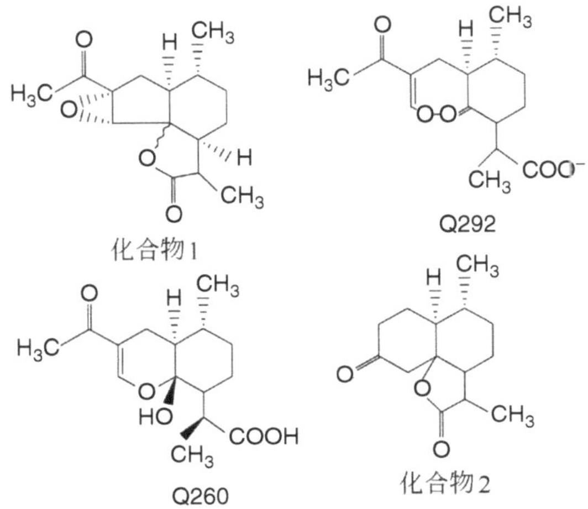

Figure 10.4: Structures of Some Acid-Base Reaction Products of Artemisinin.

electrode reaction $k_s' = 6.34 \times 10^{-6}$ cm / s, and diffusion coefficient $D = 4.3 \times 10^{-6}$ cm²/s; reaction products had adsorbability on the electrode surface. Also, they used electrochemical methods to investigate the interaction between artemisinin and hemin. For artemisinin, an irreversible reduction, in which two electrons transferred, occurred at −1.08 V on a glassy carbon electrode. Chan et al. [20] also used HPLC-ECD to investigate electrochemical properties of artemisinin.

3 Color Reactions of Artemisinin [1,6,21,22]

Color reactions are simple and feasible methods for qualitative analysis and identification of artemisinin. Test tube reactions or TLCs are commonly used, and there are several methods as follows.

3.1 Condensation Reaction of PDAB

Dissolve approximately 10 mg of artemisinin in 2 mL of ethanol, add 1 mL of PDAB reagent, and heat in water bath. The solution should be bluish violet.

3.2 Ferric Hydroxamate Test

Dissolve approximately 10 mg of artemisinin in 1 mL of methanol, add 4–5 drops of solution of 7% hydroxylamine hydrochloride in methanol, heat in water bath to boil, cool down, adjust to be acidic with hydrochloric acid, and then add 1–2 drops of 1% $FeCl_3$ in ethanol. The solution should be purplish red.

3.3 2,4-Dinitrophenylhydrazine Test

Dissolve approximately 10 mg of artemisinin in 1 mL of CCl_4, apply it to a thin layer plate, spray with 2,4-dinitrophenylhydrazine reagent, and bake in an oven at 80°C for 10 min. A yellow spot should be produced.

3.4 Alkaline m-Dinitrobenzene Test

Dissolve approximately 10 mg of artemisinin in 2 mL of ethanol, separately add several drops of 2% *m*-dinitrobenzene in ethanol and KOH in ethanol, and slightly heat in water bath. The solution should be red.

3.5 Vanillin-Sulfuric Acid Color Reaction

Apply a certain amount of artemisinin TS to the silica gel G plate, using petroleum ether–ethyl acetate (8:2) or benzene–ether (4:1) as the mobile phase, 1% vanillin in concentrated sulfuric acid as the chromogenic reagent. The spot of artemisinin should be yellow, and then turn to blue.

3.6 Acidic Potassium Iodide Solution Test

It is a color reaction of the peroxy group. After reaction of artemisinin with potassium iodide solution, add the starch solution, and bluish violet is produced immediately.

References

[1] Traditional Chinese medicine research institution. Artemisinin-based antimalarial album (1971–1978). Traditional Chinese medicine research data, 1978.3.
[2] Research Group of Artemisinin Structure (China academy of traditional Chinese medicineang The Chinese Academy of SciencesInstitute of Biophysics). A new kind of sesquiterpene lactone—artemisinin. Chin. Sci. Bull. 1977;22(3):142.
[3] Qinghaosu Antimalarial Coordinating Research Group. Antimalarial studies on Qinghaosu. Chin. Med. J. 1979;92(12):811–6.
[4] Liu JM, Ni MY, Fan JF, Tu YY. Structure and reaction of arteannuin. Acta Chim. Sin. 1979;37(2):129–43.
[5] Klayman DL. Qinghaosu (artemisinin): an antimalarial drug from China. Science 1985;228(4703):1049–55.
[6] Wang ZD, Sun FH. The progress of studies on physical, chemical characters and determination of artemisinin. Acta Agric. Univ. Jiangxiensis 1999;21(4):606–11.
[7] Blasko G, Cordell GA, La Kin DC. Definitive 1H and 13C-NMR assignments of artemisnin (Qinghaosu). J. Nat. Prod. 1988;51(6):1273–6.
[8] Wang ZS, Nakashima TT, Kopecky KR, et al. Qinghaosu: 1H and 13C nuclear magnetic resonance spectral assignments and luminescence. Can. J. Chem. 1985;63:3070–4.
[9] Prakash Q, Roy R, Kapil RS. Two-dimensional NMR studies of arteether. Indian J. Chem. 1990;29B(2):103–7.
[10] Luo D, Yeh HJC, Brossi A, et al. Configuration of antimalarials derived from qinghaosu: dihydroqinghaosu, artemether, and artesunic acid. Helv. Chim. Acta 1984;67:1515–22.

[11] Leppard DG, Rey M, Dreiding AS, et al. The structure of arteannuin B and its acid product. Helv. Chim. Acta 1974;57(3):602–15.
[12] Jeremi D, Joli A, Behbud A, et al. A new type of sesquiterpene lactone isolated form Artemisia annua L. arteannuin B. Tetrahedron Lett. 1973;14(32):3039–42.
[13] An YL. Phytochemistry. Harbin: Northeast Forestry University Press; 1996. 359.
[14] Sy LK, Hui SM, Cheung KK, et al. A rearranged hydroperoxide from the reduction of artemisinin. Tetrahedron 1997;53(22):7493–500.
[15] Zhao SS, Zeng MY. Deteminating artemisinin in southernwood by high-performance liquid chromatograph. Chin. J. Pharm. Anal. 1986;6(1):3–4.
[16] Brown GD. Secondary metabolism in tissue culture of Artemisia annua. J. Nat. Prod. 1994;57(7):975–7.
[17] Li Y, Yu PL, Chen YX, Zhang JL, Wu YL. Studies on analogs of artemisinine. Chin. Sci. Bull. 1985;30(7):1313–5.
[18] Chen Y, Zhu SM, Chen HY, Li Y. Study on electrochemical behaviors of artemisinin (Qinghaosu) and its derivatives. I. Reduction of artemisinin at Hg electrode. Acta Chim. Sin. 1997;55(9):925–9.
[19] Chen Y, Zhu SM, Chen HY. Study on electrochemical behaviors of artemisinin (Qinghaosu) and its derivatives II. Reduction mechanism of artemisinin in the presence of hemin. Acta Chim. Sin. 1998;56(9):921–5.
[20] Chan KL, Yuen KH, Jinadasa S, et al. A high performance liquid chromatography analysis of plasma artemisinin using a glassy carbon electrode for reductive electrochemical detection. Plant Med. 1997;63(1):66–9.
[21] Zhang JJ. Chinese Medicine Chemistry. Beijing: People's Medical Publishing House; 1994. 168–170, 319–321.
[22] Kan YM, Huang Tk. Chinese Medicine Chemistry Experiment Operating Technology. Beijing: China Medical Science Press; 1988. 193–197.

CHAPTER 11

Research on Assays of Artemisinin

Chapter Outline

The Institute of Chinese Materia Medica, China Academy of Chinese Medical Sciences (ICMM-CACMS) discovered artemisinin from Herba Artemisiae Annuae (*Artemisia annua* L.) in 1972; after confirmation of its chemical structure in 1975, further research had been carried out on assays of artemisinin, and quantitative analysis methods of artemisinin, including iodometry and ultraviolet (UV) spectrophotometry [1,2], were established for assays of Herba Artemisiae Annuae, its extracts and active pharmaceutical ingredients, and preparations of artemisinin. With the progress of modern analytical techniques, scholars, at home or abroad, continue to develop new methods to make quantitative analysis of artemisinin more accurate, efficient, and simple to meet demands for assays of artemisinin from different sources of samples. This chapter briefly introduces quantitative analysis methods of artemisinin and its derivatives.

From *Artemisia annua* L. to Artemisinins
http://dx.doi.org/10.1016/B978-0-12-811655-5.00011-8

1 Quantitative Reactions of Artemisinin and Its Analysis Methods [1–4]

Artemisinin is characterized by no absorption at near UV band, simply weak absorption at end (203 nm), no fluorophore, and decomposition at high temperature, leading to a certain difficulty in direct analysis and determination. Thus, characteristic groups of artemisinin can be used to carry out a series of quantitative reactions to establish corresponding quantitative analysis methods. Zeng et al. [3,4] studied and summarized quantitative reactions during analysis of artemisinin (Figs. 11.1 and 11.2).

1.1 Reaction of Iodometry

Iodometry is used for assay of active pharmaceutical ingredients of artemisinin. In acid medium consisting of absolute ethanol/sulfuric acid, one artemisinin molecule, by its peroxy

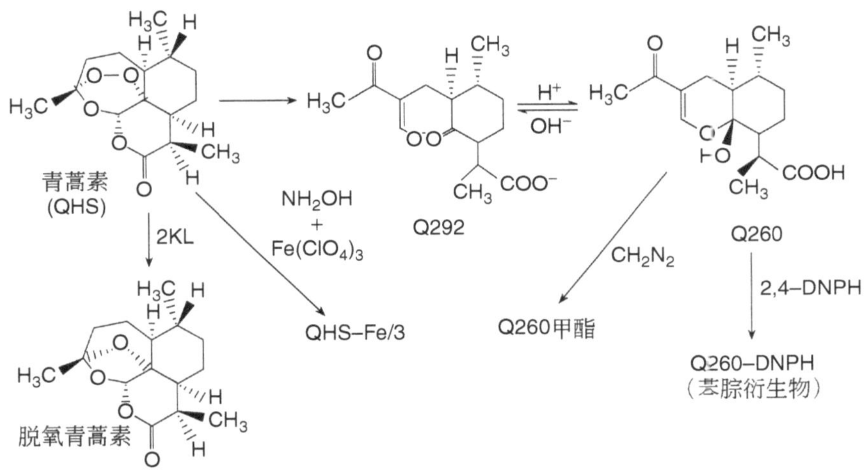

Figure 11.1: Quantitative Reactions of Artemisinin.

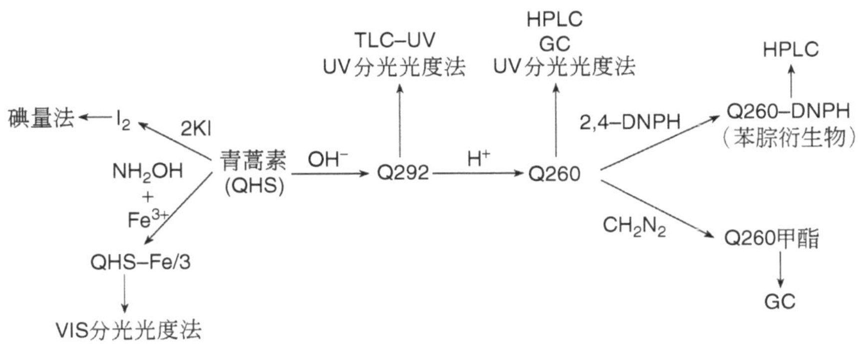

Figure 11.2: Quantitative Reactions and Determination Methods of Artemisinin.

group, oxidizes two KI molecules to one free iodine molecule, while artemisinin is reduced to one deoxyartemisinin molecule; the resulting free iodine is determined by classical iodometry. The improvement of this method is to use 2.5 mol/L sulfuric acid (as acidic reagent) and absolute ethanol as an acid medium, instead of glacial acetic acid which is prone to oxidize iodide ions in conventional iodometry. This can reduce the occurrence of self-oxidation of iodide ions and increase the concentration of iodide ion in the reaction solution, thereby improving the accuracy of the method and the reaction rate. The recovery rate of this method (up to 99.9%) will be ideal if it is performed at a constant temperature of 42°C for 1 h. This method is used for assay of artemisinin, with a variation coefficient of 0.9%.

1.2 Ferric Hydroxamate Test [1–3]

Reaction of the saturated lactonic group of artemisinin with the alkaline hydroxylamine produces a derivative of hydroxamic acid. In the acidic nonaqueous system, this derivative and ferric ion produce a purplish red complex. The maximum absorption peak is at 520 nm. The specific absorptivity is approximately 45. Spectrophotometrically, the absorbance–concentration curve is linear. The linear regression equation is $Y = 0.084X - 0.019$. The correlation coefficient is 0.993. The method detection limit is 2×10^{-5} g/mL.

1.2.1 Methods

Weigh artemisinin sample accurately, and prepare a 0.00250 g/mL solution with absolute ethanol. Apply 3.0 mL of this solution into a 25 mL ground round-bottom flask, dilute with absolute ethanol to 5 mL. Add 3.0 mL of alkaline hydroxylamine solution (mix 12.5% hydroxylamine hydrochloride in methanol and 12.5% NaOH in methanol equally, filter; newly prepared when used), reflux on water bath for 5 min, cool down, apply 20 mL of ferric perchlorate solution (dissolve 5 g of ferric perchlorate in 10 mL of water and 10 mL of 70% perchloric acid) into a 500 mL volumetric flask, add 6 mL of 70% perchloric acid. Slowly dilute with absolute ethanol to volume, quantitatively transfer the reaction solution into a 50 mL volumetric flask, and dilute to volume. Using absolute ethanol reaction solution as a blank, absorbances at 250 nm are measured on a spectrophotometer.

This method, combined with silica gel column chromatography and using petroleum ether (60–90°C) and acetone (95:5) as eluent, can be used for assay of artemisinin from *A. annua* L. The variation coefficient is below 2%.

1.3 Base Reaction [1–3]

The mechanism of alkali treatment is to convert artemisinin into a compound with apparent absorption peaks at 292 nm for further determination. Artemisinin, in 0.2% NaOH solution, quantitatively produces an α, β-unsaturated keto acid salt, that is, Q292, which has the maximum absorption peak at 292 nm, with an absorptivity of 1.65×10^4 L/(mol·cm).

The absorbance–concentration curve has a good linear relationship by using UV-spectrophotometric method. The regression equation is $Y = 0.005 + 1.17X$; the correlation coefficient is 1.00; the method detection limit is at 10^{-6} g/mL.

Shen et al. [5] used this reaction for assay of artemisinin. When this method was modified as treatment of artemisinin in ethanol with 0.2% NaOH solution at 50°C for 30 min, the result was ideal and the correlation coefficient was below 0.7%. This method, combined with silica gel thin-layer chromatography (TLC) and using a mixture of petroleum ether (60–90°C) and ether (6:4) as mobile phase, can be used for assay of artemisinin from *A. annua* L. plant and its callus. The correlation coefficient is generally below 5%.

Gordi et al. [6] established a method of high-performance liquid chromatography (HPLC)-postcolumn derivatization and successfully applied it to the determination of plasma artemisinin. Three columns were used in this experiment: anterior, analytical, and posterior columns. Of them, the analytical column was followed by a high pressure pump, which pumped in the 0.3 mol/L KOH/ethanol in water (9:1, v/v) solution to allow artemisinin to carry out derivation reaction in the posterior column. The anterior column was an ADS-V18 column, which was used to remove biomacromolecules from the plasma; the analytical column was an RP-C18 column (100 × 4.6 mm, 3 μm), with a mixture of water and acetonitrile (50:50) as mobile phase, at a flow rate of 0.75 mL/min; the posterior column was a winding 5 m PTFE column, 0.46 mm in inner diameter, immersing in the water bath at 70°C; the artemisinin-containing fraction carried out derivation reaction in the 0.3 mol/L KOH/ethanol in water (9:1, v/v) solution, and produced Q292; then UV detection was performed, with a detection limit of 10 ng. Elsohly et al. [7] reported analysis of artemisinin in *A. annua* L. by HPLC with postcolumn derivatization and UV detection.

1.4 Acid Conversion Reaction After Base Reaction [1–3]

Artemisinin produces Q292 via base reaction. At pH 5.58–6.64, Q292 can be quantitatively converted into a compound Q260, with its maximum absorption peak at 260 nm. The absorptivity of Q260 is 1.12×10^4 L/(mol·cm). The absorbance–concentration curve has a good linear relationship by using UV-spectrophotometric method. The linear regression equation is $Y = 0.032X - 0.003$. The correlation coefficient is 1.00. The method detection limit is at 10^{-6} g/mL.

1.4.1 Methods

Accurately apply 10 mL of artemisinin in ethanol solution to a 50 mL volumetric flask. Dilute with 0.2% NaOH solution to volume, and react in the water bath at 50°C for 30 min. Cool it to ambient temperature, then apply 4.0 mL of this solution to a 10 mL volumetric flask, and add 0.5 mL of 0.04 mol/L acetum. Dilute with ethanol to volume, shake well. The absorbance at 260 nm is measured on an UV spectrophotometer, using 95% ethanol reaction solution, plus isometric acetum and ethanol, as blank.

This method, in combination with gas chromatography (GC) and liquid chromatography, can be used for analysis of various artemisinin samples.

1.5 High-Performance Liquid Chromatography (HPLC)

Shishan et al. [8–10] used acid conversion of Q292, the base reaction product, into Q260, on a Lichrosorb RP-18 (10 μm), using a mixture of 0.01 mol/L NaH_2PO_4–Na_2HPO_4 buffer and methanol solution (55:45) as mobile phase. Absorbance was detected at 260 nm by UV detector. Signals were linearly related to concentrations. The detection limit was at 10^{-7} g/mL. When using the direct on-column enrichment technique, the detection limit could be at 10^{-9} g/mL. This method can be available for assays of *A. annua* L. plant, artemisinin preparations, and biological fluid samples.

1.5.1 Methods

Apply 2.00 mL of artemisinin extract to a 25 mL volumetric flask, add 10 mL of 0.20% NaOH solution, and react in the water bath at 50°C for 30 min. Then, cool it in the running water to ambient temperature, dilute with 0.08 mol/L acetum to volume, shake well. HPLC–UV analysis is performed for this derivatization reaction solution. Chromatographic conditions: Waters HPLC system, including a Waters 510 HPLC pump, a Waters 717 Plus HPLC autosampler injector, an RP-C18 column (4.6 mm × 25 cm, 5 μm stationary phase), and a 486 HPLC UV detector. Mobile phase: Na_2HPO_4 (0.9 mmol/L)–NaH_2PO_4 (3.6 mmol/L) buffer (methanol–water–acetonitrile = 45:45:10, v/v, pH 7.76). Flow rate: 0.5 mL/min. Sample size: 10 μL. Detection wavelength: 260 nm. Column temperature: 30°C. Linear dependence of experiment $r = 0.9996$, RSD: 1.1%, spike-and-recovery: 98.9%.

1.6 Gas Chromatography (GC) [2,3]

Artemisinin cannot be directly detected by GC because it can break down into several products at high temperatures. However, artemisinin needs no further derivatization after its conversion into Q260. Using nitrogen as a carrier gas, high-temperature (345°C) vaporization for injection, it can be detected on a 5% poly(methylsiloxane-*co*-phenylsiloxane) stationary phase fused-silica capillary column by splitless low-temperature enrichment method. Temperature programming: the programmed temperature is elevated (at a rate of 10°C/min) from an initial temperature of 80°C to a final temperature of 200°C (for 15 min) immediately after injection. The peak temperature is 188°C. Flame ionization detector is used for detection, with a detection limit of 3×10^{-8} g.

1.7 Methylation of Base Reaction–Acid Conversion Product [2,3]

Q260 is methylated by diazomethane to produce a methyl ester derivative of Q260.

1.7.1 Methods

As described in the "base reaction," 10 mL of 0.100 mg/mL artemisinin in ethanol solution is treated. Adjust the pH value to 3 with dilute hydrochloric acid, and extract Q260 with ethyl acetate. Volatilize ethyl acetate, and then dissolve the residue in 1.2 mL of methanol. Place in an ice bath at −5°C, dropwise add excessive diazomethane in ether solution, and volatilize the solvent after 0.5 h. Dissolve the methylation product in 100 μL of methanol.

This reaction can be used in GC: 5% poly(methylsiloxane-*co*-phenylsiloxane) stationary phase fused-silica capillary column; nitrogen as a carrier gas, splitless injection at 260°C, low-temperature enrichment; temperature programming from an initial temperature of 50°C (5 min) to a final temperature of 200°C (for 15 min). The peak temperature of the methyl ester derivative of Q260 is 180°C. Flame ionization detector is used for detection, with a detection limit of 2×10^{-8} g. The gas chromatography–mass spectrometry (GC–MS) method can identify the consistency between its mass spectrum and process of GC, suggesting the temperature stabilization of this methyl ester derivative.

This method is available for the determination of animal plasma concentration. Each mouse is administered orally at a dose of 2.0 g/kg, and approximately 1 mL of blood is drawn. Other blood components do not interfere with the determination.

1.8 2,4-Dinitrophenylhydrazine (2,4-DNPH) Derivatization of Base Reaction–Acid Conversion Product [2,3]

After conversion of artemisinin into Q260, phenylhydrazone derivative can be formed by condensation of the keto group in the Q260 molecule with 2,4-DNPH in the presence of an acidic catalyst. Its maximum absorption peak is at 391 nm. The absorptivity is 2.02×10^4 L/(mol·cm). Reaction reagents and derivatives are separated on an HPLC column (ODS, Du Pont 7 μm) using a mixture of water and methanol (22.5:77.5) as mobile phase, and detected at 347 nm by the UV detector; signals are linearly related to concentrations. The correlation coefficient is 0.992. The method detection limit is 2×10^{-5} g/mL.

1.8.1 Methods

Accurately weigh Q260, and prepare a 0.10 mg/mL methanol solution. Apply 1.0 mL of this solution to a 10 mL volumetric flask, and dilute with 0.15% 2,4-DNPH in methanol solution (containing 1% acetic acid) to volume. React at room temperature for 24 h. This solution is separated from the ODS column, and the peak height (or peak area) is measured.

2 Advances in Analytical Methods of Artemisinin

2.1 Pulse Polarography

Zhang et al. [11] used pulse polarography to determine the artemisinin content of *A. annua* L. plant based on the reducibility of the peroxy group of the artemisinin molecule. Using mercury electrode as working electrode, 0.05 mol/L ammonium sulfate as base solution, artemisinin showed a good peak profile at 0.0 V (vs SCE); the concentration ranged from 2 to 16 μg/mL; peak heights showed a good linear relation to concentrations; the detection limit was 2 μg/mL; and the error was within 2%. This method has high sensitivity and resolution, and is available for assay of microsample.

2.2 Gas Chromatography (GC)

A study of thermal stability of artemisinin by Lin et al. [12] found that artemisinin was thermostable below 150°C, but it was decomposed into several products when heated to 180–200°C. Sipahimalani et al. [13] established a fast GC method, by which determination of two decomposition products quantitatively produced by artemisinin could indirectly determine the artemisinin content of *A. annua* L. plant and its tissue cultures, with the minimum detection limit of 1 μg. Woerdenbag et al. [14] established a GC–MS method simultaneously determining the content of artemisinin, artemisitene, arteannuin II, and artemisinic acid, and determined the content of each constituent from different parts of *A. annua* L. plant, showing that the minimum detection limit of each constituent could be at 10^{-12} g (ng).

The gas chromatography with electron-capture detection (GC–ECD) developed by Liu et al. [15] requires only 100 mg of plant material, and can be easily applied to determine the content of artemisinin in an individual leaf.

2.3 Thin-Layer Chromatography (TLC) Scanning

Luo et al. [16] established a quantitative dual-wavelength TLC-scanning method for assay of artemisinin in leaves of *A. annua* L.: silica gel G plate as stationary phase, a mixture of petroleum ether and ether (6:4) as mobile phase, dimethylaminobenzaldehyde as chromogenic reagent, detection wavelength at 580 nm, reference wavelength at 750 nm. The findings showed that the standard deviation (SD) was within 4% and that the spike-and-recovery of artemisinin was more than 95%.

Chen et al. [17] determined the content of dihydroartemisinin, artemether and its metabolite dihydroartemisinin, artesunate and its metabolite dihydroartemisinin in plasma using the quantitative TLC-scanning method. The linear range was 0.1–4.0 μg/mL; the detection limit was 0.01–0.03 μg; the recovery rates of dihydroartemisinin, artemether versus dihydroartemisinin, and artesunate versus dihydroartemisinin were 96%, 82.8% versus 92.2%, and 95.5% versus 89.5%, respectively.

A simple TLC, reversed-phase TLC, and high-performance TLC have been developed for the rapid and accurate analysis of artemisinin in a large number of *A. annua* plantlets cultured in vitro [18–20].

2.4 High-Performance Liquid Chromatography (HPLC)

2.4.1 Direct HPLC–UV determination

Wang et al. [21] used an internal standard method for HPLC assays of artemisinin and its derivatives. The chromatographic column was Silica A (amorphous full-pore microparticle silica gel, particle size 13 μm, specific surface area 400 m^2/g); the mobile phase was isooctane solution containing 1% methanol and ethanol; cholesterol was an internal standard; the detection wavelength was 216.5 nm; the absorption coefficient of artemisinin was 6.5; the sensitivity under experimental conditions was approximately 0.5 μg.

Bharati and Sabat [22] described a sensitive and rapid UV method, which is highly specific to artemisinin with a single spectral resolution at 291 nm. Identification of impurities in artemisinin was chromatographed using HPLC [23].

2.4.2 HPLC–evaporative light scattering detection (HPLC–ELSD)

Christen and Veuthey [24] established an HPLC–ELSD method for simultaneous determination of artemisinin and artemisinic acid, which was carried out on a C18 silica column using an isocratic elution with a mobile phase consisting of acetonitrile and water (60:40) adjusted to pH 3 with trifluoroacetic acid. The conditions of ELSD were gas pressure 0.05 MPa; flow rate 6 L/min; nebulization chamber and heating chamber temperature 40°C. This method can also be used to determine the samples of *A. annua* L. plant and its SFE-CO_2 extract. Zhang et al. [25] used HPLC–UV–ELSD for determination of the content of artemisinin, arteannuin B, and artemisinic acid in Herba Artemisiae Annuae. Avery et al. [26] applied HPLC–ELSD to analyze and determine the artemisinin and its analogues.

2.4.3 HPLC–electrochemical detection (HPLC–ED)

Artemisinin can be determined by electrochemical methods because electrochemical reduction can be carried out in its peroxy group. Acton et al. [27] established a method for the determination of artemisinin in *A. annua* L. by HPLC with Au/Hg electrode-based electrochemical detector; this method was fast, sensitive, and well selective, but the Au/Hg electrode needed to frequently wash, polish, and prepare. Charles et al. [28] and Chan et al. [29] used a glassy carbon electrode for ED of artemisinin content in *A. annua* L. and human plasma, respectively, and the findings showed not only high sensitivity but also higher stability. Vandenberghe et al. [30] established an HPLC–UV–ED method, which could simultaneously determine artemisinin and its bioprecursors in *A. annua* L., of which arteannuin B and artemisinic acid were detected at 228 nm using a UV detector due to lack of peroxy group, whereas artemisinin and artemisitene were electrochemically detected using a glassy carbon electrode at a potential of −1.0 V. There were more reports regarding studies of HPLC–ED for the simultaneous determination of artemisinin-based drugs (artesunate, artemether, and arteether) and their major metabolite (dihydroartemisinin) in biological fluids [31–34].

HPLC–ED can determine the artemisinin without derivatization and is available for determination of low-content samples due to its high sensitivity and good selectivity. However, due to its determination at a reduction potential and higher demand for the system, both sample solution and mobile phase need strict deoxidation treatment, and the device is complex and cumbersome.

2.4.4 HPLC–MS

Filip et al. [35] established a HPLC-ESI Q-TOF MS/MS method for the simultaneous quantification of artemisinin, artemisitene, arteannuin B, and artemisinic acid in *Artemisia annua* L., showing a recovery of >97%, so this method is fast and highly sensitive, with excellent selectivity. Naik et al. [36], Souppart et al. [37], Rajanikanth et al. [38], and Huang et al. [39] reported the determination of artesunate, artemether, arteether, and their metabolite dihydroartemisinin in vivo by HPLC–MS, respectively.

A rapid HPLC–MS method was also used to determine biosynthetic precursors in *A. annua* L. crude extracts such as 9-epi-artemisinin, which is reported for the first time [40,41].

2.5 Supercritical-Fluid Chromatography (SFC)

In recent years, SFC has also been used for the quantification of artemisinin. Kohler et al. [42] established a method for the determination of artemisinin and artemisinic acid by SFC–ELSD. When using silica gel column or Silica Gel C18 as a chromatographic column and carbon dioxide containing 3% methanol as mobile phase, artemisinin and artemisinic acid could be separated within 3 min; this method is available for the quality control of pure compounds, but not for plant extracts, because some other substances interfere with the determination of artemisinic acid. When using ZORBAX NH2 column as a chromatographic column and carbon dioxide/methanol for gradient elution, the analysis process can be finished within 8 min; this method has good selectivity and is available for the determination of extracts of *A. annua* L. Mount et al. [43] established a method for the quantification of artemisinin by SFC with an electron-capture detector.

2.6 Enzyme-Linked Immunosorbent Assay (ELISA)

Jaziri et al. [44] reported an ELISA method for the quantitative determination of artemisinin. Ferreira and Janick [45] used ELISA for the determination of artemisinin, by which artemisinin was derivatized to dihydroartemisinin carboxymethyl ether in three steps, and then linked to bovine serum albumin (BSA), and finally artemisinin–BSA conjugates were injected in female New Zealand rabbits and generated polyclonal antibodies. This assay could detect artemisinin concentrations as low as 1.5 ng/mL and was ε 400-fold more sensitive than the HPLC–ED.

The mAb 3H2 was used to develop a highly sensitive and specific indirect competitive enzyme-linked immunosorbent assay for artemisinin, which can recognize the artemisinin analogs artesunate, dihydroartemisinin, and artemether with cross-reactivity of 650, 57, and 3%, respectively [46]. Zehnacker et al. [47] developed a sensitive direct chemiluminescent enzyme immunoassay for the determination of dihydroartemisinin.

2.7 Radioimmunoassay (RIA)

Song et al. [48], using RIA, linked the C-12 hydroxyl group of dihydroartemisinin to acetic acid–BSA antigen to produce an artemisinin-12-O-acetic acid–BSA antigen complex, and then intramuscularly injected to the sheep to obtain an antiserum antibody, which could be used in RIA for artesunate and artemisinin. This method is well reproducible and highly sensitive, up to 2–10 ng. RIA also has some problems despite its high sensitivity and specificity, for example, demands for some special substances, instability, high cost, and certain harm to the health.

In addition, there were also research reports regarding titrimetry [49], infrared spectroscopy [50], circular dichroism [51], and capillary electrophoresis [52] for the quantitative analysis of artemisinin and its derivatives.

References

[1] 中国中医研究院中药研究所.青蒿抗疟研究 (1971–1978). 中药研究资料, 1978, 3. [Available only in Chinese: Traditional Chinese Medicine Research Institution. Artemisinin-based antimalarial album (1971–1978). Traditional Chinese medicine research data 1978, p. 3].

[2] 中国中医科学院中药研究所. 常用中药材品种整理与质量研究——青蒿, 1990, 104–109 (内部资料). [Available only in Chinese: Traditional Chinese Medicine Research Institution. Quantitative analysis and identification of traditional Chinese medicine: artemisinin 1990, pp. 104–109].

[3] 曾美怡, 赵世善, 付桂兰. 青蒿素分析中的各种定量反应. 药物分析杂志. 1986;6(3): 183–5. [Available only in Chinese: Zeng MY, Zhao SS, Fu GL. Various quantitative reaction in artemisinin analysis. Chin J Pharmaceut Anal 1986;6(3):183–5].

[4] 曾美怡. 桥式有机过氧化物碘量法的改进研究——碘量法测定青蒿素. 药物分析杂志, 1984;4(6):327–9. [Available only in Chinese: Zeng MY. A modified iodometric method in determination of organic bridged peroxides: iodometric determination of Qinghaosu. Chin J Pharmaceut Anal 1984;4(6): 327–9].

[5] 沈旋坤, 严克东, 罗泽渊等. 紫外分光光度法测定青蒿素含量. 药物分析杂志, 1983;3:24–26. [Available only in Chinese: Shen XK, Yan KD, Luo ZY, et al. Determination of arteannuin in element content by the ultraviolet spectrophotometry. Chin J Pharmaceut Anal 1983;3:24–6].

[6] Gordi T, Nielsen E, Yu ZX, et al. Direct analysis of artemisinin in plasma and saliva using coupled-column high-performance liquid chromatography with a restricted-access material pre-column. J Chromatogr B Biomed Sci Appl 2000;742(1):155–62.
[7] Elsohly HN, Croom EM, Elsohly MA. Analysis of the antimalarial sesquiterpene artemisinin in *Artemisia annua* by high-performance liquid chromatography (HPLC) with postcolumn derivatization and ultraviolet detection. Pharm Res 1987;4(3):258–60.
[8] 赵世善，曾美怡．高效液相色谱法测定青蒿植物中的青蒿素．药物分析杂志，1986;6(1):3–4. [Available only in Chinese: Zhao SS, Zeng MY. Deteminating artemisinin in southernwood by high-performance liquid chromatograph. Chin J Pharmaceut Anal 1986;6(1):3–4].
[9] Shishan Z, Meiyi Z. Application of high-performance liquid chromatography of Qinghaosu in animal plasma. Anal Chem 1986;58:289–93.
[10] Shishan Z. High performance liquid chromatography determination of artemisinin (qinghaosu) in human plasma and saliva. Analyst 1987;112(5):661–4.
[11] 张秀琴，徐礼桑．中药青蒿中青蒿素的脉冲极谱测定法．药学学报，1985;20(5):383–86. [Available only in Chinese: Zhang XQ, Xu LS. Determination of Qinghaosu (arteannuin) in *Artemisia annua* L. by pulsepolarography. Acta Pharmaceut Sin 1985;20(5):383–6].
[12] Lin AJ, Klayman DL, Hoch JM, et al. Thermal rearrangement and decomposition products of artemisinin (qinghaosu). J Org Chem 1985;50:4504–8.
[13] Sipahimalani AT, Fulzele DP, Heble MR. Rapid method for the detection and determination of artemisinin by gas chromatography. J Chromatogr 1991;538(2):452–5.
[14] Woerdenbag HJ, Pras N, Bos R, et al. Analysis of artemisinin and related sesquiterpenoids from *Artemisia annua* L. by combined gas chromatography/mass spectrometry. Phytochem Anal 1991;2:215–9.
[15] Liu S, Tian N, Liu Z, Huang J, Li J, Ferreira JF. Affordable and sensitive determination of artemisinin in *Artemisia annua* L. by gas chromatography with electron-capture detection. J Chromatogr A 2008;1190(1-2):302–6.
[16] 罗亨明，赵萍萍，余朝菁等．青蒿中青蒿素的薄层扫描定量法．药学通报，1980;5(8):344–46. [Available only in Chinese: Luo HM, Zhao PP, Yu LJ. Quantitative method of artemisinin in *Artemisia annua* by TLC scanning method. Chin Pharmaceut J 1980;15(8):344–6].
[17] 陈来舜，曾衍霖．蒿甲醚、蒿琥酯及其代谢产物双氢青蒿素在血浆中的薄层 扫描定量法．中国医药工业杂志，1989;20(2):75–8. [Available only in Chinese: Chen LS, Zeng YL. Thin-layer chromatographic densitometric method for simultaneous determination of artemether or artesunate and their metabolite in rat plasma. Chin J Pharmaceut 1989;20(2):75–8].
[18] Marchand E, Atemnkeng MA, Vanermen S, Plaizier-Vercammen J. Development and validation of a simple thin layer chromatographic method for the analysis of artemisinin in Artemisia annua L. plant extracts. Biomed Chromatogr 2008;22(5):454–9.
[19] Bhandari P, Gupta AP, Singh B, Kaul VK. Simultaneous densitometric determination of artemisinin, artemisinic acid and arteannuin-B in Artemisia annua using reversed-phase thin layer chromatography. J Sep Sci 2005;28(17):2288–92.
[20] Khan S, Ali A, Ahmad S, Abdin MZ. Affordable and rapid HPTLC method for the simultaneous analysis of artemisinin and its metabolite artemisinic acid in Artemisia annua L. Biomed Chromatogr 2015;29(10):1594–603.
[21] 王仲山，朱耀华，张叔良等．青蒿素及其衍生物的高压液相色谱含量测定 研究．药学学报，1981;16(6):466–69. [Available only in Chinese: Wang ZS, Zhu YH, Zhang SL, Lv XL. Determination of artemisinine and its derivatives by high pressure liquid chromatography. Acta Pharmaceut Sin 1981;16(6):466–9].
[22] Bharati A, Sabat SC. A spectrophotometric assay for quantification of artemisinin. Talanta 2010;82(3):1033–7.
[23] Stringham RW, Pennell M, Cabri W, Carzana G, Giorgi F, Lalli S, Marazzi G, Torri M. Identification of impurities in artemisinin, their behavior in high performance liquid chromatography and implications for the quality of derived anti-malarial drugs. J Chromatogr A 2011;1218(38):6838–42.
[24] Christen P, Veuthey JL. New trends in extraction, identification and quantification of artemisinin and its derivatives. Curr Med Chem 2001;8:1827–39.

[25] 张东，杨岚，杨立新等HPLC-UV-ELSD法同时测定青蒿中青蒿素、青蒿乙素和青蒿酸的含量. 药学学报,2007;42(9):978–81. [Available only in Chinese: Zhang D, Yang L, Yang LX, et al. Determination of artemisinin in arteannuin B and artemisinic acid in herba *Artemisiae annuae* by HPLC-UV-ELSD. Acta Pharmaceut Sin 2007;42(9):978–81].

[26] Avery BA, Venkatesh KK, Avery MA. Rapid determination of artemisinin and related analogues using high-performance liquid chromatography and an evaporative light scattering detector. J Chromatogr B Biomed Sci Appl 1999;730(1):71–80.

[27] Acton N, Klayman DL, Rollman IJ. Reduction electrochemical HPLC assay for artemisinin (Qinghaosu). Plant Med 1985;51(7):445–6.

[28] Charles DJ, Simon JE, Wood KV, et al. Germplasm variation in artemisinin content of Artemisia annua using an alternative method of artemisinin analysis from crude plant extracts. J Nat Prod 1990;53(1):157–60.

[29] Chan KL, Yuen KH, Jinadasa S, et al. A high performance liquid chromatography analysis of plasma artemisinin using a glassy carbon electrode for reductive electrochemical detection. Plant Med 1997;63(1):66–9.

[30] Vandenberghe DR, Vergauwe AN, Montagu MV, et al. Simultaneous determination of artemisinin and its bioprecursors in Artemisia annua. J Nat Prod 1995;58(5):798–803.

[31] 杨树德. 还原型电化学极谱检测高液相色谱法测定人血浆中青蒿酯和双氢青蒿素. 药学学报, 1985;20(6):457–62. [Available only in Chinese: Yang SD. Determination of artesunate and dihydroqinghaosu in human plasma by high performance liquid chromatography with reductive mode electrochemical detection. Acta Pharmaceut 1985;20(6):457–62].

[32] Na-Bangchang K, Congpuong K, Hung LN, et al. Simple high-performance liquid chromatographic method with electrochemical detection for the simultaneous determination of artesunate and dihydroartemisinin in biological fluids. J Chromatogr B Biomed Sci Appl 1998;708(1–2):201–7.

[33] Navaratnam V, Mordi MN, Mansor SM. Simultaneous determination of artesunic acid and dihydroartemisinin in blood plasma by high-performance liquid chromatography for application in clinical pharmacological studies. J Chromatogr B Biomed Sci Appl 1997;692(1):157–62.

[34] Sandrenan N, Sioufi A, Godbillon J, et al. Determination of artemether and its metabolite, dihydroartemisinin, in plasma by high-performance liquid chromatography and electrochemical detection in the reductive mode. J Chromatogr B Biomed Sci Appl 1997;691(1):145–53.

[35] Filip CW, Sofie RF, Lies M, et al. Quantitation of artemisinin and its biosynthetic precursors in Artemisia annua L. by high performance liquid chromatography electrospray quadrupole time of flight tandem mass spectrometry. J Chromatogr A 2006;1118(3):180–7.

[36] Naik H, Murry DJ, Kirsch LE, et al. Development and validation of a high-performance liquid chromatography mass spectroscopy assay for determination of artesunate and dihydroartemisinin in human plasma. J Chromatogr B 2005;816(1–2):233–42.

[37] Souppart C, Gauducheau N, Sandrenan N, et al. Development and validation of a high-performance liquid chromatography-mass spectrometry assay for the determination of artemether and its metabolite dihydroartemisinin in human plasma. J Chromatogr B 2002;774(2):195–203.

[38] Rajanikanth M, Madhusudanan KP, Gupta RC. Liquid chromatographic-mass spectrometric method for the determination of α-, β-atreether in rat serum. J Chromatogr B 2003;783(2):391–9.

[39] Huang L, Jayewardene AL, Li X, Marzan F, Lizak PS, Aweeka FT. Development and validation of a high-performance liquid chromatography/tandem mass spectrometry method for the determination of artemether and its active metabolite dihydroartemisinin in human plasma. J Pharm Biomed Anal 2009;50(5):959–65.

[40] Suberu J, Song L, Slade S, Sullivan N, Barker G, Lapkin AA. A rapid method for the determination of artemisinin and its biosynthetic precursors in Artemisia annua L. crude extracts. J Pharm Biomed Anal 2013;84:269–77.

[41] Carrà A, Bagnati R, Fanelli R, Bonati M. Fast and reliable artemisinin determination from different Artemisia annua leaves based alimentary products by high performance liquid chromatography-tandem mass spectrometry. Food Chem 2014;142:114–20.

[42] Kohler M, Haerdi W, Christen P, et al. Supercritical fluid extraction and chromatography of artemisinin and artemisinic acid. An improved method for the analysis of Artemisia annua samples. Phytochem Anal 1997;8:223–7.

[43] Mount DL, Todd GD, Navaratnam V. Packed-column supercritical fluid chromatography of artemisinin (qinghaosu)with electron-capture detection. J Chromatogr B Biomed Sci Appl 1995;666(1):183–7.
[44] Jaziri M, Diallo B, Vanhaelen M, et al. Immunodetection of artemisinin in Artemisia annua cultivated hydroponic conditions. Phytochemistry 1993;33(4):821–6.
[45] Ferreira JFS, Janick J. Immunoquantitaive analysis of artemisinin from Artemisia annua using polyclonal antibodies. Phytochemistry 1996;41(1):97–104.
[46] He SP, Tan GY, Li G, Tan WM, Nan TG, Wang BM, Li ZH, Li QX. Development of a sensitive monoclonal antibody-based enzyme-linked immunosorbent assay for the antimalaria active ingredient artemisinin in the Chinese herb *Artemisia annua* L. Anal Bioanal Chem 2009;393(4):1297–303.
[47] Zehnacker L, Nevers MC, Sinou V, Parzy D, Créminon C, Parzy D, Azoulay S. Development of sensitive direct chemiluminescent enzyme immunoassay for the determination of dihydroartemisinin in plasma. Anal Bioanal Chem 2015;407(25):7823–30.
[48] 宋振玉，赵凯存，梁晓天等. 青蒿琥酯和青蒿素的放射免疫测定法. 药学学报, 1985;20(8):610–14. [Available only in Chinese: Song ZY, Zhao KC, Liang XT, Liu CX, Yi MG. Radioimmunoassay of Qinghaosu and artesunate. Acta Pharmaceut Sin 1985;20(8):610–4].
[49] 朱汉松. 青蒿素的滴定分析法. 药学通报, 1980;15(8):342–44. [Available only in Chinese: Zhu HS. Titrimetric analysis of Qinghaosu. Chin Pharmaceut J 1980;15(8):342–4].
[50] 刘炳玉，田惠君，崔进等. 青蒿素的红外光谱定量测定. 药物分析杂志, 1994;14(4):17–18. [Available only in Chinese: Liu BY, Tian HJ, Cui J, Liu ZJ. Infrared spectrum quantitative analysis of Qinghaosu. Chin J Pharmaceut Anal, 1994;14(4):17–8].
[51] 沈春镒. 圆二色谱在药物分析中的应用. 青蒿素和美多巴中左旋多巴的定 量测定. 分析化学, 1992;20(10):1150–52. [Available only in Chinese: Shen CY. Application of circular dichroism in pharmaceutical analysis I. Quantitative analysis of artemisinine and laevodopa in "laevodapa et benserazide". Chin J Anal Chem, 1992;20(10):1150–2].
[52] Chen HL, Wang KT, Pu QS, et al. On-line conversion and determination of artemisinin using a flow-injection capillary electrophoresis system. Electrophoresis 2002;23:2865–71.

CHAPTER 12

Research on Extraction Methods for Artemisinin

Chapter Outline

Since 1980s, research on the chemical synthesis and biosynthesis of artemisinin has been carried out in succession, wherewith chemical synthesis, biosynthesis, or both increase the resources of artemisinin. However, it is difficult to solve practical application problems because of difficulties in preparation and constraints of such conditions as low yield and high cost. Thus, production of artemisinin-based drugs still relies on natural resources so far.

Since the advent of artemisinin, no putting it into action has been further reported although individual and sporadic literature mentioned that there were also other plants containing artemisinin other than *Artemisia annua* L. For example, in 1986, Liersch et al. [1] reported that the congener *Artemisia apiacea* Hance from Japan contained 0.08% of artemisinin; in 1991, Luo et al. [2] reported that artemisinin was separated from the congener *Artemisia lancea* Van from Yunnan; in 1998, Yi et al. [3] reported that artemisinin was separated from the congener *Artemisia subdigitata* Mattf.

Youyou Tu's research team from the Institute of Chinese Materia Medica, China Academy of Chinese Medical Sciences (ICMM-CACMS) had investigated chemical constituents of confused Artemisia congeners in China. In particular, their systematic study of five confused Artemisia congeners confirmed that only the certified *A. annua* L. in TCM contains artemisinin. Hu et al. [4], also from ICMM-CACMS, used a TLC method for determination of artemisinin from 19 species in 4 Artemisia groups, and also carried out the antimalarial screening of 8 Artemisia congeners (*A. annua* L., *A. apiacea* Hance, *Artemisia argyi* H. Léveillé & Vaniot, *Artemisia gmelinii* Weber ex Stechmann, *Artemisia lactiflora* Wallich

From *Artemisia annua* L. to Artemisinins
http://dx.doi.org/10.1016/B978-0-12-811655-5.00012-X

ex Candolle, *Artemisia scoparia* Waldst. et Kit., *Artemisia japonica* Thunb, and *Artemisia eripoda* Bunge), demonstrating only *A. annua* L. contains artemisinin and has antimalarial activity as well.

As an annual herb, repeated study has showed that the active antimalarial monomer artemisinin in *A. annua* L. is only contained in the leaf of the adult plant (young plant contains no artemisinin), so the preanthesis and leafy stage is the preferable harvest time. Despite wide resource distribution in China, artemisinin of crude drugs varies from place to place, so there is difference in extraction method as well.

1 From the Initiation of the Extraction and Separation Technology of Artemisinin to the Formation of the First Production Line [5]

In the previous chapter, "Antimalarial Research on Herba Artemisiae Annuae," Youyou Tu's research team from ICMM-CACMS first determined the neutral antimalarial part of Herba Artemisiae Annuae in 1971, and described the separation and extraction method of the active antimalarial monomer artemisinin therefrom found in 1972. As the discovery of artemisinin began with Herba Artemisiae Annuae (*A. annua* L.) from Beijing, which had the yield of only 0.02%, according to the operational need, inferior Herba Artemisiae Annuae from Beijing could only be used since its massive extraction in 1972. In such historical context of no cooperation with pharmaceutical factories, the team overcame mountains of difficulties, and more than 100 g of artemisinin obtained for use was hardearned. This extraction and separation protocol was feasible for the extraction and separation of artemisinin from low-content resources of Herba Artemisiae Annuae and was a hardy pioneer.

In 1974, No."523" Office urged ICMM-CACMS to direct and organize a national large-scale cooperation, involving in Yunnan, Shandong, Sichuan, Guangdong, and Guangxi, and thus found quality resources of Herba Artemisiae Annuae (ppt) and explored the fact that use of gasoline as solvent could leave out chromatography to obtain a simple process.

Since 1974, a pharmaceutical factory affiliated to ICMM-CACMS had purchased a great quantity of Herba Artemisiae Annuae from Youyang, Sichuan, in successive years, and extracted a quantity of artemisinin with gasoline [5] for comprehensive and intensive scientific research. Subsequently, they cooperated with the Guilin Aromatics Unit, Guangxi (which had idle off-the-shelf equipment because of short-term extraction of sweet-scented osmanthus each year) to extract artemisinin from local Herba Artemisiae Annuae. They had also invested in an artemisinin production site in a county of Xiamen, which annually extracted dozens of kilograms of artemisinin. Its brief extracting process is illustrated in Fig. 12.1. During the Appraising Meeting of Artemisin in 1978, gasoline process had been confirmed to be available to extract Herba Artemisiae Annuae resources with artemisinin content of more than 0.3%.

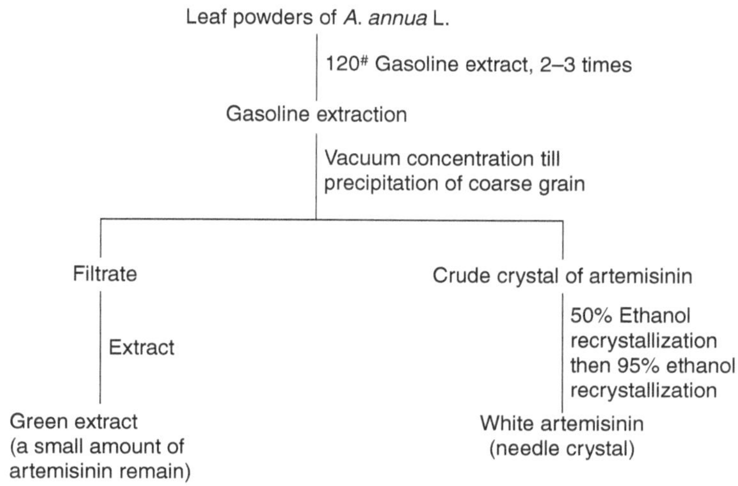

Figure 12.1: The Flow Chart of the Extraction Process.

After the ICMM obtained a new drug certificate of artemisinin in 1986, for poverty alleviation purpose, they also built a production base in Jishou, Hunan, by virtue of local quality Herba Artemisiae Annuae resource. The former "Jishou Pharmaceutical Factory, ICMM" manufactured a quantity of artemisinin for the market.

In recent years, production sites have mostly replaced gasoline with high-boiling petroleum ether due to its easy recovery and high yield. Also, current resources are in short supply and people try to make the most of every resource, so silica gel column chromatography is combined, using a mixture of petroleum ether and a certain amount of ethyl acetate as eluent. Such early pioneering work plays a pivot role.

2 Research Progress on Extraction and Separation Technology of Artemisinin

In 1975, the chemical structure of artemisinin was elucidated by ICMM-CACMS; subsequently, its physical and chemical properties were further illuminated to create conditions for the establishment of an assay. With an assay, people can discover a wide range of quality Herba Artemisiae Annuae resources; also, along with the scientific and technological progress, there will continually be new progress on extraction and separation methods of artemisinin.

There are two major problems during the extraction and separation of artemisinin. First, the artemisinin content is not high in Herba Artemisiae Annuae and there are limited Herba Artemisiae Annuae resources available for the industrial production of artemisinin. Second, active ingredients, for example, artemisinin, are mostly intracellular products, so when extracting, they are released from the cell and diffuse into the extraction medium slowly, affecting their extraction efficiency and increasing the operating cost. Scholars, at home or abroad, had carried out a series of studies on these problems and Herba Artemisiae Annuae from different places of origin. A brief overview is made as follows.

2.1 Extraction With Conventional Organic Solvents

Zhao et al. [6] extracted artemisinin with ether, chloroform, *n*-hexane, and petroleum ether (30–60°C) from the dry leaf powder of Herba Artemisiae Annuae while stirring, and the finding showed that petroleum ether was a more appropriate extraction medium. Systematic research was made on process conditions of extraction of artemisinin with petroleum ether, indicating that more appropriate extraction conditions included raw material granularity 60 mesh, extraction time 2 h, extraction temperature 50°C, solvent volume 60 mL (1 g of raw material), and stirring speed 800 r/min. The study investigated the efficiency of use of ultrasound in enhancement of petroleum ether extraction, and the result showed that the extraction rate could reach 83% when treating in a single run with 20 kHz, 90 W ultrasonic wave below 50°C for 20 min, followed by stirring to 30 min, whereas it could reach 81% when sonicating 6 times for a total of 12 min (each for 2 min), followed by stirring to 30 min. Artemisinin is mainly present in the glands of *A. annua* L. plant; if it is affected by ultrasonic cavitation, glands are easily broken and the artemisinin dissolves out to be conducive to extraction; at the moment, only a small amount of other impurity fractions enters into the extract, so the impurity content is low in the product [6,7].

Vonwiller et al. [8] used methanol–ethanol–ether extraction method. Using dried stems and leaves of *A. annua* L. as raw materials, they extracted with methanol at room temperature, filtered after 24-h mechanical agitation; after vacuum distillation of the filtrate, extracted with ethanol twice, followed by sodium carbonate solution; adjusted to pH 1 with concentrated hydrochloric acid, then extracted with ether thrice at room temperature, dehydrated the extract with anhydrous sodium sulfate, and evaporated to dryness in vacuum. The concentrate was dissolved in a solution of benzene sulfonic acid in methanol and reacted for 3 days under agitation. It was found that fatty acids present in *A. annua* L. were easily esterified by methanol at room temperature in the presence of acid catalyst, but artemisinic acid therein remained unchanged. Aforementioned method can extract not only artemisinic acid separately but also artemisinin simultaneously. However, after extraction of artemisinin and artemisinic acid with boiling ethanol and evaporation of ethanol, washing the concentrate with the solvent could remove nonpolar substances therein, but almost half of the artemisinic acid could be lost in extracting and washing while causing difficulties in its further purification.

Brown [9] reported a method to extract artemisinin with solvents from tissue cultures. Tissue cultures were crushed in liquid nitrogen, and then extracted with ether in a Soxhlet extractor; the extract was obtained by vacuum evaporation of solvents.

Klayman et al. [10] extracted *A. annua* L. introduced from China and grown in the USA with low boiling point solvents, for example, dichloromethane, chloroform, ether, acetone, and petroleum ether (30–60°C), and results showed that petroleum ether had the best effect. They extracted the dried *A. annua* L. with boiling petroleum ether for 48 h and obtained a black syrupy substance by vacuum evaporation; dissolved it in chloroform, and added acetonitrile to remove the wax therein; crude artemisinin extract was obtained by filtration and vacuum evaporation.

Singh et al. [11] investigated the pilot-scale extraction and separation. Using 50 kg of leaves of *A. annua* L. from Europe as raw materials and *n*-hexane as solvent, the powder sample was extracted at room temperature, and further separated and purified by silica gel column chromatography after vacuum concentration.

Elsohly et al. [12] also carried out a large-scale study of artemisinin extraction. Using dry leaf powders of *A. annua* L. as raw materials and *n*-hexane as solvent, the powder sample was extracted in a Soxhlet extractor for 48 h; after evaporation of the organic solvent, the concentrate was extracted thrice in the *n*-hexane-acetonitrile system; the acetonitrile extract was dehydrated with NaCl; after vacuum evaporation of acetonitrile, a yellow concentrate was obtained for further separation and refinement of various constituents. Throughout the experiment, a total of 400 kg of leaf powders of *A. annua* L. were treated to obtain 485 g of artemisinin (0.12%), 2.12 kg of artemisinic acid (0.53%), and 170 g of arteannuin B (0.04%).

2.2 Microwave-Assisted Solvent Extraction

Microwave-assisted extraction (MAE), also known as microwave extraction (ME), is an extraction method of a combination of microwave with conventional solvent extraction. The mechanism of microwave irradiation heating is a pattern of internal heating, which oscillates the interior of the substance molecule to generate transient energy to cause fierce thermal motions of intracellular molecules, and finally leads to increased cell rupture to enable the extracting objects to enter into the extraction solvent quickly and efficiently [13,14].

Hao et al. [15,16] extracted artemisinin from *A. annua* L. using MAE method. Ethanol, chloroform, cyclohexane, *n*-hexane, petroleum ether (30–60 and 60–90°C), 120# solvent oil, and 6# extraction solvent oil were compared, and effect of solvent dielectric constant on yield of artemisinin was investigated. Also, comparison was made between MAE and Soxhlet extraction, supercritical CO_2 extraction, and heating-agitating extraction. The experimental results showed that neither absolute ethanol nor chloroform was appropriate solvent because what they extracted were mostly impurities which increased the difficulty in subsequent separation. The artemisinin obtained by 6# extraction solvent oil was characterized by the highest extraction rate and low price, so it was selected as the solvent of MAE of artemisinin. Under the same extraction conditions, the extraction rate increased with the increase of the solvent dielectric constant, and an increasing tendency was prominent. To achieve the effect of MAE for 4–6 min, using conventional heat extraction and Soxhlet extraction will take several or even more than dozens of hours. MAE has a comparative solvent recovery to the heating-agitating extraction and Soxhlet extraction.

Liang et al. [17] also investigated the effect of microwave pretreatment of *A. annua* L. on extraction rate of artemisinin. Microwave pretreatment accelerated the rate of extraction of artemisinin from *A. annua* L. plant; within the same time frame, the extraction rate and yield of artemisinin by microwave pretreatment increased compared with the nonmicrowave pretreatment. They compared experiments whether to use microwave pretreatment process to dry powders of *A. annua* L. by Soxhlel extracting method with petroleum ether, cyclohexane, and 120# solvent oil as extracting solvents, and studied the influence of the pretreatment power and time and the extracting time on extraction rate and yield of artemisinin. Results showed that the yield of artemisinin by microwave pretreatment could enhance about 15% compared with the nonmicrowave pretreatment process; when the solvent was 120# solvent oil, and the pretreatment power was 450 W, and the pretreatment time was 240 s, and the Soxhlel extraction time was 6 h, the extraction rate and yield of artemisinin could attain above 88 and 0.49%, respectively.

2.3 Supercritical Fluid Extraction (SFE)

Supercritical fluid (SCF) refers to the fluid at above critical temperature (T_c) and critical pressure (P_c). SFE technology has emerged for just 2 or 3 decades, but many researchers have carried out extensive and far-ranging research due to its high performance and promising application prospect. So far, it has preliminarily been put into industrialized production and been one of the most mature and widely used SCF technologies. SCF has its particular advantages and very good industrial application prospects in extracting and separating natural drugs with low content of active ingredients due to its high diffusion mass transfer and solubility, easily adjustable process, and simple separation process [18].

He and Liang [19], using Herba Artemisiae Annuae with artemisinin content of 0.296% as raw material, crushed it into crushed aggregates with a particle size of 0.2–2 mm, and extracted for 2 h (60°C, 20 MPa, cycling quantity of CO_2 2.0–2.5 kg/h), with an extraction rate of artemisinin of 95%. The findings showed that higher extraction pressure and temperature resulted in higher extraction rate of artemisinin, but higher extraction temperature also led to higher impurity content in the extract [19].

Ge et al. [20] investigated the chemical constituents of *A. annua* L., suggesting that the extraction rate of artemisinin by supercritical CO_2 fluid extraction increased by 11–59% as compared with that by solvent method (gasoline and dilute ethanol) in the traditional industrial production, and by more than twice as compared with traditional gasoline process; the extraction time was greatly shortened; the cost was reduced; different temperatures, pressures, and time were controlled; and such constituents as octadecanol were also obtained.

Kohler et al. [21] carried out a study of extraction of *A. annua* L. with supercritical CO_2 carrying a modifier. In the experiment, 3% methanol (or ethanol) was selected as a modifier; the extraction pressure was 15 MPa; the temperature was 50°C; the flow rate of CO_2 was 2 mL/min; the extraction time was 20 min; the extracts were detected by SFC-FID method, with an extraction rate of artemisinin of about 95%. Addition of the modifier could improve the extraction kinetics of artemisinin and greatly shorten the extraction time.

Qian et al. [22] investigated the influencing factors of extraction of artemisinin from *A. annua* L. with supercritical CO_2. At pressures between 15.2 and 29.7 MPa and temperatures between 40 and 60°C, higher extraction pressure and temperature resulted in higher extraction yield and lower extraction selectivity. Supercritical extraction conditions were optimized based on the extraction yield and selectivity, and the optimum operation conditions were extraction pressure of 20 MPa, temperature of 50°C, solvent ratio of 1 kg of CO_2 per kg of raw material per hour, and particle size of 60–80 mesh. In these conditions, the extraction yield was more than 95% and the purity of the extract was above 10% for extraction time of 4 h.

SFE can significantly improve the extraction rate and is not strict for raw materials (artemisinin content of just more than 0.1%); it is also characterized by short extraction time, high product quality, and good prospect of industrial application.

2.4 High-Speed Countercurrent Chromatography

High-Speed Countercurrent Chromatography (HSCCC), a chromatographic technique developed by Dr. Y. Ito, can complete high-efficient separation and preparation in a short time [23,24]. Initially, the major aim of HSCCC was to perform preparative separation. Since 1980s, with the improvement and perfection of the counter current chromatography, its analytical ability has advanced greatly and its application range has expanded from preparative separation to microanalysis [25,26]. This technique has been used in biochemistry, bioengineering, medicine, natural products and foods, and so on. Some researchers reported studies of extraction and separation of artemisinin using this technique.

Acton et al. [27] purified artemisinin from *A. annua* L. using HSCCC, in which a mixture of isooctane and ethyl acetate (7:3) was used as stationary phase and a mixture of methanol and water (6:4) as mobile phase. Roth and Acton [28] extracted epideoxyarteannuin from *A. annua* L. by HSCCC using a mixture of isooctane, methanol, and water (10:7:3) as stationary and mobile phases. In 2000, Weiss et al. [29] investigated a method to separate artemisinin, artemisitene, and arteannuin B in a series of solvent systems using countercurrent chromatography (CCC).

In view of not big difficulties in laboratory-scale separation of artemisinin and its derivatives from the plant, and of their special structures, physicochemical properties, and good medicinal value, in 1989, an article was published in the *Journal of Chemical Education* to recommend the separation of artemisinin from *A. annua* L. as an experiment item in the textbook [30]. In brief, a lot of research has been done on the extraction and separation of artemisinin, and not a few patents have come out as well. Research on novel extraction and separation technologies will undoubtedly continue to appeal to scientists and technologists because production of artemisinin and its derivatives remains mainly dependent on natural resources to date.

References

[1] Liersch R, Soicke H, Stehr C, et al. Formation of aremisinin in *Artemisia annua* during one vegetation period. Plant Med 1986;52(5):387–90.

[2] Luo SD, Ning BM, Hu WY, et al. Studies on peroxides of Artemisia lamcea. J Nat Prod 1991;54(2):573–5.

[3] Yi P, Zhang QF, Wang HY, Fan DQ, Luo SD. Studies on the constituents of Artemisia dubia. Chin Herb Med 1998;29(1):13–4.

[4] Hu SL, Xu QC, Liu JF, Gu YX. Plant resource research on Qinghaosu. Bull Chin Mater Med 1981;6(2):13–6.

[5] Traditional Chinese medicine research institution. Artemisinin-based antimalarial album (1971–1978). Traditional Chinese medicine research data, 1978, 3.

[6] Zhao B, Wang YC, Ouyang F. The present situation of the extraction and separation of artemisinin medicinal ingredients. Chin Herb Med 1998;29(11):784–6.

[7] Zhao B, Wang YC, Wu J, Bi JX, Ouyang F. Studies on the extraction processes of artemisinin. Chin Herb Med 2000;31(6):421–3.

[8] Vonwiller SC, Haynes PK, King G, et al. An improved method for the isolation of qinghao (artemisinic) acid from *Artemisia annua*. Plant Med 1993;59(6):562–3.

[9] Brown GD. Secondary metabolism in tissue culture of *Artemisia annua*. J Nat Prod 1994;57(7):975–7.

[10] Klayman DL, Lin AJ, Acton N, et al. Isolation of artemisinin (Qinghaosu) from *Artemisia annua* growing in the United States. J Nat Prod 1984;47(4):715.

[11] Singh A, Vishwakarma RA, Husain A. Evalutation of *Artemisia annua* strains of higher artemisinin production. Plant Med 1988;54(7):475–6.

[12] Elsohly HN, Croom EM, El-Feraly JFS, et al. A large-scale extraction technique of artemisinin from *Artemisia annua*. J Nat Prod 1990;53(6):1560–4.

[13] Ganzler K, Salgo A, Valko K. Microwave extraction: a novel simple preparation method for chromatography. J Chromatogr 1986;371:299–306.

[14] Pu CT. Chemical Process and Equipment. Beijing: Chemical Industry Press; 1996. 127.

[15] Hao JY, Han W, Shi CO, Deng X. Microwave-assisted extraction of artemisinin from *Artemisia annua* L. Chin J Pharm 2002;33(8):385–7.

[16] Hao JY, Han W, Huang SD, et al. Microwave-assisted extraction of artemisinin from *Artemisia annua* L. Sep Purif Technol 2002;28(3):191–6.

[17] Liang ZS. Effect of microwave pre-treatment on the extraction rate of arlemisinin. Cent South Pharm 2004;2(6):342–4.

[18] Jiang WX, Guo JZ. Study and application of supercritical fluid technology. J Petrochem Univ 2001;14(2):15–20.

[19] He CM, Liang ZY. Studies on extraction of artemisinin from sweet wornwood (*Artemisia annua*) by supercritical carbon dioxide. Chin Herb Med 1999;37(7):497–9.

[20] Ge FH, Zhang JC, Chen L, Li J, Huang XF. Supercritical-CO_2 fluid extraction of artemisinin from *Artemisia annua* L. Chin J Pharm 2000;31(6):250–3.

[21] Kohler M, Haerdi W, Christen P, et al. Extraction of artemisinin and artemisinic acid from *Artemisia annua* L. using supercritical carbon dioxide. J Chromatogr A 1997;785(1–2):353–60.

[22] Qian GP, Yang YW, Wu CJ, Ren QL. Supercritical-CO_2 extraction of artemisinin from *Artemisia annua* L. Chem Ind Eng Prog 2005;24(3). 286–290, 302.
[23] Ito Y. Efficient preparative counter-current chromatography with a coil plantet centrifuge. J Chromatogr 1981;214(1):122–5.
[24] Ito Y. New continous extraction method with a coil planet centrifuge. J Chromatogr 1981;207(2):161–9.
[25] Zhang TY. Counter Current Chromatography, 1991. Beijing: Science and Technology Press; 2001. 267.
[26] Kong ZR, Rinehart KL, Milberg RM, et al. Application of high-speed countercurrent chromatography/ electrospray ionization mass spectrometry (HSCCC/ESIMS) in natural products chemistry. J Liq Chromatogr Relat Technol 1998;21(1–2):65–82.
[27] Acton N, Klayman DL, Rollman IJ, et al. Isolation of artemisinin (Qinghaosu) and its separation from artemisitene using the Ito multilayer coil separator-extractor and isolation of arteannuin B. J Chroamtogr 1986;355:448–50.
[28] Roth RJ, Acton N. Isolation of arteannuic acid from *Artemisia annua*. Plant Med 1987;53(5):501–2.
[29] Weiss E, Ziffer H, Ito Y. Use of countercurrent chromatography (CCC) to separate mixtures of artemisinin, artemisitene, and arteannuin B. J. Liq. Chromatogr. Relat Technol 2000;23(6):909–13.
[30] Roth RJ, Acton N. The isolation of sesquiterpenes from *Artemisia annua*. J Chem Educ 1989;66(4):349–50.

CHAPTER 13

Research on Synthetic Method of Artemisinin

Chapter Outline

As artemisinin merged as a brand new chemical structure, its outstanding efficacy and unique chemical structure of peroxy rings have drawn extensive attention of chemists at home and abroad. Total synthesis of artemisinin is a challenging research task with great technical difficulties. In 1983, two Swiss, G. Schmid and W. Hofheinz, first published the total synthetic method for artemisinin in the *Journal of the American Chemical Society* (*JACS*); later, Xu et al. also published processes of the total synthesis and semisynthesis of artemisinin in succession. Since the completion of the total synthesis and semisynthesis of artemisinin, researchers of different countries also investigated many synthetic technologies. Each of these technologies was of its own academic significance, but the yield of the key reaction introducing the peroxy group remains unsatisfactory. So far, neither chemical synthesis nor biosynthesis can solve the production problem of artemisinin; source of artemisinin still has to rely on extraction and separation from Herba Artemisiae Annuae, but in plant resources, artemisinin content is not that high. Therefore, studies of novel chemical synthetic methods and routes are of academic and practical significance. Now a brief overview is made with respect to chemical synthesis of artemisinin.

1 Total Synthesis

Total synthesis of artemisinin was finished in 1983, and Schmid and Hofheinz [1] first reported a route to total synthesis of artemisinin (Fig. 13.1). They completed the total synthesis of artemisinin via 13-step reactions using (−)-isopulegol as raw material.

In the same year, Xu et al. [2] synthesized the artemisinic acid starting out with vanillin, and then semisynthesized artemisinin via 8-step reactions, with a total of 20 steps of reactions and a yield of 0.25% (Fig. 13.2). Both synthetic methods show good agreement, and their key reactions are peroxidation and cyclization, in which the enol ether is photochemically oxidized with the action of singlet oxygen and acids.

From *Artemisia annua* L. to Artemisinins
http://dx.doi.org/10.1016/B978-0-12-811655-5.00013-1

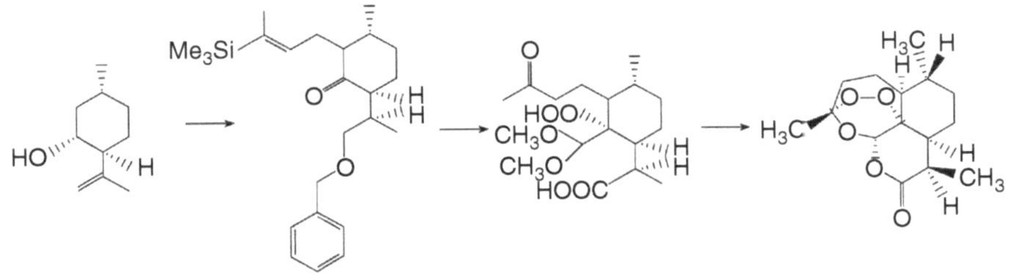

Figure 13.1: Schmid's Synthetic Route.

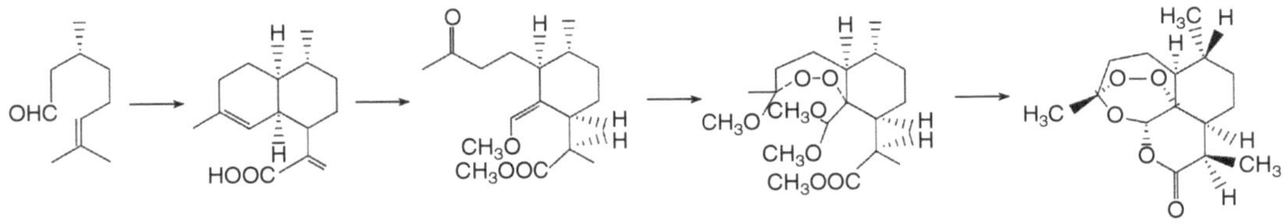

Figure 13.2: Xu's Synthetic Route.

In 1985, applying a basically similar synthetic route, Wu et al. [3] reported an approach to reorganize artemisinin from its acidic degradation product, diketo ester.

In 1987, Avery et al. [4] introduced an approach to prepare artemisinin by ozonation of vinyl silane. The reaction had a yield of 37% following rearrangement and cyclization, and its mechanism of oxidative cyclization greatly differed from the process routes previously mentioned (Fig. 13.3).

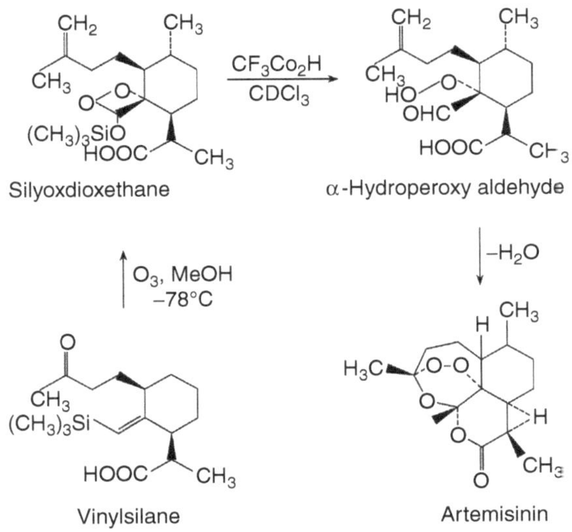

Figure 13.3: Avery's Synthetic Route I.

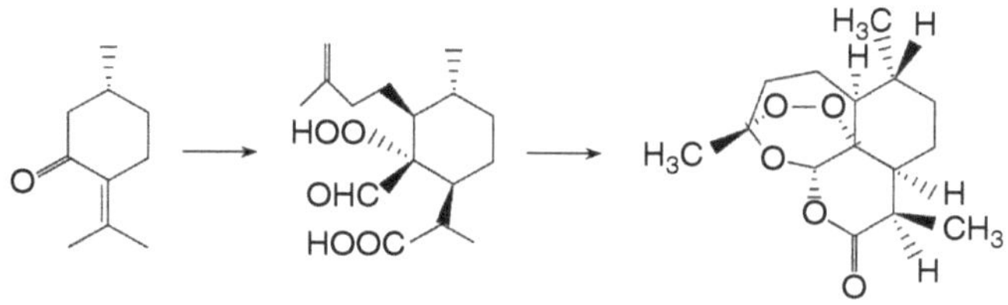

Figure 13.4: Avery's Synthetic Route II.

In 1990, Ravindranathan et al. [5] designed a synthetic route of artemisinin from (+)-isolimonene.

In 1992, Avery et al. [6] also designed a 10-step synthetic route using (R)-(+)-pulegone as raw material, with a yield of 3.6%. The intermediate was similar to the process routes of two synthetic reactions published in 1983; compared with both the previous ones, its advantage was that the product was a single stereoisomer (Fig. 13.4).

In 1993, Liu et al. [7] reported an efficient process of total synthesis of artemisinin, which used (−)-β-pinene as reaction substrate; the synthetic route greatly differed from those previously reported. The synthesis started with intramolecular Diels–Alder reaction catalyzed by zinc chloride, and introduction of a methoxyl group to the ketone group used Wittig reaction (Fig. 13.5).

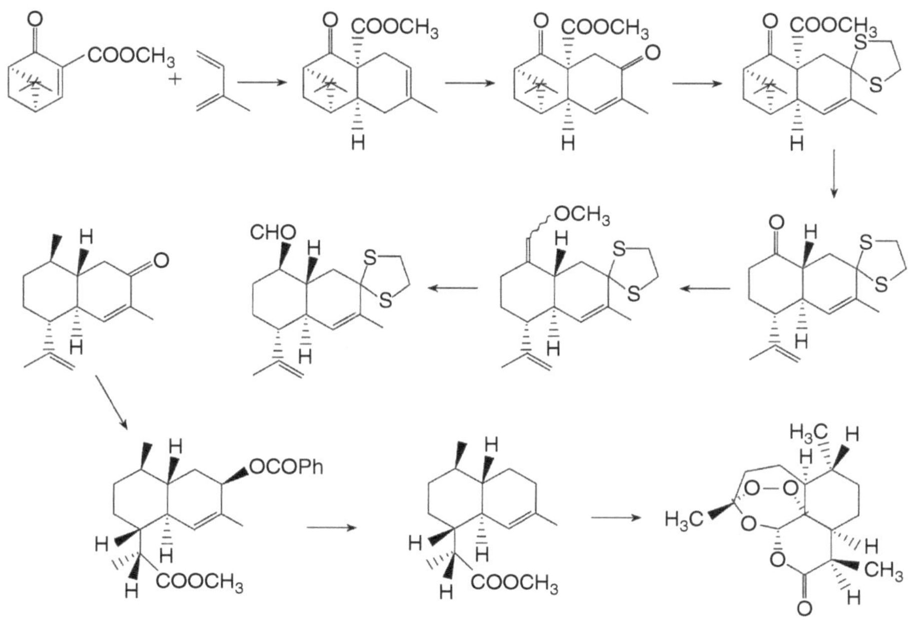

Figure 13.5: Liu's Synthetic Route.

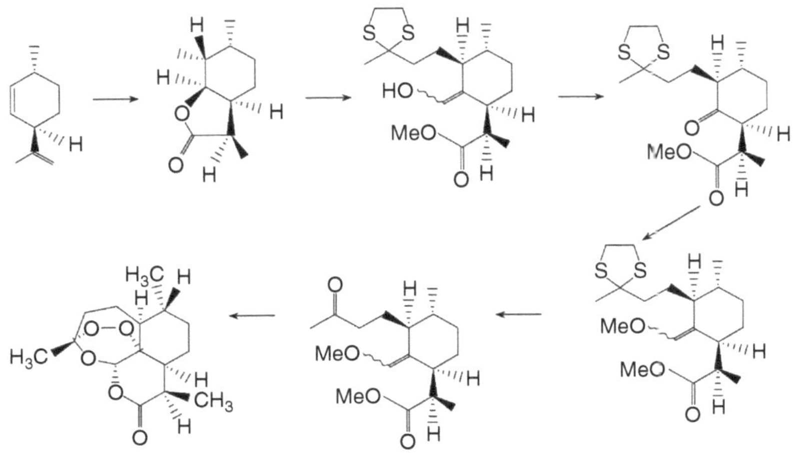

Figure 13.6: Yadav's Synthetic Route.

In 2003, Yadav et al. [8] reported another synthetic route of artemisinin starting with (+)-isolimonene. The reaction had 11 steps in total. The intermediate experienced a structure of indolyl lactone; introduction of a methoxyl group to the ketone group and formation of a structure of methyl vinyl ether used Wittig reaction; finally, the intermediate formed artemisinin through photooxidation and acidification (Fig. 13.6).

When artemisinin is finally synthesized through aforementioned reaction routes, photooxidation is mostly required, and the yield has been difficult to improve due to great technical difficulties. New total synthesis method is still an important subject to synthesis researchers.

2 Semisynthesis

Relative to total synthesis, semisynthesis technology has some potential in batch production. Artemisinic acid is a sesquiterpenoid which is highly contained in Herba Artemisiae Annuae. Many study groups have worked on the synthesis of artemisinin from the natural products in Herba Artemisiae Annuae, including artemisinic acid (the precursor of artemisinin) and arteannuin B [9–11]. In 1993, Vonwiller et al. [12] reported a process route to efficient separation of artemisinic acid from *Artemisia annua* L. Photooxidation of artemisinic acid had been well reported [13–15]. In 1991, Jung et al. [16] reported photooxidation and cyclization of artemisinic acid into desoxyartemisinin. In 1992, Acton and Roth [17] reported a two-step process of semisynthesis of artemisinin from desoxyartemisinin. In 1991, Roth and Acton [18] also reported a method of semisynthesis of artemisinin from artemisinic acid, with a yield of 30%. In 1992 and 1998, both Nowak and Lansbury [19,20] reported research efforts on the synthesis of artemisinin from higher levels of arteannuin B and artemisinic acid in *A. annua* L.

References

[1] Schmid G, Hofheinz W. Total synthesis of qinghaosu. J Am Chem Soc 1983;105(3):624–5.

[2] 许杏祥，朱杰，黄大中等.青蒿素及其一类物结构和合成的研究.X.从青蒿酸立体控制合成青蒿素和脱氧青蒿素.化学学报，1983;41(6):574–5. [Available only in Chinese: Xu XX, Zhu J, Huang DZ, Zhou WS. Studies on structure and syntheses of arteannuin and related compound X: The stereocontrolled syntheses ofarteannuin and deoxyarteannuin from arteannuic acid. Acta Chim Sin 1983;41(6):574–5.

[3] 吴毓林，张景丽，李金翠.青蒿素及其类似物的合成研究——由青蒿素降解产物重组青蒿素.化学学报，1985;43(9):901–03. [Available only in Chinese: Wu YL, Zhang JL, Li JC. Studies on the synthesis of Qinghaosu and its analogs-reconstruction of Qinghaosu from its degradation product. Acta Chim Sin 1985;43(9):901–3].

[4] Avery MA, JenningsWhite C, Chong WKM. The total synthesis of (+)-artemisinin and (+)-9-desmethylartemisinin. Tetrahedron Lett 1987;28(40):4629–32.

[5] Ravindranathan T, Kumar MA, Menon RB, et al. Stereoselective synthesis of artemisinin. Tetrahedron Lett 1990;31(5):755–8.

[6] Avery MA, Chong WKM, Jennings-White C. Stereoselective of (+)-artemisinin, the antimalarial constituent of *Artemisia annua* L. J Am Chem Soc 1992;114(3):974–9.

[7] Liu HJ, Yeh WL, Chew SY. A total synthesie of the antimalarial natural product (+)-qinghaosu. Tetrahedron Lett 1993;34(28):4435–8.

[8] Yadav JS, Babu RS, Sabitha G. Stereoselective total synthesis of (+)-artemisinin. Tetrahedron Lett 2003;44(2):387–9.

[9] Ravindranathan T. Artemisinin (qinghaosu). Curr Sci 1994;66(1):35–41.

[10] Roth RJ, Acton N. Isolation of arteannuic acid from *Artemisia annua*. Plant Med 1987;53(5):501–2.

[11] Jung M, Elsohly HN, Mc Chesney JD. Artemisinic acid: a versatile chiral synthon and bioprecursor to natural products. Plant Med 1990;56(6):624.

[12] Vonwiller SC, Haynes RK, King G, et al. An improved method for the isolation of qinghao (artemisinic) acid from *Artemisia annua*. Plant Med 1993;59(6):562–3.

[13] EL-Feraly FS, AL-Meshal IA, AL-Yahya MA, et al. On the possible role of qinghao acid in the biosynthesis of artemisinin. Phytochemistry 1986;25(12):2777–8.

[14] Jung M, Elsohly HN, Croom EM, et al. Practical conversion of artemisinic acid in desoxyartemisinin. J Org Chem 1986;51(26):5417–9.

[15] Xu XX, Zhu J, Huang DZ, et al. Total synthesis of arteannuin and deoxoarteannuin. Tetrahedron 1986;42(3):819–28.

[16] Jung M, Yu D, Bustos D, et al. A concise synthesis of 12-(3′-hydroxy-*n*-propyl)-deoxoartemisinin. Bioorg. Med. Chem. Lett. 1991;1:741–4.

[17] Acton N, Roth RJ. On the conversion of dihydroartemisinic acid into artemisinin. J Org Chem 1992;57(13):3610–4.

[18] Roth RJ, Acton N. A facile semisynthesis of the antimalarial drug qinghaosu. J Chem Educ 1991;68(7):612–3.

[19] Lansbury PT, Nowak DM. An efficient partial synthesis of (+)-artemisinin and (+)-deoxoartemisinin. Tetrahedron Lett 1992;33(8):1029–32.

[20] Nowak DM, Lansbury PT. Synthesis of (+)-artemisinin and (+)-deoxyartemisinin from arteannuin B and arteannuic acid. Tetrahedron 1998;54:319–36.

CHAPTER 14

Pharmacological Studies on Artemisinin

Chapter Outline

1 Research on Antimalarial Effect

1.1 Effect of Artemisinin on Plasmodium

1.1.1 In vivo experiments

1. Malaria in mouse models
   a. Inhibitory effect on the drug-sensitive strain of *Plasmodium berghei*
   The experiment was carried out in Swiss mice weighing between 18 and 22 g in reference to Peter's method. The mice were intraperitoneally inoculated with the drug-sensitive strain (or drug-resistant strain) of *P. berghei* 10,000,000 (1×10^7) on day 1 and were then divided into several dosing groups geometrically and one control group. There were 10 animals in each group. The dosing mice were administered with artemisinin-Tween suspension orally or via subcutaneous injection once daily, from day 1 to day 3 (or from day 1 to day 4), totally three doses (or four doses). Parasitemia in each group was detected by Giemsa staining of blood smears on day 4, and the inhibitory rate in each dosing group was calculated by comparing with the parasitemia in the control group. Otherwise, microscopic observation was made on the negative conversion of plasmodium to calculate the negative conversion rate in each group. Finally, the 50% suppression dose (SD_{50}) and the 50% effective dose (ED_{50}) were statistically calculated, both expressed in mg/kg/day.

From *Artemisia annua* L. to Artemisinins
http://dx.doi.org/10.1016/B978-0-12-811655-5.00014-3

The results of comparison of SD_{50} of artemisinin preparations are provided in Table 14.1.

b. Effect on various drug-resistant strains and comparison of resistance index
The results of resistance index showed that artemisinin had a noticeable antimalarial effect on various strains that are resistant to chloroquine, primaquine, and mefloquine, and a cross-resistance to strains that are highly resistant to chloroquine. The results are provided in Table 14.2.

2. Malaria in monkey models
   a. Therapeutic effect of artemisinin orally administered on simian malaria
   Rhesus monkeys that developed parasitemia after inoculation of sporozoites of simian malaria (*Plasmodium cynomolgi*) were orally administered 200 mg/kg/day of artemisinin-water suspension for 3 consecutive days. Chloroquine 20 mg/kg/day was used as the control. Examination of blood smears was performed daily after start of dosing, showing that all the animals turned to negative for parasitemia 2 or 3 days after dosing. The results are provided in Table 14.3.
   b. Therapeutic effect of artemisinin intramuscularly injected on simian malaria
   Rhesus monkeys were intravenously injected with red blood cells infected by simian malaria (*P. cynomolgi*) and then intramuscularly injected 20, 10, 4, or 1 mg/kg/day of

Table 14.1: Antimalarial potency of artemisinin for normal strain of *P. berghei*

Group	Experimental Dose (mg/kg) (Oral)	The Number of Animals (*n*)	Negative Conversion Rate (%)	Probability Unit	Mean Infection Rate (%)	Inhibitory Rate (%)	Probability Unit
I	5.75	10	0		18.5	23.4	4.2743
II	11.06	10	0		11.14	51.5	5.0376
III	23	10	10	3.7148	1.45	93.8	6.5382
IV	46	10	30	4.4756	0.38	98.4	7.1444
V	92	10	70	5.5244	0.05	99.8	7.8782
VI	184	10	90	6.2816	0.017	99.9	8.0902

ED_{50} = 69.3 mg/kg (38.6–124.7 mg/kg). SD_{50} = 9.97 mg/kg (6.7–14.9 mg/kg).

Table 14.2: Effect of artemisinin on various drug-resistant strains and comparison of the resistance index

Strain	SD_{50} (mg/kg/day)	SD_{90} (mg/kg/day)	Resistance Index I_{90}
Normal strain (N)	0.9	2.1	1.0
Chloroquine-resistant strain (RC high resistance)	46.5	78	37.1
Chloroquine-resistant strain (NS mild resistance)	4.4	8.4	4.0
Primaquine-resistant strain (P)	1.7	2.7	1.3
Pyrimethamine-resistant strain (PYR)	2.0	3.9	1.9
Mefloquine-resistant strain (N/1100)	1.6	2.9	1.4

Table 14.3: Therapeutic effect of artemisinin orally administered on simian malaria

Batches Tested	Drug and Dose (mg/kg/day)	Monkey ID Number	Time After Inoculation of Sporozoites (day)	Parasitism Rate Before Treatment (%)	Time to Conversion to Negative (day)	Time to Parasite Recrudescence (day)
Batch 1	Artemisinin 200	58	31	1.00	3	11
	Artemisinin 200	50	31	6.00	2	※
	Artemisinin 200	59	11	0.01	3	※
	Chloroquine 20	45	47	16.00	2	※
Batch 2	Artemisinin 200	22	18	6.0	2	23
	Artemisinin 200	23	18	2.0	2	11
	Artemisinin 200	67	18	0.10	2	11
	Artemisinin 200	66	18	13.00	2	8
	Chloroquine 20	64	18	1.00	2	※

※, Parasite recrudescence was not observed after half year of observation.

artemisinin–oil suspension, once daily for 3 consecutive days. Examination of blood smears was performed daily after start of dosing to observe negative conversion of the parasites. The results showed that parasitemia in all Rhesus monkeys treated with different dose of artemisinin–oil suspension turned to negative 2–3 days after dosing. With the exception of the large dose group, whose negative observation lasted for half year, in other dosing groups parasite recrudescence was observed 8–24 days after dosing. The results are provided in Table 14.4.

The aforementioned results show that artemisinin has a noticeable therapeutic effect on blood-stage simian malaria.

1.1.2 In vitro experiments

1. Effect of artemisinin on rodent plasmodium cultured in vitro

   P. berghei-infected mouse blood with a parasitism rate of approximately 2% was collected and centrifuged for removal of plasma and then diluted with culture medium (Medium 199 with fetal bovine serum) to the original volume; afterward, the resulting diluent was transferred to vials (0.42 mL per vial), to which 2.56 mL of culture medium

Table 14.4: Therapeutic effect of artemisinin intramuscularly injected on simian malaria

Drug	Dose (mg/kg/day)	Time to Conversion to Negative (day)	Time to Parasite Recrudescence (day)
Artemisinin–oil suspension	20	2	—
	20	2	—
	10	2	20
	10	2	23
	4	4	8
	4	2	24
	1	3	9

was added, followed by addition of such drugs as ethanol solution of artemisinin, chloroquine, and so on, and ethanol was used as control. The vials were incubated for 6 or 12 h at a temperature of 37°C in a vacuum drier that contained a certain proportion of CO_2, and then the culture harvest was inoculated into healthy mice. Microscopic examination of blood smears was performed at regular intervals to observe the occurrence of parasitemia, and the results showed that mice in the artemisinin large-dose group were not found to develop parasitemia, while mice in the corresponding control groups were all died due to the progression of parasitemia (Table 14.5).

2. Effect of artemisinin on human *Plasmodium falciparum* cultured in vitro [1]
Twenty-five percent of red blood cell suspension with a 1% parasitism rate of *P. falciparum* was treated with various drugs and cultured for 48 h at a temperature of 37°C to detect the drug sensitivity of human *P. falciparum* in vitro, as reported by Rioherds. Blood smears were prepared for staining and observation under microscope to detect the parasitism; the inhibitory rates were obtained through comparison with the control group, and then the effective dose 50% (ED_{50}) and the effective dose 90% (ED_{90}) were calculated according to the inhibitory rates in different dose groups. The results were as follows: the ED_{50} and ED_{90} of artemisinin were 1.99 and 4.52, respectively, while the ED_{50} and ED_{90} of chloroquine were 2.24 and 7.95, respectively, indicating that artemisinin is 1.13–1.76 times potent than chloroquine.

Table 14.5: Effect of artemisinin on rodent plasmodium cultured in vitro

Drug and Solvent	Batch 1			Batch 2		
	6 Days After Inoculation	8 Days After Inoculation	10 Days After Inoculation	6 Days After Inoculation	8 Days After Inoculation	10 Days After Inoculation
Artemisinin 50 μg/mL + normal saline	0	0	2.00	0	0	0
Artemisinin 250 μg/mL + normal saline	0	0	0	0	0	0
Artemisinin 25 μg/mL + ethanol absolute	0	0	0.05	0	0	0
Artemisinin 125 μg/mL + ethanol absolute	0	0	0	0	0	0
Chloroquine 450 μg/mL + normal saline	0	0.15	Dead	0	0	0
Chloroquine 225 μg/mL + normal saline	0	0.26	Dead	0	0	0.8
Chloroquine 112.5 μg/mL + normal saline	0	0.01	Dead	0	0	0
Normal saline	0.03	3.60	Dead	0.02	0.92	Dead
Normal saline + Tween	0.02	1.80	Dead	0.03	0.62	Dead
Ethanol absolute	0.02	3.15	Dead	0	0.12	Dead

The parasitism rate given in the table was the mean of the measured values for six mice.

1.2 Effect of Artemisinin on Tissue-Stage Plasmodium

1. Effect of artemisinin on *Plasmodium gollinaceum* in preerythrocytic stage
   Chicken infected with *P. gollinaceum* were allowed to be stung by *Aedes aegypti* or *Aedes albopictus* 9–11 days after development of a lot of gametocytes in the blood and then fed for 8–10 days; afterward, the salivary glands of the chicken were dissected for preparation of sporozoite-normal saline suspension, and the chicken were orally administered with the suspension 30 min after inoculation, once daily, for 6 consecutive days; blood smears were prepared for microscopic examination once daily since day 7. The result showed that artemisinin was not effective on *P. gollinaceum* in preerythrocytic stage when administered at a dose of 100–300 mg/kg/day for 6 consecutive days, and that when the dose was increased to 400 mg/kg/day, all the animals in the group died from drug toxicity within 7 days following 6 consecutive days of dosing.
2. Effect of artemisinin on *Plasmodium cyomolgi* in tissue stage
   Experimental monkeys were intravenously inoculated with a suspension that was prepared with sporozoites collected from the salivary glands of *Anopheles stephensi*, and the effect of the drug on *P. cynomolgi* in the early and late tissue stage was observed. The results showed that administration of the drug did not inhibit occurrence of parasitemia, whether started on day 1 or day 5 of inoculation, indicating that artemisinin is not effective on *P. cynomolgi* in both the preerythrocytic stage and the tissue stage.

1.3 Antirecrudescence Effect of Artemisinin

Parasite recrudescence is discovered at both clinic and laboratory after administration of artemisinin at therapeutic doses. To explore ways to prevent rescrudescence, the authors conducted studies from the aspect of rational drug use, including increasing the dose, prolonging the duration of dosing, increasing the frequency of dosing, and changing the route of administration. The preliminary results showed that when the dose increased to more than 4 times the effective dose (the total dose of 3 days was 1200 mg/kg), parasite recrudescence was not observed for 41 consecutive days following whole-blood inoculation, which was carried out on day 2 after negative conversion of parasites. When the route of administration with an effective dose of artemisinin was changed from oral administration to subcutaneous injection, that parasite recrudescence was also not detected in the 15 mice during 28–32 days of observation.

Discussion and summation

1. Artemisinin has a definite antimalarial effect on mice and monkey models of malaria Administration of artemisinin 100 mg/kg for 3 consecutive days can make plasmodium-infected mice turn to negative and administration of artemisinin 200 mg/kg can cause infected rhesus monkeys to turn to negative. The experiments have proved that artemisinin is not effective on plasmodium in the preerythrocytic stage or the tissue stage, and is

mainly effective on blood-stage plasmodium. The results of in vitro culture have provided preliminary evidence that artemisinin directly works on plasmodium.
2. Artemisinin has the problem of a high recrudescence rate in the treatment of malaria Experimental exploration from the aspects of rational drug use preliminarily discovered that in mice models of malaria, oral administration 4 times the effective dose and hypodermic injection of artemisinin-oil suspension inhibited recurrence of rodent plasmodium.

1.4 Effect of Artemisinin on the Ultrastructure of Rodent Plasmodium

To expound how artemisinin works on plasmodium, the present experiment used electronic microscope to investigate the effect of artemisinin on the ultrastructure of rodent plasmodium in the blood stage.

1.4.1 Materials and methods

Healthy mice weighing 18–22 g were intraperitoneally inoculated with *P. berghei* strain 1×10^7, and those with a parasitism rate of plasmodium >10% 3–4 days after inoculation were selected for the experiment.

The mice were divided into four groups: (1) healthy control group; (2) artemisinin minimum effective dose group (100 mg/kg); (3) artemisinin maximum effective dose group (800 mg/kg); and (4) chloroquine control group (40 mg/kg). Blood was collected at 4, 8, 12, 16, 20, and 24 h after intragastric administration of drug for preparation of specimens using the conventional ultrathin sectioning method.

The blood was collected from the retroorbital plexus, in the experimental mice in a way that 7–8 drops were allowed to fall into a flat-bottomed glass bottle and spread at the bottom; 5% glutaraldehyde (prepared with Palade buffer solution, pH 7.4) was added to cover the blood surface immediately the blood clotted; afterward, the blood was fixed for 1 h (the blood clot was taken out at the time of 30 min and cut into 1 mm pieces, which were then put back into the bottle to be fixed continuously until 1 h), followed by washing with Palade buffer solution for 1 h; the blood was fixed with barbituric isosmotic osmium tetroxide fixer for another hour and then washed with Palade buffer solution for 30 min, followed by gradual dehydration with 50–100% acetone (the aforementioned steps were all carried out at a temperature of 4°C); at the end of dehydration, a mixture of 100% acetone and prepared epoxy resin 618 embedding medium (1:1) was added to vials containing blood clot specimens to permeate the specimens, and 30 min later, the blood clots were taken out, put into capsules, embedded with epoxy resin 618, and then placed in a 37°C thermostat overnight; afterward, the blood clots were placed in a 60°C thermostat for 48 h for polyreaction. The resulting blood clots were sectioned using LKB ultramicrotome, and the sections were stained using uranyl acetate and lead citrate and then observed under JEM-100B electron microscope.

1.4.2 Results

1. Effect of artemisinin 100 mg/kg on the ultrastructure of rodent plasmodium in the blood stage
   a. 4 h: There was no noticeable change compared with the normal structure of plasmodium (Figs. 14.1 and 14.2).
   b. 8 h: Some food vacuoles within the trophozoite began to change, manifested as threaded membrane, swollen mitochondria, and shallow pigments; most trophozoites showed changes (Fig. 14.3).
   c. 12–14 h: The food vacuoles within the trophozoite were threaded, and obviously swollen threaded cytostome was seen within some of the trophozoites. Threads began to appear on the limiting membrane, and some were even multilayer threads. Mitochondrial outer membrane obviously swelled. Some of the ribosomes within the cytoplasm detached from the endoplasmic reticulua and dispersed; some of the endoplasmic reticulua were linearly arranged, some were vacuole-like, and some disintegrated to some extent. Nuclear membrane swelled, with widening of the nuclear membrane region and asteroid aggregation of the chromatin. A few trophozoites began to develop autophagic vacuoles, some of which disgorged out of parasite into the erythrocytes. Extremely few trophozoites had mild changes (Figs. 14.4–14.8).
   d. 16 h: The limiting membrane within the trophozoite swelled significantly, with widening of the outer and inner membrane region complicated by cystiform vacuoles, the food vacuoles were threaded, the mitochondria swelled to a considerable

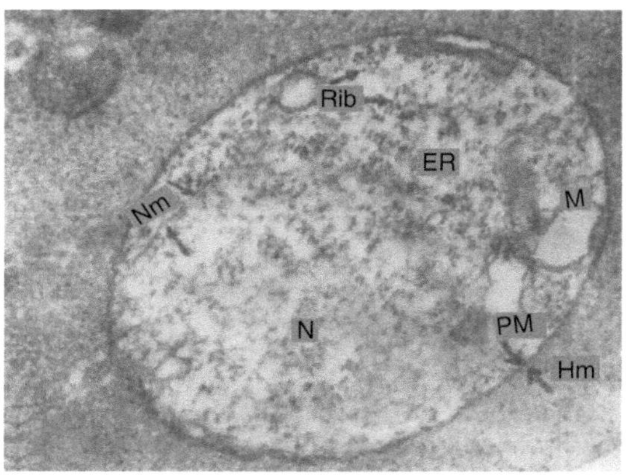

Figure 14.1: Normal Structure of Plasmodium in the Blood Stage (×24,000).
The young trophozoite of normal plasmodium has parasite membrane *(PM)*, host erythrocyte membrane *(Hm)*, nucleus *(N)*, nuclear membrane *(Nm)*, mitochondria *(M)*, ribosome *(Rib)*, and endoplasmic reticulum *(ER)*.

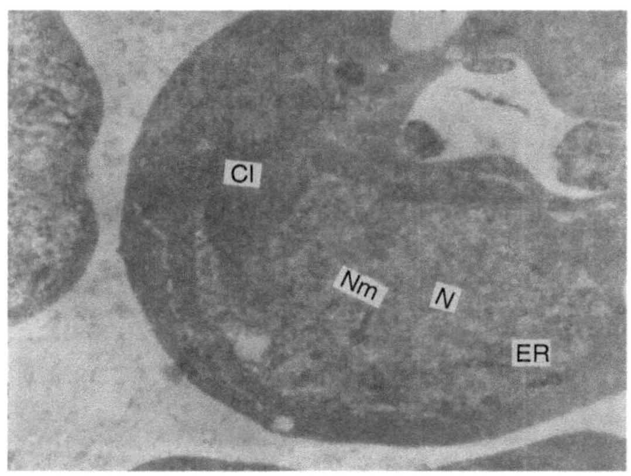

Figure 14.2: Normal Structure of Plasmodium in the Blood Stage (×15,000).
The trophozoite of normal plasmodium has cytoplasm indentation *(Cl)*, nucleus *(N)*, nuclear membrane *(Nm)*, endoplasmic reticulum *(ER)*, and so on.

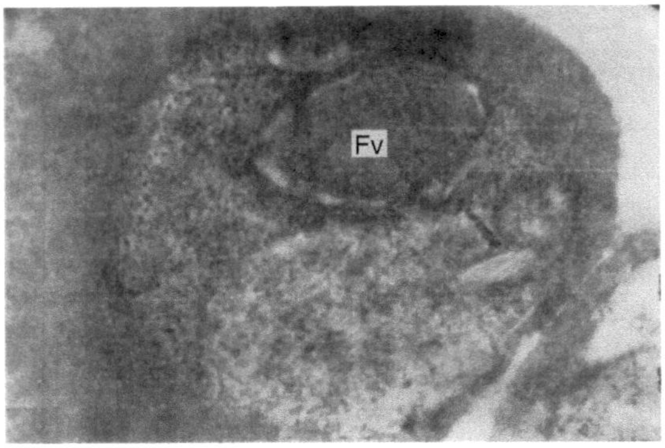

Figure 14.3: The Structure of Rodent Plasmodium in the Blood Stage 8 h After Administration of Artemisinin (100 mg/kg group).
Food vacuole *(Fv)* develops threads *(arrow)* 8 h after oral administration of artemisinin 100 mg/kg.

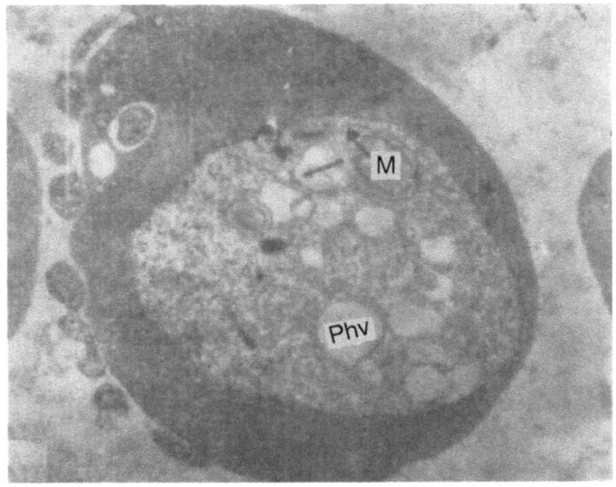

Figure 14.4: The Structure of Rodent Plasmodium in the Blood Stage 12 h After Administration of Artemisinin (100 mg/kg group) (×12,000).
The outer membrane of the mitochondria *(M)* within the trophozoite swells, with autophagic vacuoles *(Phv)*, 12 h after oral administration of artemisinin 100 mg/kg.

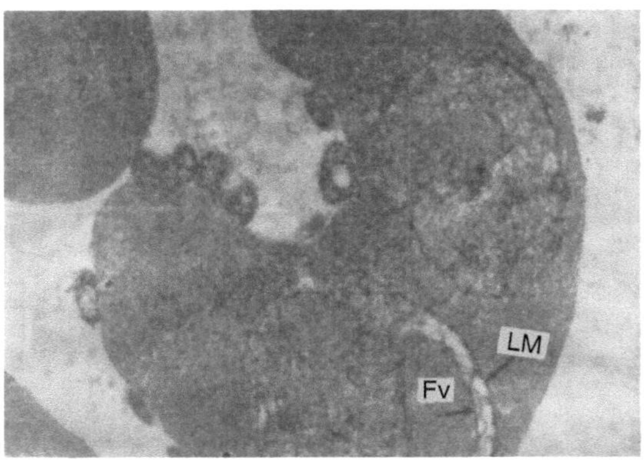

Figure 14.5: The Structure of Rodent Plasmodium in the Blood Stage 12 h After Administration of Artemisinin (100 mg/kg group) (×12,000).
Both the food vacuoles and the limiting membrane *(LM)* of the trophozoite develop threads 12 h after oral administration of artemisinin 100 mg/kg.

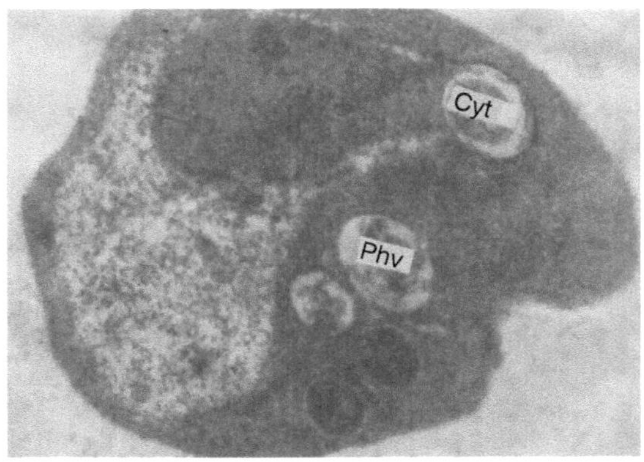

Figure 14.6: The Structure of Rodent Plasmodium in the Blood Stage 12 h After Administration of Artemisinin (100 mg/kg group) (×15,000).
The cytostome *(Cyt)* is threaded and autophagic vacuoles *(Phv)* disgorged out of parasite 12 h after oral administration of artemisinin 100 mg/kg.

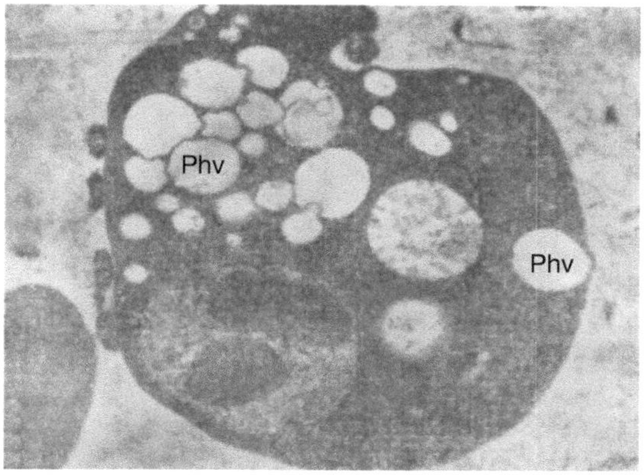

Figure 14.7: The Structure of Rodent Plasmodium in the Blood Stage 12 h After Administration of Artemisinin (100 mg/kg group) (×9000).
A lot of autophagic vacuoles are produced within the trophozoite and they disgorge out of parasite 12 h after oral administration of artemisinin 100 mg/kg.

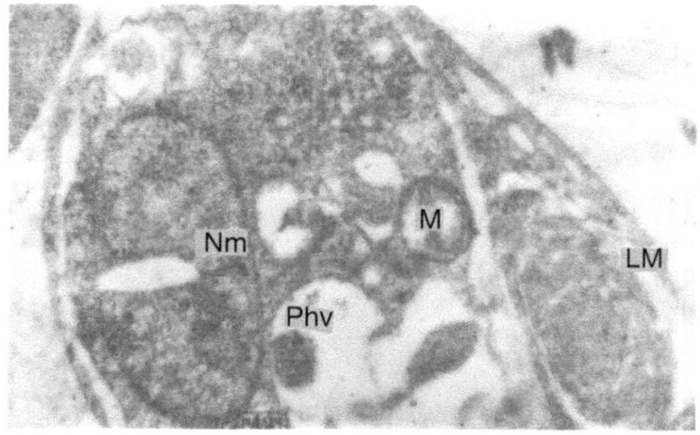

Figure 14.8: The Structure of Rodent Plasmodium in the Blood Stage 12 h and 30 min After Administration of Artemisinin (100 mg/kg group) (×15,000).
Threaded limiting membrane *(LM)*, swollen nuclear membrane *(Nm)*, asteroid aggregation of the intranuclear chromatin, autophagic vacuoles with parasite cytoplasm, and swollen mitochondria *(M)* are seen within the trophozoite 14 h and 30 min after oral administration of artemisinin 100 mg/kg.

degree, and the endoplasmic reticulua were vacuole-like or disintegrated. There were an increased number of autophagic vacuoles, some of which disgorged out of parasite. Pigmentation increased and expanded, and the nuclear membrane swelled, with asteroid or nucleol form aggregation of the chromatin. The merozoite was not significantly affected, and the nuclear membrane of the karyokinetic schizont widened (Fig. 14.9).

e. 20:h: The changes of the food vacuoles, mitochondria, nuclei, and endoplasmic reticulua within the trophozoite were similar to 16 h, with decreased ribosomes and clubbed pigments, some of which were disgorged into the parasite cytoplasm; the autophagic vacuoles enlarged, and a lot of autophagic vacuoles were disgorged out of parasites into the cytoplasm of the erythrocytes; the structure of cytoplasm in some trophozoite totally disintegrated, with only autophagic vacuoles and faint limiting membrane. Extremely few trophozoites had mild changes (Fig. 14.10).
f. 24 h: The majority of parasites within the trophozoites completely disintegrated, and there were only many autophagic vacuoles within the host erythrocytes. The changes of the schizont were mild (Fig. 14.11).

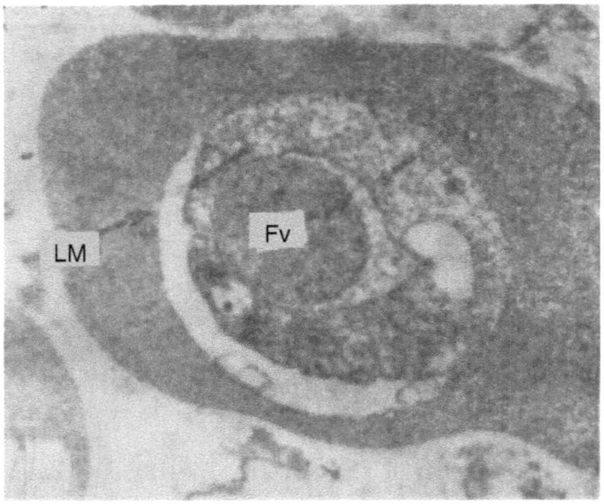

Figure 14.9: The Structure of Rodent Plasmodium in the Blood Stage 16 h After Administration of Artemisinin (100 mg/kg group) (×12,000).
Swollen food vacuoles, widening outer and inner membrane region, and swollen limiting membrane are seen within the trophozoite 16 h after oral administration of artemisinin 100 mg/kg.

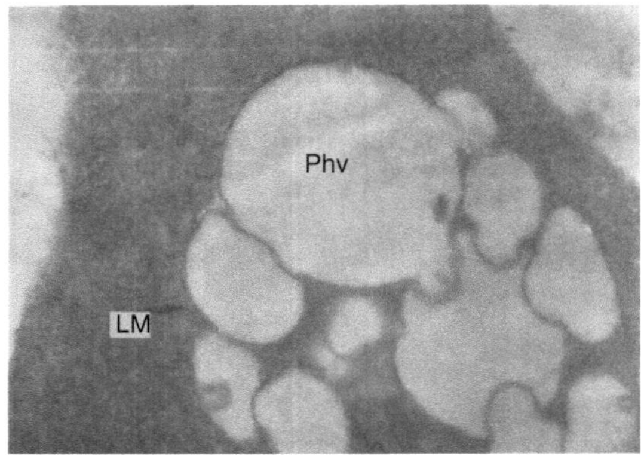

Figure 14.10: The Structure of Rodent Plasmodium in the Blood Stage 20 h After Administration of Artemisinin (100 mg/kg group) (×24,000).
The structure within the parasite cytoplasm has completely disintegrated, with only autophagic vacuoles and faint limiting membrane *(LM)* of the parasite 20 h after oral administration of artemisinin 100 mg/kg.

2. Effect of artemisinin 800 mg/kg on the ultrastructure of rodent plasmodium in the blood stage
   a. 4 h: In some trophozoites, there was already irregular accumulation of ribosomes, the endoplasmic reticulum disintegrated, the chromatin began to aggregate, and autophagic vacuoles formed.
   b. 8 h: The nuclear membrane region in the trophozoite expanded, with aggregation of the chromatin, threaded food vacuoles, decreased ribosomes, and disintegration of the endoplasmic reticulum.
   c. 16 h: Some of the limiting membranes within the trophozoite developed vacuoles, with highly threaded membrane; the nuclear membrane region expanded, with asteroid aggregation of the chromatin; the ribosomes piled up, the endoplasmic recticulum disintegrated to some extent, and the mitochondria swelled, with a dramatically increased electron density in the middle. Autophagic vacuoles appeared, some of which disgorged out of parasite (Figs. 14.12, 14.13, and 14.14).
   d. 20 h: There was an extreme aggregation of the chromatin within the nuclei of the trophozoite, with aggregation of ribosomes, disintegration of linearly arranged endoplasmic recticulum, and expanding autopahgic vacuoles, some of which disgorged out of parasite.

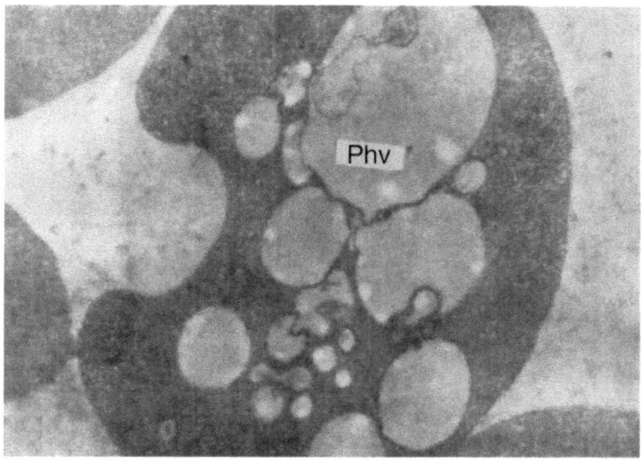

Figure 14.11: The Structure of Rodent Plasmodium in the Blood Stage 24 h After Administration of Artemisinin (100 mg/kg group) (×12,000).
The structure within the parasite cytoplasm has completely disintegrated, with residual autophagic vacuoles within the host erythrocytes 24 h after oral administration of artemisinin 100 mg/kg.

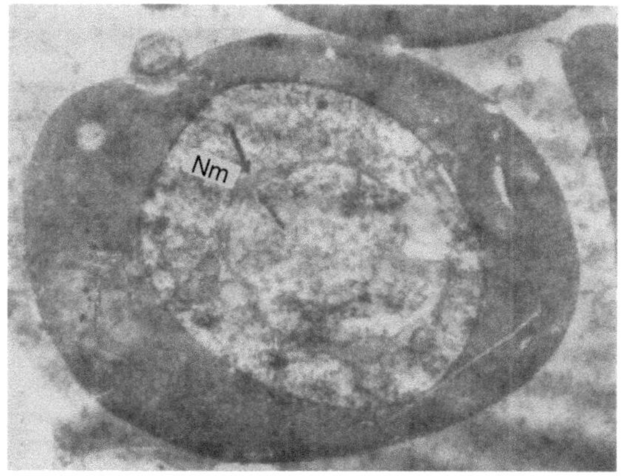

Figure 14.12: The Structure of Rodent Plasmodium in the Blood Stage 16 h After Administration of Artemisinin (800 mg/kg group) (×12,000).
Swollen nuclear membrane *(Nm)*, asteroid aggregation of the chromatin, and swollen mitochondria are seen within the trophozoite 16 h after oral administration of artemisinin 800 mg/kg.

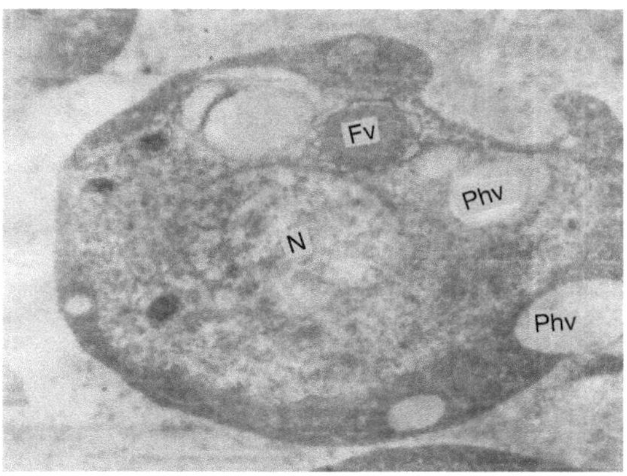

Figure 14.13: The Structure of Rodent Plasmodium in the Blood Stage 16 h After Administration of Artemisinin (800 mg/kg group).
Threaded food vacuoles, widening nuclear membrane, asteroid aggregation of the chromatin, and autophagic vacuoles are seen within the trophozoite, some of which disgorged out of parasite 16 h after oral administration of artemisinin (800 mg/kg group).

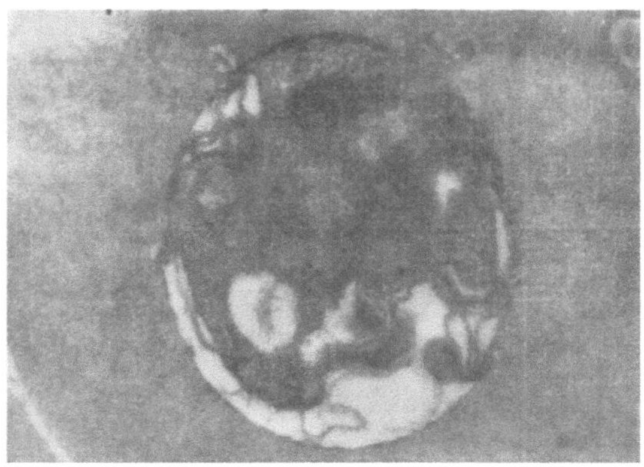

Figure 14.14: The Structure of Rodent Plasmodium in the Blood Stage 16 h After Administration of Artemisinin (800 mg/kg group) (×24,000).
Deep threads are seen on the nuclear membrane of the trophozoite 16 h after oral administration of artemisinin (800 mg/kg group).

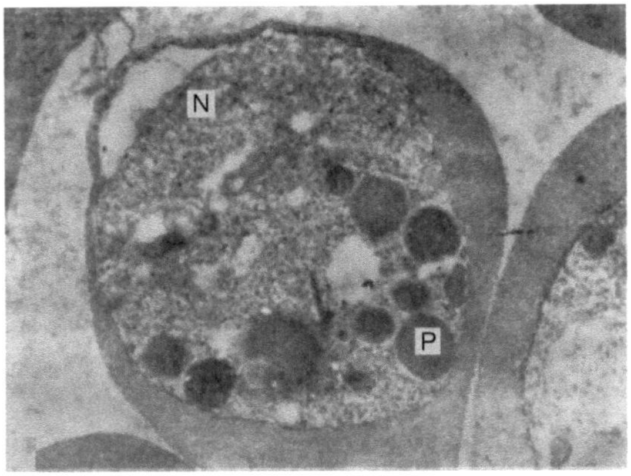

Figure 14.15: The Structure of Rodent Plasmodium in the Blood Stage 16 h After Administration of Chloroquine (40 mg/kg) (×18,000).
Clustering pigments *(P)* protrude into the parasite cytoplasm 16 h after oral administration of chloroquine4 mg/kg.

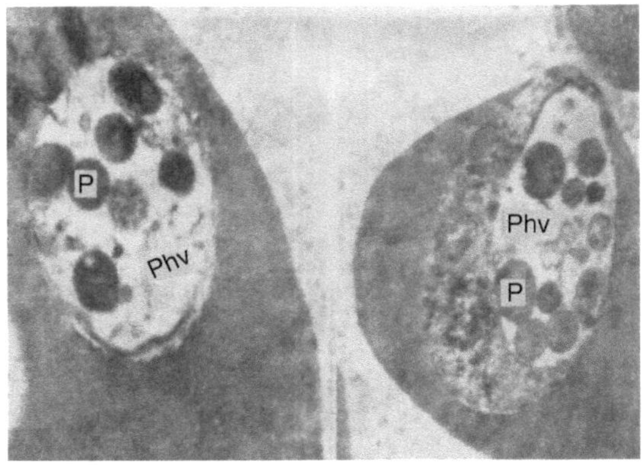

Figure 14.16: The Structure of Rodent Plasmodium in the Blood Stage 24 h After Administration of Chloroquine (40 mg/kg) (×145,000).
Clustering pigments *(P)* appear in large autophagic vacuoles *(Phv)* of the trophozoite 24 h after oral administration of chloroquine 4 mg/kg.

3. Effect of chloroquine 40 mg/kg on the ultrastructure of rodent plasmodium in the blood stage
   Observation was made 4, 16, and 24 h after intragastric administration. The most outstanding characteristics in the control group were as follows: large autophagic vacuoles formed, with massive aggregation of pigments (pigments appeared 4 h after administration and clustering pigments were seen by the time of 16 h, some of which protruded into the parasite cytoplasm), and the nuclear membrane region in some trophozoites expanded. By 24 h, some of the autophagic vacuoles protruded into the cytoplasm of the host erythrocytes together with the clustering pigments) (Figs. 14.15 and 14.16).

Discussion and summation

Artemisinin affects ultrastructure of rodent plasmodium at blood stage, mainly by working on the membrane system structure of the plasmodium. It first acts on food vacuoles, mitochondria, and limiting membrane, following which the endoplasmic recticulum and the mitochondria swell manifested as widening region between the outer and inner membranes or threads within the food vacuoles and the limiting membrane. Afterward, changes take place in the endoplasmic reticulum and the nuclear membrane, manifested in linearly arranged or vacuole-like or disintegrated endoplasmic reticulum, and swollen nuclear membrane. In addition to changes in the membrane system structure, changes of the nuclei are also seen, manifested by aggregation of the chromatin. Autophagic vacuoles were obviously seen 12 h after administration, some of which even disgorged out of parasite, and this phenomenon developed rapidly: by 20–24 h, the inner structure of some trophozoites disintegrated, only with several residual autophagic vacuoles. Different doses of artemisinin had differences in action speed rather than in action site or action characteristics.

Artemisinin differs from chloroquine in site and characteristics of action. In mice orally administered with chloroquine, the main manifestations were enlarged digestive vacuole and clustering pigments. Macomber et al. (1967) reported formation of large autophagic vacuoles with pigment 2 h after administration of chloroquine in an in vivo experiment. Ali et al. (1969) and Howells, Peters, Homewood, and Warhurst (1970) pointed out that it was the result of the plasmodium's hemoglobin metabolism being affected. However, Warhurst believed that it was the result of direct action on pigments and digestive vacuole, including changes of the digestive vacuole. The authors also observed this action characteristic of chloroquine during experiment.

Plasmodium in the blood stage is mainly nourished by hemoglobin from host erythrocytes. Electron microscope studies have proved that plasmodium is fed in a way that cytostome swallows the cytoplasm of host erythrocytes first to form a cytoplasm indentation (CI), in which the electron density is consistent to that of the host erythrocytes. CI has two layers of membranes—the outer membrane (parasite membrane) and the inner membrane (host erythrocyte membrane). After being swallowed, the erythrocyte cytoplasm, along with its outer membrane, cuts off at the cytostome to form a food vacuole that contains two layers of membranes, and the cytostome is then sealed. Afterward, the cytostome swallows the cytoplasm of host erythrocytes again to form food vacuoles again, and this cycle repeats. Therefore, food vacuoles are the earliest stage of nutrition intake by the plasmodium. Host hemoglobin in the food vacuoles is digested and enzymolyzed in the food vacuoles, during which single-membrane digestive vacuole forms and sometimes residual membrane pieces of host erythrocytes are seen in the digestive vacuoles. Insoluble pigment granules are produced during the process of digestion. Plasmodium finally digests host hemoglobin into amino acids and peptides, which are the raw materials for synthesis of protein of the protozoa.

The primary site artemisinin works on is membrane system of plasmodium. It acts on the food vacuoles directly to block the earliest stage of nutrition intake, thus causing the plasmodium to hunger for amino acids, develop autophagic vacuoles rapidly, and disgorge out of parasite continuously, in which way the plasmodium loses a lot of cytoplasm and then dies. Theakston et al. (1969) proved that the pellicle of rodent plasmodium contains cytochrome oxidase. Howelles et al. (1973) reported that the mitochondria of rodent plasmodium in the trophozoite stage mainly contained cytochrome oxidase, both chloroquine-sensitive strain and chloroquine-resistant strain. The pellicle and mitochondria of rodent plasmodium are also morphologically similar. Artemisinin mainly acts on the membrane system structure of plasmodium at such primary sites as food vacuoles, pellicle, and mitochondria. The antimalarial effect of artemisinin can be considered as that it mainly interferes with the pellicular-mitochondrial function.

The present experiment observed changes in the ultrastructure of rodent plasmodium following intragastric administration of artemisinin and compared with chloroquine control group. The results proved that artemisinin mainly acts on the membrane system structure of plasmodium; specifically, it first acts on food vacuoles, pellicle, and mitochondria, and then on the nuclear membrane and the endoplasmic reticulum. The results also proved that artemisinin has a certain effect on the chromatin within the nucleus. We believe that the mode of action of artemisinin mainly affects the pellicular-mitochondrial function, blocking nutrition supply that is mainly based on the host erythrocyte cytoplasm. Artemisinin significantly differs from chloroquine in the site of action and the development rate of action, suggesting that artemisinin can achieve rapider negative conversion of plasmodium than chloroquine in clinical treatment.

2 General Pharmacology Studies

2.1 Nervous System

2.1.1 Test in dogs

1. Two dogs were intragastrically administered a single dose of artemisinin 105 and 158 mg/kg, respectively, and were not found to have any abnormalities in general behavior, activities, and neurological status during the 14 days of observation after administration of the drug.
2. Four dogs were used for the test, each of which was intragastrically administered artemisinin 100 mg/kg daily for 5 consecutive days. The dogs were not found to have any abnormalities in appearance and activities during the 1-week observation following administration of the drug.

2.1.2 Test in rats

Twenty-four rats were equally divided into three groups and intragastrically administered artemisinin-water suspension 250, 500, and 1000 mg/kg, respectively, for 14 consecutive days. The rats in all dose groups were not observed with abnormalities in general behavior after administration of the drug.

Based on the aforementioned results of tests in dogs and rats, we conclude that artemisinin has no obvious pharmacological activity on the nervous system.

2.2 Cardiovascular System

2.2.1 Test in dogs

1. The dogs were intragastrically administered a single dose of artemisinin 105 and 158 mg/kg, respectively, and their heart rates (HR) and electrocardiograms (ECG) were measured 3, 7, and 14 days, respectively, after administration of the drug and compared with those before administration of the drug. The results showed that artemisinin had no obvious effects on the HR and ECG of the dogs, and abnormal changes in the HR and ECG were still not observed even when the dose was increased up to 1200 mg/kg.

2. Three dogs were anesthetized with pentobarbital sodium through intravenous injection, and a mercury manometer was connected to the left common carotid artery of each dog to measure the blood pressure (BP). Each of the dogs was intragastrically administered a dose of artemisinin 100 mg/kg after their BP stabilized, and then their HR and BP within 2 h after administration of the drug were measured and recorded. Comparison with the HR and BP before administration of the drug showed that there were no obvious changes. No abnormal changes in the HR and BP of the dogs were observed even when artemisinin 100 mg/kg was administered via intraperitoneal injection.

Furthermore, rabbits and guinea pigs were also used for test using the method of intraperitoneal administration, and the results obtained for HR, BP, and ECG were identical to those obtained in the aforementioned test.

2.2.2 Test in cats

Six cats were divided into two groups and intragastrically administered a dose of artemisinin 100 and 500 mg/kg, respectively, without being anesthetized, and the ECG of the cats was recorded for 2–3 h after administration of the drug and then compared with those before administration of the drug. In addition, another three cats were anesthetized with urethane and then intragastrically administered a dose of artemisinin 800 mg/kg, respectively; afterward, the ECG and BP of the three cats were recorded until 4 h and then compared with those before administration of the drug. The results showed that artemisinin at any of the aforementioned doses did not significantly affect the HR, BP, and ECG of the cats, whether the cats were anesthetized or not.

Based on the aforementioned results of test, we believe that artemisinin does not have a significant pharmacological effect on the cardiovascular systems of dogs and cats when administered at a dose slightly higher than the effective dose against plasmodium.

2.3 Respiratory System

2.3.1 Test in dogs

Four dogs were intragastrically administered artemisinin 100 mg/kg daily for 5 consecutive days and were not found to the have respiratory abnormalities during the following 1-week observation.

2.3.2 Test in rabbits

Six domestic rabbits were catheterized and connected to Marley's drum following separation of the trachea under local anesthesia with procaine and then intraperitoneally injected a dose of artemisinin 100 mg/kg. No obvious changes were found in the respiratory frequency and depth of the domestic rabbits during the 2 consecutive hours of observation after administration of the drug.

The aforementioned results lead to a conclusion that artemisinin has no obvious effects on the respiratory systems of dogs and domestic rabbits.

In summary, experimental studies on the general pharmacological effects of artemisinin have shown that artemisinin, when administered at a dose level slightly higher than its effective dose against plasmodium, has no obvious pharmacological effects on the nervous system, cardiovascular system, and respiratory system of the experimental animals.

3 Pharmacokinetic Studies [1,2]

3.1 Animals

Artemisinin is a novel compound that is totally different from known antimalarial agents, and its chemical structure has already been clearly identified. A lot of studies on the antimalarial effect of artemisinin from pharmacological and clinical aspects have proved that it is a novel antimalarial agent superior to chloroquine. It has the advantages of high effectiveness, rapid onset of action, and low toxicity. To further investigate the absorption, distribution, excretion, and metabolism of artemisinin in the body and provide reference for clinical rational use to enhance the therapeutic effect, tracer test was carried out using ^{3}H-labeled artemisinin.

3.1.1 Materials and methods

1. Drug: ^{3}H-artemisinin that was labeled by China Institute of Atomic Energy using the method of discharge exposure tritium exchange, with radiochemical purity >90% and specific radioactivity of 1 mCi/mg after identification by chromatography.
2. Animals: Adult mice, weighing between 18 and 22 g, either gender.
3. Administration: ^{3}H-artemisinin suspension was prepared with 1% methylcellulose added with an appropriate amount of carrier after being dissolved with a small volume of ethanol (total volume <10%). Each of the mice was intragastrically administered ^{3}H-artemisinin 20 mCi/2 mg. For whole-body autoradiography, ^{3}H-artemisinin was prepared into a suspension with oil for injection, and each of the mice was subcutaneously injected 20 mCi/2 mg into the axilla.
4. Preparation and determination of samples for use in absorption, distribution, and excretion tests.
   a. Test on blood concentration: There were five mice in ^{3}H-artemisinin group, and 10 μL of blood was collected from the caudal vein 0.5, 1, 2, 4, 6, 8, 12, 14, 16, and 24 h, respectively, after intragastric administration of the drug. After addition of 20 mL of 98% formic acid, 20 mL of 20% hydrogen peroxide, and a small drop of octanol, the blood specimens were warmed up in a 75°C water bath for 30 min to 1 h for digestion and decolorization, and scintillation solution was added to the specimens for the determination of radioactivity.

b. Test on tissue distribution: There were four mice in each dose group. Animals were killed by breaking the cervical cord 1, 4, 24, and 28 h, respectively, after intragastric administration of ^{3}H-artemisinin, and such organs (<100 mg) as heart, liver, spleen, lung, kidney, muscle, bone, and brain were harvested and then, after addition of 0.2 mL of 70% perchloric acid, 0.4 mL of 30% hydrogen peroxide, and one drop of octanol, were warmed up in a 75°C water bath for 30 min to 1 h to digest and decolorate the tissue. Afterward, the tissue was diluted with water to 1 mL, from which 0.1 mL was collected for radioactivity assay. Although the tissue was harvested, bile was absorbed using lens paper and then placed in a measuring bottle directly for determination after being weighed.

c. Excretion test: Four or five mice were used. Urine and fecal samples were collected 4, 8, 12, and 24 h, respectively, after intragstric administration of ^{3}H-artemisinin. The urine was absorbed using filter paper and then allowed to dry. The urine and fecal specimens were treated the same way using the method of perchloric acid digestion as the aforementioned organ tissue. FJ353G double-channel liquid scintillation counter (LSC) was used for radioactivity assay. Dimethylbenzene–glycol ether (7:3) solution of 0.4% PPO was used as the scintillation solution; 5 mL of the scintillation solution was added to each of the samples. The various tissue samples were all quenched with internal standard for calibration.

5. Thin-layer chromatography (TLC) for radioactive metabolites in blood and urine samples

   Blood was collected at different times after intragastric administration, intramuscular injection, or intravenous injection of ^{3}H-artemisinin 20 μCi, or 1 mL of anticoagulant whole blood was collected in vitro from healthy mice and incubated at a temperature of 37°C for a different period of time after addition of ^{3}H-artemisinin 4 μCi. Deionized water was added to the blood samples to break the red blood cell, and the resulting solution was extracted using triple volume of chloroform. After centrifugation, chloroform extract in the lower layer was aspirated for TLC to measure the radioactivity, with petroleum ether–ethyl ether (1:1) used as the developing solvent. After TLC, the radioactive silica gel sheet was cut off at a distance of 1 cm and the pieces were put into a measuring flask, to which 3 mL of dimethylbenzene scintillation solution of 0.4% TP was added for radioactivity assay. Radioactive and stable artemisinin samples and standards were dotted the same way for radioactivity assay, or were developed using 2% vanillin-concentrated sulphuric acid. Urine samples were dotted directly for TLC.

6. Whole-body autoradiography

   Four mice were used 1, 4, 8, and 24 h, respectively, after hypodermic injection of ^{3}H-artemisinin 2 mCi/2 mg into the axilla, and then quickly frozen and fixed using acetone–dry ice under anesthesia with ethyl ether. The body of the animal was embedded with 7.5% methylcellulose sodium and then cut into sagittal sections of the whole body

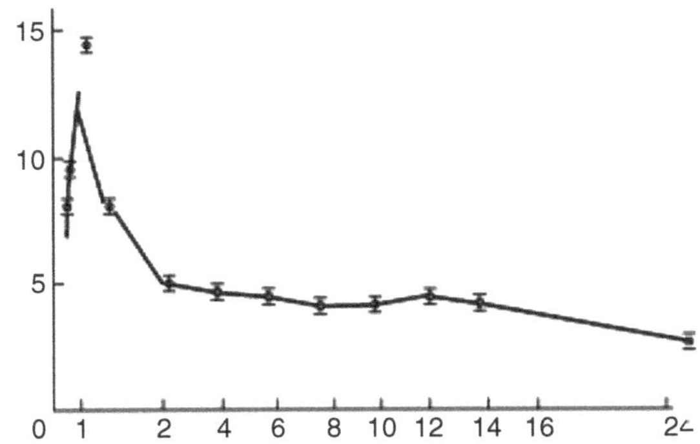

Figure 14.17: Changes of Radioactivity in the Blood of Mice Over Time After Oral Administration of ^{3}H-artemisinin Drawn Based on the Data (mean ± SD) of the Five Animals.

40 μm in thickness at a low temperature of −20°C. After lyophilization, the sections were pasted onto nuclear-4 latex dry plate tightly and exposed in the cassette for 1–2 months. Observation was made after the development and fixation.

3.1.2 Results

1. Blood concentration: After oral administration of ^{3}H-artemisinin to mice, the radioactivity in the blood increased rapidly and peaked by the time of 1 h, followed by rapid decrease; the radioactivity decreased to below 1/2 of the peak value by the time of 4 h and then plateaued (Fig. 14.17).
2. In vitro distribution: The results of intensity of radioactivity in such organs as liver, spleen, kidney, lung, muscle, bone, brain, and heart, as well as in their secretions at a different time after oral administration of ^{3}H-artemisinin to mice, are provided in Table 14.6. The results showed that the radioactivity in various organs and their secretions peaked by the time of 1 h and then decreased gradually. Specifically, bile had the highest radioactivity; as for solid organs, kidney had the highest radioactivity, followed by liver as compared, and the sequence was as follows: kidney>liver>heart>lung>spleen>muscle>bone>brain (based on 1-h data).

   The outline of whole body and some organs were developed 1 h after administration of ^{3}H-artemisinin, indicating that the radioactivity was widely distributed in the body and mostly in the gallbladder and the urinary bladders. By the time of 4 h, only the contents in the gallbladder, the urinary bladders, and the intestine were developed, but not the outline of whole body and some organs. By the time of 8 h, the gallbladder was not developed and only the contents in the urinary bladders and the intestine were developed. By the time of 24 h, only the contents in the intestine were developed. These results were consistent with the aforementioned results of quantitative radioactivity assay.

Table 14.6: Intensity of radioactivity in tissue organs and their secretions at a different time after oral administration of ^{3}H-artemisinin to mice

Tissue	Intensity of Radioactivity in Tissue at Different times (×10^5 dpm/g)			
	1 h After Administration	4 h After Administration	24 h After Administration	48 h After Administration
Liver	18.67 ± 2.77	9.85 ± 1,14	3.67 ± 0.57	1.85 ± 0.08
Spleen	5.24 ± 0.46	3.62 ± 1.03	2.63 ± 0.23	1.81 ± 0.29
Kidney	25.52 ± 2.81	11.05 ± 1.26	4.82 ± 1.04	3.04 ± 0.29
Lung	6.22 ± 0.46	4.17 ± 0.37	2.72 ± 0.41	1.55 ± 0.27
Muscle	4.99 ± 0.65	4.19 ± 0.33	3.11 ± 0.24	1.56 ± 0.13
Bone	3.42 ± 0.21	2.56 ± 0.27	1.57 ± 0.07	0.76 ± 0.07
Brain	3.18 ± 0.21	2.67 ± 0.15	2.26 ± 0.25	1.43 ± 0.21
Heart	7.07 ± 0.84	4.36 ± 0.24	3.34 ± 0.37	1.95 ± 0.19
Bile	221.67 ± 64.02	22.10	23.01	1.36

The data provided in the table were the mean values of the four animals ± SD.

3. Urinary and fecal excretions: Within 24 h after oral administration of ^{3}H-artemisinin to five mice, 56.3 ± 9.8% and 22.2 ± 2.3% of the radioactivity were excreted from urine and stool, respectively; the radioactivity excreted from urine and stool accounted for about 80% together.
4. Preliminary observation on metabolism in the body: TLC performed for determination of radioactivity in the chloroform extract of blood collected from the mice at a different time (5 min, 10 min, 30 min, and 1 h) after oral administration, intramuscular injection, or intravenous injection of ^{3}H-artemisinin showed that the radioactive peaks decreased gradually at the site of artemisinin, and increasingly bigger radioactive peaks appeared around the site where the blood samples were dotted on the silica gel plate (Fig. 14.18).

   Fig. 14.18 is the TLC map of radioactivity in blood 30 min after intravenous injection of the labeled compound. According to the map, the radioactivity was mainly around the site where the blood samples were dotted, and there was only a tiny amount of radioactivity left at the R_f site of the original compound.

   The results of TLC performed for radioactivity assay in urine samples showed that there were only single radioactive peaks at the site where the samples were dotted on the silica gel plate, indicating that there was no ^{3}H-artemisinin in the urine any more (Fig. 14.19).
5. Radiopharmaceutical agent and blood incubation test: Thirty minutes after in vitro incubation of ^{3}H-artemisinin with the anticoagulant whole blood of mice, TLC was performed for radioactivity assay in the chloroform extract of the incubated blood. The result showed that the radioactivity concentrated at R_f site of the original compound, indicating that blood did not affect the action of the drug significantly in vitro (Fig. 14.20).

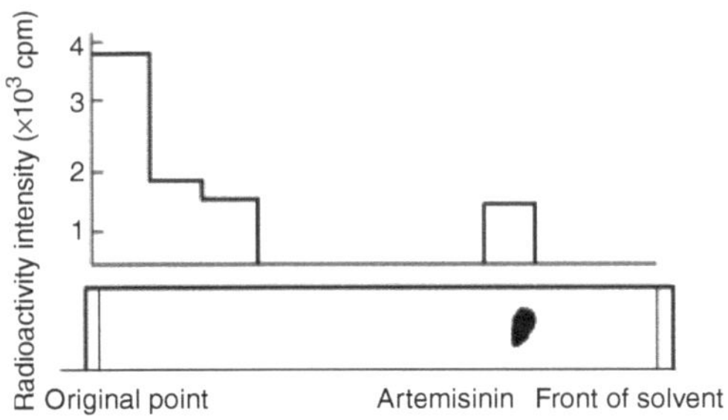

Figure 14.18: TLC Map of Radioactivity in the Standard Sample and Chloroform Extract of Blood Collected From the Mice 30 min After Intravenous Injection of ^{3}H-artemisinin.

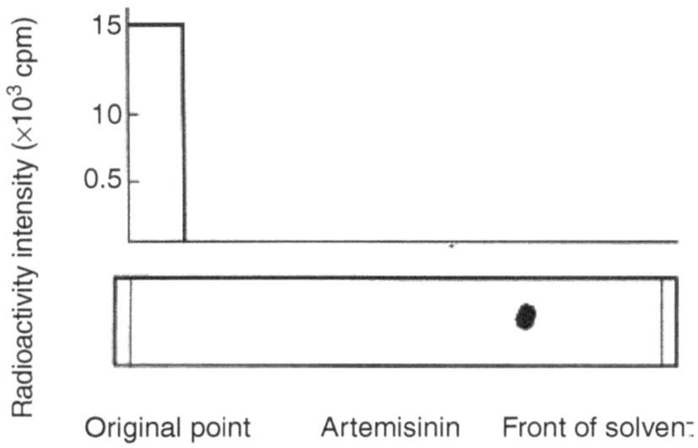

Figure 14.19: TLC Map of Radioactivity in the Standard Sample and Urine Samples.

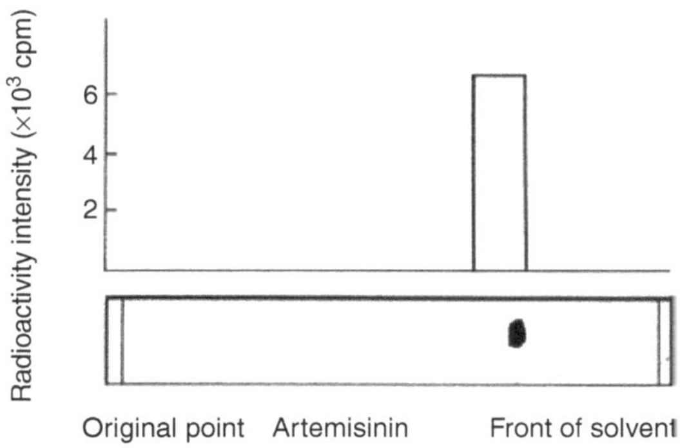

Figure 14.20: TLC Map of Radioactivity in the Standard Sample and Chloroform Extract of Blood Collected 30 min After ^{3}H-artemisinin was Incubated With Anticoagulant Whole Blood of Mice.

Discussion and summation

Performing TLC for the radioactivity in blood and urine, ^{3}H-artemisinin was incubated with blood for 30 min and then extracted chemically, the radioactivity was found to mainly concentrate at the R_f site of the parent compound, suggesting it is not easily damaged by blood in vitro. However, 30 min after intravenous injection, the radioactivity of the two labeled compounds mainly concentrated around the site where the samples were dotted, with a tiny amount of radioactivity at the R_f site of the original compound, indicating that the original compound significantly decreased in the animal blood. Almost all of the radioactivity excreted from urine concentrated at the site where the samples were dotted, indicating that the drug did not exist in the form of the original compound any more. The aforementioned results demonstrate that the drug is quickly metabolized in the body. The radioactivity of the samples measured in the experiment should be the sum of the radioactivity of artemisinin and that of its metabolites.

The process of artemisinin in the body was similar: it was quickly absorbed in the gastrointestinal tract after oral administration, and the blood concentration reached the peak 1 h after administration, thus enabling the drug to take effect rapidly. However, due to short maintenance of blood concentration resulting from quick excretion from the body, it will not help eliminate plasmodium completely and thoroughly.

Artemisinin was widely distributed in the body after oral administration and absorption, with radioactivity distributed in such tissue organs as bone and brain. The radioactivity in solid organs liver and kidney was higher, and it was probably associated with metabolism and excretion. The levels of radioactivity in various organs decreased with blood concentration, indicating that the drug will not accumulate in the body. The results of whole-body autoradiography showed that the gallbladder and the urinary bladders had the highest radioactivity levels, indicating that a lot of radioactivity was discharged within a short period of time.

To conclude, studies using ^{3}H-labeled artemisinin have proved that the drug has the properties of quick absorption, wide distribution, quick excretion, and rapid metabolism in the body. However, further in-depth studies are still required regarding whether the molecules of the drug after metabolism will produce an antimalarial effect, as well as the structure of its metabolites.

3.2 Human Bodies

3.2.1 Pharmacokinetics of artemisinin tablets in human body after oral administration

Artemisinin tablets are a newly developed oral dosage form of artemisinin and have the advantages of convenient use and reliable efficacy. With a view to promoting clinical use, we conducted a clinical pharmacokinetic study for artemisinin tablets as per requirements of new drug approval regulation.

The study was carried out at Xiyuan Hospital, China Academy of Chinese Medical Sciences (CACMS) in Beijing. The investigational product artemisinin tablets were supplied by the Institute of Chinese Materia Medica, CACMS. The subjects were three healthy adults weighing between 47 and 71 kg and were orally administered artemisinin tablets 15 mg/kg; 1.5 mL blood was collected from each of the subjects at different times after administration of the drug and centrifuged at ambient temperature for separation of serum. The obtained serum was preserved at a temperature of −20°C. Radioimmunoassay was used to determine the plasma concentration.

The plasma concentration reached the peak 1.5 h after oral administration of artemisinin tablets 15 mg/kg to the human subjects, with the peak concentration being 0.09 μg/mL. The mean retention time, half-life, and steady-state volume of distribution of the drug in human body were 3.27 ± 0.32 h, 2.27 ± 0.22 h, and 152.34 ± 35.09 L/kg, respectively.

The experimental results showed that artemisinin tablets were quickly absorbed in human body after oral administration. As compared with tests in animals, there was a significant difference in absorption of artemisinin tablets between human subjects and animals. In dogs, the drug was still not detected in blood up to a dose of 70 mg/kg of artemisinin tablets orally administered. The aforementioned results indicate that artemisinin tablets are well absorbed in human body.

3.2.2 Bioavailability of artemisinin tablets in human body

As artemisinin is a water-insoluble compound and cannot be administered via intravenous injection, we only had to compare artemisinin tablets with the approved artemisinin suppositories to obtain bioavailability relative to that of the suppositories. The relevant parameters of the two drugs are provided in Table 14.7.

Table 14.7 shows that oral artemisinin tablets are more quickly absorbed than artemisinin suppositories, with T_{max} of 1.5 h, significantly shorter than that (4.6 h) of the suppositories. In addition, artemisinin tablets are also more quickly excreted than the suppositories, with a half-life of 2.27 h, shorter than that (4.8 h) of the suppositories. Judging from the ratio of area under the curve, the bioavailability of artemisinin tablets is 1.06 times that of artemisinin suppositories.

Table 14.7: Comparison of bioavailability between artemisinin tablets and artemisinin suppositories

Relevant Parameters	Artemisinin Tablets (15 mg/kg)	Artemisinin Suppositories (15 mg/kg)
Time to maximum concentration (h)	1.5	4.6 ± 0.82
Peak plasma concentration (μg/mL)	0.09 ± 0.01	0.04 ± 0.01
Mean retention time (h)	3.27 ± 0.32	6.98 ± 3.91
Half-life (h)	2.27 ± 0.22	4.84 ± 2.71
Steady-state volume of distribution (L/kg)	152.34 ± 35.90	285.14 ± 176.62
Clearance [L/(kg·h)]	46.55 ± 9.52	54.85 ± 44.07
Area under cure [μg/(h·mL)]	0.33 ± 0.07	0.31 ± 0.22

References

[1] Institute of Chinese Materia Medica, CACMS. Research on the antimalarial effect of artemisinin (1971–1978) album. Research data of Chinese Materia Medica. 1978, 3.
[2] Pharmacology Laboratory of the Institute of Chinese Materia Medica, CACMS. J New Med 1 (1979) 24–32.

CHAPTER 15

Toxicological Studies of Artemisinin

Chapter Outline

1 Acute Toxicity

Kunming mice (male:female = 1:1) weighing 12–22 g were used.

1. The mice were fasted for about 20 h before the test and then randomized in different groups. There were 10 mice in each group.
2. Artemisinin was suspended in physiological saline containing 1% Tween, and five dose groups were set. Observation began 30 min following oral administration and lasted for 1 week. Death and poisoning manifestations of animals were recorded; the LD_{50} was calculated.
3. See Table 15.1 for the results.

Calculated by simplified probit method, the oral LD_{50} is 4721.6 mg/kg, and the 95% confidence interval was 3180–6546 mg/kg.

2 Subacute Toxicity

2.1 Subacute Toxicity Test of Artemisinin in Rats

Rats weighing 80–100 g were used. Artemisinin was intragastrically administered at doses of 250 and 500 mg/kg for 14 days. A blank control group was set. Before and after drug administrations, the rats were weighed; ECG examination, urine and blood routine tests were carried out. After animals were killed, pathological examination was carried out. No significant abnormality was found.

From *Artemisia annua* L. to Artemisinins
http://dx.doi.org/10.1016/B978-0-12-811655-5.00015-5

Table 15.1: Acute toxicity of artemisinin to mice

Dose (mg/kg)	Logarithmic Dose	Number of Deaths (Mice)	Death Rate (%)	Probit (g)
10,000	4.000	10	100	7.40
7,000	3.8451	8	80	5.84
4,900	3.6902	4	60	5.85
3,430	3.5353	3	30	4.48
2,401	3.3804	0	0	2.60

2.2 Subacute Toxicity Test of Artemisinin in Monkeys

Yunnan wild monkeys were used. Artemisinin-oil suspension was intramuscularly injected. Five groups were set and given 24, 48, 96, and 192 mg/kg for 14 consecutive days. A blank control group was set. Observation items included 8 clinical items (behavior, appetite, body temperature, body temperature, etc.), 12 urine routine items (pH, color, urine sugar, urine protein, ketone bodies, etc.), 9 blood routine items (RBC, reticular cells, platelets, hemoglobin, WBC and classification, blood sedimentation, etc.), 4 bone marrow smear items, 19 blood biochemical items (protein electrophoresis, blood urine nitrogen, creatine kinase, aspartate aminotransferase, etc.), and histopathological examination (52 specimens including organs, tissues, and glands).

2.2.1 Results

In the 24 mg/kg group, only decreased reticular cell count was found; in the 48 mg/kg group, the RBC count and hemoglobin concentration decreased, blood sedimentation increased, and blood volume mildly decreased; in the 96 mg/kg, toxic reactions intensified, and one of the six animals died 2 days after administration of the last dose; in the 192 mg/kg group, three of the four animals died 3 days after administration of the last dose. Positive toxic reactions in the 48 and 96 mg/kg group all returned to normal 22 days after drug withdrawal. Therefore, doses below 24 mg/(kg·day) × 14 days are safe, and doses above 48 mg/(kg·day) × 14 days are not. Bone marrow hemopoietic cells, particular erythroid cells, and myocardial cells are considered the target organ of toxicity.

3 Particular Toxicity

3.1 Mutation Tests

3.1.1 Ames test

Artemisinin was quantitatively analyzed using Salmonella/microsome enzyme method introduced by Ames and referring to the modified shaking incubation method proposed by Takie Yahagi.

In the first batch, *Salmonella typhimurium* TA98, TA100, TA1537, TA1535, and TA1538 were used. Every milliliter of the solution contained 300, 30, 3, 0.3, or 0.03 μg of artemisinin. In the second batch, *S. typhimurium* TA100 and TA98 were used. Artemisinin was dissolved in dimethyl sulfoxide. Every 0.1 mL contained 3000, 1500, 1000, 750, 500, or 375 μg of artemisinin. When S-9 mix or phosphate buffer was added, different degrees of precipitation occurred in solutions which contained artemisinin more than 500 μg. To each tube, 0.1 mL of different concentrations of artemisinin solution and 0.1 mL of freshly cultured bacterial solution were added; 0.5 mL of S-9 mix was added to the test tubes and 0.5 mL of phosphate buffer was added to the control tube. The tubes were incubated with shaking in a 37°C water bath for 20 min and then 2 mL of soft agar was added. The mixtures were poured to nutrient agar plates and incubated at 37°C for 48–72 h. Revertant colonies were counted. The test result was the mean value of three plates.

According to the results, with or without the presence of S-9 mix, the numbers of revertant colonies of all test strains obtained with different doses of artemisinin fluctuate around that obtained with the solvent dimethyl sulfoxide and S-9 (Tables 15.2 and 15.3).

Strains and S-9 used in the test were with high quality, ensuring validity of the test results.

3.1.2 Micronucleus test

Using the micronucleus test method proposed by Matter and Schmid, effects of artemisinin on micronuclei of bone marrow polychromatophilic erythrocytes of mice were investigated, with active cyclophosphamide as control.

Both batches of tests were carried out using Kunming mice. Water suspension of artemisinin was intragastrically twice within 48 h at doses equivalent to 1/2, 1/3, 1/5, 1/10, and so on of oral LD_{50} for mice. Six hours after administration of the second dose, bone marrow smears were prepared using femoral bones of mice. Following Giemsa staining, micronuclei in every 1000 polychromatophilic erythrocytes were counted. The results were expressed

Table 15.2: Numbers of revertant colonies of five test strains obtained with artemisinin

Group[a]	TA100	TA98	TA1535	TA1537	TA1538
Spontaneous reverse mutation	105	44	25	8	24
Solvent dimethyl sulfoxide + S-9	118	55	31	15	39
Artemisinin (0.03) + S-9	141	40	14	14	19
Artemisinin (0.03)−S-9	79	46	14	5	12
Artemisinin (0.3) + S-9	144	46	24	5	14
Artemisinin (0.3)−S-9	94	49	27	12	16
Artemisinin (3.0) + S-9	102	38	23	17	39
Artemisinin (3.0)−S-9	85	44	20	18	20
Artemisinin (30) + S-9	77	52	27	22	37
Artemisinin (30)−S-9	78	40	24	15	30
Artemisinin (300) + S-9	89	44	26	22	23
Artemisinin (300)−S-9	44	40	24	16	24

[a]The unit of figures in parentheses is μg/0.1 mL.

Table 15.3: Number of revertant colonies of test strains TA100 and TA98 obtained with artemisinin

Group[a]	TA100		TA98	
	+S-9	−S-9	+S-9	+S-9
Dimethyl sulfoxide	134	143.5	47	39
Artemisinin (3000)	104	Toxic	37	41
Artemisinin (500)	119	77	36	29
Artemisinin (1000)	185	172	63.5	58.5
Artemisinin (750)	181	124	26	32
Artemisinin (500)	228	144	40	50
Artemisinin (375)	128	74	22	27
Spontaneous reverse mutation	185	38		

[a]The unit of figures in parentheses is μg/0.1 mL.

as permillages (‰). Effects of different administration frequencies on micronuclei were compared.

U test following treatment by Poisson distribution method showed no significant difference between any dose group of artemisinin and the control group in terms of effects on micronucleus rate of bone marrow polychromatophilic erythrocytes of mice (Tables 15.4 and 15.5). Changes of administration frequency had no influence on the micronucleus rate. The active comparator cyclophosphamide could significantly increase the micronucleus rate.

Under the test conditions, artemisinin had no effects on the micronucleus rate of bone marrow polychromatophilic erythrocytes of mice.

3.1.3 Induction of sister chromatid exchange and micronuclei in V79 cells with artemisinin

The sister chromatid exchange (SCE) test method was used; 1×10^6 Chinese hamster fibroblast V79 cells were cultured in DMEM complete medium, followed by addition of artemisinin dissolved in dimethyl sulfoxide. After action for 2 h, culture medium was

Table 15.4: Effects of artemisinin on micronucleus rate of bone marrow polychromatophilic erythrocytes of mice (I)

Group	Number of Animals (Mice)	Micronucleus Rate (‰) ($\bar{X}$)		
		Administration Once	Administration Twice	Administration Three Times
Artemisinin 1/80 LD_{50}	5	2.6	2.0	0.8
Artemisinin 1/40 LD_{50}	5	1.4	3.3	0.6
Artemisinin 1/20 LD_{50}	5	3.4	3.6	1.0
Artemisinin 1/10 LD_{50}	5	2.8	3.4	2.2
Artemisinin 1/5 LD_{50}	4	1.0	3.8	2.6
Cyclophosphamide 1/2 LD_{50}	5	34.0	39.2	30.0
Tween solution as control	5	1.6	2.8	2.2

Table 15.5: Effects of artemisinin on micronucleus rate of bone marrow polychromatophilic erythrocytes of mice (II)

Group	Number of Animals (Mice)	Micronucleus Rate (‰) ($\bar{X}$)
Artemisinin 1/3 LD_{50}	10	4.1
Artemisinin 1/2 LD_{50}	10	2.9
Cyclophosphamide 1/10 LD_{50}	10	79.7
Tween solution as control	10	3.8

replaced with complete medium free of artemisinin, with 5 μg/mL of 5-bromodeoxyuridine. Colchicine (1.2 μg/mL) was added after 24 hand smears were made by the routine method after additional 2 h later. Numbers of SCE in 25 normal metaphase stages and the mean value were calculated.

The number of V79 cells with micronuclei in the same test was calculated.

The results showed that artemisinin, with or without hepatomicrosome enzyme activation system, could not induce elevation of SCE or micronuclei in V79 cells (Tables 15.6 and 15.7).

3.1.4 Conclusions

In the test, artemisinin has not significantly damaged chromosomes of V79 cells, indicating it is not a chromosome damage factor.

Aforementioned three tests indicate artemisinin is not mutagenic.

Table 15.6: Induction of sister chromatid exchange (SCE) and micronuclei in V79 cells with artemisinin (without S-9)

	Dose			
Test Item	**Control**	**100 μg**	**200 μg**	**400 μg**
SCE (cells ± SE)	5.48 ± 0.49	5.00 ± 0.41	6.72 ± 0.41	[a]
Cells with micronuclei (‰)	8	13	12	10

[a]Due to unclear differential straining of chromatids, Brdu labeling was inhibited.

Table 15.7: Induction of sister chromatid exchange (SCE) and micronuclei in V79 cells with artemisinin (with S-9)

		Dose			
Test Item	**Control**	**50 μg**	**100 μg**	**200 μg**	**400 μg**
SCE (cells ± SE)	6.40 ± 0.40	5.60 ± 0.34	5.84 ± 0.48	5.84 ± 0.34	5.56 ± 0.57
Cells with micronuclei (‰)	3	11	12	8	8

3.2 Reproductive Toxicity Test

3.2.1 Effects of artemisinin on reproductive functions of male rats

Effects of artemisinin on reproductive functions of male Wistar rats were investigated. Aqueous suspensions of artemisinin were prepared with Tween 80. Doses were 1/25, 1/50, and 1/100 of oral LD_{50} for rats. A blank control group and a Tween solution control group were set. Each group contained 30 rats. Artemisinin suspensions and controls were intragastrically administered once daily for 8 consecutive weeks. On the second day after drug administration, 1/3 of rats in each group were killed. Male reproductive organs including testicles and epididymes were isolated and histopathological examination was carried out. Another 1/3 of male rats were allowed to mate with female rats having not administrated with drug, and effects on embryo growth and development were investigated through teratogenicity tests. The remaining 1/3 of rats were retained for observation of recovery.

Test results of each indicator are given as follows.

1. Effects on reproductive organs of male rats: No significant histopathological changes of reproductive organs (including testicles, epididymes, spermatic ducts, and prostates) of male rats were found in any of the three dose groups of artemisinin; generation of sperms was not affected.
2. Effects on reproductive functions of male rats: Artemisinin at doses equivalent to 1/100 LD_{50} or 1/25 LD_{50} showed no significant effects on mating conception rate, implantation index, live fetus rate, or mean body weight of fetal rats. Values of all indicators in the 1/50 LD_{50} group were lower than in the control group, but no significant difference was found following statistical treatment (see Tables 15.8–15.12 for details).
   The results indicate artemisinin has no effects on reproductive functions of male rats.
3. Effects on bone development and soft tissues of fetal rats: Appearance of 656 fetal rats in the five groups was observed. All fetal rats developed normally without deformity. No deformity was found in bone or soft tissue. The results indicated intragastric administration of artemisinin to male rats mating with female rats having not administrated with drug had no effects on fetal rats.

Table 15.8: Effects of artemisinin on reproductive functions of male rats

Group	Mating Conception Rate (%)	Implantation Index	Live Fetus Rat (%)
Blank control	93.3	12.7	99.4
Tween solution control	86.7	11.7	100.0
Artemisinin 1/100 LD_{50}	92.9	11.8	99.3
Artemisinin 1/50 LD_{50}	66.7[a]	10.0	100.0
Artemisinin 1/25 LD_{50}	100.0	10.9	100.0

[a]No significant difference was found following statistical treatment in comparison to the control group.

Table 15.9: Effects of artemisinin on mating conception rate

Group	Number of Pregnant Female Rats/Number of Female Rats With Positive Mating	Mating Conception Rate (%)
Blank control	14/15	93.3
Tween solution control	13/15	86.7
Artemisinin 1/100 LD_{50}	13/14	92.9
Artemisinin 1/50 LD_{50}	6/9	66.7[a]
Artemisinin 1/25 LD_{50}	10/10	100.0

[a]No significant difference was found following statistical treatment in comparison to the control group.

Table 15.10: Effects of artemisinin on implantation index

Group	Number of Implantations/Number of Pregnant Female Rats	Implantation Index (%)
Blank control	178/14	12.7
Tween solution control	152/13	11.7
Artemisinin 1/100 LD_{50}	152/13	11.8
Artemisinin 1/50 LD_{50}	66/6	10.0
Artemisinin 1/25 LD_{50}	109/10	10.9

Table 15.11: Effects of artemisinin on live fetus rate

Group	Number of Live Fetuses/Number of Implantations	Live Fetus Rate (%)
Blank control	177/178	99.4
Tween solution control	152/152	100.0
Artemisinin 1/100 LD_{50}	152/153	99.3
Artemisinin 1/50 LD_{50}	66/66	100.0
Artemisinin 1/25 LD_{50}	109/109	100.0

Table 15.12: Effects of artemisinin on mean litter weight and mean body weight of fetal rats

Group	Number of Pregnant Rats (Rats)	Number of Live Fetuses (Rats)	Mean Litter Weight (g)	Mean Body Weight (g)
Blank control	14	177	52.3	4.1
Tween solution control	13	152	49.2	4.2
Artemisinin 1/100 LD_{50}	13	152	48.1	4.1
Artemisinin 1/50 LD_{50}	6	66	42.7	3.9
Artemisinin 1/25 LD_{50}	10	109	45.7	4.2

As all observed indicators were normal, 1/3 of male rats remained in each group were killed 1 month after drug withdrawal.

Following oral administration of artemisinin at doses equivalent to 1/100 LD_{50}, 1/50 LD_{50}, and 1/25 LD_{50} to male rats for 8 consecutive weeks, all observed indicators were negative, indicating that, under the test conditions, artemisinin had no influence on sperm generation and reproductive functions of male rats.

3.2.2 Studies on teratogenic action of artemisinin [1–5]

Teratogenicity or embryotoxicity of artemisinin to fetuses of rats was observed using the method reported by Wilson.

A water suspension of artemisinin was prepared with Tween 80 and orally given to Wistar rats. Adult rats were mated and pregnant female rats were grouped and given the drug in different stages of pregnancy. On day 20 of pregnancy, pregnant rats were killed. Bone and soft tissue development of live fetal rats was observed by structural teratological methods. A Tween solution control group was set.

1. Effects of artemisinin given to pregnant rats during the organogenetic period: On days 6–15 of pregnancy, rats received continual drug administration for 10 days at doses equivalent to 1/400–1/5 of oral LD_{50}. Among 200 fetuses of 27 pregnant rats in the control group, only 2% were reabsorbed fetuses; no dead fetus was found; 98% of live fetuses were normal without deformity. In the 1/400 LD_{50} group, 45% of fetuses were alive; in all of other dose groups, 100% of fetuses were reabsorbed (see Table 15.13 for the results).

   The results indicate artemisinin given to pregnant rats during the organogenetic period has significant embryotoxicity, manifested as occurrence of reabsorbed fetuses.
2. Effects of artemisinin given to pregnant rats during different stages of the organogenetic period: Artemisinin was administered during the early, middle, and late stages of the

Table 15.13: Toxicity of artemisinin to fetal rats

Group	Number of Pregnant Rats (Rats)	Total Number of Implantations	Live Fetuses: Number (Fetuses)	Live Fetuses: Proportion (%)	Reabsorbed Fetuses: Number (Fetuses)	Reabsorbed Fetuses: Proportion (%)	Remarks
Control	27	200	196	98	4	2	
Artemisinin 1/400 LD_{50}	9	71	32	45	39	55	
Artemisinin 1/200 LD_{50}	17	173	9	5.2	162	93.6	Two dead fetuses
Artemisinin 1/5 LD_{50}	6	52	0	0	52	100	

Table 15.14: Embryotoxicity of artemisinin given during the early, middle, and late stages of the organogenetic period

Administration Time	Number of Pregnant Rats (Rats)	Total Number of Implantations	Live Fetuses		Reabsorbed Fetuses	
			Number (Fetuses)	Proportion (%)	Number (Fetuses)	Proportion (%)
Days 6–8	17	224	164	73.2	60	26.8
Days 9–11	21	285	0	0	285	100
Days 12–22	22	279	0	0	279	100

organogenetic period, that is, on days 6–8, 9–11, and 12–22 of pregnancy, at a dose equivalent to 1/25 of LD_{50}. When artemisinin was given on days 6–8 of pregnancy, 26.8% of fetal rats were reabsorbed, and 6.1% of live fetuses had umbilical hernia deformities, while all other live fetuses were normal. When artemisinin was given during the middle and late stages, 100% of fetuses were reabsorbed, that is, all fetuses were reabsorbed fetuses, indicating significant embryotoxicity was exhibited (see Table 15.14 for the results).

3. Effects of artemisinin given to pregnant rats during early, middle, and late stages: Artemisinin was administered during the early, middle, and late stages, that is, on days 1–6, 7–14, and 15–19 of pregnancy, administration of a dose equivalent to 1/25 of LD_{50} to observe effects of artemisinin on early, middle, and late stages of pregnancy of rats.

When artemisinin was given on days 1–6 of pregnancy, 93.6% of fetal rats were alive, and all fetal rats developed normally without deformity, indicating early administration had minimal effects on fetal rat development. When artemisinin was given during the middle and late stages, 100% of fetuses were reabsorbed, that is, significant toxicity to fetal rats was exhibited (see Table 15.15 for the results).

Teratogenicity results indicate artemisinin orally given to rats has significant embryotoxicity. The toxicity is mainly demonstrated during the middle and late stages of the organogenetic

Table 15.15: Embryotoxicity of artemisinin orally given during early, middle, and late stages of pregnancy at a dose equivalent to 1/25 of LD_{50}

Administration Time	Number of Pregnant Rats (Rats)	Total Number of Implantations	Live Fetuses		Reabsorbed Fetuses	
			Number (Fetuses)	Proportion (%)	Number (Fetuses)	Proportion (%)
Days 1–6	7	78	73	93.6	5	6.4
Days 7–14	10	121	0	0	121	100
Days 15–19	9	104	0	0	104	100

period. The influence of administration in the early stage of pregnancy is minimal. The aforementioned test results were obtained in rats and cannot be based on for inferring its toxicity to human beings. Many drugs demonstrating teratogenic effects in animals have not proved teratogenic to human fetuses, for example, salicylates, penicillin, tetracycline, and such antimalarial drugs as pyrimethamine and chloroquine. Nevertheless, in clinical application, caution should be taken for use in pregnant women and women with suspected pregnancy (except for treating severe malaria).

Reproductive toxicity test results of artemisinin indicate artemisinin had no effects on reproductive functions of male rats, but has significant embryotoxicity if administered during the organogenetic period of pregnant rats, which is mainly manifested as occurrence of reabsorbed fetuses; it is not teratogenic.

References

[1] Institute of Chinese Materia Medica, China Academy of Chinese Medical Sciences. A Collection of Antimalarial Studies of *Artemisia annua* (1971–1978). Research on Chinese Materia Medica 1978, 3.

[2] Pharmacology Research Laboratory, Institute of Chinese Materia Medica, China Academy of Chinese Medical Sciences. Pharmacological studies of *Artemisia annua*. J New Med 1979;(1):23–33.

[3] China Coperatice Research Group on Qinghaosu and its Derivatives as Antimalarials. J Tracit Chin Med 1982;2(1):31–38.

[4] 矢作多貴江.環境中のがン物質な使つてスクリ.ニングすろ驗法.てついて.蛋白質 核酸 酵素, 1975;20(13):1178. [Available only in Japanese: Yahagi T. Experimental method of screening carcinogen in the environment with microorganism. Protein Nucleic Acid Enzyme 1975;20:1178].

[5] Yang LX, Wu BF, Dai BQ. Effects of artemisinin on micronucleus rate of bone marrow polychromatophilic erythrocytes of mice. Tradit Chin Med J 1984;9(3):36–8.

CHAPTER 16

Clinical Studies of Artemisinin

Chapter Outline

Clinical studies of artemisinin: Institute of Chinese Materia Medica of China Institute of Traditional Chinese Medicine verified the effectiveness of the active part for the first time in all of 30 cases infected with *Plasmodium vivax* and *Plasmodium falclparum* in the malarial epidemic area in Changjian of Hainan Province in 1972 [1] and observed the efficacy of Artemisinin Monomer Capsules in the same malarial epidemic region in 1973 and verified that artemisinin is the active antimalarial ingredient in *Artemisia annua* [2]. Over several consecutive years until 1978, when the identification meeting was held, the efficacy of artemisinin had been validated by the Institute of Chinese Materia Medica of China Institute of Traditional Chinese Medicine in 529 patients [3] and collaboratively validated nationwide in 2099 patients, including 1511 with *P. vivax* infection and 588 with *P. falclparum* infection. All patients were clinically recovered except four patients who also responded. The mean time to fever reduction and time to negative conversion were significantly shorter than those treated with chloroquine. In addition, significant effects were obtained in treating 143 patients with chloroquine-resistant *P. falclparum* malaria and 141 with cerebral malaria [4]. With artemisinin, malaria parasites can be quickly killed, so the time to negative conversion is shorter than that with chloroquine or quinine. It has been confirmed that artemisinin has "potent efficacy, fast action, and low toxicity," and is a new specific antimalarial drug that is effective in treating chloroquine-resistant *P. falclparum* malaria.

1 Acquisition of the First IIPR Class I New Drug Certificate for Artemisinin

Institute of Chinese Materia Medica of China Institute of Traditional Chinese Medicine obtained the class I new drug certificate [No.: (86) Weiyaozhengzi X-01] for artemisinin in 1986 [5] and successively developed artemisinin suppositories [6] and artemisinin tablets [7].

From *Artemisia annua* L. to Artemisinins
http://dx.doi.org/10.1016/B978-0-12-811655-5.00016-7

After artemisinin was discovered, the administration route of injection was recommended to further highlight its advantages of "potency efficacy and rapid action." After dihydroartemisinin was discovered and the structure–activity relationship was established, studies on derivatives of dihydroartemisinin were conducted after 1976 in China. In 1987, Guangxi Guilin Pharmaceutical Plant and Shanghai Institute of Materia Medica successively developed water-soluble sodium artesunate for intravenous injection and oil-soluble artemether for intramuscular injection. They have their own advantages, but products prepared directly from *A. annua* have advantages of convenient preparation and low prices.

2 Artemisinin Tablets

According to the clinical opinion that drugs should be convenient for patients to use, oral tablets were developed. Artemisinin tablets, according to three administration regimens with total doses of 2.5 g/2 days, 3 g/3 days, and 4 g/5 days, could all eliminate plasmodia and achieve quick recovery.

More than 100 cases were observed on clinic [8], as briefly described later (Tables 16.1–16.3).

According to the aforementioned results, within 24 h, the mean time needed for decreasing the survival rate of plasmodia (more than 95%) was 24 h with artemisinin tablets 1000 mg and 20 h with artemisinin tablets 1500 mg.

Table 16.1: Comparability of patients in different dose groups of artemisinin tablets

Group	Number of Patients (*n*)	Local Adults (*n*)	Local Children (*n*)	Nonlocal People (*n*)	Patients With History of Malaria	Patients With Splenomegaly (*n*)	Mean Course of Disease (days)	Mean Body Temperature (°C)	Mean Geometric Density of Plasmodia (μL)
2.5 g/2 days	38	15	10	13	4	10	3.5 ± 1.9	39.5 ± 0.7	16,563
3 g/3 days	39	16	6	17	3	6	4.1 ± 2.0	39.7 ± 0.9	18,462
4 g/5 days	31	16	5	10	5	5	3.7 ± 1.9	39.3 ± 0.7	18,098

Table 16.2: Comparison of time to fever reduction and time to plasmodium negative conversion in different dose groups of artemisinin tablets

Group	Time to Fever Reduction (h)		Time to Plasmodium Negative Conversion (h)	
	Patients (*n*)	$\bar{x} \pm SD$	Patients (*n*)	$\bar{x} \pm SD$
2.5 g/2 days	31	289 ± 167	31	354 ± 115
3 g/3 days	38	317 ± 177	39	390 ± 146
4 g/5 days	37	310 ± 144	38	337 ± 128

Table 16.3: Comparison of asexual plasmodia killing rate in different dose groups of artemisinin tablets

Dose (mg)	Number of Observed Patients (n)	Survival Rate of Asexual Plasmodia at Different Time Points After Drug Administration						
		0 h	4 h	8 h	12 h	16 h	20 h	24 h
1000	2	100	82.4	54.2	34.3	13.8	7.8	3.4
1000 + 500	2	100	84.6	51.7	32.1	8.6	2.8	0.6

Clinical validation results showed that all of the three regimens of artemisinin tablets for treating *P. falclparum* (patients with established chloroquine-resistant plasmodia and high parasitemia) yielded high effectiveness and rapid action. The mean time to fever reduction was 28.9–31.7 h. The mean time to plasmodium negative conversion was 33.7–39.0 h. The mean time needed for eliminating more than 95% of asexual plasmodia was 20–24 h with 1.0–1.5 g administered within 24 h. Oral tablets produced more definite effects than rectal administration of suppositories. The efficacy in critical patients with high parasitemia was also significant. Eleven patients with plasmodium densities of 116,000–265,700 μL^{-1} (geometric mean: 167,240 μL^{-1}) were treated with artemisinin tablets ($n = 4$ in 2.5 g/2 days group, $n = 5$ in 3 g/3 days group and $n = 2$ in 4 g/5 days group). The mean time to fever reduction was 43.6 ± 17.4 h, and the mean time to negative conversion of asexual plasmodia was 56.4 ± 18.9 h. Although the time to fever reduction and time to plasmodium negative conversion were longer than in general patients, the results still indicated that artemisinin can quickly control clinical symptoms and clear parasitemia.

It is considered that artemisinin tablets have definite clinical efficacy and low toxic and side effects with convenient use and compete absorption. Among these dose regimens, the recrudescence rate with 4 g/5 days is low. Prolongation of the administration period can lower the recrudescence rate.

The instructions for use are attached as follows.

- **Drug Name**
  Generic name: Artemisinin tablets
  English name: Qinghaosu tablets
  Chinese phonetics: Qing Hao Su Pian
  Active ingredient of this product and its chemical name: The active ingredient of this product is artemisinin. The chemical name of artemisinin is octahydro-3,6,9-trimethyl-3,12-epoxy-12H-pyrano[4,3-*J*]-1,2-benzodioxepin-10(3H)-one.

Its structural formula is:

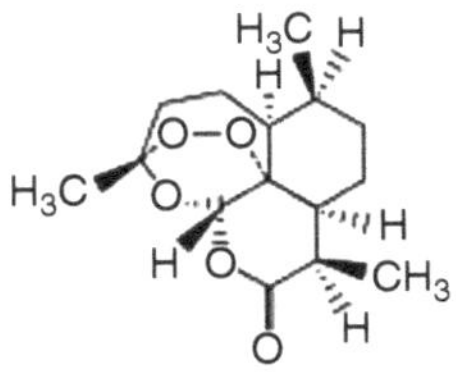

Molecular formula: $C_{15}H_{22}O_5$
Relative molecular weight: 282

- **Description**
  White round tablets.
- **Pharmacology and Toxicology**
  This product has strong action to kill asexual plasmodia and can quickly control symptoms and kill asexual plasmodia. It has particularly salient efficacy against chloroquine- and piperaquine-resistant malaria. It is a high-efficacy and low-toxicity antimalarial agent. With regard to reproductive toxicity, administration in the sensitive period of pregnancy to mice increased occurrence of fetal absorption, but no teratogenic effects were found.
- **Pharmacokinetics**
  Following oral administration, artemisinin can be quickly absorbed, extensively distributed, and rapidly excreted and metabolized; the time to peak plasma concentration is 1.5 h; the half-life is 2.27 h.
- **Indications**
  Various types of malaria. It has specific efficacy against chloroquine- and piperaquine-resistant malaria.
- **Dosage and Administration**
  Take orally. The total dose is 4 g divided for use on five consecutive days q.d., prim.vic No2. Doses in children should be decreased in a gradual manner depending on the age.
- **Adverse Reactions**
  No significant adverse reaction has been found at the clinical dose.
- **Use in Pregnant and Lactating Women**
  Exercise caution for use in early pregnancy.
- **Strength**
  250 mg/tablet.
- **Storage**
  Preserve in a dry, shady, and cool place.
- **How Supplied**
  Aluminum plastic sheets in packs.

- **Shelf Life**
  Tentatively set as 2 years (artemisinin is well stable).

3 Artemisinin Suppositories

Intending to avoid the liver first-pass effect and considering that absorption by mucosa is fast, artemisinin suppositories were developed for the sake of convenient use in children and unconscious patients. The product was granted with the new drug certificate in the same year as artemisinin (1986). The instructions for use are briefly introduced as follows.

- **Indications**
  Artemisinin suppositories are for rectal use. The product may be used for treating *P. falclparum* and *P. vivax* malaria and can rapidly control clinical symptoms. It is particularly suitable for children patients, patients with vomiting and unconsciousness, and critical patients with *P. falclparum* malaria. It has favorable efficacy in treating chloroquine-resistant malaria.
- **Dosage and Administration**
  Insert into the anus to a depth beyond the anal sphincter (see Table 16.4 for the dosage).
- **Precautions**
  - Use in early pregnancy (the first trimester) is not recommended.
  - To avoid poor absorption, caution should be taken for use in patients with frequent diarrhea and shock.
  - It is preferential to pass the stool before taking the medicine. Patients who passed the stool within 2 h after drug administration should take an additional dose to ensure the drug quantity is sufficient.
- **Storage**
  Preserve in a shady and cool place, and protect from light.

Table 16.4: Dosage of artemisinin suppositories

Age Group (years)	Total Dose (mg)	Day 1 (mg)		Day 2 (mg)		Day 3 (mg)	
		Initial Dose	4 h After Initial Dose	a.m.	p.m.	a.m.	p.m.
≥16	2800	600	600	400	400	400	400
11–15	2200	600	400	300	300	300	300
7–10	1400	300	300	200	200	200	200
3–6	1000	200	200	200	100	200	100
1–2	600	100	100	100	100	100	100
<1	400	100	100	50	50	50	50

On days 2 and 3, the interval between the dose in the morning and the dose in the afternoon is 8 h.

References

[1] Institute of Chinese Materia Medica, China Institute of Traditional Chinese Medicine. Clinical data of 30 malarial patients initially treated with *Artemisia annua* in 1972. Science and Technology Documents of Institute of Chinese Materia Medica, China Institute of Traditional Chinese Medicine.

[2] Institute of Chinese Materia Medica, China Institute of Traditional Chinese Medicine. Summary of treatment of malaria with artemisinin for the first time. Science and Technology Document 000892-000910 of Institute of Chinese Materia Medica, China Institute of Traditional Chinese Medicine.

[3] Institute of Chinese Materia Medica, China Institute of Traditional Chinese Medicine. Album of anti-malarial studies with *Artemisia annua* (1971–1978). Studies of Traditional Chinese Medicine, 1978, 3.

[4] National Leading Group on Prevention and Treatment of Malaria. Identification Certificate of Artemisinin.

[5] Institute of Chinese Materia Medica, China Institute of Traditional Chinese Medicine. New Drug Certificate of Artemisinin (No.: [86] Weiyaozhengzi X-01).

[6] Institute of Chinese Materia Medica, China Institute of Traditional Chinese Medicine. New Drug Certificate of Artemisinin Suppositories (No.: [86] Weiyaozhengzi X-04).

[7] Institute of Chinese Materia Medica, China Institute of Traditional Chinese Medicine. New Drug Certificate of Artemisinin Tablets (No.: Guoyaozhengzi H2003014).

[8] Cai XZ, et al. Efficacy of artemisinin tablets in treating 108 patients with pernicious malaria. Chin. J. Parasitic Dis. Control 1994;7(3):175–7.

PART 3

Dihydroartemisinin

CHAPTER 17

Overview of Dihydroartemisinin

The antimalarial research group of the Institute of Chinese Material Medica, China Academy of Chinese Medical Sciences, successfully extracted artemisinin from Chinese traditional medicine—*Artemisia annua* L.—in 1972. They identified artemisinin as sesquiterpenes lactones through chemical reaction, elemental analysis, and spectral data analysis in the researches on chemical structure that was conducted very soon. To confirm the presence of the carbonyl group in the structure, the research group further developed researches on artemisinin derivatives in 1973. The compounds that were derived by reduction of artemisinin with sodium borohydride were artemisinin derivatives—dihydroartemisinin [1]—which were first developed by Youyou Tu in 1973, thereby confirming the presence of a carbonyl group in the structure of artemisinin.

Upon successfully development of dihydroartemisinin, the research group launched the studies on the structure–activity relationship of artemisinin. They found that peroxide group played a vital role in the structure of artemisinin, which lost antimalarial activity after peroxide destruction (Table 17.1). By contrast, if peroxide group was retained, the antimalarial activity may be enhanced via the generation of hydroxyl group by the reduction of carbonyl group in lactone ring followed by introduction of acetyl group [2,3], suggesting that structural modification can improve the biological activity when peroxide group was retained. This structure–activity relationship offered new ideas for the development of new medicines. After being disclosed in internal antimalarial research meeting held in Henan in 1975, the aforementioned findings opened up a new prospect for domestic researches on the development of derivatives in the following year [1–3].

Owing to the absence of functional groups which can be used for direct preparation of derivatives in the structure of artemisinin, dihydroartemisinin is generated by the introduction of hydroxyl group into the structure. This approach can not only improve antimalarial activity, but also become the only way to synthesize artemisinin derivatives with antimalarial activity. Such antimalarial drugs as artesunate, artemether, arteether, and the like have been successively developed and marketed. The research and development of artelinic acid are underway (Fig. 17.1).

After further in-depth studies on the structure–activity relationship, the research group led by Youyou Tu confirmed that dihydroartemisinin had a great value to the development of a new generation of antimalarial drugs. In 1992, "Dihydroartemisinin and Its Tablets" (Trade Name: Ketaixing and Ketaixing Tablet), developed by the Institute of Chinese Material

From *Artemisia annua* L. to Artemisinins
http://dx.doi.org/10.1016/B978-0-12-811655-5.00017-9

Table 17.1: The study of the relationship between structure and efficacy

Derivative	Structural Formula	Dose (mg/kg)	Antimalarial Effect
Artemisinin	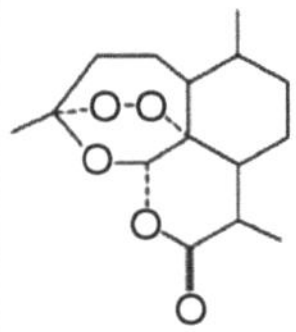	50~100	Became negative (all plasmodia became negative)
Reduction (dihydroartemisinin)	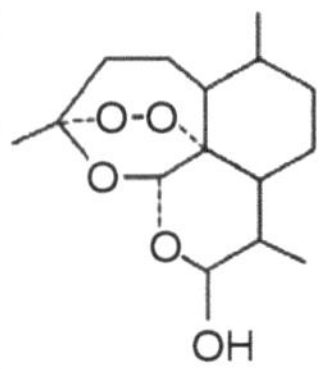	12.5	Became negative
Acetylation following reduction	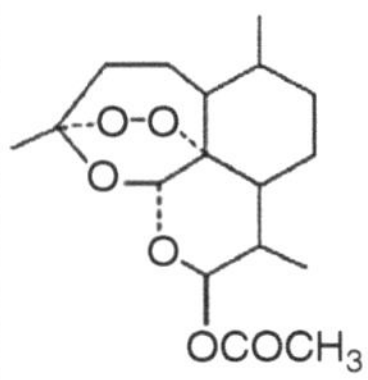	6	Became negative
Hydrogenation	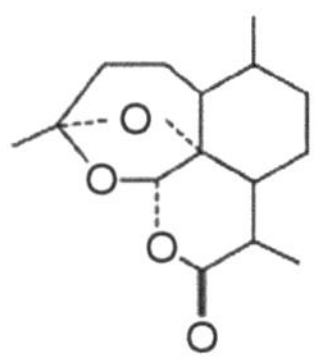	100	Ineffective
Hydrogenation following reduction	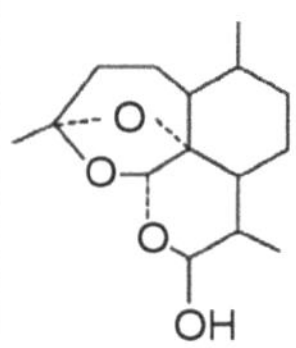	100	Ineffective
Acid treatment	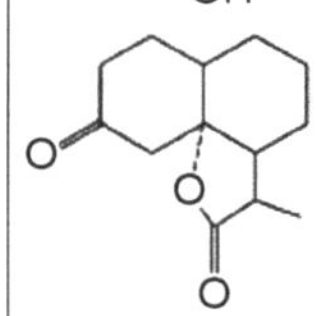	100	Ineffective
Alkali treatment	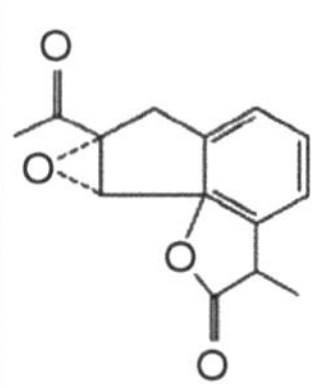	100	Ineffective

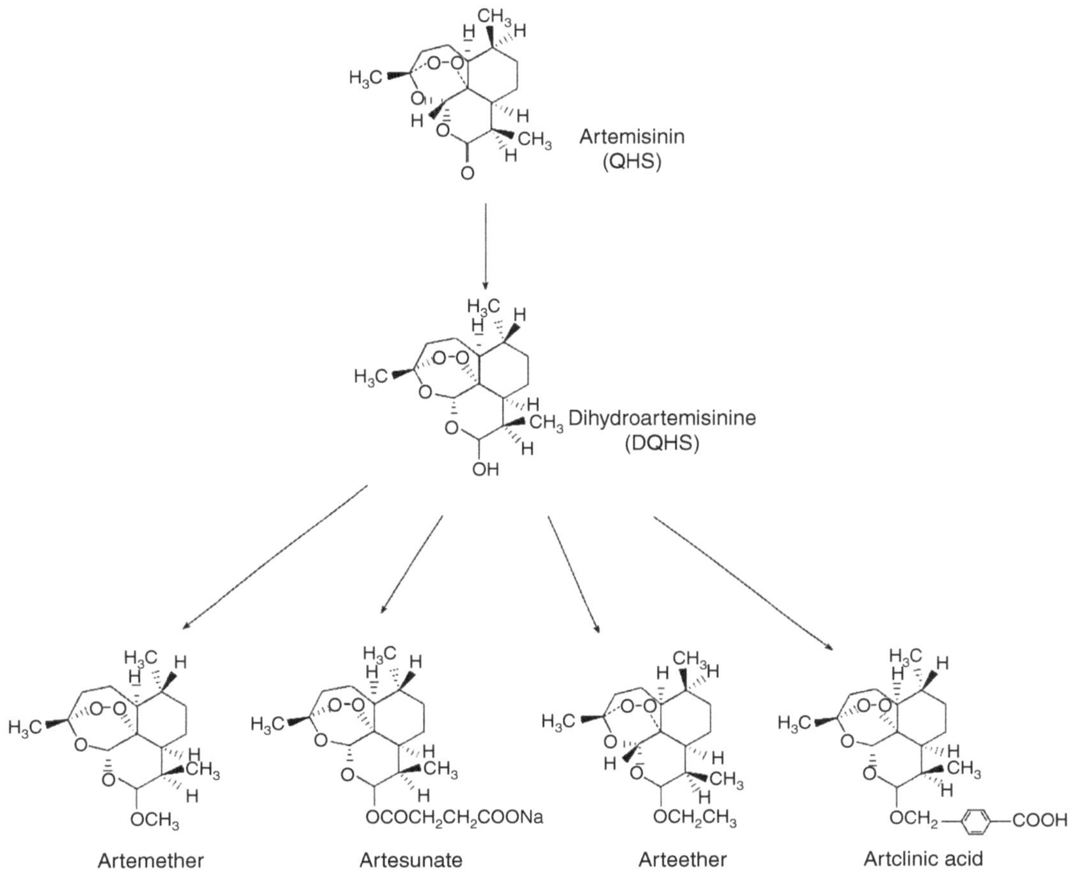

Figure 17.1: Artemisinin and Its Main Derivatives.

Medica, China Academy of Chinese Medical Sciences, was granted with "New Drug Certificate" [4] (Class 1). Its efficacy was 10 times higher than artemisinin, but recrudescence rate was reduced to 1.95%. It was awarded with the "Prize for China's Top 10 Scientific Achievements" (1992), and appraised by the Ministry of Health as the "New China's Top 10 Health Achievements" (1997). Compared with commercially available artesunate sodium injection and artemether injections (Table 17.2), it was characterized by "high efficacy, quick action, high safety, small dose, easy-to-take, and simple and inexpensive preparation." So dihydroartemisinin was accepted as the first choice among available artemisinin drugs.

Over the past decade, the artemisinin research center of China Academy of Chinese Medical Sciences has made new progress in the serial studies on dihydroartemisinin: (1) developed a "dihydroartemisinin suppositories" suitable for children patients and those with cerebral malaria presenting severe coma based on clinical needs; (2) to solve the spread of multidrug-resistant falciparum malaria around the word—the World Health Organization advocated

Table 17.2: Comparisons of efficacy and safety (according to clinical data)

Drug Name	Administration Route	Dose	Clinical Efficacy	Plasmodium-Killing Rate (95% of Plasmodium-Killing Time) (h)	Recurrence Rate (%)	Safety	
						LD_{50} (mg/kg)	Chemotherapeutic Index (LD_{50}/SD_{50})
Dihydroartemisinin tablet	p.o.	480 mg (240 mg is also curable, but with high recrudescence rate)	Cured	16	1.95	834.5	843.0
Sodium artesunate	i.v.	480 mg	Cured	16	5.6	711.1	792.8
Artemether	i.m.	480 mg	Cured	20	6.7	268	447

artemisinin-based combination therapy. The significant synergistic effect of the compound consisting of dihydroartemisinin and pyronaridine has been demonstrated in experiments. Its synergy index was 7.6, higher than that of the widely used artemether composite consisting of artemether and lumefantrine (synergy index of 6). In addition, this compound was also characterized by multiple target sites and short course; (3) more and more pharmacological activities of artemisinin-like drugs in nonmalaria areas have been identified (seen in relevant sections of the book). In the studies, dihydroartemisinin has shown an excellent two-way adjustment effect in the field of immunization, which can not only lower B cell hyperresponsiveness so as to reduce autoimmune disease caused by the deposition of immune complexes, but also improve the immune function of T cells. It can be clinically used in the treatment of lupus erythematosus and photosensitivity disease. The preliminary clinical observation of 100 cases showed that the total effective rate was 94% and the response rate was 44%. The latter two won the national invention patent.

This chapter is based on the new drug application materials for "dihydroartemisinin and dihydroartemisinin tablet" that were completed by the Institute of Chinese Material Medica, China Academy of Chinese Medical Sciences, in 1992. To reflect historic original appearance, we reserved original research data, maps, and information, as more as possible.

References

[1] 1973-9-20, Science and technology archives in Traditional Chinese medicine research institution 000647.

[2] Traditional Chinese medicine research institution. Artemisinin-based antimalarial album (1971–1978). Traditional Chinese medicine research data, 1978, 3.

[3] Tu YY. Chemical studies on qinghaosu. Tradit. Chin. Med. 1982;2(1):3–8.

[4] Traditional Chinese medicine research institution. New drug certificates of dihydroarteminisin and dihydroartemisinin tablets, (92) Wei and drug administration X No. 66 No. 67.

CHAPTER 18

Preparation and Identification of Dihydroartemisinin

Chapter Outline

1 Preparation of Dihydroartemisinin

1.1 Chemical Equation

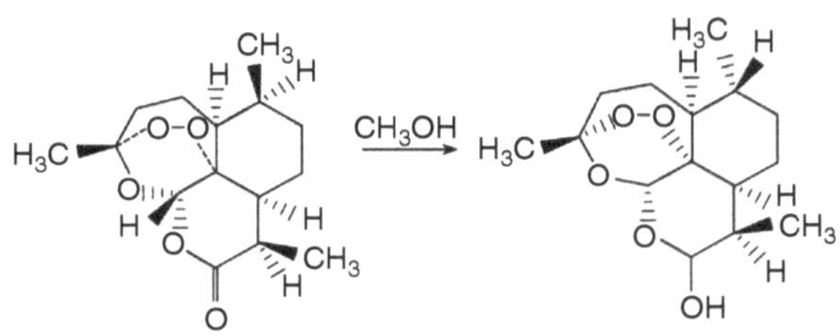

1.2 Preparation

Dissolve artemisinin in methanol, cool to 0–5°C, and add sodium borohydride that is four-tenths of artemisinin. After reaction was complete, neutralize with acid, precipitate crystals with ice water, filtrate, and dry, the yield of dihydroartemisinin was about 85%.

Dissolve dihydroartemisinin crude product in acetone by heating, precipitate crystals after hot filtration, and then filtrate. The yield of pure product was about 75%. The acetone mother liquor was reused or recovered.

From *Artemisia annua* L. to Artemisinins
http://dx.doi.org/10.1016/B978-0-12-811655-5.00018-0

Table 18.1: A list of mass spectrum data and analysis

Mass-to-Charge Ratio (*m/z*)	Relative Intensity (%)	Note
43	100.0	CH_3-C≡O^+
41	32.4	CH_2=CH_2-$CH_2^{\pm}$
55	31.6	C_3HO^+
194	17.6	$[M-C_3H_6O]^+$
137	13.2	$[M-C_4H_6O-CH_2O-CH_2]^+$
152	11.8	$[M-C_4H_6O-CH_2O]^+$
71	11.6	$C_4H_7O^+$
69	11.3	$C_4H_5O^+$
252	5.8	$[M-O_2]^+$
		$[M-H_2O]^+$
266	3.2	(Ion source temperature is reduced to 150°C)

2 Identification of Dihydroartemisinin

Molecular structure:

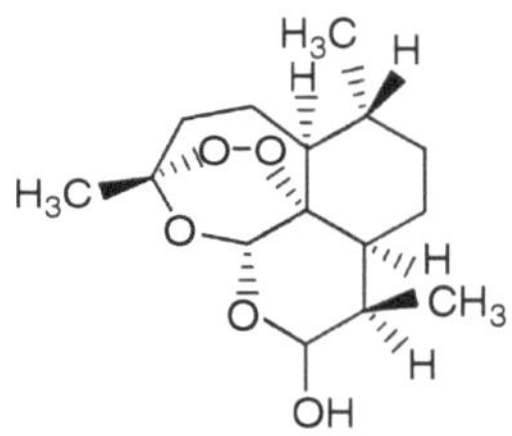

Molecular formula: $C_{15}H_{24}O_5$

2.1 Elemental Analysis

Calcd: C 63.43%; H 8.25%.

Found: C 63.31%; H 8.60%.

2.2 Mass Spectrometry

Main fragments *m/z* (%): 43 (100.0), 41 (32.4), 55 (31.6), 194 (17.6), 137 (13.2), 152 (11.8), 71 (11.6), and 69 (11.3) (Table 18.1).

Although molecular and ion peaks are not displayed in spectrum, peroxide bridge remaining in molecules frequently results in loss of O_2 in the mass spectrometry process as dihydroartemisinin is generated by reduction of artemisinin, thus the peak at *m/z* of 252 is actually the fragment of [M−O_2]+. As a result, the relative molecular mass of reduced artemisinin is estimated to be 284.

2.3 NMR Analysis

NMR analysis: (CDCl3 TMS 90 Hz) (Fig. 18.1).

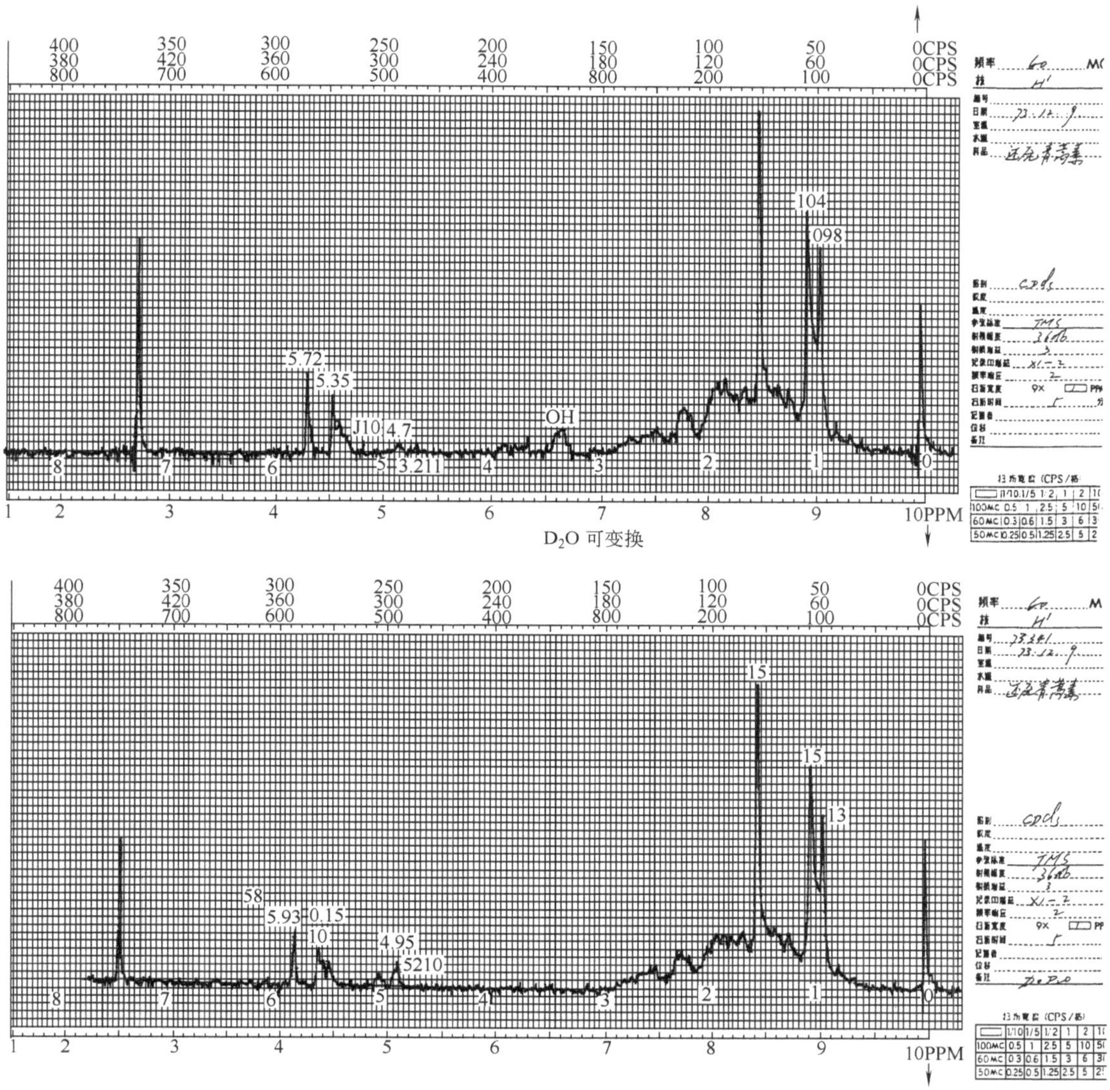

Figure 18.1: NMR Spectra of Dihydroartemisinin (Also Known as Reduced Artemisinin) (the Peak Disappeared After an Exchange With D_2O).

δ: 0.96 (6H, d, *J* = 7.2, C14, and C13-CH3), 1.44 (3H, s, C15-CH3), 4.83 (0.5H, d, *J* = 9.00, C12-Hα), 5.40 (0.5H, d, *J* = 36, C12-Hβ), 5.44 (0.5H, s, C5-Hα), 5.70 (0.5H, s, C5-Hβ), 3.0 (1H, brs, C12-OH-D2O disappeared after exchange), suggesting the difference between dihydroartemisinin and artemisinin lies in that the former has an additional hydroxyl group.

2.4 Infrared Spectroscopic Analysis

Comparing with artemisinin, 1745 peak (σ lactone) disappeared, while 3356 peak appeared in the structure of dihydroartemisinin, which provided the evidence supporting the existence hydroxyl group. There were still peaks of peroxide groups at wavelengths of 831, 880, and 1100 in the IR spectral map (Fig. 18.2).

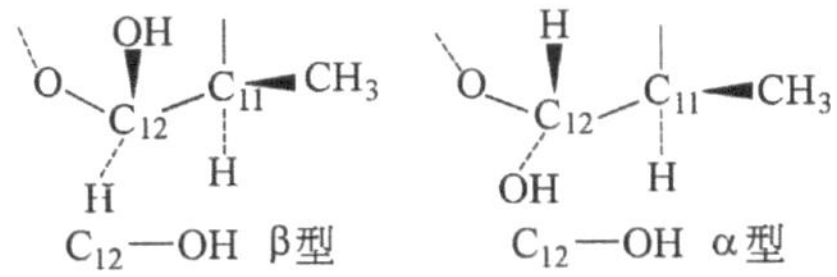

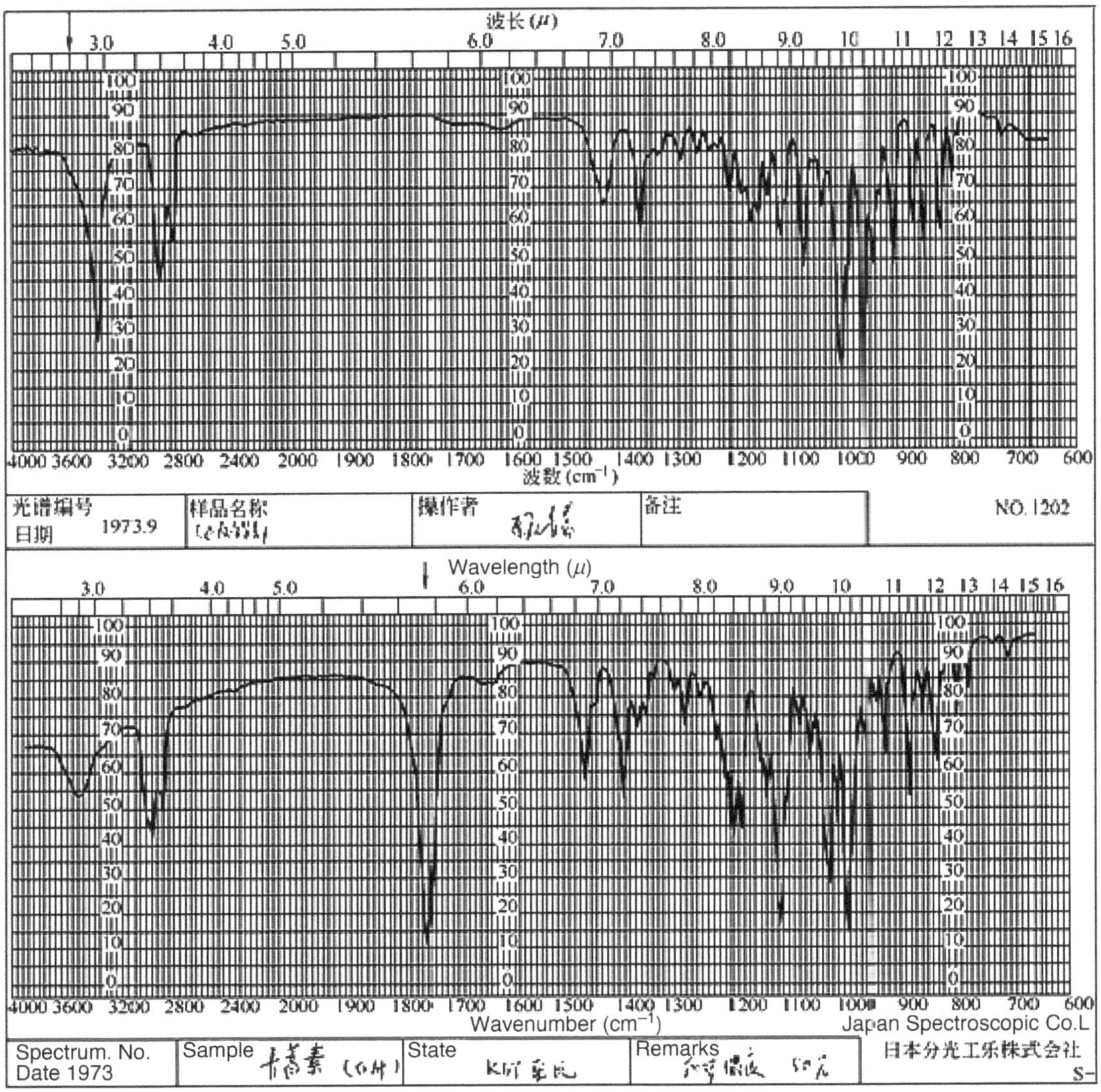

Figure 18.2: IR Spectra of Dihydroartemisinin (Also Known as Reduced Artemisinin) Versus Artemisinin.

2.4.1 X-ray diffraction

The crystal belongs to the orthorhombic system, space group: D42-P212121, cell parameter: $a = 5.659$, $b = 14.270$, $c = 18.954$, $Z = 4$. C12 in the molecule of dihydroartemisinin is chiral center, which gives rise to different configuration of C12-OH (β, α).

2.4.2 X-ray diffraction findings

The crystal exists in β-configuration but exists in α and β configuration in the solution. The changes in the 1H-NMR spectrum clearly showed that the crystal first existed in the form of β configuration followed by the form of the mixture of α and β isomers in the solution. The change rate and equilibrium position of α and β configuration depend on the types of solvent, and, for example, α and β configuration reached equilibrium in the CDC13 solution at the ratio of 1:1 in 8 h, but in the CD3OD solution only within a few minutes at the ratio of 2:1. Thus, the measurement of specific rotation is affected by the solvent, temperature, and storage time (before equilibrium is reached) to some extent.

CHAPTER 19

Analytical Methods and Quality Standards for Dihydroartemisinin

Chapter Outline

1 Physical and Chemical Parameters of Dihydroartemisinin

1.1 Description

White needle-like crystal, odorless, with a bitter taste.

Easily soluble in chloroform, soluble in dimethyl sulfoxide, ether, and acetone, slightly soluble in methanol and ethanol, and practically insoluble in water.

Melting point: Melting and decomposition of crystals occur simultaneously at 145–150°C.

Specific rotation: [α] 20D = +140° to +146° (*c* = 1.0026, CHCl3)

1.2 Identification

1. Dissolve 5 mg in 0.5 mL of dehydrated ethanol, add 2.5 mL of acidic potassium iodide solution (dissolve 0.4 g of potassium iodide in 2.5 mL of sulfuric acid at a concentration of 2.5 mol/L), add 1–2 drops of 2% starch solution. The blank control solution is prepared with anhydrous ethanol following the same procedures. The blue and violet color in test solution is immediately produced, while only light violet color is produced in blank control solution.
2. Add a few drops of chloroform solution and evaporate to dryness, add a drop of 2% vanillin sulfuric acid solution, the red color is produced immediately and become brown after placement.
3. The infrared absorption spectrum is concordant with the reference spectrum.

From *Artemisia annua* L. to Artemisinins
http://dx.doi.org/10.1016/B978-0-12-811655-5.00019-2

1.3 Inspection

Loss on drying: Dry over phosphorus pentoxide R under reduced pressure (not exceeding 2.67 kPa); it loses not more than 0.5%.

Residue on ignition: <0.1%.

2 Thin-Layer Chromatography (TLC)

Dissolve the product in chloroform to produce 10 mg/mL solution (test solution), and dissolve dihydroartemisinin in chloroform to produce 10 mg/mL solution (control solution). Apply 10 μL of aforementioned two solutions in the same silica gel G plate, respectively. Develop with benzene–ethyl acetate (8:2), dry the plate, and spray with 2% vanillin sulfuric acid solution. The spot of dihydroartemisinin in sample is not darker than the reference spots, and no other individual impurity spot is observed.

Dihydroartemisinin contains not less than 98.0% and not more than the equivalent of 102.0% of $C_{15}H_{24}O_5$.

3 High-Performance Liquid Chromatography (HPLC)

Preparation of internal standard solution: Accurately weigh about 10 mg biphenyl (spectroscopic ally pure), transfer into a 10 mL volumetric flask, dissolve in methanol and dilute to the mark, and shake thoroughly.

Preparation of reference solution: Accurately weigh about 10 mg dihydroartemisinin reference, transfer into a 10 mL volumetric flask, dissolve in methanol, add 20 mL of above internal standard solution, dilute to the mark, and shake thoroughly.

Preparation of test solution: The same as the preparation of reference solution.

Assay: Waters-201 high performance liquid chromatography, variable wavelength UV monitor Mode 1481, wavelength: 210 nm, sensitivity: 0.05AUFS, column: μ-Bondapa KC13 (4.0 mm × 300 mm), mobile phase: methanol–water (8:2), flow rate: 1.0 mL/min, and injection volume: 10 μL. Transfer about 10 mL of methanol into a 10 mL volumetric flask, add 20 μL of internal standard, and dilute with methanol to the mark (as the blank control). Conduct the assay. Peak time is within 15 min, and the content is calculated based on peak area and response factor (or calibration factor) with the following equation (Fig. 19.1):

1. Correction factor method

$$C\% = Fi \times AiAis \times WisWi \times 100\%$$

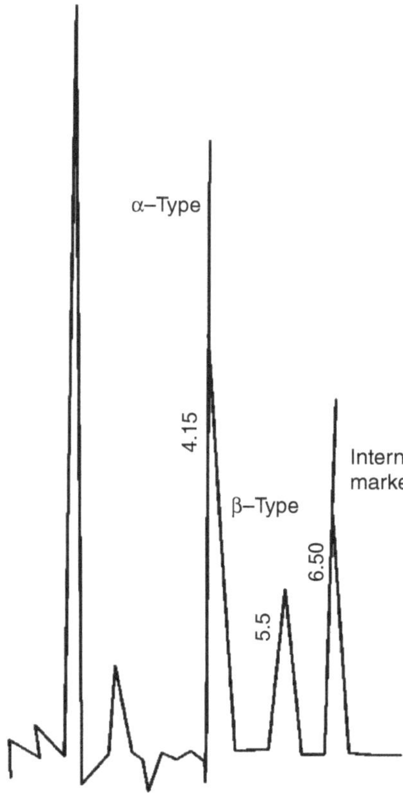

Figure 19.1: A Chromatogram.

where Fi (correction factor) = $W'iW'is \times A'isA'i$; A′is and Ais are the peak area of internal standard in reference solution and test solution, respectively; A′i and Ai are the peak area of certain substance in reference solution and test solution, respectively; W′is and Wis are the weight of internal standard in reference solution and test solution, respectively (μg/μL); W′i and Wi are the weight of certain substance in reference solution and test solution, respectively (μg/μL).

2. Response coefficient method (the spike amount in reference solution is the same as that in internal standard)

$$C\% = A'isAis \times A'iWi \times rF1000 \times 100\%$$

where rF (response factor) = $W'iA'i \times 1000$.
Other symbols labels are the same as (1).

Laboratory samples and pilot trial samples are measured with the aforementioned methods, and the results are shown in Table 19.1.

Table 19.1: Laboratory and pilot trial results of dihydroartemisinin content

Batch No.	Measured Value (%)		
	I	II	Mean
Laboratory 1	99.38	101.18	100.53
Laboratory 2	99.51	99.74	99.63
Laboratory 3	98.22	96.87	98.79
Laboratory 4	98.78	98.58	98.68
Laboratory 5	99.32	100.10	99.71
Pilot trial 1	98.82	98.11	98.97
Pilot trial 2	99.85	99.85	99.85
Pilot trial 3	98.29	99.82	99.06

4 UV Spectrophotometry

Preparation of control solution: Accurately weigh about 10 mg of dihydroartemisinin reference substance, place into a 50 mL volumetric flask, add ethanol solution to dissolve and dilute to the mark, shake, and then stand for 2 h.

Preparation of test solution: Accurately weigh 10 tablets, porphyrize, transfer appropriate amount (about equivalent to 10 mg of dihydroartemisinin) into a 50 mL volumetric flask, add ethanol, and shake to dissolve, dilute to the mark, stand for 2 h, and filtrate. Discard the primary filtrate, and use the subsequent filtrate.

Measuring procedures: Accurately transfer each of 1 mL of the reference solution and the test solution into a 10-mL volumetric flask, respectively, add 1 mL of ethanol, shake thoroughly, add 2% sodium hydroxide solution to the mark, shake, allow it to react in a water bath at 60°C for 30 min, and then cool to room temperature. With 2% sodium hydroxide–ethanol (4:1) as the blank, the absorbance was measured at a wavelength of 238 nm using UV–visible spectrophotometer.

CHAPTER 20

Pharmacological Studies of Dihydroartemisinin

Chapter Outline

1 Studies on Antimalarial Pharmacodynamics

1.1 In Vivo Tests

1.1.1 Pharmacodynamic test against murine plasmodia [1]

1. Test materials
   a. Animals: Kunming mice weighing 18–23 g (male:female = 1:1) were used.
   b. Plasmodia: Murine plasmodia included normal strains (N) and chloroquine-resistant (RC) strains of *Plasmodium berghei kerberg 173* (K173). Human *Plasmodium falciparum* is the FCC-1/HN strain.
2. Methods and results
   a. Methods: The 4-day inhibitory test method reported by W. Peters was used. Animals were randomized. Six dose groups were set by a geometric sequence, with 10 mice each group. A plasmodium infection group was set as a control. Chloroquine-sensitive strains (normal strains, N) or chloroquine-resistant strains were intraperitoneally injected at a dose of 1×10^7 infected erythrocytes/mouse. In the administration group, Tween-suspended drug solution was orally given or hypodermically injected from the day of drug administration to day 3, once daily, for 4 times in total. On day 4, blood

From *Artemisia annua* L. to Artemisinins
http://dx.doi.org/10.1016/B978-0-12-811655-5.00020-9

was collected for Giemsa staining, followed by observation under a microscope. The suppression rate and negative conversion rate in each dose group were calculated. The 50% suppression dose (SD_{50}) and the 50% effective dose (ED_{50}) were calculated, both expressed in milligrams per kilograms of body weight (mg/kg). The chemotherapeutic index (LD_{50}/SD_{50}) and the resistance index (SD_{50} of resistant strains/ SD_{50} of sensitive strains) of dihydroartemisinin were calculated.

b. Results and discussion: Calculated according to the aforementioned data (Tables 20.1–20.3), the chemotherapeutic index of dihydroartemisinin is 834.5/0.99 = 842.9; the resistance index of dihydroartemisinin is 22.3/0.99 = 22.5; the resistance index of chloroquine is 50.9/0.68 = 74.9.

It can be known from the aforementioned test data that the antimalarial potency of dihydroartemisinin is lower than that of chloroquine against normal strains of murine plasmodia, while the potency of the former is higher than that of the latter against chloroquine-resistant strains. The resistance index values of dihydroartemisinin and chloroquine are 22.5 and 74.9, respectively. Dihydroartemisinin and chloroquine have cross-resistance, but the resistance index of the former is lower than that of the latter. The SD_{50} and chemotherapeutic index of dihydroartemisinin are approximately equivalent to those of sodium artesunate (the SD_{50} is 0.97; the chemotherapeutic index is 792.8). Of the two administration routes of dihydroartemisinin, hypodermic injection has higher efficacy than oral administration. The antimalarial potency of dihydroartemisinin tablets is equal to that of dihydroartemisinin API. Comparison of ED_{50} against murine plasmodia indicates that the potency of oral dihydroartemisinin is 5.7 times higher than that of artemisinin.

Table 20.1: Antimalarial potency against normal strains of murine plasmodia

Drug	Route of Administration	ED_{50} [mg/(kg·day)]	SD_{50} [mg/(kg·day)]
Dihydroartemisinin	Per os	12.2	0.99
	Hypo	5.8	
Artemisinin	Per os	69.3	9.97
Chloroquine	Per os		0.63

Table 20.2: Antimalarial potency against chloroquine-resistant strains

Drug	Route of Administration	ED_{50} [mg/(kg·day)]	SD_{50} [mg/(kg·day)]
Dihydroartemisinin	Per os	151	22.3
Chloroquine	Per os		50.9

Table 20.3: Antimalarial potency of preparations

Drug	Route of Administration	SD_{50} [mg/(kg·day)]
Dihydroartemisinin API	Per os	1.55
Tablets	Per os	1.44

1.1.2 Pharmacodynamic test against simian plasmodia [2]

1. Test materials
   a. Animals: Male and female rhesus monkeys weighing 2–5.5 kg were used.
   b. Plasmodia: *Plasmodium knowlesi*.
   c. Preparation of drugs: Tween 80 was added to dihydroartemisinin, and the mixture was grinded. Then the distilled water was added to prepare suspensions of desired concentrations. Sodium Artesunate Injection (prepared by dissolving artesunate and sodium bicarbonate in a solvent and adding distilled water) was used as a control drug.
2. Methods and results
   a. Methods: The 7-day therapeutic regimen reported by D.E. Davidson et al. (1976) was used. Three dose groups of dihydroartemisinin, 10.0, 3.16, and 1.0 mg/kg, were set, with each group containing three monkeys; all doses were intragastrically administrated (i.g.). One dose group (3.16 mg/kg) was set for the control drug, sodium artesunate, which was intravenously administrated (i.v.) to a rear leg. The group also contained three animals. Each monkey received intravenous (i.v.) inoculation of 5×10^8 plasmodium-infected erythrocytes to a rear leg. Drug administration started when parasitemia reached more than 2.5% and conducted once per day for 7 days. Parasitemia was detected every 6 h after administration of the first dose and changed to once daily for 15 days after negative conversion of plasmodia for 3 times. Afterward, parasitemia was detected once every other day, and after 2 months, changed to twice per week until day 105. With the plasmodium parasitizing rate before drug administration as 100%, the decrease rates of plasmodia at different time points following administration of the first dose were calculated, and the times (in hours) to 50 and 90% decreases of plasmodia were calculated using regression equations. Negative conversion of plasmodia in test monkeys was considered when negative parasitemia was found in 200 vision fields of thick films in 3 consecutive observations. The time of the first negative conversion observation was the time of plasmodium negative conversion. The parasite killing rate of the drug was determined according to times when 50 or 90% decreases and negative conversion of plasmodia. Cure effects of the drug were determined by whether negative conversion was potentiated and recrudescence occurred following negative conversion. The criteria are as follows:
      - Ineffective (I): Parasitemia had no significant difference, compared with parasitemia before drug administration or continued increase.
      - Slight suppression (SS): Parasitemia was temporarily suppressed but intensified later.
      - Marked suppression (MS): Negative conversion of plasmodia lasted for at least 2 days, but recrudescence occurred within 30 days.
      - Complete suppression (CS): Recrudescence occurred within 30–105 days after negative conversion.
      - Cure (C): No recrudescence occurred within 105 days after negative conversion.

b. Results and discussion: PC50, PC90, and PCT. First, the speed of parasite clearance increased with the dose of dihydroartemisinin among the test groups, that is, PC50, PC90, and PCT were reduced accordingly. However, there was no significant difference in PC50, PC90, and PCT among the three dose levels of i.v. sodium artesunate. The medium and small doses were even found to have faster parasite clearance compared with large dose (which was based on the primary pharmacodynamic studies by Academy of Military Medical Sciences on sodium artesunate). The speed of parasite clearance of dihydroartemisinin was clearly correlated with the dose level. Second, the speed of parasite clearance for i.g. dihydroartemisimin at 3.16 mg/kg was slightly faster compared with i.v. sodium artesunate of the same dose level, indicating that oral dosage form of dihydroartemisimin will be an alternate choice for the patients with *P. falciparum* malaria who are still able to swallow. In addition, it can be found in Table 20.4 that PC50 and PC90 of i.g. dihydroartemisimin at 10.0 and 3.16 mg/kg were faster than those of i.g. artemisimin of the same dose, and the PCT of dihydroartemisimin was more than 10 h faster at the same dose (according to the primary pharmacodynamic studies done on artesunate oral tablets). It reveals that the speed of parasite clearance of i.g. dihydroartemisinin is faster than that of i.v. sodium artesunate and i.g. artesunate at the same dose level. Dihydroartemisinin will be an ideal choice as a new antimalarial agent with fast action.

The results of the study on monkey malaria reveal that the speed of *P. knowlesi* clearance of i.g. dihydroaratemisinin is superior to that of i.v. sodium artesunate and i.g. artesunate.

Table 20.4: Speed of *Plasmodium knowlesi* clearance for i.g. dihydroaratemisinin and i.v. sodium artesunate at asexual blood stage

Agent	Dosage (mg/kg/day)	Number of Animals	Average Parasite Infection Rate Before Dosing (%)	Average PC50 (h)	Average PC90 (h)	PCT (h)
Dihydroartemisinin (DQHS)	10.0	3	50.7	7.04 ± 0.96	11.94 ± 0.74	40.0
	3.16	3	51.6	7.11 ± 0.95	12.05 ± 1.63	42.7
	1.0	3	45.3	10.99 ± 0.53	22.77 ± 3.67	Negative parasitemia was not observed in one animal. The average PCT of rest two animals was 52 h
Sodium artemisinin	3.16	3	56.7	8.06 ± 1.07	14.27 ± 1.25	45.3

1.2 Pharmacodynamic Properties Against Human Plasmodium falciparum *In Vitro [3]*

1.2.1 Test materials

Human *P. falciparum* FCC-1/HN strain, which was conserved and tested by Parasite Research Laboratory of China Medical University. The culture medium was RPMI-1640 plus 15% rabbit serum.

1.2.2 Methods and results

1. Methods: The parasites with an infection rate of 0.8 and 0.61% were used for in vitro culture. The parasites were aliquoted into a 96-well plate at 200 μL per well. The parasites were incubated at 37°C, 5% CO_2 for 4 h. After blood cell sedimentation, the supernatant was withdrawn. Drugs were prepared freshly. The drugs were diluted with RPMI-1640 containing 0.05% ethanol to the desired concentrations. The dilutions were added to the wells (three wells per concentration, 200 μL per well). RPMI-1640 containing 0.05% ethanol was added to the control group. The plate was placed in a CO_2 incubator for 48 h. Subsequently, Giemsa-stained blood smears were prepared; 5000 red blood cells were counted for each blood smear to calculate the infection rate of parasites. The infection rate of the control group was considered to be 100. The parasite clearance rate of each dose level was calculated.
   Parasite clearance rate = (percentage of infected erythrocytes in control group − percentage of infected erythrocytes in test groups)/percentage of infected erythrocytes in control group × 100.
   Curves were plotted based on different drug concentrations and parasite clearance rates (Figs. 20.1–20.3).
2. Results and discussion: See Figs. 20.1–20.3 and Table 20.5.
   It is found from the aforementioned results that the antimalarial activity of dihydroartemisinin against human *P. falciparum* is superior to artemisinin. The median effective doses (ED_{50}) for in vitro antimalarial activity are 0.385 and 5.65 ng/mL,

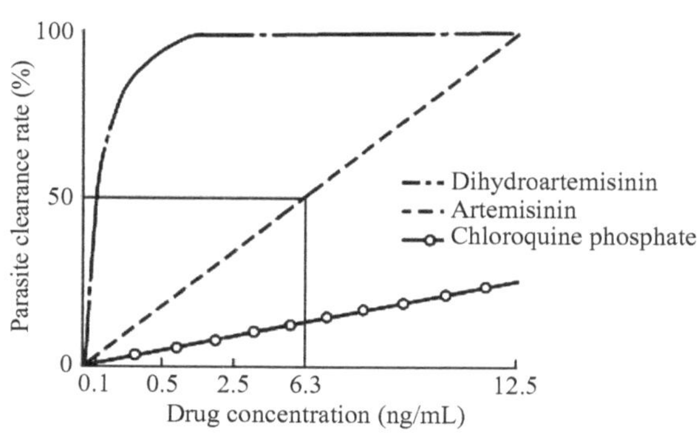

Figure 20.1: In Vitro Pharmacodynamics of Three Agents (With Parasite Infection Rate of 0.8%).

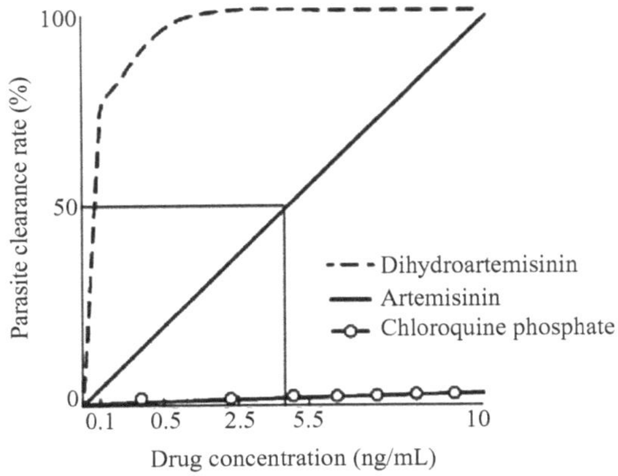

Figure 20.2: In Vitro Pharmacodynamics of Three Agents (With Parasite Infection Rate of 0.61%).

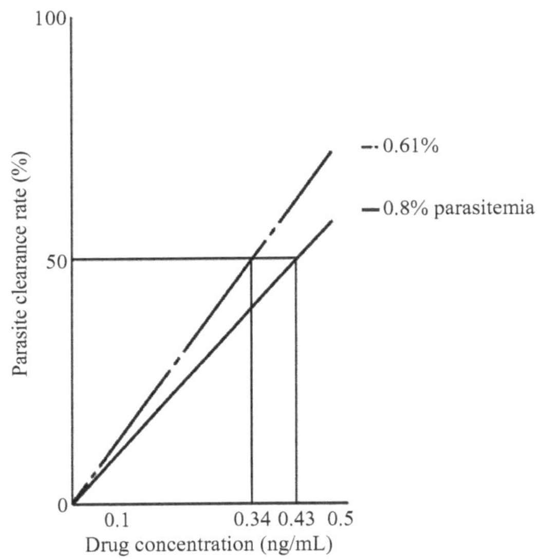

Figure 20.3: In Vitro Pharmacodynamics of Dihydroartemisinin.

Table 20.5: In vitro ED_{50} determination for three agents against human *P. falciparum*

Starting Parasite Infection Rate (%)	ED_{50} of Three Agents Against Human *P. falciparum*		
	Dihydroartemisinin	Artemisinin	Chloroquine
0.8	0.43	6.3	Undetectable (even at a dose up to 25 ng/mL)
0.61	0.34	5.0	Undetectable
$\bar{X} \pm SD$	0.385 ± 0.045	5.65 ± 0.65	

respectively. The former compound exhibits a 14-fold superior efficacy over the latter compound, even more over chloroquine. The ED_{50} of chloroquine could not detect even at a dose level up to 25 ng/mL.

2 General Pharmacology Studies [4]

2.1 Nervous System

The effects of dihydroartemisinin on the locomotor activity, behavioral changes, and sedative and hypnotic activities of mice were investigated. Considering that the ED_{50} of dihydroartemisinin against murine parasites was 12.2 mg/kg, the doses of 10, 20, and 40 mg/kg were chosen for the study.

2.1.1 Effects on the locomotor activity of mice by righting reflex

Kunming mice weighing 18–20 g were randomized into four groups, according to Irwin behavior grading method [1]. Dihydroartemisinin was dosed at 10, 20, and 40 mg/kg, respectively. One hour after dosing, the animals were tail-lifted and rotated for three circles, then thrown up 1 m into the air. The abnormal landing gestures and behavioral changes were observed. The procedure was repeated 5 times for each animal for grading. The results are listed in Table 20.6. There is no significant difference between each dose group and the control group, indicating that dihydroartemisinin does not influence the righting reflex and general activities of mice at the investigated dose levels.

2.1.2 Writhing test in white mice (analgesic test)

Kunming mice weighing 18–20 g were randomized into six groups, and the room temperature was maintained at 20°C. Dihydroartemisinin was dosed at 10, 20, and 40 mg/kg, respectively. Aspirin was used as positive control at doses of 0.1 and 0.2 g/kg. One hour after dosing, the mice were i.p. injected with 0.2 mL of 1.2% acetic acid. After 20-min observation, the number of typical writhing symptoms was counted; *t* test was used to compare the number

Table 20.6: Effects of dihydroartemisinin on the righting reflex of mice

Experiment times	Group	Dose Level (mg/kg)	Number of Animals	Grading					
				0	1	2	3	4	5
1	Control group	0	10	10					
	Dihydroartemisinin	10	10	10					
		20	10	9	1				
		40	10	8	1			1	
2	Control group	0	10	10					
	Dihydroartemisinin	10	10	9					
		20	10	10					
		40	10	9	1				

Table 20.7: Analgesic action of dihydroaretemisinin

Experiment Times	Group	Dose Level (mg/kg)	Number of Animals	Episodes of Writhing Behavior ($\bar{X} \pm SD$)	Reduction Rate (%)	P
1	Control	1	9	36.33 ± 5.21		
	Dihydroartemisinin	10	10	33.30 ± 6.22	8.34	
	Dihydroartemisinin	20	9	31.44 ± 6.24	13.46	
	Dihydroartemisinin	40	10	20.60 ± 5.72	43.30	<0.1
	Aspirin	0.1 g/kg	10	21.10 ± 4.39	41.92	<0.05
	Aspirin	0.2 g/kg	10	14.60 ± 4.12	59.81	<0.005
2	Control	0	11	29.18 ± 3.96		
	Dihydroartemisinin	10	12	21.67 ± 4.09	25.74	
	Dihydroartemisinin	20	12	23.08 ± 3.79	20.90	
	Dihydroartemisinin	40	10	18.90 ± 6.80	35.23	<0.2
	Aspirin	0.1 g/kg	11	16.91 ± 4.84	42.05	<0.05
	Aspirin	0.2 g/kg	12	11.83 ± 2.50	59.46	<0.005

of writhing behavior between each test group and control group. The reduction in writhing behavior in test groups was also calculated (see Table 20.7 for the results). It is found that the positive control reduces the number of writhing behavior significantly ($P < .05$), and the large dose of dihydroartemisinin is also effective to reduce the writhing behaviors.

2.1.3 Sedative acitivity

The effects of dihydroartemisinin on hypnosis induced by subthreshold dose of pentobarbital sodium and sleep duration of mice were investigated.

Kunming mice weighing 18–20 g were randomized into five groups: blank control group, dihydroartemisinin groups (three dose levels), and positive control group (chlorpromazine). The mice were i.p. injected with 30 mg/kg pentobarbital sodium 1 h after oral dosing. The animals were observed immediately to detect the animal number with loss of righting reflex for up to 1 min. If 1 min was achieved, it was considered that sleep onset occurred and sleep duration was also observed. The sleep rate and sleep duration were compared between test and control groups. The study was performed at a quiet environment at 24°C (see Tables 20.8 and 20.9 for the test results).

Table 20.8: Effects of dihydroartemisinin on the subthreshold hypnotic dose of pentobarbital sodium

Group	Dose Level (mg/kg)	Number of Animals	Number of Animals in Sleep	Sleep Rate (%)	P Value for X^2 Test
Blank control	0	20	7	35	
Dihydroartemisinin	10	20	4	20	
Dihydroartemisinin	20	20	6	30	
Dihydroartemisinin	40	20	12	60	>0.5
Chlorpromazine	30	20	20	100	<0.05

Table 20.9: Effects of dihydroartemisinin on the sleep duration of mice induced by pentobarbital sodium

Group	Dose Level (mg/kg)	Number of Animals in Sleep	Sleep Duration (min) ($\bar{X} \pm SE$)	Prolongation of Sleep Duration (%)	*P* Value
Blank control	0	7	17.14 ± 8.26		
Dihydroartemisinin	10	4	32.50 ± 7.31	89.73	>0.5
Dihydroartemisinin	20	6	31.50 ± 8.82	66.76	>0.5
Dihydroartemisinin	40	12	68.48 ± 18.04	299.70	<0.05
Chlorpromazine	30	20	180	950.80	<0.005

Aforementioned results show that two dose levels of dihydroartemisinin do not influence the sleep rate of mice. The large dose of dihydroartemisinin is able to prolong the sleep duration of mice for more than 4 times. It indicates that the agent has relative strong sedative-hypnotic activities. (All the mice in chlorpromazine control group achieved a sleep duration of more than 3 h.)

2.2 Cardiovascular and Respiratory Systems

Five cats, weighing 3.20 ± 0.32 kg, including both males and females, were anesthetized intravenously with 30 mg/kg pentobarbital sodium. The skin was incised at the center of neck, and the trachea and one common carotid artery were isolated. The common carotid artery was cannulated. Systolic blood pressure (SBP) and diastolic blood pressure (DBP) were measured at the common carotid artery with an electrosphygmomanometer. A trachea cannula was connected to a modified Marey tambour. A wire was attached to a force transducer from the center of the tambour. The respiratory signal was amplified through a carrier preamplifier, and respiratory rate (RR) and respiratory amplitude (RA) were recorded. Heart rate (HR) and lead-II ECG were measured with a cardiotachometer. The esophagus was cannulated for intragastric administration. SBP, DBP, HR, ECG, RR, and RA were recorded on an 8-lead physiological recorder at a paper speed of 25 mm/s. The change of indicators was observed predosing and at 15, 30, 60, and 90 min after i.g. dosing. The test results were expressed with $\bar{X} \pm SD$; *t* test was used to determine the significance of difference for each indicator before dosing, at 15, 30, 60, and 90 min after dosing. Each dihydroartemisinin tablet contained 20 mg. Vehicle blank of the same amount was used as a control.

It can be found from Table 20.10 that the BP, HR, ECG, RR, and RA did not have significant difference within the 90-min observation period after i.g. administration of dihydroartemisinin (40 mg/kg) and equal amount of vehicle blank to the anesthetized cats, indicating that oral administration of 40 mg/kg dihydroartemisinin will not have adverse effects on the BP, respiration, ECG, and HR of anesthetized cats.

Table 20.10: Effects of dihydroartemisinin on the BP, HR, ECG, and respiration of cats

Group	Time	SBP	DBP	HR (beats/min) ($\bar{X} \pm SE$)	RR (breaths/min) ($\bar{X} \pm SE$)	RA (mm) ($\bar{X} \pm SD$)	Lead-II ECG
Dihydroartemisinin (40 mg/kg)	Predosing	158.0 ± 43.8	116.0 ± 45.1	166.0 ± 47.7	19.6 ± 4.6	4.0 ± 1.4	Normal
	15 min post dosing	145.0 ± 42.0	110.0 ± 45.6	164.0 ± 47.7	19.6 ± 4.9	4.0 ± 1.4	Normal
	30 min post dosing	152.0 ± 39.6	118 ± 46	162.0 ± 48.1	19.8 ± 4.1	4.0 ± 1.4	Normal
	60 min post dosing	150.0 ± 45.3	100.0 ± 46.9	162.0 ± 48.1	19.4 ± 4.4	4.20 ± 1.09	Normal
	90 min post dosing	154.0 ± 39.7	118 ± 46	162.0 ± 48.1	19.6 ± 4.6	4.0 ± 1.4	Normal
Vehicle blank	Predosing	138.0 ± 51.2	110.0 ± 48.9	168 ± 46	20.0 ± 5.1	3.9 ± 1.4	Normal
	15 min post dosing	127.0 ± 51.2	110.0 ± 48.4	164.0 ± 48.7	20.2 ± 5.1	4.0 ± 1.4	Normal
	30 min post dosing	148.0 ± 49.6	120 ± 50	164.0 ± 45.0	19.6 ± 3.9	4.0 ± 1.4	Normal
	60 min post dosing	156.0 ± 53.1	120.0 ± 51.1	162 ± 48	21.0 ± 5.3	4.20 ± 1.09	Normal
	90 min post dosing	154.0 ± 50.3	120.0 ± 48.4	162 ± 48	20.2 ± 5.1	3.9 ± 1.4	Normal

$n = 5$, $P > .05$ for all the indicators, cats weighing 3.20 ± 0.32 kg.

3 Pharmacokinetic Studies

3.1 Animals

1. Determination of dihydroartemisinin serum concentrations after oral administration. Radioimmunoassay was used for the pharmacokinetic study of dihydroartemisinin tablet administered orally in rabbits [5].
   Rabbits weighing 1.5–2.0 kg were divided into three groups. Each group contained four male rabbits and one female rabbit. The animals received dihydroartemisinin tablet orally at 10, 20, and 30 mg/kg, respectively. Blood samples were collected at different times after dosing. The serum was isolated for determination of drug concentration. The pharmacokinetic parameters were calculated with statistical moment method.
   The results revealed that dihydroartemisinin tablet was well absorbed. The serum concentration reached peak value within 1–2 h of dosing. The peak concentrations were 0.03 ± 0.01, 0.05 ± 0.02, and 0.13 ± 0.11 μg/mL for 10, 20, and 30 mg/kg dose levels, respectively. The mean retention times (MRTs) in rabbits were 1.73 ± 0.56, 1.36 ± 0.17, and 1.53 ± 0.21 h, and $T_{1/2}$ were 1.19 ± 0.37, 1.00 ± 0.13, and 1.10 ± 0.26 h, respectively. There was no significant difference in pharmacokinetic behavior among different dose groups. The serum concentration–time curve for dihydroartemisinin tablet administered orally in rabbits at 10, 20, and 30 mg/kg is given in Fig. 20.4.
2. Study on the absorption, distribution, excretion, and metabolism of labeled dihydroartemisinin in mice [6]. The study used liquid scintillation and radioactive TLC techniques to investigate the absorption, distribution, excretion, and metabolism of tritium-labeled dihydroartemisinin in mice.

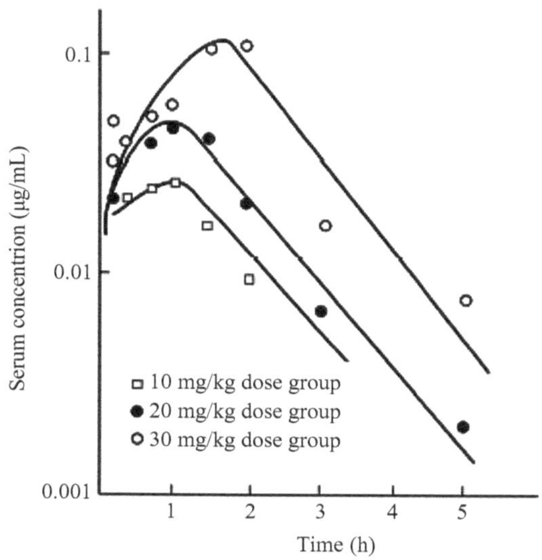

Figure 20.4: Serum Concentration–Time Curve for Dihydroartemisinin Tablet Administered Orally in Rabbits at 10, 20, and 30 mg/kg.

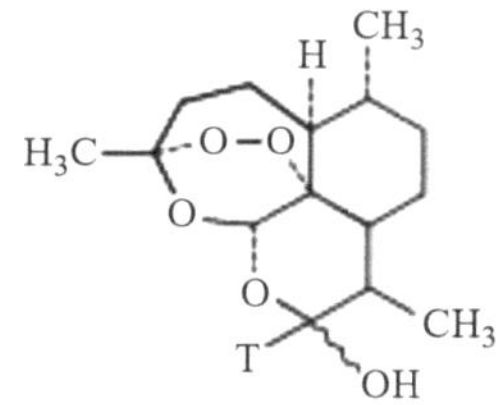

Figure 20.5: Dihydroartemisinin.

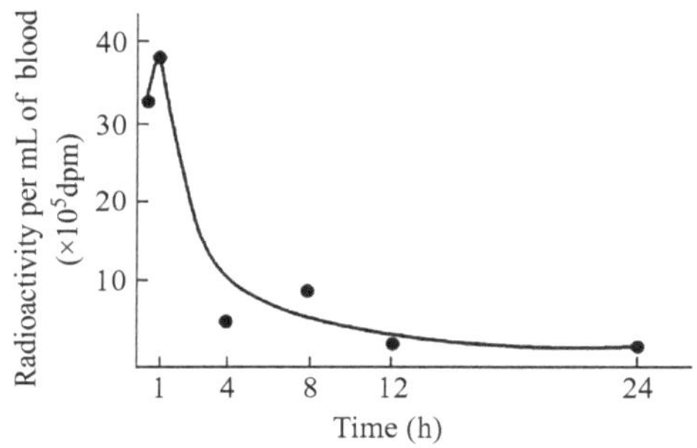

Figure 20.6: Change of Radioactivity Level in Blood With Time.

> ^{3}H-labeled dihydroartemisinin was used in the study. The hydrogen atom bonded to C12 was replaced by ^{3}H (Fig. 20.5). The radiochemical purity was >90%. Kunming mice weighing 18–22 g were used, four to five animals in each group. Each animal was i.g. administrated with 50 mCi/mg ^{3}H-labeled dihydroartemisinin (the dose in excretion experimental group was 10 mCi/mg). Blood samples were collected from the tail vein, and feces and urine were collected at different time points after dosing. Tissues and organs were isolated after animals were sacrificed. The samples were digested with acid. The quench calibration for liquid scintillation counting adopted internal standard method. Radioactive TLC technique was used to investigate the stability of labeled compound and the compound metabolism in blood and urine samples.

The results revealed that oral ^{3}H-labeled dihydroartemisinin was absorbed quickly. The level of radioactivity in blood reached peak value at 1 h after dosing, and decreased to less than half of its peak value at 4 h after dosing (Fig. 20.6). With regard to the radioactivity distribution in tissues and organs, peak levels were observed in the liver and heart at 0.5 h after dosing, while 1 h for other tissues and organs. The radioactive intensity at 1 h after dosing was compared among different tissues and organs. The intensity was as follows: kidney > lung > liver > spleen > bone > heart > muscle > brain (Table 20.11). When the blood concentration of drug decreased, the radioactive intensity in various tissues and organs decreased accordingly. The biliary radioactivity excretion also reached peak level at 1 h after dosing. The radioactivity excreted in urine and feces reached 82.7% within 24 h. The majority was excreted in urine, accounting for 67.1 ± 1.7% of total radioactivity (Table 20.12). Radioactive TLC analysis revealed that ^{3}H-labeled dihydroartemisinin was relatively stable in in vitro environment. It was metabolized quickly in in vivo environment. Parent compound was undetectable in urine (Figs. 20.7 and 20.8).

Table 20.11: Radioactive intensity of multiple tissues and organs at four stages after oral ^{3}H-labeled dihydroartemisinin administration

Tissue/Organ	Radioactive Intensity at Four Stages (×10^5 dpm/g)			
	0.5 h	1 h	4 h	24 h
Liver	63.56 ± 7.33	57.77 ± 5.63	40.36 ± 1.13	3.85 ± 2.06
Spleen	42.64 ± 3.31	53.36 ± 2.83	25.24 ± 3.49	11.93 ± 4.82
Kidney	53.00 ± 2.82	78.01 ± 21.56	20.16 ± 0.79	6.72 ± 0.37
Lung	31.43 ± 2.83	60.68 ± 21.56	13.35 ± 1.55	4.96 ± 0.24
Muscle	23.39 ± 0.81	23.99 ± 3.49	9.42 ± 0.72	3.70 ± 0.28
Bone	31.23 ± 3.45	38.45 ± 4.02	13.44 ± 0.90	6.37 ± 0.78
Brain	16.52 ± 1.67	20.06 ± 0.85	5.84 ± 0.35	3.12 ± 0.37
Heart	32.95 ± 1.43	30.18 ± 1.02	12.59 ± 0.52	4.81 ± 0.41
Gallbladder	997.83 ± 383.62	1777.41 ± 686.84	966.74 ± 320.98	2.87

The data in the table are the mean value ±SD of four mice.

Table 20.12: Radioactivity excreted after oral ^{3}H-labeled dihydroartemisinin administration in mice[a]

Source of Sample	Radioactivity of 1% Dose Excreted at Each Stage				
	0.4 h	4–8 h	8–12 h	12–24 h	0–24 h
Urine	56.5	8.7	1.3	0.6	67.1 ± 1.7[b]
Feces	7.6	4.1	2.2	1.6	15.6 ± 1.3

[a]Each mouse was given 10 mCi/mg^{3}H-labeled dihydroartemisinin.
[b]$\overline{X}$ ± SD (mean value of four mice) ± SE.

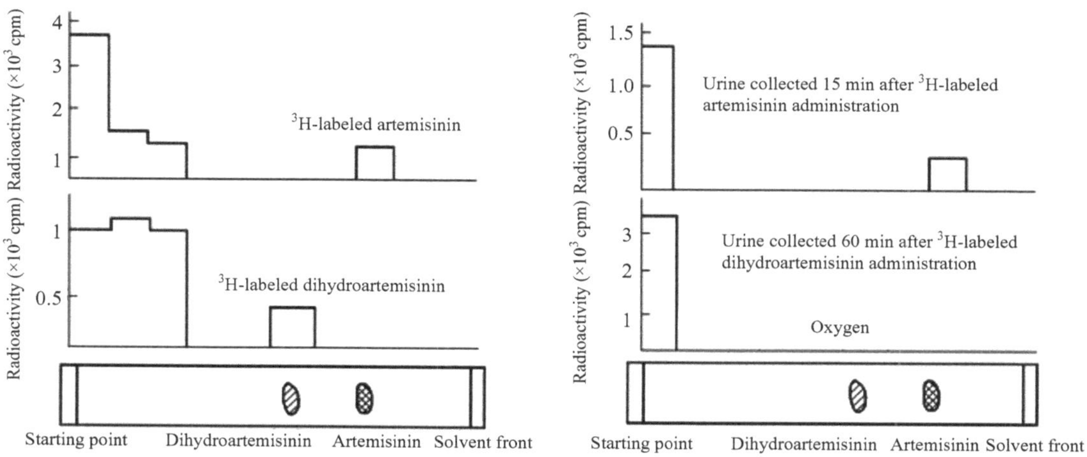

Figure 20.7: Isolation of ^{3}H-labeled Artemisinin or ^{3}H-labeled Dihydroartemisinin With TLC for 30 min.

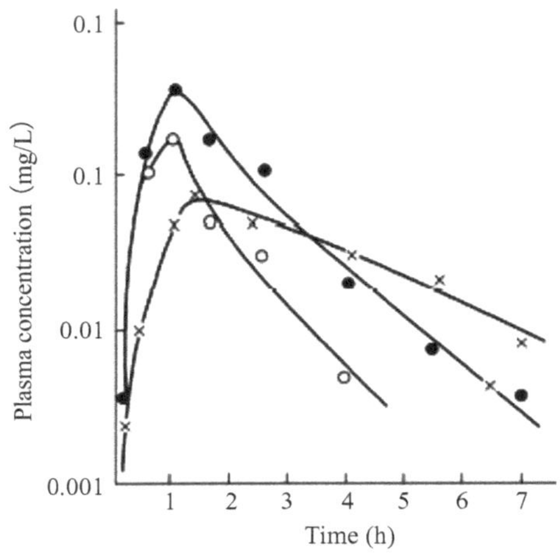

Figure 20.8: Plasma Concentration–Time Curves for Oral Administration of 1.1 mg/kg (°—°) and 2.2 mg/kg (·—·) Dihydroartemisinin and 15 mg/kg (×— ×) Artemisinin by Human.

In conclusion, dihydroartemisinin has a quick absorption, extensive distribution, quick excretion, and metabolism.

3.2 Human [7]

The pharmacokinetics of artemisinin and dihydroartemisinin in human were investigated with radioimmunoassay. Peak plasma concentration of 0.09 μg/mL was achieved at 1.5 h after artemisinin tablet was administered orally at 15 mg/kg. MRT was 3.27 h. However, peak plasma concentrations of 0.13 and 0.71 μg/mL were achieved at 1.33 h after oral administration of dihydroartemisinin tablet at 1.1 and 2.2 mg/kg, respectively. The MRTs were 2.36 and 2.26 h, respectively. It indicates that the bioavailability of artemisinin tablet only accounts for 1.62–10.08% of dihydroartemisinin bioavailability. Therefore, dihydroartemisinin is more appropriate for oral administration.

Peak plasma concentrations of 0.04 and 0.11 μg/mL were achieved at 4.6 and 4.7 h after rectal administration of artemisinin and dihydroartemisinin suppositories at 15 and 8 mg/kg, respectively. The MRTs were 6.98 and 6.96 h, respectively. It reveals that the bioavailability of artemisinin suppository only accounts for 29% of dihydroartemisinin bioavailability. Therefore, dihydroartemisinin can also be used as a suppository.

3.2.1 Materials and methods

Each artemisinin or dihydroartemisinin tablet contains 20 mg of drug. There were 4 suppository dosage forms, including 80, 60, 40, and 20 mg. The agents were all provided by Institute of Chinese Materia Medica, China Academy of Chinese Medical Sciences. The investigation center was Xiyuan Hospital, China Academy of Chinese

Medical Sciences. The subjects were adult male patients, weighing 50–70 kg. The liver and kidney functions and reticulocyte count of the subjects were examined strictly. Only the patients with normal results were enrolled in the study. The subjects did not take any medicine 2 weeks before the study. They were divided into five groups. Of which, three groups consisted of three patients per group, and the doses were 1.1 and 2.2 mg/kg dihydroartemisinin tablets and 15 mg/kg artemisinin tablet, respectively. The remaining two groups consisted of five patients per group. The patients were given 8 mg/kg dihydroartemisinin suppository and 15 mg/kg artemisinin suppository, respectively, via rectal administration. Two milliliter of blood samples were collected from upper extremity vein at different times after dosing. The serum was isolated and kept at −30°C for determination.

Plasma and urine drug concentration were determined with radioimmunoassay [4]. The data of plasma concentration were processed with statistical moment method. The pharmacokinetic parameters were calculated.

3.2.2 Results

1. Pharmacokinetics of dihydroartemisinin tablet and oral artemisinin
   The plasma concentration–time curve for artemisinin tablet administered orally in human was given in Fig. 20.8. Peak plasma concentration of 0.09 μg/mL was achieved at 1.5 h after dosing. The MRT was 3.27 h and $T_{1/2}$ was 2.77 h. For other pharmacokinetic parameters, see Table 20.13. Fig. 20.8 shows that dihydroartemisinin tablet was absorbed quickly after oral administration. Peak plasma concentration was obtained at 1.33 h after dosing. The peak concentrations were 0.13 and 0.71 μg/mL for 1.1 and 2.2 mg/kg dose levels, respectively. The MRTs were 2.36 and 2.26 h, respectively. $T_{1/2}$ were 1.63 and 1.57 h, respectively. It is found that there was no significant difference in pharmacokinetic profile between the two dose groups except for peak concentration. Table 20.13 gives the remaining pharmacokinetic parameters of the two dose groups.

Table 20.13: Pharmacokinetic parameters of oral dihydroartemisinin and artemisinin in volunteers (n = 3)

Agent	Dihydroartemisinin		Artemisinin
Dosage (mg/kg)	1.1	2.2	15
Time to peak value (h)	1.33 ± 0.29	1.33 ± 0.29	1.50 ± 0.32
Peak concentration (μg/mL)	0.13 ± 0.03	0.71 ± 0.30	0.09 ± 0.01
MRT (h)	2.36 ± 0.98	2.26 ± 0.50	3.27 ± 0.32
$T_{1/2}$ (h)	1.63 ± 0.68	1.57 ± 0.34	2.27 ± 0.22
Volume of distribution at steady state (L/kg)	10.85 ± 3.08	4.12 ± 2.38	152.34 ± 35.90
Clearance rate (L/kg/h)	4.98 ± 1.74	1.97 ± 1.47	46.55 ± 9.52
Area under curve (μg/h/mL)	0.24 ± 0.08	1.49 ± 0.82	0.33 ± 0.07
Variance of mean retention time (h^2)	0.93 ± 0.31	2.34 ± 1.38	2.92 ± 0.22

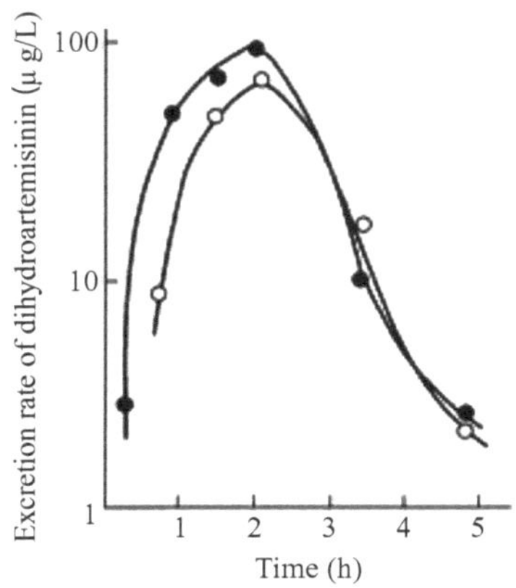

Figure 20.9: Urinary Excretion Rate–Time Curves for Oral Administration of 1.1 mg/kg (°—°) and 2.2 mg/kg (·—·) Dihydroartemisinin by Human.

Fig. 20.9 is the urinary excretion rate-time curve for oral dihydroartemisinin. It can be found that urinary excretion rate reached peak value at 2.5 h after dosing. Table 20.14 reveals that little parent compound was excreted in urine after oral administration of dihydroartemisinin tablet by human. The excreted amount within 7 h after dosing only accounted for 0.10–0.15% of the administered dose.

The study revealed that little parent compound was excreted in urine after oral administration of dihydroartemisinin tablet by human, which was consistent with the excretion of intravenous sodium artesunate [6], indicating that urinary excretion was not the major elimination route of artemisinin-based drugs.

2. Pharmacokinetics of dihydroartemisinin and artemisinin suppositories

The plasma concentration–time curves for rectal administration of 15 mg/kg artemisinin and 8 mg/kg dihydroartemisinin suppositories were given in Fig. 20.10. Peak plasma concentrations of 0.04 and 0.11 μg/mL were achieved at 4.6 and 4.7 h, respectively, after

Table 20.14: Parent compound excreted in urine after oral administration of dihydroartemisinin by human volunteers

Group	Dosage (mg/kg)	Total Amount of Excretion (mg)	Excretion Rate (%)
1	1.1	0.05	0.08
2	1.0	0.07	0.09
3	1.2	0.23	0.29
$\overline{X} \pm SD$	1.10 ± 0.10	0.12 ± 0.09	0.15 ± 0.12
4	2.0	0.07	0.06
5	2.2	0.17	0.11
6	2.3	0.22	0.14
$\overline{X}$	2.16 ± 0.15	0.15 ± 0.08	0.10 ± 0.04

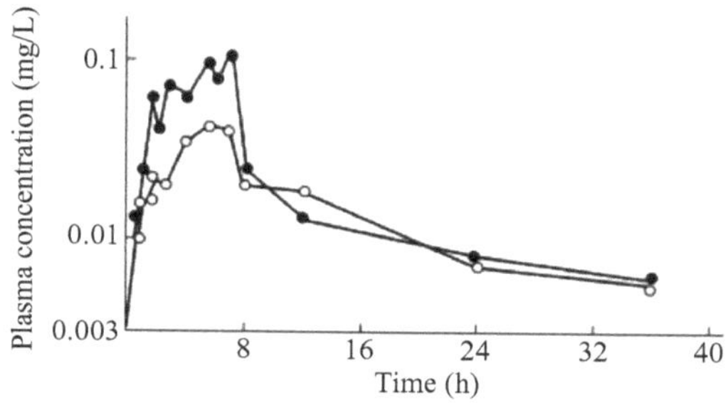

Figure 20.10: Plasma Concentration–Time Curves for Rectal Administration of 15 mg/kg (°—°) Artemisinin and 8 mg/kg (·—·) Dihydroartemisinin.

administration of artemisinin and dihydroartemisinin. The MRTs were 6.98 and 6.96 h, and $T_{1/2}$ were 4.84 and 4.82 h, respectively. For other pharmacokinetic parameters of the two agents, see Table 20.15. It is found from the table that the two agents have similar pharmacokinetic behavior.

It is found from Fig. 20.10 and Table 20.15 that both artemisinin and dihydroartemisinin suppositories are absorbed well. Peak plasma concentrations were achieved at about 5 h after administration. The dose of dihydroartemisinin was about half of artemisinin dose, while its peak concentration increased by 2.5-fold. It can be found from the results that dihydroartemisinin has a higher bioavailability compared with artemisinin. The latter only accounts for 29% of the former bioavailability. By comparison of Figs. 20.9 and 20.10, it is also found that the bioavailabilities of tablet dosage form of the two compounds are both higher than theirsuppository dosage form, especially for dihydroartemisinin. It should be pointed out that the MRT for both compounds in tablet dosage form was only about 3 h, while 7 h for suppository dosage form. It is considered that, despite its delayed time to peak concentration, suppository has a prolonged MRT which will benefit the extension of efficacy.

Table 20.15: Pharmacokinetic parameters of rectal artemisinin and dihydroartemisinin suppositories in volunteers (n = 5, $\bar{X} \pm SE$)

Agent	Dihydroartemisinin	Artemisinin
Dosage (mg/kg)	8	15
Time to peak value (h)	4.70 ± 1.44	4.60 ± 0.82
Peak concentration (μg/mL)	0.10 ± 0.04	0.04 ± 0.01
MRT (h)	6.96 ± 3.88	6.98 ± 3.91
$T_{1/2}$ (h)	4.82 ± 2.69	4.84 ± 2.71
Volume of distribution at steady state (L/kg)	116.18 ± 87.47	285.14 ± 176.62
Clearance rate (L/kg/h)	17.66 ± 9.09	54.85 ± 44.07
Area under curve (μg/h/mL)	0.57 ± 0.30	0.31 ± 0.22
Variance of mean retention time (h^2)	33.76 ± 44.85	22.09 ± 33.74

4 Effects of Dihydroartemisinin on the Development of Plasmodium yoelii in Mosquitoes [8]

Dihydroartemisinin (DQHS) is a newly synthesized artemisinin derivative in China, which exhibits a high efficacy, quick onset of action, safe use, low recrudescence rate, low dose, and convenience for oral dosing. There are studies on the influence of artemisinin and its derivatives on the infectivity of parasite gametocytes and its development in mosquito. Dutta initially used light microscope and scanning electron microscope to observe the mosquito infectivityin *Plasmodium cynomolgi* B–*Anopheles stephensi* model. The results revealed that artemisinin was able to influence plasmodium gametocyte infectivity; however, it did not directly inhibit the proliferation of sporozoites. It was reported by Chen Peiquan that artemisinin partially inhibited the development of gametocytes in mosquitos in the treatment of patients with *P. falciparum* malaria. There have been some domestic reports on the study of dihydroartemisinin in this field. Considering that *P. yoelii*–*A. stephensi* model in mice is convenient for repeating the experiment, the study specially investigated the effects of dihydroartemisinin on the infectivity of *P. yoelii* gametocytes and effects of different doses on the development of parasites in mosquitoes. We also observed if there was a direct influence on the proliferation of sporozoites.

4.1 Materials and Methods

4.1.1 Parasites

P. yoelii BY265 strain, provided by Academy of Military Medical Sciences. Second generation of parasites was used in the experiment.

4.1.2 Animals

Kunming mice, weighing 18–20 g were used in the experiments. One drop of blood was used for preparing Giemsa-stained thin blood smear. The mice with about 50 gametocytes per 100 white blood cells in a smear were used as donor mice, feeding the mosquitoes. The mice were divided into 3.5-day (with mature gametocytes achieving peak q.d. value) and 2.5-day infection experimental groups.

4.1.3 Test drug

The compound dihydroartemisinin was provided by Institute of Chinese Materia Medica, China Academy of Chinese Medical Sciences. The medicinal solution was prepared by dissolution of dihydroartemisinin and Tween-80 in distilled water. The infected mice were dosed at 20, 60, 120, 180, and 240 mg/kg, respectively. The infected mouse was orally administrated with a single dose of 0.2 mL drug solution.

4.1.4 Anopheles stephensi

A. stephensi were provided by Academy of Military Medical Sciences. They were fed and infected with plasmodia according to the methods previously used in the laboratory. After blood feeding, the mosquitoes were dissected and observed under light microscope to calculate the positive rate (ratio of number of infected mosquitoes to number of all dissected mosquitoes) and density (number of oocysts and sporozoites in the stomach and salivary glands of each mosquito) of oocysts and sporozoites. Some infected mosquitoes were treated for transmission electron microscopy (TEM) to observe the change in ultrastructure with a Hitachi H-600 TEM.

4.1.5 Gametocyte morphology observation

Blood samples were collected from the tail vein of infected mice to prepare Giemsa-stained blood smears. The smears were observed under light microscope (oil immersion lens) for the change of gametocytes in morphology.

4.1.6 Statistical processing

$\bar{X} \pm \mathrm{SD}$ was calculated, and t test (or variance analysis) was performed.

4.2 Results

4.2.1 Effects of test dugs on the infectivity of gametocytes

On day 2.5 or day 3.5 after mice were infected with merozoites, mice were administrated with 60 mg/kg of dihydroartemisinin and fed to mosquitoes 4 h later. The mosquitoes were dissected at 7–8 and 14–16 days after blood feeding. The positive rate and density of oocysts and sporozoites were observed (Table 20.16). The positive infection rates of oocysts on 2.5- and 3.5-day were both 100%. There was a difference in the positive infection rate of sporozoites in 2.5-day infection experimental group. The level in control group was 40% above the test group. Density comparison: in 2.5-day infection experimental group, there were significant differences between control group and test group both in oocyst and sporozoite ($t_{\mathrm{oocyst}} = 4.34$, $t_{\mathrm{sporozoite}} = 7.30$, $P < .001$); in 3.5-day inflection experimental group, the difference in oocyst was significant ($t = 3.80$, $P < .05$), while not significant in sporozoite ($t = 0.78$, $P > .05$).

4.2.2 Effects of different doses of dihydroartemisinin on the development of malaria parasites in mosquitoes

The same batch of infected mice were divided into different groups. On day 3.5, mice were orally administrated with a single dose of dihydroarteminisin at 60, 120, 180, or 240 mg/kg, respectively, and 16 h later fed to the mosquitoes. The mosquitoes were dissected for observation of the positive rate and density of oocysts and sporozoites 7–8 and 14–17 days after blood feeding. The results revealed that the positive rate of oocysts and sporozoites in each test group decreased in a drug concentration-dependent manner. The density of oocysts and sporozoites in each test group also decreased gradually (Table 20.17). The number of oocysts in the stomach and sporozoites in the salivary glands of each mosquito was calculated

Table 20.16: Infection of mosquitoes after feeding with infected mice 4 h after dosing with 60 mg/kg of dihydroartemisinin

Group	2.5-Day Infection Mice				3.5-Day Infection Mice			
	Oocyst Infection		Sporozoite Infection		Oocyst Infection		Sporozoite Infection	
	Positive Rate (%)	Average Density ($\bar{X} \pm SE$)	Positive Rate (%)	Average Density ($\bar{X} \pm SE$)	Positive Rate (%)	Average Density ($\bar{X} \pm SD$)	Positive Rate (%)	Average Density ($\bar{X} \pm SD$)
Control group	100 (10/10)	4.90 ± 1.11	90 (9/10)	4.20 ± 1.75	100 (10/10)	5.36 ± 0.50	90 (9/10)	3.04 ± 2.05
Test group	100 (10/10)	2.75 ± 1.10	50 (10/20)	0.80 ± 0.83	100 (10/10)	3.65 ± 1.34	90 (18/20)	2.47 ± 1.86
t	4.34		7.30		3.80		0.7	
P	<0.001		<0.001		<0.05		>0.05	

Table 20.17: Infection of mosquitoes after blood feeding of with infected mice 16 h after dosing of different levels of dihydroartemisinin

Group	Dosage	Oocyst Infection		Sporozoite Infection	
		Positive Rate (%)	Average Density ($\bar{X} \pm SD$)	Positive Rate (%)	Average Density ($\bar{X} \pm SD$)
Control group		100 (30/30)	5.51 ± 1.25	85 (17/20)	3.34 ± 2.25
Test group	60	100 (20/20)	4.25 ± 1.47	55 (11/20)	2.29 ± 2.61
	120	90 (18/20)	3.28 ± 1.48	22.2 (2/9)	0.84 ± 2.00
	180	70 (21/30)	0.88 ± 0.85	0 (0/18)	0
	240	74 (26/35)	1.24 ± 1.30	5 (1/20)	0.08 ± 0.36

and averaged. The results were as follows: the average number of oocysts in the control group was 411.77 (3–1100), and those in each test group were 143.0 (4–485), 54.6 (0–23), 3.17 (0–11), and 9.03 (0–90), respectively. The average number of sporozoites in the control group was 129.75, and those in each test group were 113.75, 45.0, and 0.25, respectively. After comparison of differences between each test group and control group and the differences among each test group using variance analysis, it should be specially pointed out that the difference in sporozoite density was not significant between 180 and 240 mg/kg dose groups.

4.2.3 Effects of dihydroartemisinin on the morphology and structure of gametocytes and sporozoites

Abnormal morphology in gametocytes were observed in the blood smears from 10 mice with 3.5-day infection and 16 h after dosing (60 mg/kg). The abnormalities included gametocyte shrinkage, cytoplasmic hypervacuolization, and clustering of haemozoin. The ratio of normal to abnormal gametocytes in morphology in the blood smears from each mouse was calculated. The results were as follows: 73:37, 14:25, 102:18, 15:30, 21:74, 43:76, 33:14, 54:12, 42:21, and 31:43. The number of gametocytes with abnormal morphology accounted for 43% (350/814). The mouse with largest difference had an abnormal gametocyte count (74), 3.5-fold more than the normal gametocyte count. In the meantime, the same batch of mice were fed to the mosquitoes. The stomach of mosquitoes were dissected after 12–13 days. The changes in ultrastructure of late oocysts were observed with TEM. Compared with control group, the mosquitoes in test groups had many oocysts and sporozoite buds with abnormal development and morphology in stomach. The abnormalities were manifested in membrane system, malformation and damage in organelles (including endoplasmic reticulum and mitochondria), even cytoplasmic hypervacuolization.

4.2.4 Effects of dihydroartemisinin on sporozoite proliferation and development after direct action on the oocysts in mosquitoes

The mosquitoes were divided into control and test groups 4 days after blood feeding with infected mice (3.5-day infection), when oocysts of 3-day age occurred in the stomach of mosquitoes. The control mosquitoes bit naive mice and the test ones bit the mice 6 h after dosing (120 mg/kg). Six days after first blood feeding, 10 mosquitoes from each group were dissected to observe the infection of oocysts and 30 mosquitoes were dissected on days 12 and 14 from each group to observe the infection of sporozoites. The infection of oocysts was positive in both groups. Fourteen and sixteen mosquitoes were dissected on days 12 and 14 in control group. The number of positive mosquitoes were 12 and 13, respectively. The oocyst (10 mosquitoes per group) and sporozoite (30 mosquitoes per group) density were observed for both groups. The oocyst density was 5.7501 ± 0.7750 and 5.5520 ± 0.9310 for the control and test groups, respectively. The difference between the groups was of no significance when analyzed with *t* test ($t = 0.33$, $P > .05$). In addition, when observed with TEM, the difference in the ultrastructure of oocysts in stomach of mosquitoes in both groups (16 mosquitoes dissected for each group) was not found obvious significance. It indicated that the test drug had no direct inhibition on the development of *P. yoelii* oocysts.

4.3 Discussion

The study revealed that dihydroartemisinin had certain inhibition on the gametocytes of *P. yoelii*. The inhibition was related to the maturity of gametocytes and dosage. Dihydroartemisinin had a stronger effect on immature gametocytes than on mature gametocytes, that is, lower dose (20 mg/kg) was also able to affect the positive rate of sporozoites. The results were consistent with the report by Mehra and Bhasin in 1993. However, it is still required to observe the ratio of damaged immature to mature gametocytes in the blood of dosed mice with TEM to obtain further conclusion. The infection and density of oocysts and sporozoites decreased gradually, as the increase of dose. However, when the malaria parasites in the mosquitoes developed to oocysts of 3-day age, the development of oocyst to sporozoite could not be inhibited by dihydroartemisinin (16 h, 120 mg/kg), that is, the product was unable to inhibit sporozoite proliferation. Therefore, it is inappropriate to suppose that artemisinin-based drugs might be used to prevent malaria, considering that such drugs have common antimalarial active groups.

References

[1] Zhou Zhang, Liu Zhongwan. Dihydroartemisinin NDA Dossier.

[2] Shi Yunlin, Yang Junde, NingDianxi. Dihydroartemisinin NDA Dossier.

[3] Liu Er'xiang, Fan Rugong, et al. Dihydroartemisinin NDA Dossier.

[4] Institute of Chinese MateriaMedica China Academy of Chinese Medical Sciences. Dihydroartemisinin NDA Dossier.

[5] Kaicun Zhao, Zhenyu Song. Pharmacokinetic study of oral dihydroartemisinin in rabbits and dogs. Acta Pharm Sin 1990;25(2):147–9.

[6] Pharmacology Laboratory of Institute of MateriaMedica, Chinese Academy of Medical Sciences. Pharmacological studies of wormwood herb (Qing Hao). China N Med 1979;(1):23–33.

[7] Kaicun Zhao, Zhenyu Song. Pharmacokinetics of dihydroartemisinin in human and comparison with artemisinin. Acta Pharm Sin 1993;28(5):342–6.

[8] Peihui Chen, Youyou Tu, Fengyun Wang, et al. Effects of dihydroartemisinin on the development of *Plasmodium yoelii* in mosquitoes. Chin J Parasitol Parasitic Dis 1998;16(16):421–4.

CHAPTER 21

Toxicology Studies of Dihydroartemisinin

Chapter Outline

1 Acute Toxicity

The animals were administered via two routes. The toxic reactions occurred after dosing were observed, and the median lethal dose (LD_{50}) was determined.

1.1 Animals

Kunming mice, weighing 18–23 g, half males and half females.

1.2 Test Methods

1. The animals were fasted for about 20 h before the study. They were randomized into different groups by 10 animals in each group.
2. Dihydroartemisinin was suspended in saline with 1% Tween. Five dose groups were set according to a geometric progression. Intragastric and intraperitoneal routes of administration were applied.
3. One control group was set.
4. The animals were observed 30 min after dosing and observation lasted for 1 week. Toxic reactions and deaths were recorded. LD_{50} was calculated with Karber's method.

From *Artemisia annua* L. to Artemisinins
http://dx.doi.org/10.1016/B978-0-12-811655-5.00021-0

1.3 Results

Three hours after dosing, the animals started to show toxic reactions. The intoxicated mice had low spirit and finally died.

LD_{50} of dihydroartemisinin (mg/kg): The oral LD_{50} was 834.5 mg/kg, with a 95% confidence interval of 705.2–987.5 mg/kg. The intraperitoneal LD_{50} was 288.5 mg/kg, with a 95% confidence interval of 256.9–324.2 mg/kg. Chemotherapy index LD_{50}/SD_{50} = 843.0

2 Subacute Toxicity [1]

2.1 Subacute Toxicity in Rats

Eighty Wistar outbred rats were used. Twenty animals were divided into each group, half males and half females. Three doses of dihydroartemisinin were set at 20, 60, and 180 mg/kg, respectively. The animals were i.g. administered once daily for 15 consecutive days. The body weight increased in medium- and low-dose groups compared to that in control group, and no significant difference was observed. In large-dose group, the body weight decreased, manifested with reduced food intake and low spirit. One female rat had diarrhea on the 10th day. One male and one female died immediately due to inappropriate intragastric procedure on day 10. The tracheas of dead rats were found to have filled with drug solution after dissection, while other organs and tissues were normal. After drug withdrawal, no abnormalities were observed during the pathological examination of tissues and organs from all groups. Peripheral blood biochemical indices also fluctuated in normal range. Serum glutamic pyruvic transaminase (SGPT) of each group was within normal range before and after dosing. The weight coefficient of organs had no significant changes except for heart, spleen, and kidney with certain deviation.

2.2 Subacute Toxicity in Dogs

Sixteen dogs, half males and half females, were randomized into four groups, of which three were dose groups (11, 33, and 99 mg/kg/day, respectively) and the remaining one was control group.

During the i.g. dosing period, there was no obvious change in weight in each dose group (weight change: −0.4 to 1.2 kg). On day 3, there was a slight decrease in weight (weight change: −0.1 to 0.4 kg) except for the medium-dose group. Subsequently, body weight increased gradually. On day 14, slight decrease was seen again in large-dose group (weight change: −0.38 kg); however, the change was of no significance. Decreased food intake and low spirit were found in large-dose group during the treatment period. One dog even vomited on day 12. Two dogs with prolonged observation started taking food 3 days after drug withdrawal with recovered spirit and body weight increased to 17 and 18 kg, respectively.

No damages were found during the pathological examination of organs from all groups. The results of blood tests revealed that dihydroartemisinin had no significant influence on total RBC count, total WBC count, and hemoglobin of dogs. The values of all groups were all within normal range before and after dosing. No obvious change was observed for peripheral blood reticulocyte count within the investigated dose range after dosing. The absolute value in large-dose group decreased compared with control group; however, the difference was of no statistical significance. The reticulocyte count of the two dogs with prolonged observation returned to normal 1 week after drug withdrawal (0.712%).

Dihydroartemisinin also had no obvious influence on the SGPT of dogs. SGPT fluctuated within normal range. The value decreased in both control group and each dose group after dosing. However, there was of no significance intergroup. The influence on creatinine and urea nitrogen was also minimal.

The ECGs obtained before, during, and after dosing revealed that dihydroartemisinin has no obvious influence on the ECG of dog.

2.2.1 Conclusions

The subacute toxicity studies with intragastric administration with dihydroartemisinin for half month revealed that dihydroartemisinin had no obvious influence on the body weight, peripheral blood biochemical indices, liver function, and kidney function of rats and dogs both at medium and low doses. The large dose in rats was 180 mg/kg/day, about 150-fold the recommended human dose. The large dose in dogs was 33 mg/kg/day, about 27.8-fold the recommended human dose. The large doses were able to induce decreased food intake and body weight in rats or dogs. One female rat had diarrhea. One dog had vomited. The dogs in large-dose group had aforementioned reactions, including the two dogs with prolonged observation for 15 days, no deaths or other serious outcomes were found. The ECGs obtained before, during, and after dosing revealed that no abnormalities in all groups were observed. Pathological examination showed that the three doses of dihydroartemisinin did not have any damage on the organs of animals. The aforementioned results indicate that dihydroartemisinin does not have obvious adverse effects. The clinical dose is relatively safe in human.

3 Special Toxicities

3.1 Mutagenicity Tests [2]

3.1.1 Micronucleus test

The test was carried out by detecting the micronuclei in bone marrow polychromatic erythrocytes of male NIH mice. A preliminary study of the micronucleus test was performed by acridine orange fluorescent staining to investigate the dose and sample collection time (24, 48, and 72 h). The preliminary study did not find increased micronucleation induced by

Table 21.1: Results of micronucleus test in mice administrated with dihydroartemisinin

Group	Agent	Dosage (mg/kg)	Number of Animals	Micronucleation Rate (%) ($\bar{X} \pm SD$)	*P*	Result
Solvent control	5% starch paste	—	6	2.50 ± 1.38		
Positive control	Cyclophosphamide	120	6	57.02 ± 10.00	<.01	+
Test group A	Dihydroartemisinin	38.625	8	3.51 ± 1.38	>.05	–
Test group B	Dihydroartemisinin	77.25	6	2.50 ± 1.52	>.05	–
Test group C	Dihydroartemisinin	154.5	6	1.83 ± 1.17	>.05	–
Test group D	Dihydroartemisinin	309	6	2.77 ± 1.36	>.05	–

the compound. Therefore, the formal study used routine 24-h sample time. The mice were i.g. administered with dihydroartemisinin, and the dosage was set with 1/2 LD_{50} as the maximum dose (i.e., 1/2 LD_{50}, 1/4 LD_{50}, 1/8 LD_{50}, and 1/16 LD_{50}). One thousand polychromatic erythrocytes were counted for each animal. The result showed negative under the test conditions (Table 21.1).

3.1.2 Chromosome aberration test with Chinese Hamster Lung cells

Chinese Hamster Lung (CHL) cells were used for the chromosome aberration test. Appropriate doses were decided according to a growth inhibition test (50% inhibition). The cells were treated for 24 and 48 h in the absence of S9 mix (−S9) and 48 h with S9 mix (+S9). One hundred metaphases per dose were observed to calculate the aberration percentage. It revealed a negative result for dihydroartemisinin in chromosome aberration test under the test conditions (Table 21.2).

3.1.3 Ames test

Salmonella typhimurium strains with histidine deficiency, TA97, TA98, TA100, and TA102, were used as the test bacteria. The bacteria were maintained in culture medium after qualified in biological identification. A preliminary study was used to select the solvents and strain TA100−S9 was used to decide the highest dose. The results revealed that a dose of 1000 μg/dish was of slight toxicity, while doses above 1000 μg/dish were of obvious toxicity. Therefore, 1000 μg/dish was used as the highest dose in formal study. Five dose groups, including one positive control and one negative control, were set. Each test was performed both with and without S9 mix and repeated twice. There was no significant difference between each dose group and negative control in the number of revertant colonies. However, the number of revertant colonies in positive control was twofold above the negative control. Therefore, a negative result was obtained for dihydroartemisinin in Ames test.

Table 21.2: Chromosome aberration test results for dihydroartemisinin with CHL cells

Group	Agent	Dosage (μg/mL)	S9 mix	Time (h)	Aberration Rate (%)	Result
Solvent control	DMSO	–	–	24	4	
	DMSO	–	+	24		
Positive control	MMC	0.3125	–	24	41.51	+++
	DMN	0.625	–	24	58.0	+++
		1000	+	24	19.0	+
Test group	Dihydroartemisinin group A	0.406	–	24	6	–
		0.406	–	48	5	–
	Dihydroartemisinin group B	0.8125	+	24	4	–
		0.8125	–	24	6	–
		0.8125	–	48	5	–
	Dihydroartemisinin group C	1.625	+	24	1	–
		1.625	–	24	5.3	–
		1.625	–	48	4	–
	Dihydroartemisinin group D	3.25	+	24	2	–
		3.25	–	24	No metaphase	
		3.25	–	48	No metaphase	
	Dihydroartemisinin group E	6.5	+	24	2	–
		6.5	–	24	No metaphase	
		6.5	–	48	No metaphase	

3.2 Reproductive Toxicity Test [1]

Kunming mice, aged 6–11 weeks and weighing 28–30 g, were used in the study. There were 100 female and 50 male mice. Dihydroartemisinin was suspended with Tween-80 into aqueous solutions at dose levels of 38.44, 9.61, and 2.4 mg/kg. A normal control group was also arranged. Oral route of administration was used. The mice were dosed through the sensitive period of pregnancy, that is, days 6–15.

3.2.1 Effects on mother mice

Body weight of mother mice was detected on days 0, 6, 10, 15, and 18 of pregnancy. Average body weight was calculated. The results revealed that there was significant difference in body weight of dosing pregnant mice, compared with control mice. The average body weight on day 18 of pregnancy was 26.66 g for the large-dose group, while 33.44, 35.59, and 45.65 g for the medium- and low-dose groups and control group, respectively. The P values were all $<.001$ after statistical analysis. It is considered that the difference is caused by absorbed fetus induced by dihydroartemisinin.

3.2.2 Effects on fetal mice

The number of pregnant mice in each group was 16, while 17 in the control group. The number of completely absorbed fetal mice increased with the dosage, 0 for the control group, 7 for the low-dose group (43.75%), 12 for the medium-dose group (75%), and 16 for the large-dose group (100%). Similarly, the incidence of absorbed fetus also increased with the dosage. The overall incidence of absorbed fetus was 16.02% in control group, while 46.75, 82.09, and 100% in the low-, medium-, and large-dose groups, respectively. The live birth rate was inversely proportional to the dosage, with 92.77, 53.25, 17.91, and 0% in the control group, low-, medium-, and large-dose groups, respectively. In low-dose group, the appearance, body weight, and weight of placenta of live fetal mice were normal, except for the absorbed fetuses.

To further investigate if there was abnormity in viscera and soft tissues, the fetal mice soaked in Bouin's solution were prepared with 2-mm-thick sections by hand. The sections were observed by eyes for checking the presence of abnormality in the viscera. No abnormality was observed in trachea, esophagus, heart, liver, stomach, kidney, intestines, bladder, testis, uterus, and fallopian tubes in each dose group. In addition, 11 indices were set for the study of skeleton system in fetal mice, to investigate the skeletal development of fonticulus, occipital bone, cervical vertebrae, thoracic vertebrae, sternum, ribs, lumbar vertebrae, bones of the extremities, tailbone, and phalanx. The results showed that the skeletons did not show any abnormalities in each dose group, compared with the control group. Incomplete ossification was observed in the fonticulus and thoracic vertebrae of fetal mice in each dose group. However, the difference was of no statistical significance compared with the control group.

The aforementioned results indicate that the fetal toxicity of oral dihydroartemisinin in mice is similar with artemisinin toxicity when administered in the sensitive period of pregnancy. The toxicity was manifested as the formation of absorbed fetuses. The number of absorbed fetuses was related to the dosage. The appearance, body weight, and weight of placenta of live fetal mice were normal.

References

[1] Dihydroartemisinin NDA Dossier from Fu Hangyu Research Team of Institute of Chinese Materia Medica China Academy of Chinese Medical Sciences.

[2] Lin Fei, Feng Xinlu, et al. Dihydroartemisinin NDA Dossier.

CHAPTER 22

Clinical Studies of Dihydroartemisinin

Chapter Outline

1 Dihydroartemisinin—Another Class One New Drug Certificate

In 1985, the registration of artemisinin as a new drug was drawing to a close. After serious study and analysis on the structure–activity relationship, Youyou Tu firmly believed that dihydroartemisinin was with high value to be further developed into another new drug, to overcome the shortages of artemisinin, such as relatively large dose and high recrudescence rate, and to be a better new drug for malaria patients. Therefore, she reorganized personnel to complete the 26 studies required by Provisions for New Drug Approval. After another 7-year effort, two class one New Drug Certificates on dihydroartemisinin and its tablets were obtained in 1992. The total dose is 480 mg and recrudescence rate is as low as 1.95%, with 10-fold increase in clinical efficacy. The achieved outcomes further reflected the characteristics of artemisinin-based drugs, such as high efficacy, quick onset of action, low toxicity, and convenience for use. The drug was honored as Chinese "magic drug" after practical use in Africa for years. The achievement was awarded China Top 10 Science and Technology Achievement Awards in 1992 and Health Achievement of New China in 1997 by Ministry of Health of China.

From *Artemisia annua* L. to Artemisinins
http://dx.doi.org/10.1016/B978-0-12-811655-5.00022-2

Original copy

New Drug Certificate Original copy

Code: (92) WeiYaoZhengZi X-66

Institute of Chinese Materia Medica, China Academy of Chinese Medical Sciences:

This is to certify that the new drug, dihydroartemisinin, developed by your institute according to Drug Administration Law of the People's Republic of China, complies with the stipulations of Provisions for New Drug Approval issued by Ministry of Health.

Ministry of Health of the People's Republic of China (Stamped)
July 20, 1992

Original copy

New Drug Certificate

Code: (92) WeiYaoZhengZi X-67

Institute of Chinese Materia Medica, China Academy of Chinese Medical Sciences:

This is to certify that the new drug, dihydroartemisinin tablets, developed by your institute according to Drug Administration Law of the People's Republic of China, complies with the stipulations of Provisions for New Drug Approval issued by Ministry of Health.

Ministry of Health of the People's Republic of China (Stamped)
July 20, 1992

2 Dihydroartemisinin Tablets (Brand Name: COTECXIN Tablets)

2.1 Phase I Pharmacokinetic Study

Dihydroartemisinin is the first effective derivative of artemisinin, a new antimalarial drug. Animal studies reveal that the drug is of low toxicity, high efficacy, and quick absorption. Therefore, the drug is developed into an oral tablet. Before launching into clinical use and large-scale production, a phase I pharmacokinetic study was performed in Xiyuan Hospital, China Academy of Chinese Medical Sciences, as required by Provisions for New Drug Approval.

Six male subjects, aged between 25 and 47 years, were enrolled in the study. The subjects were divided into two groups, orally administrated with 1.1 and 2.2 mg/kg dihydroartemisinin, respectively, with three subjects each group. The physiological and biochemical parameters of the subjects, including alkaline phosphatase and reticulocyte count, were examined strictly. Only the subjects with normal results in all parameters were enrolled in the study.

Blood samples (1–1.5 mL) were collected from upper extremity vein at different times after dosing. The serum was isolated and serum drug concentrations were determined with radioimmunoassay. Urine was also collected to determine the excretion amount and excretion rate in urine.

After oral administration of dihydroartemisinin tablet, the compound was absorbed quickly; peak serum concentration was achieved at 1.33 h after administration, reaching 0.13 and 0.71 μg/mL when administered at 1.1 and 2.2 mg/kg, respectively. MRT were 2.36 and 2.26 h, and $T_{1/2}$ were 1.63 and 1.57 h, respectively.

Little parent compound was excreted in urine after oral administration of dihydroartemisinin tablet. The cumulative excretion in urine within 7 h after dosing was 0.12–0.15 mg, only accounting for 0.10%–0.15% of administered dose. There was no significant difference between the two dose groups.

The results revealed that dihydroartemisinin was absorbed well and appropriate to be dosed orally.

2.2 Phase I Clinical Study

Twent-four healthy adult volunteers were enrolled in the study, with 13 males and 14 females. The subjects were randomized to double-blind treatment. All the volunteers stayed in the hospital for 5 days (days 0–4). They came to the hospital for reexamination on days 7, 14, 21, and 28. ECG, hematology, blood biochemistry, urine, and stool routines were examined on days −4, −2, 1, 3, 7, 14, and 28. For dose arrangement, please refer to Table 22.1. The

Table 22.1: Dose arrangement in phase I clinical trial of dihydroartemisinin

Dosage (mg/kg)					Progressive Increase Rate (%)	Number of Test Subjects	Number of Control Subjects
Total	First Dose	Day 1	Day 2	Day 3			
0.5	0.1	0.2	0.15	0.15		1	0
1.5	0.3	0.6	0.45	0.45	200.0	1	0
3.0	0.6	1.2	0.90	0.90	100.0	2	1
5.0	1.0	2.0	1.50	1.50	66.7	5	2
7.0	1.4	3.0	2.00	2.00	40.0	3	1
9.0	1.8	4.0	2.50	2.50	28.6	3	1
11.0	2.2	5.0	3.00	3.00	18.2	3	1

drug was administered 3 times per day for 3 consecutive days. The first dose was doubled, followed with the remaining doses.

No adverse reactions were found in each dose group. No abnormal changes were found in ECG, hematology, blood biochemistry (including serum proteins, glucose, bilirubin, urea nitrogen, creatinine, cholesterol, triglyceride, potassium, sodium, calcium, phosphorus, nitrogen, alkaline phosphatase, cholinesterase, glutamate pyruvate transaminase, glutamic oxalacetic transaminase), and urine and stool examination in all dose groups, except for reticulocyte. Mean reticulocyte count decreased slightly on days 1 and 3 in 9.0 and 11.0 mg/kg dose groups; however, the fluctuation was within normal range. The reticulocyte count in 11.0 mg/kg decreased to the lower limit of normal range on day 1. Both groups almost returned to the levels before dosing on day 7. It might be considered that the maximum tolerable dose is 11.0 mg/kg (equal to total dose of 550 mg in adults) for 3 consecutive days of dihydroartemisinin dosing in healthy subjects.

2.3 Phase II Clinical Trial

2.3.1 Stage I study

1. Explorative study of dihydroartemisinin dose in the treatment of *Plasmodium falciparum* malaria. Previous studies with artesunate injection and tablet indicate that there is no significant difference in the efficacy against *P. falciparum* malaria between once daily and multiple dosing daily. However, appropriate increase of total dose and extension of treatment duration is a feasible method to improve cure rate and lower recrudescence rate. Therefore, 3- and 5-day treatment, with total doses of 240 and 360 mg, respectively, were used and compared in the explorative study. The drug was administered once daily (Table 22.2). The doses in pediatric patients decreased by age in both groups.
   There were 30 patients with *P. falciparum* malaria, 22 adult patients and 8 pediatric patients in each group, treated with a 3- or 5-day regimen, respectively, with each.
   All the patients were cured. The average fever subsidence time was 18.5 ± 3.9 h and 23.4 ± 15.1 h, respectively. The average PCT was 68.6 ± 11.8 h and 64.2 ± 14.8 h,

Table 22.2: Dosing regimens in two groups

Group	Number of Subjects Observed	Total Dose (mg)	Daily Dose (mg)				
			Day 0	Day 1	Day 2	Day 3	Day 4
3-Day treatment group	30	240	120	60	60	—	—
5-Day treatment group	30	360	120	60	60	60	60

respectively. The recrudescence rates within 28 days were 52.0% in 3-day group and 5.6% in 5-day group. The difference in recrudescence rate was of obvious significance ($P < .01$). Recrudescence rate of 5% was a relatively ideal level of *P. falciparum* malaria treatment at that time. As the sample size was small, an expanded study was required to compare with 7-day regimen.

2. Observation of the parasite clearance rate of dihydroartemisinin. Seven patients with *P. falciparum* malaria were orally administrated with dihydroartemisinin once, at a dose of 120 mg for adult patients, and reduced dose for pediatric patients. The average PC95 was 16 h, which was similar with that of i.v. artesunate and i.m. artemether, but obviously faster than i.v. infusion of quinine dihydrochloride (with an average of 82 h).
3. Comparison of 5- and 7-day regimens in the treatment of *P. falciparum* malaria. Fifty patients with *P. falciparum* malaria were enrolled in each group, treated with 5-day dihydroartemisinin (total dose of 360 mg) regimen, 7-day dihydroartemisinin (total dose of 480 mg) regimen and domestic treatment standard at that time, piperaquine phosphate in the control group, respectively. There were 36, 35, and 37 adult patients and 14, 15, and 14 pediatric patients were enrolled in the 3 treatment groups, respectively. The results revealed that the average fever subsidence time and average PCT were 16.3 ± 6.6 h and 19.7 ± 13.2 h as well as 64.0 ± 12.0 h and 66.5 ± 10.4 h, respectively, for both dihydroartemisinin groups. The differences were of no significance. The recrudescence rate was 6.3% (2/32) for 5-day regimen group and 0% for 7-day regimen group (0/42). In piperaquine phosphate control group, the clinical symptoms were not controlled in 11 patients after treatment. No obvious reduced or even increased parasite density was observed (grade III resistance). Symptoms were controlled while PCT was not observed within 7 days in 13 patients (grade II resistance). Parasite recrudescence was observed within 28 days of treatment in another 10 patients (grade I resistance). Only 17 patients were cured, accounting for 33.3%. It was found that the efficacy of dihydroartemisinin in both 5- and 7-day regimens were both superior to piperaquine phosphate (Table 22.3).

Two patients in 7-day dihydroartemisinin group had mild morbilliform rash without itching and pain. The symptom disappeared spontaneously in one patient about 12 h later. For the other patient, the symptom disappeared on day 9 (2 days after the regimen was completed).

Table 22.3: Efficacy comparison between dihydroartemisinin and piperaquine phosphate

Group	Dosage (mg)	Number of Patients	Fever Subsidence Time (h)	PCT (h)	Recrudescence Rate (%)
Dihydroartemisinin	360	50	16.3 ± 6.6	64.0 ± 12.0	6.3 (2/32)
Dihydroartemisinin	480	50	19.7 ± 13.2	66.5 ± 12.0	0 (0/42)
Piperaquine phosphate	1500	51	37.3 ± 16.3	104.0 ± 16.8	55.26 (21/38)

In piperaquine phosphate group, the 11 patients without PCT within 7 days of treatment and 13 patients without obvious decrease or with increase in parasite density within 48 h of treatment are not included in the table. *P* is <.01 when compared with two dihydroartemisinin groups.

2.3.1.1 Reticulocyte examination

There were 13 (26.0%) and 19 (38.0%) patients in 5- and 7-day dihydroartemisinin groups with slight decrease in reticulocyte count to below normal level 3 days after treatment. On sixth day after treatment, all patients in 5-day regimen group returned to normal, while three patients in 7-day regimen group still had abnormal reticulocyte count while the average value in that group had returned to the level before dosing.

2.3.1.2 ECG examination

One patient in 5-day dihydroartemisinin group ($n = 29$) had occasional atrial premature beats on sixth day after treatment, while in the 7-day regimen group (totally 32), one patient on day 3 and one patient on day 6 after treatment had occasional atrial premature beats, and three patients had grade I atrioventricular block on day 6 after treatment, and one recovered to normal during 14-day reexamination. Another two patients recovered to normal during 1-month reexamination. In piperaquine phosphate group (totally 31 patients), one patient showed frequent atrial premature beats and one patient had occasional atrial premature beats on day 3 and day 6 after treatment, respectively. And one patient had grade II atrioventricular block on day 3 after treatment.

There was no statistically significant difference regarding parasite recrudescence rate between 5- and 7-day dihydroartemisinin groups. However, considering the significant difference between 5- and 3-day regimens, the insignificance between 7- and 5-day regimens might be due to small sample size. In addition, when the total dose was increased to 480 mg with treatment duration extended to 7 days, no obvious toxic effects were observed. Therefore, 7-day regimen was recommended in stage II of phase II clinical trial.

2.3.2 Stage II study

During 1988–1989, 7-day dihydroartemisinin regimen was used to treat 189 patients with *P. falciparum* malaria in Dongfang and Sanya of Hainan Province, with 119 adults and 70 pediatric patients. There were two pregnant patients with gestation of 5–6 months.

1. Efficacy. The clinical symptoms were controlled and parasites were cleared quickly in all the 189 patients. The recrudescence rate was relatively low. For the detail, please refer to Table 22.4. No abortion or other adverse reactions were observed in the two mid-term pregnant patients.
2. Observation of toxic reactions. One patient had morbilliform rash at the chest, abdomen, and both lower extremities on day 2 after treatment. After antiallergic treatment, the rash faded away on day 4. No other adverse reactions were observed.

The reticulocytes were examined in 113 patients. The count in 10 patients (8.8%) decreased to a level slightly lower than the normal range on day 3 after treatment and it returned to the level before dosing on day 6 after treatment. It indicated that transient decrease of reticulocytes was mainly related to doubling of first dose. The thereafter maintaining doses had no obvious influence on reticulocytes.

ECG was examined in 43 patients. Sinus bradycardia was observed in four patients on days 3 and 6 after treatment. Two patients were found with atrial premature beats at hospital admission. Another five patients were found with atrial premature beats on days 3 and 6 after treatment. At stage I of phase II clinical trial, atrial premature beats also occurred in two patients in piperaquine phosphate control group. It was still unclear whether the occurrence or increase of atrial premature beats during treatment was related to the two drugs and further investigation was required.

Routine blood tests were performed for 178 patients on days 0, 3, and 6. Thirty-nine patients were examined for glutamic-pyruvic transaminase, glutamic oxalacetic transaminase, alkaline phosphatase, bilirubin, and creatinine. Urinalysis was conducted for 50 patients. No obvious abnormalities were found after treatment.

Table 22.4: Efficacy of dihydroartemisinin in 189 patients with falciparum malaria

Institution	Sample Size	Fever Subsidence Time ($\bar{X} \pm SD$)	PCT (h) ($\bar{X} \pm SD$)	Recrudescence	
				Sample of Observation	Recrudescence Rate
Guangzhou University of Chinese Medicine	59	19.5 ± 10.2	69.2 ± 16.1	55	1.8
Guangba Farm Hospital	44	24.7 ± 13.0	66.4 ± 9.5	34	2.9
Nanxin Farm Hospital	30	18.8 ± 12.9	65.3 ± 10.0	22	0
Nanan Health Center	46	14.5 ± 7.6	64.9 ± 15.9	44	4.5
Dongfang County Hospital	10	15.2 ± 10.3	66.8 ± 8.8	8	0

2.3.3 Brief summary

1. Phase I clinical trial revealed that the maximum tolerable dose for 3 consecutive days of oral dihydroartemisinin dosing was 11.0 mg/kg in healthy volunteers.
2. A total of 239 patients with *P. falciparum* malaria were treated with dihydroartemisinin with 7-day regimen with total dose of 480 mg. Of which, 154 were adults and 85 were pediatric patients. All the patients were cured. The average fever subsidence time was 14.5–24.7 h. The average PCT was 64.9–69.2 h; 205 patients were followed up for 28 days; 4 were found with recrudescence (1.95%); 80 patients with *P. falciparum* malaria were treated with 5-day regimen with total dose of 360 mg. Of which, 58 were adults and 22 were pediatric patients. The average fever subsidence time and PCT were similar with 7-day regimen, with recrudescence rate of 6.0% (3/50).
3. During the treatment with 5-day (total dose of 360 mg) and 7-day (total dose of 480 mg) dihydroartemisinin regimens, 104 patients had ECG examination. Of which, 2 (1.9%) had atrial premature beats and 1 (1.0%) had grade I atrioventricular block before treatment, 4 (3.8%) had atrial premature beats, 1 (1.0%) had ventricular premature beats, and 3 (2.9%) had grade I atrioventricular block after treatment. Of the three patients with atrioventricular block, one patient was found normal during the reexamination on day 14, and the other two patients returned to normal 1 month later. However, phase I clinical trial of dihydroartemisinin revealed that the 3-day regimen with total dose of 550 mg in healthy volunteers was equal to 2.29-fold the dose of 3-day, 1.53-fold the dose of 5-day, and 1.15-fold the dose of 7-day in phase II clinical trial. No adverse effects on the heart were found. In addition, changes related to heart were also found in piperaquine phosphate control group. Of which, two patients (6.5%) were found with atrial premature beats and one patient (3.2%) was found with grade I atrioventricular block. Considering that abnormal changes of heart were found in both dihydroartemisinin and control groups before treatment and there were no obvious changes in heart in healthy volunteers administrated with dihydroartemisinin, it does suggest that the disease itself might induce the abnormal change in heart or dihydroartemisinin has certain influence on heart in malaria. These issues will be further investigated in the expanded study of dihydroartemisinin.
4. The total doses of 3-, 5-, and 7-day dihydroartemisinin regimens were 240, 360, and 480 mg, respectively, and in total 349 patients with *P. falciparum* malaria were enrolled. Three patients (0.9%) had rash during the treatment. Blood biochemical tests were performed for 96 patients. Urinalysis was conducted for 80 patients. No abnormal changes were found after treatment; 162 patients in 7-day regimen group were examined for reticulocytes. Of which, 29 (17.9%) patients had slight decrease in reticulocyte count to below the lower limit of normal range on day 3 after treatment, and the count almost returned to normal on day 6 after treatment. It was found that the toxic reactions of dihydroartemisinin were mild in 7-day regimen with total dose of 480 mg and the dose was relatively safe.

5. Dihydroartemisinin was convenient for use. It was an antimalarial drug with high efficacy and quick onset of action. For the *P. falciparum* malaria with resistance to chloroquine and piperaquine, dihydroartemisinin had favorable efficacy. Based on the evaluation on the efficacy and toxic reactions, 7-day dihydroartemisinin regimen with total dose of 480 mg might be used as recommended clinical regimen.

Based on current situation of *P. falciparum* malaria treatment in the world, it was thought that dihydroartemisinin was a relatively ideal drug for oral administration.

2.4 Phase III Clinical Trial

Relatively favorable efficacy was achieved in dihydroartemisinin phase II clinical trial with 7-day regimen with total dose of 480 mg in the treatment of *P. falciparum* malaria in 239 patients. To further investigate the efficacy, adverse reactions and indications of dihydroartemisinin at an expanded scale, a phase III clinical trial was conducted for confirmative evaluation of the safety and efficacy of the product. The study was organized by Science & Technology Development Center, Chinese Medical Association. It was conducted in 19 hospitals from Hainan Province, Yunnan Province, Guangzhou, and the south of Vietnam, including Tropical Medicine Institute, Guangzhou University of Chinese Medicine, Hainan Provincial Institute of Tropical Diseases, Beijing Tropical Medicine Research Institute, and Yunnan Institute of Malaria Control. In total, 585 malaria patients were treated with dihydroartemisinin using 5-day regimen with total dose of 480 mg, or 7-day regimen with total dose of 480 mg, or 7-day regimen with total dose of 640 mg. Of which, 442 patients were with *P. falciparum* malaria, 142 patients were with *Plasmodium vivax* malaria, and 1 patient was with the combination of *P. falciparum* and *P. vivax* malaria. Artesunate and Fansimef were used as control agents. The data are summarized as follows.

2.4.1 Population and methods

1. Population: There were 585 malaria patients, 411 males, and 174 females, with an age of 2–73 years; 136 patients were <16 years and 449 were ≥ 16 years. The time after disease onset ranged from 1 day to 2 months. There were 217 patients infected with malaria for the first time and 360 patients with previous malaria history. The malaria history of 8 patients was unknown; 96 patients were in the subsiding stage of fever at hospital admission, and 489 patients were with various degrees of fever. All the patients were found with asexual parasites in peripheral blood examination at hospital admission. The parasite density was 37–274, 105 μL^{-1}; 442 patients were infected with *P. falciparum* malaria. Of whom, 59 patients were from Hainan province, 54 patients from Yunnan province, and 29 patients from Guangdong province. One patient, from Hainan province, was infected with *P. falciparum* malaria in combination with *P. Vavex* malaria. Of the 585 patients, 22 patients had a parasite density over 10×10^4 μL^{-1}, and 1 patient was associated with blackwater fever, and 2 patients were with mental confusion.

2. Dosing regimen: Each dihydroartemisinin tablet contained 20 mg, provided by Beijing COTEXIN Technology Corporation and Beijing Sixth Pharmaceutical Factory. The study included 5-day regimen with total dose of 480 mg, 7-day regimen with total dose of 480 mg, or 7-day regimen with total dose of 640 mg. The first dose was doubled and followed by the maintenance dose. The batch number of the drug was 931,103.
3. Control study: The control drugs were artesunate and Fansimef. Artesunate tablet was given through a 5-day regimen with total dose of 600 mg, that is, daily oral dosing of 100 mg plus the first double dose, and a 7-day regimen with total dose of 640 mg, that is, daily oral dosing of 80 mg plus the first double dose. Artesunate tablets were provided by Guilin Pharmaceutical Co., Ltd, and the batch number was 940,102. Fansimef was taken three tablets as a single dose for the adults. Each tablet contained 250 mg of mefloquine, 500 mg of sulfadoxine, and 25 mg of pyrimethamine. The total dose of three tablets was 2325 mg. Fansimef was produced by F. Hoffmann-La Roche Ltd, and the batch number was 2C358/27.
4. Efficacy evaluation: The efficacy was assessed with reference to the 4-week observation method recommended by WHO in 1973 for chloroquine sensitivity.
   a. S (cured): Temperature gradually returned to normal after dosing. No asexual parasites were found on day 7. No recrudescence occurred within 28 days.
   b. R I (recrudescence): Temperature gradually returned to normal after dosing. No asexual parasites were found on day 7. However, recrudescence occurred within 28 days.
   c. R II (No PCT within 7 days): Parasitemia decreased ≥15% within 48 h of dosing; however, no PCT was found on day 7.
   d. R III (Ineffective): Parasitemia decreased <15% or increased within 48 h of dosing.

2.4.2 Results

1. Efficacy: Totally 585 malaria patients were treated with dihydroartemisinin using 5-day regimen or 7-day regimens. Of which, 442 patients were with *P. falciparum* malaria (including 2 patients with mental confusion, 1 patient with blackwater fever, and 22 patients with high-density parasitemia) and 142 patients were with *P. vivax* malaria. All the patients were cured. The average fever subsidence time was 9.3–34.8 h and the average PCT was 18.7–65.9 h. It indicated that clinical symptoms were controlled and parasites in blood were cleared promptly by dihydroartemisinin. The product had better efficacy for *P. falciparum* malaria.

   One hundred seventy-six patients were treated with a 7-day regimen with total dose of 480 mg. The recrudescence rate was 6.8% (the data for the 63 patients observed in Vietnam were 2.3%); 20 patients in the multidrug-resistant area of Vietnam were treated with a 7-day regimen with total dose of 640 mg. No recrudescence was found. It revealed that 7-day regimens of dihyroartemisinin had a relatively lower recrudescence rate than

5-day regimen in the treatment of *P. falciparum* malaria, and dose increase might achieve an even lower recrudescence rate.

2. Toxic and adverse reactions: Of the 585 patients treated with dihydroartemisinin, 1 patient (0.17%) had rash. However, it did not affect the completion of treatment. No obvious adverse reactions related to clinical symptoms were found in the remaining patients. There was no obvious influence on hematology, reticulocytes, glutamic-pyruvic transaminase (GPT), and serum creatinine or urea nitrogen. Three patients had elevated GPT after treatment. As that malaria might induce the elevation of GPT due to liver impairment, it was still unable to identify whether the reaction was related to the product. Two patients (0.3%) had grade I atrioventricular block after treatment. It was considered that this reaction was possibly related to the product to certain extent when combined with the data from phase II clinical trial. Nevertheless, the incidence was extremely low and was of little clinical significance.
3. Efficacy comparison with positive controls of artesunate and Fansimef: In 1994, 30 patients were treated with 7-day dihydroartemisinin regimen with total dose of 480 mg, 7-day artesunate regimen with total dose of 640 mg, and Fansimef as a single dose with total dose of 2325 mg, respectively. The patients were from two hospitals in southern Vietnam, Xuan Loc Hospital and Qui Nhon Hospital. The results were averaged and compared.

It revealed that dihydroartemisinin was an antimalarial drug with quick onset of action and was able to control the clinical symptoms of *P. falciparum* malaria quickly. The average fever subsidence time was about 26 h, which was similar with artesunate ($P > .05$) but obviously shorter than Fansimef (38 h, $P < .01$). The PCT of dihydroartemisinin was about 63 h, which was similar with artesunate ($P > 0.05$) but obviously shorter than Fansimef (75 h, $P < 0.01$).

The cure rates of the three groups were similar. During the treatment with Fansimef, R I, RII, and RIII resistance each occurred in 1 patient; 4 patients each in dihydroartemisinin and artesunate groups had R I resistance, but without R II and R III resistance. It indicated that 7-day dihydroartemisinin regimen with total dose of 480 mg was similar with 7-day artesunate regimen with total dose of 640 mg in efficacy in the treatment of *P. falciparum* malaria in southern Vietnam, and the short-term efficacy was superior to Fansimef.

The incidence of adverse reactions in both dihydroartemisinin and artesunate groups was lower than that in Fansimef group during treatment. The 7-day dihydroartemisinin regimen with total dose of 480 mg was well tolerated in patients.

In conclusion, dihydroartemisinin is convenient for use (once daily) and with definite efficacy and low adverse effects. It is a relatively ideal choice as an oral antimalarial agent. Dihydroartemisinin is recommended to be supplied as a 7-day regimen with total dose of 480 mg.

2.5 Package Insert of Dihydroartemisinin Tablets

Generic name: Dihydroartemisinin tablets

English name: Dihydroartemisinin tablets

Chinese Pinyin: shuāngqingqinghāosupiàn

Active ingredient of the product and its chemical name: dihydroartemisinin

The chemical name is octahydro-3,6,9-trimethyl-3,12-epoxy-12H-pyrano[4,3-*j*]-1,2-benzodioxepin-10(3H)ol

Structural formula:

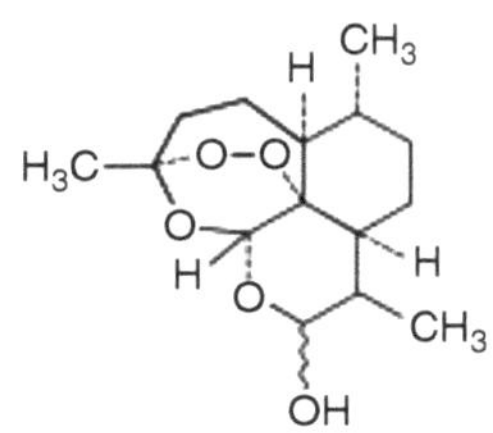

Molecular formula: $C_{15}H_{24}O_5$

Relative molecular weight: 284.35

- **Appearance**
  White tablets.
- **Pharmacology and toxicology**
  The product has powerful antimalarial effect against the asexual parasites. It is able to control the symptoms and clear the parasites promptly. The product has similar efficacy toward chloroquine- and piperaquine-resistant *P. falciparum* malaria. It is of relatively low toxicity. With regard to reproductive toxicity, the product might increase the incidence of absorbed fetuses when administered to mice during the sensitive period of pregnancy; however, no teratogenic effect was observed.
- **Pharmacokinetics**
  The product is absorbed well and has quick onset of action after oral administration. Peak plasma concentration was achieved at 1.33 h after oral administration of 2 mg/kg dihydroartemisinin in human. The peak level was 0.71 μg/mL and serum $T_{1/2}$ was 1.57 h. The characteristics of the product in vivo are quick absorption, extensive distribution, and quick excretion and metabolism.
- **Indications**
  The product is indicated for the treatment of various malaria, especially for chloroquine- and piperaquine-resistant *P. falciparum* malaria and deadly cerebral malaria.

- **Dosage and administration**
  The product should be taken orally. Dosage for adults: dosing should be taken once daily (60 mg) for 5 or 7 consecutive days with first dose doubled. Dosage for pediatric patients: the dose should be decreased by age.
- **Adverse reactions**
  No obvious adverse reactions were observed at the recommended clinical dose. A small number of patients were observed with transient slight decrease of reticulocytes.
- **Contraindications**
  The product should be used with caution in pregnant patients.
- **Precautions**
  The product should be kept appropriately in a refrigerator.
- **Strength**
  20 mg/tablet.
- **Shelf life**
  The tentative shelf life is 2 years.
- **Storage**
  Store in a cool, dark place, and sealed in a container.

3 Dihydroartemisinin Suppositories

In consideration of the characteristics of high efficacy and low dose, dihydroartemisinin is appropriate to be developed into suppositories for the malasia patients who are inconvenient to take oral tablets, such as infants and coma patients. Therefore, the suppository form of dihydroartemisinin is developed after tablets. New drug certificate has been issued and the use of the suppository is briefly described in the certificate.

Generic name: Dihydroartemisinin suppositories

English name: Dihydroartemisinin suppositories

Hanyu Pinyin: shuāngqingqinghāosushuān

Active ingredient of the product and its chemical name: dihydroartemisinin

The chemical name is octahydro-3,6,9-trimethyl-3,12-epoxy-12H-pyrano[4,3-*j*]-1,2-benzodioxepin-10(3H)ol

Structural formula:

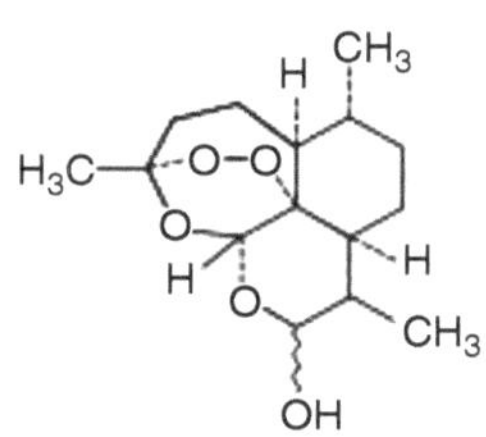

Molecular formula: $C_{15}H_{24}O_5$

Relative molecular weight: 284.35

- **Appearance**
  White bullet shaped suppository.
- **Pharmacology and Toxicology**
  The product has powerful antimalarial effect against the asexual parasites. It is able to control and clear the parasites promptly. The product has similar efficacy toward chloroquine- and piperaquine-resistant *P. falciparum* malaria. It is of relatively low toxicity.
- **Pharmacokinetics**
  The product is absorbed well, has quick onset of action and metabolism after administration. Peak plasma concentration was achieved at 7 h after dosing. The average MRT was 6.96 h in human, and $T_{1/2}$ was 4.82 h.
- **Indications**
  The product is indicated for the treatment of various malaria, especially for chloroquine- and piperaquine-resistant *P. falciparum* malaria and deadly cerebral malaria. The product is convenient for the use in infants, critically ill patients with vomiting or in coma.
- **Dosage and Administration**
  Dosing should be taken once daily (60 mg) for 5 or 7 consecutive days (with total dose of 360 mg or 480 mg) with first dose doubled.
- **Adverse Reactions**
  No obvious adverse reactions were observed at the recommended clinical dose. A small number of patients were observed with transient slight decrease of reticulocytes.
- **Contraindications**
  The product should be used with caution in pregnant patients.
- **Precautions**
  The product should be kept appropriately in a refrigerator.
- **Pediatric Use**

	Total Dose (mg)		Daily Dose (mg)	
Age	5 Days	7 Days	First Day	Days 1–6
11–15	240	320	80	40
7–10	180	240	60	30
4–6	120	120	40	20
1–3	90	120	30	15

- **Strength**
  100 mg/suppository, 200 mg/suppository, 60 mg/suppository.

- **Shelf Life**
  The tentative shelf life is 2 years.
- **Storage**
  Store in a cool, dark place, and sealed in a container.

Acknowledgments

We would like to express our thanks to the assistance of the following institutions in dihydroartemisinin tablet-related studies:

Phase I pharmacokinetic study and phase I clinical trial were both contracted by Institute of Chinese Materia Medica, China Academy of Chinese Medical Sciences and conducted by Xiyuan Hospital, China Academy of Chinese Medical Sciences and Pharmacology Laboratory of Institute of Materia Medica, Chinese Academy of Medical Sciences.

Phase II clinical trial was contracted by Institute of Chinese Materia Medica, China Academy of Chinese Medical Sciences and conducted b y Tropical Medicine Institute, Guangzhou University of Chinese Medicine.

The test agents used in Phase III clinical trial were provided by Beijing Sixth Pharmaceutical Factory.

PART 4

Advancement in Research on Artemisinins

CHAPTER 23

Artemisinins

Chapter Outline

In the aforementioned chapters, newly marketed antimalarial drugs, artemisinin, and dihydroartemisinin that have been developed by the Institute of Chinese Material Medica, China Academy of Chinese Medical Sciences are described in details. This chapter will present a brief introduction to other marketed artemisinin-like drugs which have been developed based on previously mentioned mechanism.

1 Artesunate [1–5]

Chinese name: 青蒿琥酯

English name: Artesunate

Chemical structure:

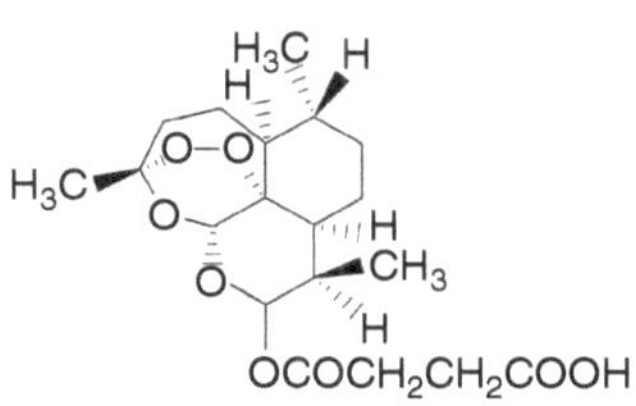

Chemical name: (3R,5aS,6R,8aS,9R,10S,12R,12aR)-decahydro-3,6,9-trimethyl-3,12-epoxy-12H-pyrano[4,3-*j*]-1,2-benzodioxepin-10-ol, hydrogen succinate

Molecular formula: $C_{19}H_{28}O_8$

Relative molecular mass: 384.43

http://dx.doi.org/10.1016/B978-0-12-811655-5.00023-4

- **Synthetic Method**
  Dissolve artemisinin in methanol, cool, add sodium borohydride, and neutralize with acetic acid after the end of the reaction. Add saturated brine into reaction mixture to precipitate crystals, separate, and dry to obtain dihydroartemisinin. The dihydroartemisinin reacts with succinic anhydride in chloroform in the presence of pyridine and then neutralized with dilute hydrochloric acid. Afterward, the chloroform layer is separated from the reaction mixture, and then the solvent is recovered to yield artesunate.
- **Antimalarial Activity**
  In vitro EC_{50} and ED_{90} values for artesunate in the treatment of human *Plasmodium falciparum* FCC-1 strain were 0.14 and 1.14 ng/L. In vitro, six isolates of *P. falciparum* (including chloroquine-resistant strains) were sensitive to artesunate, and there was no significant difference among strains. Compared with five commonly used antimalarial drugs (including chloroquine and piperaquine), the artesunate was effective at the lowest concentration of 0.018 μmol/L for complete inhibition of maturation of *P. falciparum* schizont.
  SD_{50} and SD_9 values of artesunate that was intravenously administered for the treatment of mice infected with sensitive strains of *Plasmodium berghei* were 0.97 ± 0.19 and 5.25 ± 0.25 mg/(kg·day); SD_{50} and SD_9 values of chloroquine-resistant strains were 15.65 ± 7.4 and 90.7 ± 33.25 mg/(kg·day). After intravenous injection of this product at a dose of 3.16 mg/(kg·day) or quinine dihydrochloride at a dose of 31.6 mg/(kg·day) for 7 consecutive days in the treatment of monkeys treated with *Plasmodium knowlesi*, the plasmodium parasite rate has dropped by 90% 13.2 or 50.5 h after administration, and the negative conversion time was 40 or 104 h.
- **Toxicology**
  LD_{50} of artesunate that was administered intravenously in mice was 552.82 mg/kg; and oral LD_{50} was 1409.0 ± 44.4 mg/kg. Intravenously chemotherapeutic index was 792.8, and the safety index was 79.9. A single intravenous injection of artesunate at a dose of 33 mg/kg was given in dogs, and no any side effects were seen.
  There is no significant effect of artesunate on the heart rate and coronary blood flow after administration in dogs and monkeys. The cardiovascular toxicity of artesunate was only 1/53.6 of that of chloroquine. After intravenous injection of artesunate at a dose of 11.25 kg/kg was administered in dogs once a day for 14 consecutive days, no significant side effects were observed. Reduction in blood reticulocytes is the most sensitive indicator for side effect of this drug. In Beagle dogs which were orally given with artesunate once a day for 28 consecutive days, the toxic dose was 82.5 mg/kg, and the safe dose was 15 mg/(kg·day).
  Artesunate was toxic to rat embryo, but had no effect on reproductive function in male rats. Artesunate might cause miscarriage in Hamsters and guinea pigs and embryo absorption in mice and rabbits. When the dose of 1/809 LD_{50} (0.5 mg/kg) was completely

harmless to parent body, there were still 32.8% of fetal rats absorbed, indicating that embryos were significantly more sensitive to artesunate than female rats.

In Ames test, micronucleus test, and mammalian cell chromosome aberration test, no genetic toxicity of artesunate was observed.

Artesunate had no significant effect on the nervous system of mice when it was intravenously administered at dose of 40 times ED_{90} for anti-*P. berghei*. Artesunate also had no effect on the nervous system of guinea pigs, rabbits, dogs, and monkeys when it was intravenously administered at a dose of 16 times ED_{90}. It had no significant effect on the breathing, blood pressure, heart rate, and ECG when it was intravenously administered 32 times ED_{90} in rabbits and dogs.

- **Pharmacokinetics**

  Artesunate was rapidly absorbed in the human body, and the peak serum concentration was reached in 1.5, 2, and 0.5 h after oral, rectal, and intramuscular injection, which was almost completely converted to the product with antimalarial activity—dihydroartemisinin in the metabolic process. Artesunate was also eliminated quickly, the antimalarial activity of which depended on the elimination rate of dihydroartemisinin ($T_{1/2}$ = 45 min). In thin-layer chromatography and immunoassays of artesunate in dogs, in vivo pharmacokinetic parameters were fitted to one-compartment model. The blood concentration was maintained longer after skin liniments of artesunate were applied to rabbits. Clinically, the concentration time curve was fitted to two-compartment model after intravenous injection of artesunate at a dose of 2.0 and 3.8 mg/kg, and the elimination half-life was 0.5 and 0.6–0.8 h, the steady-state volume of distribution was 0.1–0.6 and 0.3–0.4 L/kg, and the cumulative urinary drug amount was only 0.1%–6.8% of the total dose 7 h after administration.

- **Clinical Research**

  A total of 288 cases of falciparum malaria were treated intravenously with artesunate for 4 times within 3 days (60 mg/time, a total dose of 240 mg), and the fever clearance time was 14.6 ± 27.0 h, the negative conversion time was 32.3 ± 58.8 h, and the recrudescence rate was 51.3% on day 28. Another 40 cases were treated with 7-day regimen with the total dose of 480 mg, and the recrudescence rate dropped to 5.6%. Artesunate was administered by intramuscular injection into 30, 100, and 50 cases of falciparum malaria, respectively, at an initial dose of 120 mg, followed by a total dose of 240 mg within 3 days, a total dose of 360 mg within 5 days, and a total of dose of 480 mg within 7 days, and the mean fever clearance time of three doses was 19.5 h, and the mean negative conversion time was 64.1 h, and the recrudescence rate was 52% in 3-day group, 9.8% in 5-day group, and 2.5% in 7-day group.

  In drug-resistant falciparum malaria endemic area in Hainan, an open and randomized comparison was conducted in patients with uncomplicated falciparum malaria receiving artesunate tablets for 5 days (total 600 mg) and 7 days (total 800 mg). Clinical symptoms were quickly controlled in two groups (n = 50). The fever clearance time was 24.0 ± 13.6 and 20.0 ± 10.0 h in 5- and 7-day group, respectively; the negative conversion time of plasmodium was 61.7 ± 10.1 h and 55.8 ± 15.8 h, respectively, and there was no significant difference between the two groups. The 28-day plasmodium recrudescence rate was 4.2% in the 7-day group and 19.1% in the 5-day group, and there was no adverse reaction between the two groups. From April to June in 1993, 50 uncomplicated falciparum malaria patients without cerebral symptoms were randomly assigned into two treatment groups in Tanzania, of which the artesunate group received the treatment for 5 days with a total dose of 550 mg, and the intravenous artesunate group received the treatment for 4 days with a total dose of 7.2 mg/kg. The results showed that the clearance time of plasmodium and the fever clearance time after oral administration were consistent with those after intravenous administration. The recrudescence rate was 20 and 16%, respectively, after oral and intravenous administration.

A total of 30 cases of severe falciparum malaria were treated with 200 mg artesunate suppositories at 0, 4, 8, 12, 36, 48, and 60 h followed by 750 and 50 mg mefloquine tablets at 72 and 84 h, respectively, by Looareesuwans et al. The parasitologic cure rate at 28 days was 92%, and no adverse reaction was observed in the whole process.

- **Adverse Reaction**
  No adverse reactions were observed with recommended dose up to now. Transient reticulocytopenia may occur in peripheral blood when overdose of artesunate injection (more than 2.75 mg/kg) is given.
- **Indication**
  It is used to treat falciparum malaria and tertian malaria. The injection is suitable for rescue of patients with cerebral malaria and severe malarias.
- **Precautions**
  - The solution should be used immediately after the powder is dissolved. It should not be used for intravenous infusion if the solution appears cloudy or sediment occurs.
  - Doubling the initial dose is recommended for patients with serious infection.
  - Patients with malaria should be cured with other antimalarial drugs after their symptoms are controlled.
  - The speed for intravenous infusion is 3–4 mL/min.
- **Usage and Dosage**
  IV: Dissolve in 5% sodium bicarbonate injection which is supplied in the product package before use, and then add 5% glucose injection or 5.4 mL glucose sodium chloride solution to make the concentration of artesunate in 10 mg/mL for slow intravenous infusion, 60 mg/time (or 1.2 mg/kg), children <7 years: 1.5 mg/kg, double the initial dose, and repeat injection once at 4, 24, and 48 h. The initial dose is 120 mg for extremely severe cases, and a course of 3 consecutive days with a total dose of 240–300 mg for slow administration is recommended.

- **Formulation**
    - Artesunate tablets: 50 mg or 200 mg/tablet.
    - Intramuscular or intravenous injection of artesunate: Each ampoule contains 60 mg artesunate powder and 5% sodium bicarbonate solution, respectively.
    - Each suppository contains 100 or 400 mg sodium artesunate.
- **R&D Facilities**
  Guilin Pharmaceutical Factory in Guangxi, China.

2 Artemether [6–8]

Chinese name: 蒿甲醚

English name: Artemether

Chemical structure:

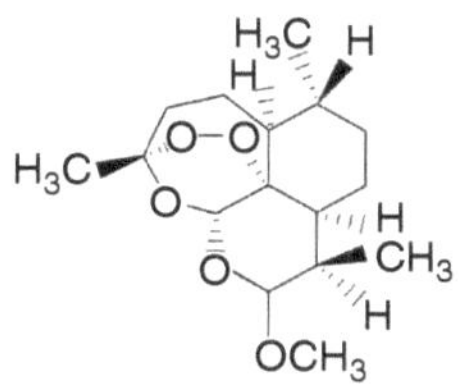

Chemical name: [3R-(3R,5aS,6S,8aS,9R,10R,12S,12aR)]-decahydro-10-methoxy-3,6,9-trimethyl-3,12-epoxy-12H-pyrano[4,3-*j*]-1, 2-benzodioxepin

Molecular formula: $C_{16}H_{26}O_5$

Relative molecular mass: 298.4

- **Synthetic Method**
  Artemisinin is reacted with sodium borohydride to produce dihydroartemisinin as a mixture. The mixture is treated with methanol and an acid catalyst to provide a pair of epimers: β-dihydroartemisinin methyl ether which was predominant, accounting for 98%, and α-dihydroartemisinin methyl ether accounting for 2%.
- **Antimalarial Activity**
  Artemether exhibits strong killing effect against blood schizont of in vitro cultured chloroquine-sensitive strains and chloroquine-resistant strains of *P. berghei*, *Plasmodium inui* (*Plasmodium cynomolgi*), and *P. falciparum*. The negative conversion time was significantly better in mice infected with plasmodium receiving artemether than that in those receiving oral chloroquine phosphate. In vitro EC_{50} and ED_{90} values for artemether in the treatment of human plasmodium falciparum FCC-1 strains were 2.19

and 4.12 ng/L. SD_{50}, ED_{50}, and SD_{90} values of artesunate that was intramuscularly administered for the treatment of mice infected with strains of *P. berghei* were 0.6, 0.72, and 1 mg/(kg·day); in cynomolgus monkeys with malaria that were treated with artemether oil-based injection via intramuscular administrated, the number of plasmodium can be reduced by more than 90% within 1 day, and the negative conversion was achieved within 2 days.

- **Toxicology**
  In mice, LD_{50} for artemether oil-based injection was 263 mg/kg when administered intramuscularly and 391 ± 60) mg/kg when administered subcutaneously, and LD_{50} for artemether suspension was 977 ± 114 mg/kg when administrated via oral gavage. LD_{50} for intramuscular injection of artemether oil-based solution was 579 mg/kg in rats, the therapeutic index was about 493, and the safety factor was about 138. After artemether was injected intramuscularly once a day into rats at a dose of 36 mg/kg, for 15 consecutive days, no toxicity was observed except slightly decrease in reticulocytes. After artemether was injected intramuscularly once a day into dogs at a dose of 6 mg/kg, for 15 consecutive days, no toxicity was observed. When the dose was increased to 10 mg/(kg·day), loss of appetite and other gastrointestinal reactions occurred.
  Some animal studies have shown that intramuscular injection of large doses of artemether or arteether may cause neurotoxicity which was manifested by scattered chromatin dissolution and necrosis in some brain stem and cerebellar nuclei, leading to depression, ataxia, spontaneous muscle spasm, and even death.
  The negative results of Ames test, mouse micronucleus test, and V79 cell chromosome aberration test results indicated that this product had no mutagenic effect. No teratogenic effects were observed in mice, rats, and rabbits in the early stage of pregnancy. However, the embryo toxicity was observed in mice in the mid-trimester stage of pregnancy when the dose was greater than 1/28 of LD_{50}.
- **Pharmacokinetics**
  Animal experiments showed that the absorption, transformation, and excretion of artemether were relatively faster and its half-life was about 17 min in rats and 9.8 h in monkeys. It is widely distributed in animals, predominantly in liver, lung, adrenal gland, and kidney.
  The peak plasma concentrations of oral artemether was reached within 2–3 h. The absorption of artemether intramuscular injection can vary greatly from individual to individual. Artemether is absorbed slowly and unstably, and its peak plasma concentration is usually achieved within 6 h, 18 h, or even longer. Artemether is slowly and incompletely metabolized in vivo to produce active product—dihydroartemisinin. Artemether plays a leading role when intramuscularly injected, while dihydroartemisinin plays a major role when administered orally.

- **Clinical Research**
  In chloroquine-resistant falciparum malaria endemic area, 308 patients with falciparum malaria were treated with artemether intramuscular injection, including 13 cases of pernicious malaria and 8 cases of pregnant women. All patients were clinically cured. The mean fever clearance and negative conversion time of plasmodium was 20–30 and 32–72 h, significantly better than oral chloroquine phosphate. The recrudescence rate was about 6.7% within 1 month.
- **Adverse Reaction**
  Artemether was injected intramuscularly into 12 healthy adult volunteers 5 times daily for 5 days, and the initial dose was doubled. The maximum tolerated dose was 1.6 mg/(kg·times) and the total dose was 9.6 mg/kg. In terms of toxicity, only transient reduction in reticulocytes was observed but returned to normal 1 week after discontinuation.
  The toxic and side effects were observed in 204 cases receiving 5-day intramuscular injection of artemether with a total dose of 480 mg, including 4 cases of transient fever after medication and 8 cases of transient and slight reduction in reticulocytes. Eight pregnant women were treated with artemether in mid- and last-trimester. During 5–8 years of follow-up, mothers were healthy, and their children grew and developed normally.
- **Indication**
  It is used to treat falciparum malaria and tertian malaria, especially suitable for critically ill patients with malaria, including chloroquine-resistant malaria and cerebral malaria.
- **Precautions**
  - If solidification occurs at low temperature, Artemether injection may be warmly dissolved before use.
  - As its embryo-fetotoxic effects were observed in animal experiments, this product should be used with caution in pregnant woman.
- **Usage and Dosage**
  Oral administration. Adults: the dose is 160 mg on the first day followed by 80 mg from the second day, once a day for 5–7 consecutive days. Children: the dose is 3.2 mg/kg on the first day followed by 1.6 mg/kg from the second day, once a day for 5 consecutive days. Intramuscular injection: once a day for 5 consecutive days as a course of treatment; for adults: 80 mg daily, the initial dose is double; for children: 1.6 mg/kg.
- **Formulation**
  Capsules: Each contains 40 mg artemether.
  Tablets: Each contains 50 mg artemether.
  Intramuscular injection containing 80 mg of artemether for adults or 40 mg of artemether in 1 mL sterile injectable oil for pediatric use.
- **R&D Facilities**
  Shanghai Institute of Materia Medica, Chinese Academy of Sciences and Kunming Pharmaceutical Factory in Yunnan, China.

3 Arteether [9–11]

Chinese name: 蒿乙醚

English name: Artemotol, arteether

Chemical structure:

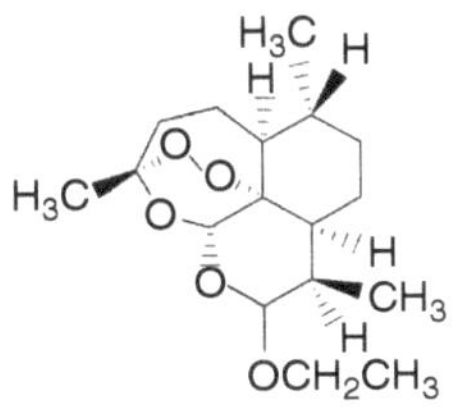

Chemical name: [3R-(3R,5aS,6S,8aS,9R,10R,12S,12aR)]-decahydro-10-ethoxy-3,6,9-trimethyl-3,12-epoxy-12H-pyrano[4, 3-*j*]-1, 2-benzodioxepin-10-ol

Molecular formula: $C_{17}H_{28}O_5$

Relative molecular mass: 312.4

- **Synthetic Method**
  Artemisinin is reacted with sodium borohydride to produce dihydroartemisinin as a mixture. The mixture is treated with ethanol and an acid catalyst to provide a pair of epimers: β-dihydroartemisinin ethyl ether and α-dihydroartemisinin ethyl ether. β-Dihydroartemisinin ethyl ether is separated from the mixture by chromatographic method.
- **Antimalarial Activity**
  This product that is a semisynthetic derivative of artemisinin is a DNA synthesis inhibitor. This product is an antiplasmodium infection drug which can kill blood schizont of falciparum malaria plasmodium. It can be used to effectively kill gametocytes of plasmodium falciparum in early stages. Furthermore, the results of in vitro studies show that this product is an effective agent against strains of plasmodium falciparum which are resistant to chloroquine, mefloquine, halofantrine, quinine, pyrimethamine, cycloguanil, and amodiaquine.
- **Toxicology**
  Five of six dogs died after receiving intramuscular injection of Arteether [20 mg/(kg·day)] for 8 consecutive days. Main adverse events included nerve damage that was dose related and site specific in located pons and medulla. In clinical trials, the cumulative dose of this product was 11.2 mg/kg, and no neurotoxicity was reported in healthy volunteers and patients.

- **Pharmacokinetics**
  The sesame oil solution of this product is slowly released into the systemic circulation after intramuscular injection. The peak plasma concentration was reached within 3–12 h after injection, and its plasma half-life was 1–2 days, and the plasma protein binding rate was 98–99%.
  This product is mainly dealkylated into dihydroartemisinin after CYP3A4-mediated oxidation. The dihydroartemisinin after glucuronidation is mainly excreted from biliary ducts, while only a very small amount (20%–30%) was excreted in the urine in the form of dihydroartemisinin glucuronide.
- **Clinical Research**
  In a positive-controlled, randomized, double-masking, parallel-group, and two-center clinical study that was conducted in 127 children with severe or complicated falciparum malaria, the survival rate was comparable between the treatment group and artemether group. After continuous treatment, the parasites were completely removed from the body and the fever subsided within 1–4 days in all patients.
  In an open multicenter and randomized parallel-group clinical study, 200 children patients with cerebral falciparum malaria received intravenous arteether or quinine injection. The survival rate was comparable between the treatment group (82%) and control group (72%). Patients woke up from a coma 4 h to 8 days after treatment, and the plasmodium was completely cleared while the fever subsided within 1–7 days.
  In five small open randomized clinical studies involving 200 adult patients with severe falciparum malaria, the survival rate was similar to that of the artemether treatment group, and the plasmodium was completely removed within 60 h.
- **Adverse Reaction**
  Adverse reactions that have been reported include weakness, loss of appetite, fever, and headache, pain at the injection site, upper respiratory tract infection, pneumonia, cough, dizziness, diarrhea, nausea, itching, and myalgia. Prolonged Q-Tc occasionally occurred in patients who showed ECG abnormalities before treatment.
- **Indication**
  It is used to treat falciparum malaria plasmodium infections.
- **Precaution**
  - Patients who are allergic to this product or sesame should not use this product, or those with liver or renal impairment should be treated with this product with caution.
  - This product must be administered via intramuscular route. In general, patients with severe infection with plasmodium falciparum also need adjuvant therapy, including intravenous or rectal administration of glucose—saline solution and acetaminophen. Within 4 weeks following falciparum malaria infection crisis, monitoring the patient's blood plasmodium once a week is recommended. Patients with relapsed plasmodium falciparum malaria must be treated with another antimalarial drug.

- Patients who are diagnosed or suspected with concurrent infection with plasmodium tertian must be treated with different antimalarial medication.
- This product does not prolong Q-Tc interval within the recommended dosage range. Patients who have experienced ECG abnormalities should be treated with caution as prolonged Q-Tc interval is occasionally observed.
- The potential neurotoxicity of this product has not yet been ruled out. The maximum tolerated dose of the product in healthy volunteers and patients have not been confirmed.
- The concomitant use of mefloquine or quinine exerts synergic effects which have not been yet confirmed in clinical studies.
- This product is metabolized to dihydroartemisinin via cytochrome P450 enzyme CYP3A4. The concomitant use of strong inhibitors of CYP3A4, such as human immunodeficiency virus protease inhibitors, may affect the metabolism of this product, resulting in elevated plasma level. By contrast, the concomitant use of strong inhibitors of CYP3A4, such as phenobarbital, may result in reduced plasma concentration.
- Pregnant women should not be treated with the product; lactating women should stop breastfeeding within 2 weeks after the first treatment.

- **Usage and Dosage**

  Intramuscular injection: initial dose of 4.8 mg/kg, followed by 1.6 mg/kg dose at 6, 24, 48, and 72 h. The initial dose must be divided into two equal doses and then injected into the anterior sides of two thighs, and the subsequent dose must be injected alternately on two thighs.

  The same patient cannot receive repeat medication within 2 weeks after the first treatment.
- **Formulation**

  Intramuscular injections: 2 mL ampoule contains 50 or 150 mg in sesame oil.
- **R&D Facilities**

  Developed by ARTECEF BV Zeewolde (the Netherlands) Company and manufactured by Rotexmedica GmbH Trittau (Germany) GmbH.

References

[1] Yang QC, Shi WZ, Li R, et al. The antimalarial and toxic effect of artesunate on animal models. J. Tradit Chin Med 1982;2(2):99.

[2] 管惟滨,黄文锦,周元昌等.青蒿素及其衍生物对体外培养恶性疟原虫的作用.中国药理学报, 1982;3(2):139. [Available only in Chinese: Guan WB, Huang WJ, Zhou YC, et al. Effect of artemisinin and its derivatives on *Plasmodium falciparum* in virto. Acta Pharmacol Sin 1982;3(2):139].

[3] 黄家章,高徐生,任娜.青蒿琥酯在体外对六株恶性疟原虫的作用.中国药理学与毒理学杂志, 1987;1(4):283. [Available only in Chinese: Huang JZ,Gao XS, Ren N. Effect of sodium artesunate on six isolates of *Plasmodium falciparum* in vitro. Chin J Pharmacol Toxicol 1987;1(4):283].

[4] Li GQ, Guo XB, Fu LC, et al. Clinical trials of artemisinin and its derivatives in the treatment of malaria in China. Trans R Soc Trop Med Hyg 1994;88(1):522.
[5] World Health Organization. WHO Guidelines for the Treatment of Malaria. WHO, Geneva. 2006;98.
[6] 管惟滨,黄文锦,周元昌等.青蒿素及其衍生物对体外培养恶性疟原虫的作用.中国药理学报 1982;3(2):139–41. [Available only in Chinese: Guan WB, Huang WJ, Zhou YC, et al. Effect of artemisinin and its derivatives on *Plasmodium falciparum* in vitro. Acta Pharmacol Sin 1982;3(2):139–41].
[7] 顾浩明,刘明章,吕宝芬等.蒿甲醚在动物的抗疟作用和毒性.中国药理学报 1981;2(2): 138–44. [Available only in Chinese: Gu HM, Liu MZ, Lv BF, et al. Antimalarial action and toxicity of Artemether in animals. Acta Pharmacol Sin 1981;2(2): 138–44].
[8] World Health Organization. WHO Guidelines for the Treatment of Malaria. WHO, Geneva. 2006;96.
[9] Bross A, Venugopalan B, Dominguez Gerpe L, et al. Arteether, a new antimalarial drug: synthesis and antimalarial properties. J Med Chem 1988;31:645–50.
[10] 李继安.细菌耐药性机制的研究与开发 II. 国外医药(合成药, 生化药, 制剂分册) 2001; 22(3): 132–3. [Available only in Chinese: Li JA. Study and development of the bacterial drug resistance mechanism II. World Clin Drugs (Volume: Synth Drug Biochem Drugs Prep) 2001;22(3):132–3].
[11] World Health Organization. WHO Guidelines for the Treatment of Malaria. WHO, Geneva. 2006;100.

CHAPTER 24

The Pharmacological Research Progress of Artemisinin for Antimalarial Treatment

Chapter Outline

Artemisinin is completely different from conventional antimalarial drugs in structure, thus is called a milestone by international peers in antimalarial research history. It has opened up a new direction for the study of antimalarial drugs, and become the main weapon for fighting multidrug-resistant falciparum malaria. This chapter gives an overview on the studies about the antimalarial characteristics, metabolism in vivo, and mechanism of action of artemisinins.

1 Pharmacodynamics Features

Artemisinins are powerful to kill all *Plasmodium*, including chloroquine-resistant *Plasmodium falciparum* malaria and multidrug-resistant *P. falciparum* malaria. In vitro, nanomolar concentrations of this drug can effectively suppress and kill *P. falciparum*. Compared with other antimalarial drugs, these drugs can inhibit the main metabolic process of *Plasmodium* more quickly and kill them. Under the action of artemisinins, the *Plasmodium* will decrease up to 10,000 times at each asexual reproductive cycle [1–3].

Artemisinins mainly act on plasmodia at erythrocytic stage. Relatively, the plasmodia of big ring form, trophozoites, and schizonts with more active metabolism are more sensitive to the drugs, but artemisinins can also inhibit the small ring form [4,5]. They have no effects on the plasmodia in preerythrocytic and exoerythrocytic cycles, so they cannot be used to prevent malaria.

Artemisinins can significantly inhibit the formation of gametocytes. Laboratory studies have also proved that artemisinins can inhibit gametocyte formation through their effects on ring forms and early gametocytes [1,4]. This discovery has been confirmed in the field studies. From 1992 to 1995, we carried out a blood gametocyte carrying rate assessment study in more

From *Artemisia annua* L. to Artemisinins
http://dx.doi.org/10.1016/B978-0-12-811655-5.00024-6

than 5000 adults and children on the Thai–Myanmar border, which showed that artesunate or artemether treatment for uncomplicated malaria can greatly reduce blood gametocyte carriage rate. This suggests that artemisinins can reduce the spread and proliferation of resistant plasmodia strains. In recent studies, although artemisinins were not found to effectively reduce the spread of malaria in regions with high malaria incidence, they are indeed helpful to reduce the spread of malaria in areas with low malaria incidence [6].

In addition, for the pathophysiological process caused by plasmodia, such as erythrocyte surface stickiness lesions caused by *P. falciparum* infection and the cells adhesion between them and microvascular system, the effect of artemisinins treatment is better than the other antimalarial drug [1,2,7].

2 Pharmacokinetics

Oral artemisinins can be absorbed quickly, and reach peak plasma concentration in 1–2 h. However, they cannot be completely absorbed, and the bioavailability is low [8]. In rats, the bioavailability of orally administered artemisinins is only 19%–35%. For intramuscular injection, the bioavailability of different derivatives is quite different [9]. The bioavailability of artesunate is the highest, up to 105%, followed by artelinic acid (95%) and dihydroartemisinin (85%). The bioavailability of artemether and arteether is lower, 54 and 34%, respectively, which is related to the slow and sustained release of the oil-soluble formulation at the injection site. Taking arteether as an example, at 24 h after the injection of the first dose and the continuous seven doses (one dose daily), the drugs that remained in the muscles were 38 and 98% of the total dose, respectively, and presented bimodal absorption.

The derivatives, but not artemisinin itself, are hydrolyzed in vivo into dihydroartemisinin, then metabolized through hepatic cytochrome P450 metabolic enzymes and other metabolic systems, and then cleared with a short half-life (1.6–2.6 h) [1,7–10]. During the metabolism, artesunate can be converted within minutes to dihydroartemisinin, and the latter plays the major role in antimalaria. The animal studies indicate that the conversion rates of different artemisinin derivatives converting into dihydrotestosterone are different. When calculated by the percentage to total dose administered, the conversion rate of artesunate is up to 25.3%–72.7%, while those of artemether and arteether are only 3.7%–12.4% and 3.4%–15.9%.

The serum protein binding rate of artemisinins is high. The serum protein binding rate of dihydroartemisinin is as high as 93% [11]. The binding rates of artemisinin, artemether, arteether, sodium artesunate, and artelinicacid to human serum or serum protein can also reach 64, 76, 78.7, 59, and 81.5%, respectively. In addition, artemisinins seem to be able to selectively bind to erythrocytes infected by plasmodia. The concentration in the infected erythrocytes is tens to hundreds of times higher than in normal erythrocytes [8,12].

Quantitative analysis of artemisinin and its derivatives is difficult, which limited the pharmacokinetic study of artemisinins. The combination of high-performance liquid chromatography and ultraviolet spectroscopy, or reducing electrochemical detection is the most widely used method. In recent years, the application of liquid chromatography–mass spectrometry, gas chromatography–mass spectrometry (GC–MS), supercritical fluid chromatography, and chemiluminescence and fluorescence assays have developed. However, all methods can only be used for serum samples, there is no suitable method for whole blood samples [13,14]. In addition, the biological activity assays can measure the total quantity of active antimalarial substances in serum samples, but cannot distinguish between the original drug and the active metabolites [15].

When single dose of artesunate is orally given to patients with uncomplicated malaria, the absolute bioavailability of antimalarial active ingredient is about 60%, but there is a large individual difference [10]. Dihydroartemisinin, which is the active metabolite, will reach peak plasma concentration in generally 1–2 h. Studies have shown that, compared with patients in recovering stage, the artesunate clearance rate in patients with acute infection is lower, which may be related to the impacts of disease on the pharmacokinetics or the self-induction of metabolic enzymes.

Although the bioavailabilities of oral artemisinin, dihydroartemisinin, and artemether are relatively low, given the plasmodia clearance rate, their pharmacodynamic activity is still satisfactory [7,9,10]. Studies have confirmed that the efficacy of oral dosage form is reliable in uncomplicated malaria patients. The clearance rate of oral multiple-dose artemisinin or artemether is increased, and is similar with artesunate. The clearance rate in convalescent patients is higher than that in acute infection patients.

The relative bioavailability of intramuscular artemether is lower than oral administration. The absorption is slow. The peak plasma concentration will be reached at a few hours after administration. To make things worse, the relevant studies found that a considerable number of patients (5/26 cases and 7/97 cases), after artemether injection, cannot detect the presence of antimalarials with the traditional bioassays [16,17]. This suggests that, as for the pharmacokinetics, the intramuscular injection of artemether is not applicable for dangerous malaria. The absorption of intramuscular arteether is slower than artemether.

For the treatment of dangerous malaria, regardless of intramuscular or intravenous injections, artesunate is superior to artemether in pharmacokinetics. Compared with artemether, artesunate can be rapidly absorbed from the intramuscular site. Within 1 h, plasma concentration of active metabolite dihydroartemisinin can reach peak, and the bioavailability of dihydroartemisinin will reach more than 80%. The severity of malaria infection seems to have no significant impact on the pharmacokinetics of artesunate, but age may have some impact.

The study of artesunate suppositories in treating pediatric patients with moderate malaria showed that absorption after rectal administration is also faster [18]. Plasma concentration of the active metabolite dihydroartemisinin will reach peak within 2 h, with a bioavailability of 20%–60%. Absorption of artemisinin suppositories in healthy people and patients with uncomplicated malaria is relatively slow. The pharmacokinetic behavior of dihydroartemisinin suppositories is similar with artemisinin [10].

Although it is known that there are no significant interactions between artesunate and mefloquine, artemether and lumefantrine, there are few studies on the interactions between artemisinin and other antimalarials or other drugs. Studies showed that artesunate and artemether have no impacts on carboxyprimaquine and carboxymefloquine. Halofantrine and mefloquine have a marginal impact on the efficiency of arteether transformed into dihydroartemisinin in vitro, but not clinically relevant [19,20].

In addition, studies on drug-metabolizing enzyme systems showed artemisinins are mainly metabolized in the liver, and its phase I metabolism is mainly mediated by the cytochrome P450 isoenzymes CYP2B6 and CYP3A4 [8]. Both mediated the conversion of artemisinin derivatives into dihydroartemisinin. The latter is mainly eliminated by glucuronidation process. The relevant uridine gluconate transferases are mainly 1A1, 1A8 to 9, and 2B7. In addition, artemisinins have different effects on human drug-metabolizing enzyme systems. Artemisinin has inducing activity on CYP2C19 and at least one other cytochrome P450 isozyme, which can lead to accelerated elimination of omeprazole, but has no effect on CYP3A4. Dihydroartemisinin and artemether have moderate or mild inhibitory activity on CYP1A2 in rats, but have no effects on CYP2E1 or 2D6 in rats. Although the biotransformation of artelinicacid is not mediated by CYP2C9, it can inhibit the activity of human diclofenac-4-hydroxylase.

The pharmacokinetic difference of artemisinins in different patient populations, especially children and pregnant women, has not been studied in depth. However, animal studies suggest that the metabolism of artemisinins is gender correlated.

3 Mechanism of Action

The early structure–activity relationship (SAR) studies conducted by Chinese scientists confirmed that peroxide bridge in artemisinin is the essential structure of its antimalarial activity, and thus it was presumed that the antimalarial effect is related to the physiological and biochemical changes caused by peroxide bridges. As peroxides can produce free radicals or other reactive oxygens and then damage biological macromolecules, it is suggested that the mechanism of artemisinins' action may be associated with free radicals.

To date, there have been many studies confirming the important relationship between free radicals and artemisinin antimalarial activity, and heme or other ferrous biological macromolecules play a key role in drug activation. For example, drugs that elevate erythrocytes oxygen tension and other drugs that can produce free radicals can enhance the antimalarial activity of artemisinins, while antioxidant (free radical scavenger) catalase, tocopherol, ascorbic acid, glutathione, and dithiothreitol can reduce the antimalarial activity of artemisinin [21]. In the presence of heme, artemisinin can lead to cell membrane lipid peroxidation and ATP activity inhibition, while some radical scavengers and iron chelators could antagonize these effects of artemisinin [21,22]. The irreversible iron-mediated reductive cleavage of artemisinins has been confirmed by electrochemical studies [23,24]. FT-IR studies also confirmed the peroxide bridge cleavage upon ferrous reduction [25,26].

In 1992, Posner and Oh [27] took ^{18}O labeled 1,2,4-trioxane molecule as a model, first brought forward that the ferrous ion-mediated cleavage of a peroxide bridge would produce a pair of oxygen anion radicals 2 and 3, which would become primary carbon radical 11 and secondary carbon radical 4 at C4 site through β-cleavage and 1,5-H migrating rearrangement, and then they are further converted to different products (Fig. 24.1).

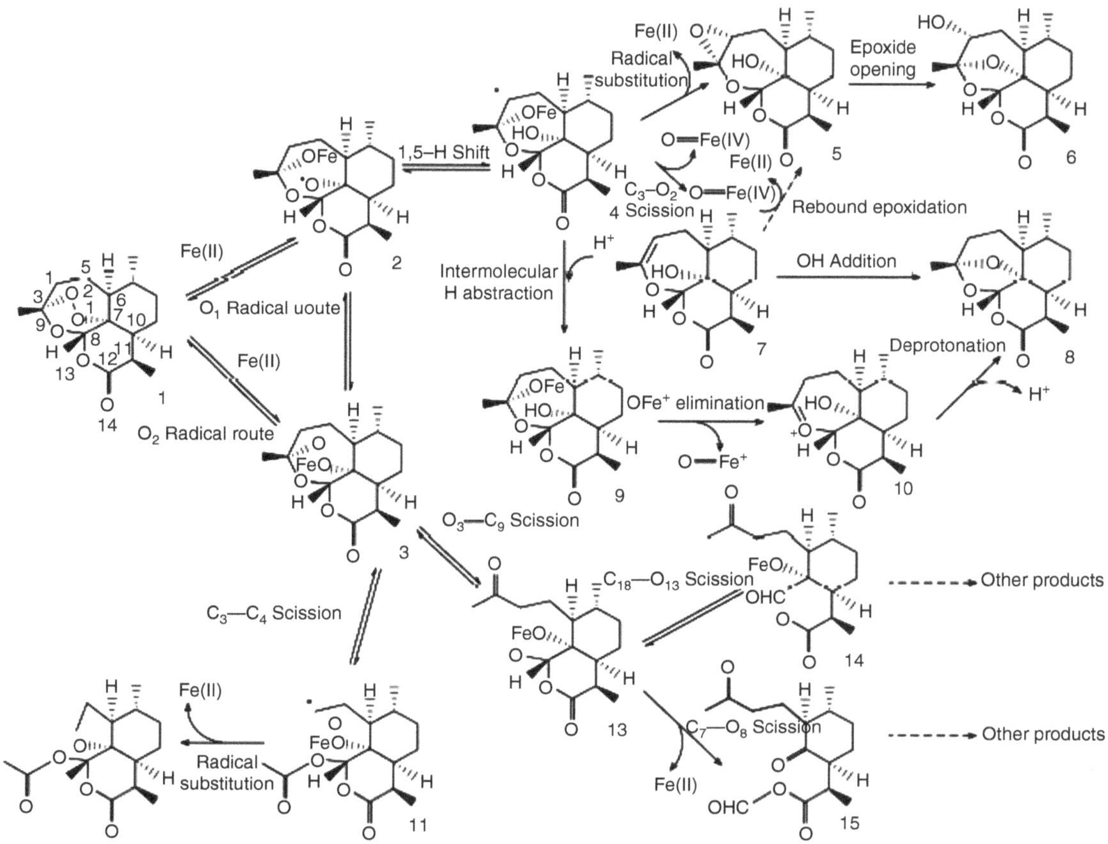

Figure 24.1: The Cleavage of Artemisinin-Induced Ferrous Ion.

After that, chemists continued using various chemical model systems to study the ferrous-mediated reductive cleavages of artemisinin and its derivatives in vitro. some of the presumed existed intermediates and final products have also been identified [27–31]. Although the products under different reaction conditions were not of the same composition, all the results have shown the important status of radicals 11 and 4 at C4 site in the reductive cleavage reactions of artemisinin derivatives. Moreover, after the oxygen atoms at different sites of the trioxane molecule model are isotopically labeled, the mass spectrometry of reductive cleavage products [26] and FT-IR absorption peak [25] isotope migration both confirmed that ethyl furan 12 is C3–C4 cleavage product. More directly, the covalent adducts of radical 11 alkylated heme and metal porphyrin have also been identified [32–34]; spin trapping experiments also obtained signals demonstrating the existence of radicals 4 and 11 [31,35].

Some scholars once raised doubts about the hypothesis that 1,5-H migration can generate 4, although spin trapping experiment and the obtaining of epoxide 5, which is the product of further evolution of 4 (Fig. 24.1), have provided solid evidence [31,35]. Wu Yulin et al. [31] believed that the seven-ring structure of 2 is close to the boat conformation, a slight distortion can meet 1,5-H migration requirement, so the energy of its 1,5-H migration may not be as high as required by open-ring structure. This view was subsequently supported by the quantum theory study of Cu et al. [36], which took 6,7,8-tri-oxygen bicyclo[3,2,2] nonane as the model molecule. Recently, the similar theory calculation performed directly on artemisinin further supported this view, which demonstrated that both primary and secondary carbon radicals are more stable than their precursors 2 and 3 or the corresponding transition states, the formation of the two is an irreversible thermodynamic process [37].

There may be several ways in the reactions from 4 to 6 and 8 (Fig. 24.1). First, direct epoxidation, or β-cleavage released oxygen hyperiron ($O = Fe^{2+}$) epoxidized into intermediate enol 7, which formed 5, the latter can be catalyzed by acids (including proton acid and Lewis acid) and rearranged into 6 [27,31]. Second, 7 can also be rearranged through the internal addition of double-bond molecules into 8 [31]. Finally, if there are suitable hydrogen donors, then 4 (possibly 2 and 3) can be turned by intermolecular dehydrogenation into 9, while 9 can also be hydrolyzed into 8 [31,38]. The harvest of epoxide 5 confirmed the reaction pathways of 4→5→6. $O = Fe^{2+}$ related reaction mechanisms have also been supported by several studies [27,39,40], but there are also scholars who presented theoretical questions on the interpretation of associated trial results [31,38,41]. More reliable evidence is needed to determine $O = Fe^{2+}$ status in artemisinin derivatives reductive cleavage and relevant trial results still need more reasonable explanations.

There is close correlation between the aforementioned ferrous ion-mediated reductive cleavage and the mechanism of drug action. However, among the radicals, epoxides and possible $O = Fe^{2+}$ produced in the reaction, exactly which one or ones are the true plasmodia killers?

Related information originally came from studies on the C4 substitution of artemisinin-like trioxane [42]. It was found C4 methyl or benzyl substitution can significantly improve their in vitro antimalarial activity, while the corresponding C4α-substitution and C4-dimethyl substitution may lead to reduced antimalarial activity, or even complete loss. The author believed that C4β-substitution was beneficial for 1,5-H migration to generate more stable tertiary carbon radicals, while C4α-substitution and C4-dimethyl substitution may hinder 1,5-H migration, therefore the high antimalarial effects of artemisinins are closely related to the pathway of 1,5-H migration rearrangement that generated carbon radical 4. Further SAR studies basically confirmed the effect of C4 substitution on antimalarial activity, only the activity of trioxane with C4β-silyl methyl and stannyl substitutions is lower than the corresponding unsubstituted trioxane [43]. It should be noted that although C4α-substitution hampered 1,5-H migration, its decreased activity may also be the result of the hindrance of the binding of peroxide bridges to corresponding ferrous ion due to α-substitution and thus hindering the drug activation. C4β-substitution can stabilize free radicals produced by rearrangement of C4 without hindering ferrous binding, thus it can enhance the antimalarial activity. The theory that α-substitutions of carbon atoms adjacent to peroxide bridges or other carbon atoms can reduce activity due to steric hindrance has been proved by a large number of studies [44–47]. Therefore, the results of the aforementioned C4 substitution studies suggest that C4 radicals are essential to the antimalarial activity of artemisinins, but the conclusion that secondary carbon radical 4 is more important still cannot be drawn.

Some scholars have questioned the importance of C4 radicals. One of the main foundations is that C4 radical generated by the reductive cleavage of carbaartemisinin derivative of O9 substituted by C is more stable, while its in vitro antimalarial activity is only 4% of that of artemisinin [48]. In this regard, the previous common explanation is that the antimalarial activity is not only related to the stability of C4 radical, but also related to its reactivity, while it seems that carbaartemisinin cannot reach a balance between the two [38,49]. However, recent artemisinin molecular electrostatic potential (MEP) study provided new clues for this [50]. PM3 semiempirically optimized artemisinin MEP diagram shows the maximum negative potential on the surface of artemisinin molecule is located on its carboxyl oxygen O13 and O14; the negative potential on the molecular surface of peroxide bridge is only half of the former. However, O9 also focused some negative potential, and bound with the peroxide bridges O1 and O2, so that the overall negative potential exceeds the carboxyl oxygen. This means that the electrostatic binding of ferrous ion and peroxide bond is closely related to the electrostatic potential distribution on the surface of drug molecule. The substitution of O9 by carbon atom is clearly not favorable to this binding. The antimalarial activity of 10-deoxy-carbaartemisinin can be increased by 4 times (the increase magnitude is similar to 10-deoxy-artemisinin) [48], and the relationship between the three-dimensional electronic properties of the artemisinin-like drugs and their neurotoxicity [51] also indicated this. Moreover, many studies have confirmed that C4 radical can be bound covalently with biological macromolecules via alkylation [32–34,52].

Due to the strong reactivity, $O = Fe^{2+}$ and epoxide 5 are also considered to be the real plasmodia killers. The rationale is when the β cleavage that is assumed to generate $O = Fe^{2+}$ is preceded by the β-cleavage of C3 that is easier to eliminate the aromatic substitution (such as phenyl), the corresponding trioxane will completely lose its antimalarial activity [27]. Silyl groups, such as methyl C4 β-substituted trioxane, have no antimalarial activity, which is also attributed to the hindrance of the pathway that β-cleavage producing oxygen hyperiron [43]. However, the author believed it should be noted that the electron-attracting effect of C3 aromatic substitution is in favor of the formation of O2 free radical 3, while as groups that are easier to be eliminated, their cleavages are clearly prior to the β-cleavage of C3–C4. The composition of their ferrous reduced products also indicated this point [27]. Although the substitution of C4 β-silyl methyl groups did not affect the formation of C4 free radicals, silyl and stannyl groups can be easily eliminated and cause the intramolecular quenching of C4 free radicals. The assumption that β-cleavage can generate $O = Fe^{2+}$ is still questionable in theory. Although epoxide 12 has been broken down, the epoxides obtained from the ferrous reduced products of high activity 10-deoxy-carbaartemisinin derivatives (two-third of artemisinin antimalarial activity) are completely free of antimalarial activity [49].

Some scholars indicated that the hydroperoxides formed by ring opening of C2–C3 cleavage mediated by electrophilic reagents, such as acid, aniline, and oxophlic metal ions, as well as their generated electrophilic oxides may also be the antimalarial active substances [29,49]. However, this hypothesis has not been generally recognized. First, their reaction rate is much slower than reductive cleavage, so even if there is such reaction, it cannot compete with the reductive cleavage reaction and generate enough electrophilic oxides. Second, the lactone ring is more sensitive to the attack of electrophilic reagents, such as acid than peroxide bridges [53]. Moreover, the majority of hydrogen peroxides that cannot generate carbon radicals by reductive cleavage only have slight antimalarial activity.

Different from typical free radical drugs, the antimalarial action of artemisinins is the result of a specific free radical acts on a specific target. This has been recognized by most scholars. A large number of data indicate that C4 free radicals 4 and 11 may be the true plasmodia killers (see earlier). But what are their specific targets? How do C4 free radicals act on them and cause the death of plasmodia?

Due to the importance of ferrous ion to the action of these drugs, the hemoglobin metabolism and heme detoxification pathway are thought to be the targets of artemisinins. Hong et al. [54] found that, when 14C-labeled artemisinin treated plasmodia-infected erythrocytes, 75% of the drugs were focused in malaria pigments, and formed covalent adducts with them. Subsequently, the alkyl metal porphyrin of C4 free radicals or the heme covalent adducts were identified [32–34,55]. SAR studies have also shown that the antimalarial activity of artemisinin derivatives is closely related to their binding affinity to heme [56,57]. Further studies found that artemisinin and dihydroartemisinin have inhibitory effects on heme polymerization and β-hemin formation in vitro [58]. Recent studies have reported that artemisinin and β-artemisia ether can inhibit the hemoglobin hydrolysis catalyzed by plasmodia's food vacuole lysates (mainly inhibiting cysteine protease activity) and heme polymerization, and the heme polymerization catalyzed by *P. falciparum* histidine-rich protein (PfHRPII) and cause the decomposition of malaria pigment [59,60], while the covalent adducts of drug and heme may be the key in inhibiting heme polymerization [61].

The plasmodia can digest 75% of the host's hemoglobin, and take them as the main source of raw material for their own protein synthesis. Under the catalysis of hemoglobin enzyme, hemoglobin is hydrolyzed and releases free amino acids and heme. The former is utilized by plasmodia, while the heme with high-reactivity aggregates into inactive hemozoin through plasmodia's detoxification pathway and accumulates in the food vacuoles. Therefore, the inhibitory effects on hydrolysis of hemoglobin and heme polymerization are exactly the primary antimalarial mechanism of quinoline drugs, such as chloroquine and quinine [62]. However, artemether does not affect hemozoin formation in vivo [60]. Even outside the cells, 10-deoxy-artemisinin with high antimalarial activity has not shown the inhibitory effects in heme polymerization inhibitory activity and β-hematin inhibitory activity tests in vitro, which was different from artemisinin and dihydroartemisinin [63]. Moreover, for the small ring form plasmodia that do not digest a lot of hemoglobin to form hemozoin, artemisinins also have powerful antimalarial activity [5]. Some inactive trioxane derivatives can also form covalent adducts with metal porphyrin by the same free radical alkylation [44], and the discovery that using plasmepsin inhibitors to block hemoglobin hydrolysis and heme release can significantly antagonize chloroquine activity without any influence on the activity of artemisinin confirmed this point [64].

A large number of studies seem to have confirmed that heme is the ferrous source required by drug activation, and a large number of action mechanism-related model studies and drug structural optimizations are also based upon this understanding. Recently, however, this consensus has been challenged. First, Wu et al. [65,66] found that when there is cysteine, a trace amount of nonheme iron can quickly cleave artemisinin, and produce a previously unknown aldehyde. Meanwhile, they also found that C4 free radical can be covalently combined under intramolecular attack with iron-complexed cysteine sulfur atoms. Therefore, they thought that *Plasmodium* function protein or ferrous ions complexed by enzymes are necessary for drug activation; the reductive cleavages mediated by other unnecessary ferrous ion, such as monomer ferrous heme or ferrous complex, are not drug related. They further proposed that the targets of artemisinins are the plasmodia's key enzymes or functional proteins that contain iron. Olliaro et al. [49] also raised questions about the roles of heme in drug activation. Recently, using fluorescence-labeled artemisinin derivatives and high-resolution confocal microscope, Krishna et al. [64] confirmed artemisinins are distributed in the membrane system of plasmodia's cytoplasm, but not accumulated in the food vacuoles like people formerly believed, and the use of plasmepsin inhibitors to block the release of heme from hemoglobin hydrolysis also does not affect the action of artemisinin. However, there is no more evidence to support the importance of nonhemoglobin ferrous ion, and there is also no direct evidence that heme is not related to the actions of these drugs.

It has been confirmed that artemisinin can cause the alkylation of plasmodia-specific proteins. When the radiolabeled artemisinin derivatives of treatment concentration were incubated with infected erythrocytes, there were six *Plasmodium* proteins specifically labeled [67], one of them was identified as translationally controlled tumor protein (TCTP) homolog. Further studies have found that *P. falciparum* recombinant TCTP could combine with dihydroartemisinin under heme catalysis. Immunoprecipitation also confirmed the their interaction inside *P. falciparum* parasites [68,69]. However, the functions of TCTP are still unknown, and there is no direct evidence that its combination with the drug is related to drug antimalarial actions. Under complex experimental conditions, the combination of the two may be accidental, but does not relate to drug actions.

The study carried out by Krishna et al. [64] brought new findings. They found that artemisinin can selectively inhibit the activity of *P. falciparum*'s heterologously expressed PfATP6 (the sarco/endoplasmic reticulum calcium-dependent ATPase, SERCA), while chloroquine, quinine, and deoxy-artemisinin have no such effect. Moreover, the antimalarial action of artemisinin can be significantly antagonized by potent SERCA inhibitor, thapsigargin. Deferoxamine can also antagonize the inhibition of artemisinin on PfATP6. In the same model, the PfATP6 inhibiting activity of some artemisinin derivatives and internal peroxide compounds and their in vitro antimalarial activity are linearly correlated. Further experiments of receptor binding competition showed that the targets of artemisinin inside plasmodia are the same with those of thapsigargin, and the binding with the targets is irreversible. PfATP6 can regulate intracellular calcium levels by excreting calcium ions to the outside. After ferrous activation, artemisinin can irreversibly bind with PfATP6 and inhibit its activity, and increase intracellular calcium levels, which may be the cause of plasmodia death. However, just as the authors suggested, there is still no direct evidence that the activated artemisinin can bind with PfATP6 at treatment concentration.

Recently, in yeast model studies, it was found that by the action between artemisinin and mitochondrial electron transport chain, it can generate regional reactive oxygen, destruct mitochondrial membrane potential, and finally result in mitochondrial dysfunction [70]. This finding is consistent with the findings in early related studies on the neurotoxicity of high-dose artemisinins [71].

In summary, the mechanism has been established that artemisinins, mediated by ferrum from heme or other sources, via reductive cleavage and rearrangement, can generate two C4 free radicals. C4 free radicals with no steric hindrance can irreversibly bind with *Plasmodium* key proteases by alkylation, inhibit their activities, and lead to parasite death, which seems a reasonable result. However, the direct targets have not yet been finally determined. Although Krishna et al. provided substantial evidence that PfATP6 may be drug targets, as previously described, this view still needs other evidence for support. The role of TCTP and other proteins in the life cycle of plasmodia that can bind with active metabolites of artemisinin, as well as the biological effects caused by their alkylation also need to be further elucidated.

4 Side Effects

No drug-related side effects were found in the prospective clinical studies with more than 1,000 cases and the postmarketing survey of more than 46,000 patients in Thailand [1]. The most common side effects after administration are headache, nausea, abdominal pain, vomiting, and occasional diarrhea. However, all these symptoms are common in patients with malaria and they will disappear after appropriate treatment. In addition, noncontrolled clinical trials once reported that artemisinins can transiently inhibit nonanemic reticulocyte reaction and cause alcaptonuria in some patients with *P. falciparum* malaria (the probability of incidence is similar with quinine). The postmarketing survey in Thailand also found two cases of hemolysis that may be related to the administration of artesunate. By far the most serious side effects reported were two cases of anaphylaxis caused by oral artesunate in 2001. Notably, although it was found in animal experiments that artemisinins do not impact reproductive function in rats, or have teratogenic effects, they have embryo toxicity when given to pregnant rats at organogenesis stage, manifesting increased fetal absorption. Therefore, it is not recommended that pregnant women take artemisinins during their first 3 months of pregnancy [1].

It is widely noted that animal experiments have shown long-term repeated injections of high doses of artemether or artemisia ether may cause neurotoxicity [1]. In trials with rats, dogs, and rhesus monkeys, both the drugs have caused selective neurotoxic symptoms of brainstem nuclei. Pathological studies found that the related neuropathological damages mainly located in the brain stem, fastigial nucleus of cerebellum, and auditory system, where neuronal chromatin dissolving and dispersed neuronal necrosis occurred. Auditory sense and vestibular nucleus are the most sensitive parts. In vitro studies confirmed that artemisinins can induce the release of lactate dehydrogenase from neurons and inhibit neuronal cells' leucine intake. This toxic effect is neuron specific and dose and time related. The studies carried out by the author's laboratory have found that high doses of artemether can cause mitochondrial swelling, reduce cristae, break or disappear in the brain cortex neurons of parent rats, and inhibit the activity of complex enzymes I and IV in neuronal mitochondrial respiratory chain, so that their capacity for NADH oxidation and cytochrome *c* reduction decreases. It is demonstrated that the inhibition of electron transport in mitochondrial respiratory chain resulting in cell oxidative metabolism and energy metabolism disorders may be the mechanism of artemisinins' neurotoxicity [71].

However, it can be confirmed that the possibility of clinical neurological toxicity is very low. A prospective study of more than 10,000 cases has found no clinical evidence of severe neurological toxicity of artemisinins. In central Vietnam, where artemisinins are widely used, the prospective study has not found any drug-induced hearing ability changes in children and adults who have taken artemisinin for more than 1 year [1].

The neurotoxicity difference between animals and humans is related to the relatively higher doses and longer period of administration in animal experiments. In addition, the pharmacokinetic studies of artemether and artemisia ether injections showed that in both animals and humans, the release of oil formulation is slow, which resulted in longer drug exposure. The toxic effects of artemisinins may be caused by long drug exposure, rather than determined by maximum drug concentration. When administered orally, even high concentration of artemisinins cannot have the aforementioned consequences.

5 Drug Resistance

Although several laboratories have successfully cultivated the artemisinin-resistant strains of *Plasmodium yoelii* and *P. falciparum*, so far there are no clinical reports of artemisinin resistance. Low artemisinins sensitive clinical isolates of *P. falciparum* were found during drug sensitivity monitoring in vitro, but most are not clinically relevant [1,2,10,50,72].

The resistance to artemisinins generated by *Plasmodium* parasites is slow for several reasons. (1) The half-lives of artemisinins are short, therefore plasmodia cannot be exposed to the drug pressure of subtherapeutic concentration for long time. (2) Artemisinins can inhibit gametocyte formations. Hence when patients were treated with artemisinins, the parasites taken ingested by mosquitoes cannot spread easily to other patients. (3) Artemisinins are often used with other antimalarial drugs in clinical practice, which can delay the development of drug resistance.

Recent reports suggest the possibility of plasmodia developing resistance to artemisinins. (1) The artemisinin resistant strains of *P. yoelii*, cultivated under selective drug pressure, can retain their stable resistance after being transmitted by mosquitos [73]. (2) The in vitro artemether-resistant *P. falciparum* was isolated in Guiana [74]. (3) Clinical studies confirmed the relationship between reduced sensitivity to artemisinin in vitro and decreased clinical efficacy of artemether monotherapy [75].

In fact, numerous studies have already confirmed that the change of *P. falciparum*'s sensitivity to artemisinins is related to its genetic change [1,10,76]. For example, studies on both *P. yoelii* (pymdr1) and *P. falciparum* (pfmdr1) have found the increase of multidrug resistance gene mdr1 copies in *P. falciparum* is related to the decrease of artemisinin sensitivity. The studies on *P. yoelii* have found that the drug intake of artemisinin-resistant plasmodia strains is significantly reduced, and the expression of TCTP homologs is increased twice. Although it may not necessarily be clinically relevant, the multidrug resistance gene of *P. falciparum* (pfmdr1), the mutation of chloroquine transporter gene (pfcrt), and polymorphism of *P. falciparum* PfATP6 gene are all related to the sensitivity to artemisinins in vitro.

However, it is interesting that in recently reported stable artemisinin-resistant strains of *P. yoelii*, no changes were found in the corresponding genes of PfATP6, pfm-dr1 and pftctp, including the sequence and copy number [73]. This may be due to large biological and physiological differences between *P. yoelii* and human *Plasmodium*, so that their sensitivity mode to artemisinins is very different from *P. falciparum*. There may also be the presence of unknown targets of artemisinins or drug transport-associated proteins, but the genetic change is the key of the developing resistance to artemisinin.

At present, artemisinins are still at the stage of being widely used. Therefore, people still have opportunities to assess the clinical resistance to artemisinins and try to reduce that risk. Thus, it is essential to closely monitor the changes in drug sensitivity of *P. falciparum* and deeply explore the mechanisms of artemisinins' actions and their resistance. More recently, specific mutations in the Kelch 13 (K13)–propeller domain were reported to be associated with delayed parasite clearance [77]. This molecular marker of artemisinin resistance provides new possibilities for tracking resistance to artemisinins.

The following comrades participated in the preparation of this chapter: Wang Jingyan, Bei Zhuchun (Institute of Microbiology and Epidemiology, Academy of Military Medical Sciences).

References

[1] World Health Organization. The use of artemisinin and its derivatives as anti-malarial drugs. Geneva. 1998; WHO/MAL/98.1086.
[2] Krishna S, Uhlemann AC, Haynes RK. Artemisinins: mechanisms of action and potential for resistance. Drug Resist Updat 2004;7:233–44.
[3] Chen PQ, Li GQ, Guo XB, et al. The infectivity of gametocytes of *Plasmodium falciparum* from patients treated with artemisinin. Zhonghua Yi Xue Za Zhi 1994;74(4):209–10.
[4] Chimanuka D, Francois G, Timperman G, et al. A comparison of stage-specific efficacy of chloroquine, artemether and diocophylline B against the rodent malaria parasite *Plasmodium chabaudi chabaudi* in vivo. Parasitol Res 2001;87:795–803.
[5] Skinner TS, Mannig LS, Johnston WA, et al. In vitro stage-specific sensivity of *Plasmodium falciparum* to quinine and artemisinin drugs. Int J Parasitol 1996;26(5):519–25.
[6] Price RN, Nosten F, Luxemburger C, et al. Effects of artemisinin derivatives on malaria transmissibility. Lancet 1996;347:1654–8.
[7] White NJ. Clinical pharmacokinetics and pharmacokinetics and pharmacodynamics of artemisinin and derivatives. Trans R Trop Med Hyg 1994;88(11):S41–3.
[8] Navaratnam V, Mansor SM, Sit NW, et al. Pharmacokinetics of artemisinin-type compounds. Clin Pharmacokinet 2000;39(4):255–70.

[9] Li GQ, Peggins JO, Fleckenstein LL, et al. The pharmacokinetics and bioavaliability of dihydroartamisinin, arteether, artemether, artesunic acid and artelinic acid in rats. J Pharm Pharmacol 1998;50(2):173–82.
[10] Woodrow CJ, Haynes RK, Krishna S. Artemisinins. Postgrad Med J 2005;81:71–8.
[11] Batty KT, Ilett KF, Davis TM. Protein binding and alpha: beta anomer ratio of dihydroartemisinin in vivo. Br J Clin Pharmacol 2004;57:173–82.
[12] Akompong T, Van Wye J, Ghori N, et al. Artemisinin and its derivatives are transported by a vacuolar-network of *Plasmodium faiciparum* and their anti-malarial activities are additive with toxic sphingolipid analogues that block the Net work. Mol Biochem Parasitol 1999;101:71–9.
[13] Xing J, Yan H, Zhang S, et al. A high-performance liquid chromatography/tandem mass spectrometry method for the determination of artemisinin in rat plasma. Rapid Commun Mass Spectrom 2006;20:1463–8.
[14] Lihua Chen, Hong Yin, Zhaoxia Yang, et al. Artemisinin detection using cytochrome C catalytic fluorescence with Pyronine B as indicator. Anal Chem 2006;34(2):173–4.
[15] Teja-Isavadham P, Peggins JO, Brewer TG, et al. *Plasmodium falciparum*-based bioassay for measurement of artemisinin derivatives in plasma or serum. Antimicrob Agents Chemother 2000;44:972–7.
[16] Silamut K, Newton PN, Teja-Isavadham P, et al. Artemether bioavailability after oral or intramuscular administration in uncomplicated faiciparum malaria. Antimicrob Angents Chemother 2003;47:3795–8.
[17] Murphy SA, Mberu E, Muhia D, et al. The disposition of intramuscular artemether inchildren with cerebral-malaria; apreilminary study. Trans R Soc Trop Med Hyg 1997;91:331–4.
[18] Barnes KI, Mwenechanya J, Tembo M, et al. Efficacy of rectal artesuante copmared with parenteral quinine in initial treatment of moderately severe malaria in African children and adults: a randomized study. Lancet 2004;363:1598–605.
[19] Mithwans S, Aarons L, Kokwaro GO, et al. Population pharmacokinetics of artemether and dihydroartemisinin following single intramuscular dosingofartemetherin African children with severe falciparum malaria. Br J Clin Pharmacol 2004;57:146–52.
[20] Svensson USH, Alin MH, Karlsson MO, et al. Population pharmacokineticand pharmacodynamic modeling of artemisinin and mefloquine enantiomersinpatients whth falciparum malaria. Eur J Clin Pharmacol 2002;58:339–51.
[21] Dhingra V, Vishweshwar Rao K, Lakshmi Narasu M. Current status of artemisinin and its derivatives as antimalarial drugs. Life Sci 2000;66(4):279–300.
[22] Wei N, Sadrzadeh SM. Enhancement of hemin-induced membrane damage by artemisinin. Biochem Pharmacol 1994;48(4):737–41.
[23] Berman PA, Adams PA. Artemisinin enhances heme-catalysed oxidationoflipid membranes. Free Radic Biol Med 1997;22(7):1283–8.
[24] Zhang F, Gosser DK Jr, Meshnick SR. Hemin-catalyzed decomposition of artemisinin (qinghaosu). Biochem Pharmacol 1992;43(8):1805–9.
[25] Chen Y, Zheng JM, Zhu SM, et al. Evidence for hemin inducing the cleavage of peroxide bond of artemisinin (qinghaosu): cyclic voltammetry and in situ FTIR sector electrochemical studies on the reduction mechanism of artemisinin in the presence of hemin. Electrochim Acta 1999;44(14):2345–50.
[26] Kapetanki S, Varotsis C. Fourier transform investigation of non-heme Fe(III) and Fe(II) decomposition of artemisinin and of a simplified trioxane alcohol. J Med Chem 2001;44(19):3150–6.
[27] Posner GH, Oh CH. A regiospecificaly oxygen-18 labeled 1,2,4-trioxane: a uncomplicated chemical model system to probe the mechanism(s) for the antimalarial activity of artemisinin (qinghaosu). J Am Chem Soc 1992;114(21):8328–9.
[28] Posner GH, Gumming JN, Ploypradith P, et al. Evidence for Fe(IV) ¼ O in the molecular mechanism of action of the rtioxane antimalarial artemisinin. J Am Chem Soc 1995;117(21):5885–6.
[29] Haynes RK, Vonwiler SC. The behaviour of qinghaosu (artemisinin) in the presence of hemeiron (II) and (III). Tetrahedron Lett 1996;37(2):253–6.
[30] Haynes RK, Vonwiler SC. The behaviour of qinghaosu (artemisinin) in the presence of non-hoemiron (II) and (III). Tetrahedron Lett 1996;37(2):257–60.
[31] Jeford CW, Vicete MGH, Jacquier Y, et al. The deoxygenation and isomerization of artemisinin and artemether and their relevance to antimalarial action. Helv Chim Acta 1996;79(5):1475–87.

[32] Jefford CW. Why artemisinin and certain synthetic peroxides are potent antimalarial. Implications for the mode of action. Curr Med Chem 2001;8(15):1803–26.
[33] Robert A, Meunier B. Characterization of the first covalent adduct between artemisinin and a heme model. J Am Chem Soc 1997;119(25):5968–9.
[34] Robert A, Cazeles J, Meunier B. Characterization of alkylation product of heme by the antimalarial drug artemisinin. Angew Chem 2001;113(10):2008–11.
[35] Rober A, Coppel Y, Meunier B. NMR characterization of covalent adducts obtained by alkylation of heme with the antimalarial drug artemisinin. Inorg Chim Acta 2002;339:448–96.
[36] Butler AR, Gilbert BC, Hulme P, et al. EPR evidence for the involvement of free radicals in the iron-catalysed decomposition of qinghaosu (artemisinin) and some derivatives; antimalarial action of some ploycyclic endoperoxides. Free Radic Res 1998;28(5):471–6.
[37] Gu J, Chen K, Jiang H, et al. The radical transformation in artemisinin: a DFT study. J Phys Chem A 1999;103(46):9364–9.
[38] Wu WM, Wu Y, Wu YL, et al. Unified mechanistic framework for Fe(II)-induced cleavage of qinghaosu and derivatives/analogues. The first spin-trapping evidence for the previously postulated secondary C-4 radical. J Am Chem Soc 1998;120(14):3316–25.
[39] Taranto AG, Carneiro JWM, de Olivveira FG, et al. The role of C-centered radicals on the mechanism of action artemisinin. J Mol Struct 2002;580:207–15.
[40] Posner GH, Wnag D, Gonzalez L, et al. Mechanism-based design of uncomplicated, symmetrical easily prepared, potent antimalarial endoperoxides. Tetrahedron Lett 1996;37(6):815–8.
[41] Kapetanaki S, Varotsis C. Ferryl-oxo heme intermediate in the antimalarial mode of action of artemisinin. FEBS Lett 2000;474(2–3):238–41.
[42] Posner GH, Oh CH, Wang D, et al. Mechanism-based design, synthesis, and in vitro antimalarial testing of new 4-methylated trioxanes structurally related to artemisinin: the importance of a carbon-centered radical for antimalarial activity. J Med Chem 1994;37(9):1256–8.
[43] Posner GH, Wang D, Cumming JN, et al. Further evidence supporting the importance of and the restrictions on a carbon-centered radical for high antimalarial activity of 1,2,4-trioxanes like artemisinin. J Med Chem 1995;38(13):2273–5.
[44] Cumming JN, Wang D, Park SB, et al. Design, synthesis, derivatization, and structure–activity relationships of simplified, tricyclic, 1,2,4-trioxane alcohol analogues of the antimalarial artemisinin. J Med Chem 1998;41(6):952–64.
[45] Provot O, Camuzat-Dedenis B, Hamzaoui M, et al. Structure–activity relationship of synthetic tricyclic trioxanes related to artemisinin: the unexpected alkylative property of 3-(methoxymethyl)anaolg. Eur J Org Chem 1999;(8):1935–8.
[46] Wang DY, Wu Y, Wu YL, et al. Synthesis, iron(II)-induced cleavage and in vivo antimalaral efficacy of 10-(2-hydroxy-1-naphthyl)-deoxoqinghaosu-(-deoxoartemisinin). J Chem Soc Perkin Trans 1999;(13):1827–31.
[47] Jefford CW, Burger U, Milasson-Schmidt P, et al. Epiartemisinin, a remarkably poor antimalarial: implications for the mode of action. Helv Chim Acta 2000;83(6):1239–46.
[48] Avery MA, Fan PF, Karle JM, et al. Structure–activity relationships of the antimalarial agent artemisinin. 3. Total synthesis of (+)-13-carbaartemisinin and related tetra- and tricyclic structures. J Med Chem 1996;39(9):1885–97.
[49] Olliaro PL, Haynes RK, Meunier B, et al. Possible modes of action of the artemisinin-type compounds. Trends Parasitol 2001;17(3):122–6.
[50] Taranto AG, Carneiro JWD, Oliveria FG. MNDO/d Caculati on soninteraction between artemisinin and heme. J Mol Struct 2001;539:267–72.
[51] Bhatacharjee AK, Karle JM. Stereoelectronic properties of antimalarial artemisinin analogues in relation to neuro-toxicity. Chem Res Toxicol 1999;12(5):422–8.
[52] Rodriguez M, Claparols C, Robert A, et al. Alkylation of microperpoxidase-11 by the antimalarial drug artemisinin. ChemBiochem 2002;3(11):1147–9.
[53] Jung M, Lees. Stability of acetal and non acetal type analogs of artemisinin in simulated stomach acid. Bioorg Med Chem Lett 1998;8(9):1003–6.

[54] Hong YL, Yang YZ, Meshnick SR. The interaction of artemisinin with malarial heomozoin. Mol Biochem Parasitol 1994;63(1):121–8.
[55] Bentoit-Vical F, Robert A, Meunier B. In vitro and in vivo potentiation of artemisinin and synthetic endoperoxide antimalarial drugs by metaloporphyrins. Antimicrob Agents Chemother 2000;44(10):2836–41.
[56] Paitayatat S, Tarnchompoo B, Thebtaranonth Y, et al. Correlation of antimalarial activity of artemisinin derivatives with binding affinity with ferroprophyrin IX. J Med Chem 1997;40(5):633–8.
[57] Tonmunphean S, Parasuk V, Kokpol S. QSAR study of antimalarial activities and atremisinin-heme binding properties obtained from docking calculations. Quant Struct Act Relat 2000;19(5):475–83.
[58] Parapini S, Basilico N, Pasini E, et al. Standardization of the physicochemical parameters to assess in vitro the beta-hematin inhibitory activity of antimalarial drugs. Exp Parasitol 2000;96(4):249–56.
[59] Pandey AV, Tekwani BL, Singh RL, et al. Artemisinin, an endoperoxide antimalarial, disrupts the hemoglobin catabolism and heme detoxifyication systems in malarial parasite. J Biol Chem 1999;274(27):19383–8.
[60] Haynes RK, Monti D, Taramelli D, et al. Artemisinin antimalarials do not inhibit hemozion formation. Antimicrob Agents Chemother 2003;47(3):1175.
[61] Kannan R, Sahal D, Chauhan VS. Heme-artemisinin adducts are crucial mediators of the ability of artemisinin to inhibit heme polymerization. Chem Biol 2002;9(3):321–32.
[62] Foley M, Tilley L. Quinoline antimalarials: mechanisms of action and resistance and prospects for new agents. Pharmacol Ther 1998;79(1):55–87.
[63] Basilico N, Pagani E, Itarat I, et al. A microtitre-based method for measuring the haem polymerization inhibitory activity (HPIA) of antimalarial drugs. J Antimicrob Chemother 1998;42(1):55–60.
[64] Eckstein-Ludwing U, Webb RJ, van Goethem IDA, et al. Artemisinins target the SERCA of *Plasmodium falciparum*. Nature 2003;424(6951):957–61.
[65] Wu Y. How might qinghaosu (artemisinin) and relatedcompunds kill the intraerythrocytic malaria parasite? A chemist's view. Acc Chem Res 2002;35(5):255–9.
[66] Wu Y, Yue ZY, Wu YL. Interaction of qinghaosu (artemisinin) with cysteine sulfhydryl mediated by traces of non-heme iron. Angew Chem Int Ed Engl 1999;38(17):2580–2.
[67] Asawamahasakda W, Itarat I, Pu YM, et al. Reaction of antimalarial endoperoxides with specific parasite proteins. Antimicrob Agents Chemother 1994;38(8):1854–8.
[68] Bhisutthibhan J, Pan XQ, Hossler PA, et al. The *Plasmodium falciparum* translationally controlled tumor protein homolog and its reaction with the antimalarial drug artemisinin. J Biol Chem 1998;273(26):16192–8.
[69] Bhisutthibhan J, Meshnick SR. Immunoprecipitation of [^{3}H]-dihydroartemisinin translationally controlled tumor protein (TCTP) adducts from plasmodium falciparum infect erythrocytes by using anti-TCTP antibody. Antimicrob Angents Chemother 2001;45(8):2397–9.
[70] Li W, Mo W, Shem D, et al. Yeast model uncovers dual roles of mitochondria in the action of artemisinin. PLoS Genet 2005;1(3):36.
[71] Zhao Yanhong, Wang, Jingyan. The impacts of artemether on the morphology and function of neuronal mitochondria. Chin J Pharmacol Toxicol 2003;17(3):196–201.
[72] Deleted in review.
[73] Afonso A, Hunt P, Chesman S, et al. Malaria parasites can develop stable resistance to artemisinin but lack mutations in candidate genes apt6 (encoding the sarcoplasmic and endoplasmic reticulum Ca^{2+} ATPase), tctp, mdrl, and cg10. Antimicrob Angents Chemother 2006;50(2):480–9.
[74] Jambou R, Legrand E, Niang M, et al. Resistance of *Plasmodium falciparum* field isolates to in-vitro artemether and point mutations of the SERCA-type PfAT Pase 6. Lacent 2005;366:1960–3.
[75] Krishna S, Woodrow CJ, Staines HM, et al. Re-evaluation of how artmeisinins work in light of emerging evidence of in vitro resistance. Trends Mol Med 2006;12(5):199–205.
[76] Msshnick SR. Artemisinin: mechanisms of action, resistance and toxicity. Int J Parasitol 2002;32(13):1655–60.
[77] Straimer J, Gnädig NF, Witkowski B, Amaratunga C, Duru V, Ramadani AP, Dacheux M, Khim N, Zhang L, Lam S, Gregory PD, Urnov FD, Mercereau-Puijalon O, Benoit-Vical F, Fairhurst RM, Ménard D, Fidock DA. Drug resistance. K13-propeller mutations confer artemisinin resistance in *Plasmodium falciparum* clinical isolates. Science 2015;347(6220):428–31.

CHAPTER 25

Advancement in Research on Artemisinin Derivatives and Relevant Compounds

Chapter Outline

The Institute of Chinese Material Medica, China Academy of Chinese Medical Sciences, discovered artemisinin in 1972 and synthesized dihydroartemisinin in 1973. Shortly afterward, it started the exploration of the relationship between artemisinin derivatives and antimalarial activity. The research staff found that the antimalarial activity of the product was enhanced after the reduction of the carbonyl group in the lactone ring to a hydroxyl group (namely dihydroartemisinin) provided that the peroxide bridge of antimalarial activity groups in artemisinin was retained. After acetylation of hydroxyl group, the activity of the product was further improved [1–3]. In 1975, the coding sequence diagram of artemisinin-like compounds as shown in Fig. 25.1 was reported in the "523" internal meeting, thus promoting the development of domestic and foreign research on artemisinin derivatives. The new drugs that have been derived and developed include sodium artesunate, artemether, dihydroartemisinin, and arteether. Based on in-depth study on the structure–activity relationship, pharmaceutical chemists have conducted extensive research on other artemisinin derivatives and artemisinin-related compounds to seek for newer antimalarial therapy.

There are two systems for coding of artemisinin-like compounds, as shown in Fig. 25.1. In this chapter, only the sample on the left is used. As derivatives which have been marketed as

From *Artemisia annua* L. to Artemisinins
http://dx.doi.org/10.1016/B978-0-12-811655-5.00025-8

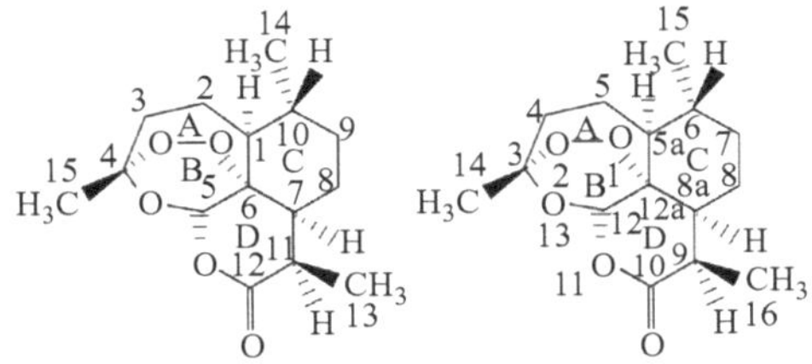

Figure 25.1: Serial Numbers of Arterminin and Its Derivatives.

new drugs are informatively described in previous chapters, this chapter mainly outlines the progress of research on other typical derivatives and their structurally related compounds.

1 Artemisinin Derivatives

Numerous studies were conducted mainly aiming at C4, C12, and C13 groups, especially great progress has been achieved in systematic research on C12 derivatives. *Plasmodium berghei* model or in vitro tests of some derivatives showed that its antimalarial activity was superior to that of artemisinin.

1.1 C12 Derivatives

C12 derivatives are the first compounds which have attracted attention at home and abroad. They account for the largest part of the artemisinin drugs, consisting of the following three main types.

1.1.1 Ether derivatives of artemisinin

After 1976, Li et al. [4–7] in the Institute of Materia Medica, Chinese Academy of Sciences, synthesized various ether derivatives of artemisinin C12. US Walter Reed Army Institute Research focused on the ether-like compound—artesunate benzyl ether acid—which was designed to overcome the instability and easy hydrolysis of sodium artesunate. A large number of ether derivatives of artemisinin C12 have been reported successively by Haynes et al. [8], O'Neil et al. [9,10], and Venugopalan et al. [11] (Fig. 25.2).

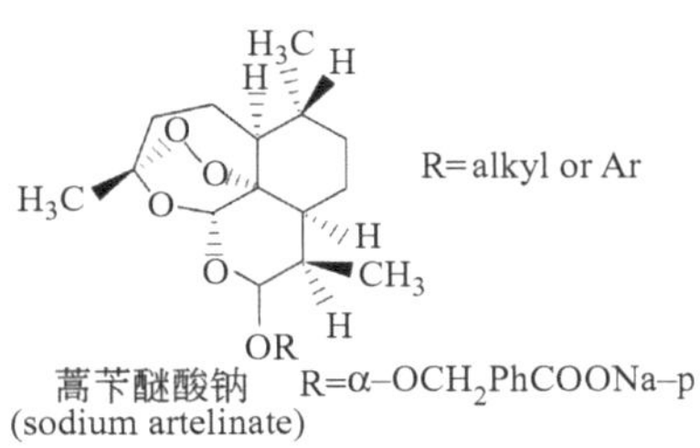

Figure 25.2: Structures of Ether Derivatives of Artemisinin C12 (I).

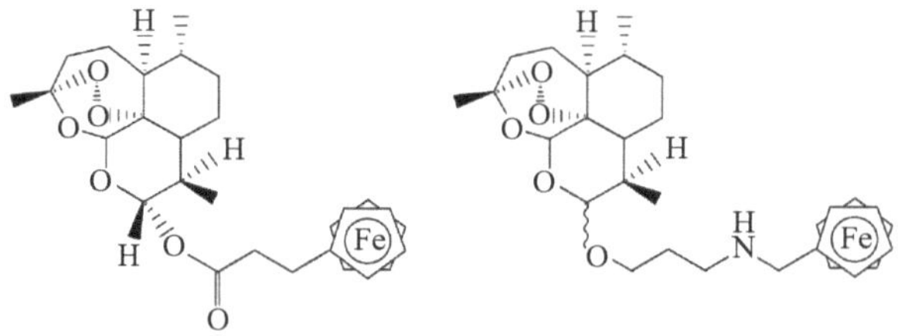

Figure 25.3: Structures of Ether Derivatives of Artemisinin C12 (II).

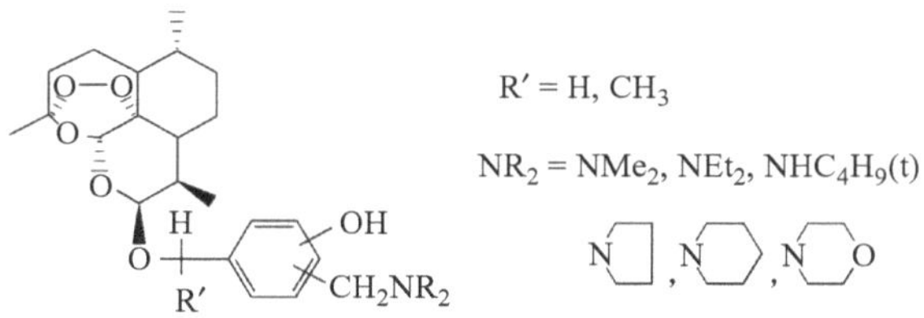

Figure 25.4: Structures of Ether Derivatives of Artemisinin C12 (III).

Delhaes et al. [12] in France reported metallocene derivatives with compounds listed in Fig. 25.3 as representatives in 2000.

In 2003, Li et al. [13] reported derivatives of ether-like compounds with Mannich base structure (Fig. 25.4).

1.1.2 Artemisinin–carboxylate and carbonate derivatives

Li et al. [4–7] synthesized a series of carboxylic acid ester and carbonate ester compounds in the study on artemisinin derivatives (Fig. 25.5). The efficacy was evaluated among some of these compounds. The *P. berghei* model showed that the antimalarial activity of some compounds was higher than artemisinin.

The ester-like derivatives derived in early stage was very unstable, and its crystals were easily decomposed to produce dihydroartemisinin after longtime storage. In 1995,

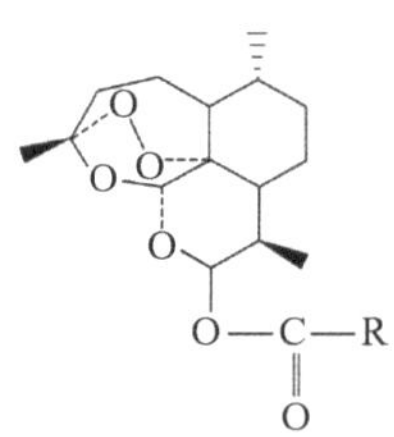

青蒿琥酯钠 (sodium artesunate)
R=—CH_2CH_2COONa

Figure 25.5: Structures of Artemisinin–Carboxylate and Carbonate Derivatives.

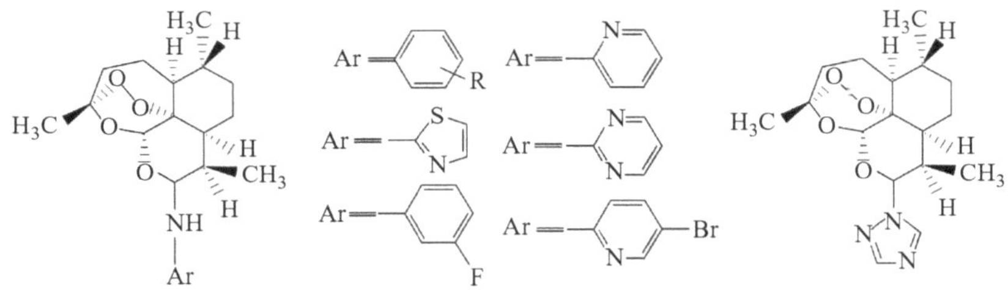

Figure 25.6: Structures of Aromatic Amine and *N*-Heterocyclic Derivatives of Artemisinin.

Li et al. [14] synthesized 11-hydroxy dihydroartemisinin ester derivatives with dehydrated dihydroartemisinin. After *P. berghei* screening, these compounds exhibited poor antimalarial activity when compared with artemisinin. This finding was supported by the results obtained by Pu et al. [15] in 1994, suggesting that the introduction of 11-hydroxy l group may decrease the antimalarial activity. The presence of bulky groups in ester derivatives contributes to their stability in vivo.

1.1.3 Aromatic amine and N-heterocyclic derivatives of artemisinin

In 1990, a series of aromatic amine artemisinin derivatives were generated by addition of bromine to dehydrated dihydroartemisinin followed by condensation with the corresponding aromatic amine in the study conducted by Lin et al. [16] (Fig. 25.6). These compounds exhibited significantly antimalarial activity in vitro, especially fluorine-substituted phenyl derivatives with highest activity. However, they failed to exert satisfactory antimalarial effect in *P. berghei* in vivo tests. Some of these compounds have also been synthesized by Yang et al. [17].

1.2 C13 Derivatives

Is C13-methyltransferase the essential group to antimalarial activity? What is the effect on antimalarial activity that can result from structural modification? Scientists have also carried out some research on this type of derivatives (Fig. 25.7).

In 1997, Paitayatat et al. [18] in Thailand synthesized a series of C13-substituted artemisinin derivatives by Michael addition between intermediates and nucleophilic reagents with artemisinin as raw materials after the elimination of organic selenium, and reported all these

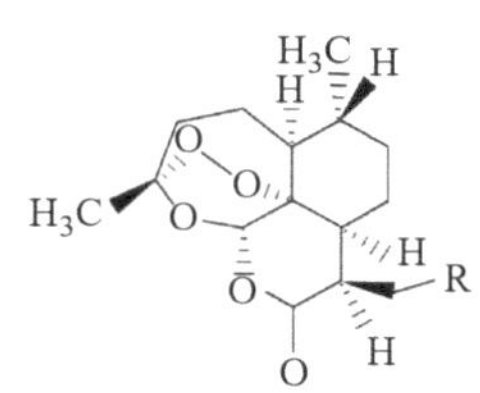

Figure 25.7: The Structure of Artemisinin C13 Derivatives.

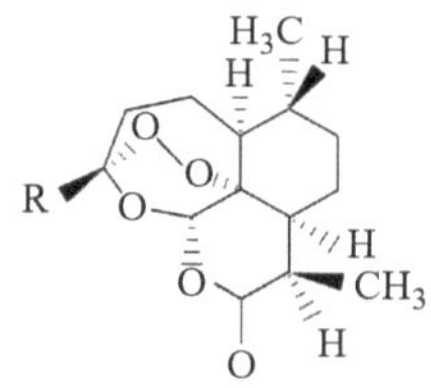

Figure 25.8: Structures of Artemisinin C4 Derivatives.

derivatives had significant antimalarial activity. They focused on the antimalarial activity of artemisinin and their derivatives, as well as the relationship between antimalarial activity and hemoglobin affinity, and proposed the complexes formed by interaction between artemisinin-like compounds and hemes may interfere with the normal metabolism of *Plasmodium*, thereby killing the *Plasmodium*. They believed that this can provide an easy way for screening of antimalarial activity of artemisinin derivatives (containing peroxide bridge structure).

In 2002, Grellepois et al. [19] reported a high yield of 12 trifluoromethyl-substituted–C13-substituted artemisinin derivatives based on anhydrodihydroartemisinin.

1.3 C4 Derivatives

A number of research groups explored C4 methyl artemisinin derivatives (Fig. 25.8). Several C4 derivatives of artemisinin synthesized by Jung and Wu and other groups [20–22] showed higher in vitro antimalarial activity than artemisinin. In 1996, four alkylated analogs of artemisinin synthesized by Avery et al. [23] showed stronger antimalarial activity than the parent compounds artemisinin.

2 Artemisinin Structure-Related Compounds

In research on artemisinin structure-related antimalarial compounds, a series of artemisinin structure-related compounds were typically derived from artemisinin or *Artemisia annua* L. acid by changing certain chemical groups by means of chemical reaction (substitution, addition, elimination, ring opening, polymerization, and so on). Some academics have also studied compounds with essential activity which had simpler structure than artemisinin, but retained necessary antimalarial groups. A brief description of progress of research on artemisinin structure-related compounds is as follows.

2.1 Artemisinin Structure-Related Compounds With Peroxide Bridge

In 1989, Ye et al. [24] synthesized deoxy artemisinin analogs A with the arteannuic acid as raw materials according to the procedures: arteannuic acid was hydrogenated by $NaBH_4$ in the

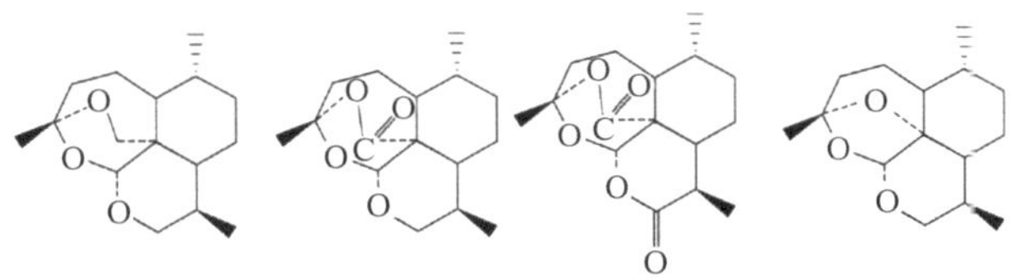

Figure 25.9: Structures of Artemisinin Structure-Related Compounds With Peroxide Bridge.

presence of $NiCl_2$ after CH_2N_2 esterification (Fig. 25.9). Then, the compound was reduced by $LiAlH_4$ to alcohol. After O_3 oxidation, the ring closure occurs to generate cycloalkenyl ether by means of Me_2S. Finally, deoxy artemisinin analogs A were generated after reaction with paraformaldehyde in the presence of $BF_3 \bullet OEt_2$ (Fig. 25.9A). Two other deoxy analogs B and C were derived from A via RuO_4 oxidation (for structure, see Fig. 25.9B–C). The antimalarial activity test showed that the compounds A–C had no antimalarial activity.

In 1990, Jung et al. [25] also synthesized deoxy artemisinin analog D (Fig. 25.9) by conversion of intermediates cycloalkene ether via three-step reaction, followed by ring closure and epoxidation in the presence of *m*-CPBA with arteannuic acid compounds as raw materials. The results of anti-*Plasmodium falciparum* activity test showed that the compound had no antimalarial activity.

2.2 Demethylated Artemisinin Structure-Related Compounds

In 1989, Avery et al. [26] synthesized 10,11-demethylated artemisinin analogs with pyrrolidine cyclohexene as raw materials (Fig. 25.10). The results of anti-*Plasmodium falciparum* activity test showed that the compound had significant antimalarial activity, suggesting that 10,11-methyl groups may not be essential to antimalarial activity of artemisinin.

2.3 Decarbonylated Artemisinin Structure-Related Compounds

In 1990, Jung et al. [25] successfully synthesized decarbonylated artemisinin analogs A by reduction of $NaBH_4$ in the presence of $BF_3 \bullet Et_2O$ during the synthesis of deoxy analog

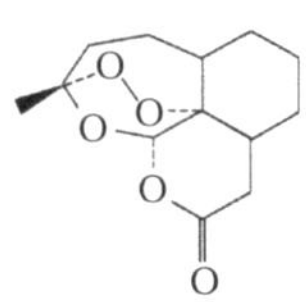

Figure 25.10: The Structure of Demethylated Artemisinin Structure-Related Compound.

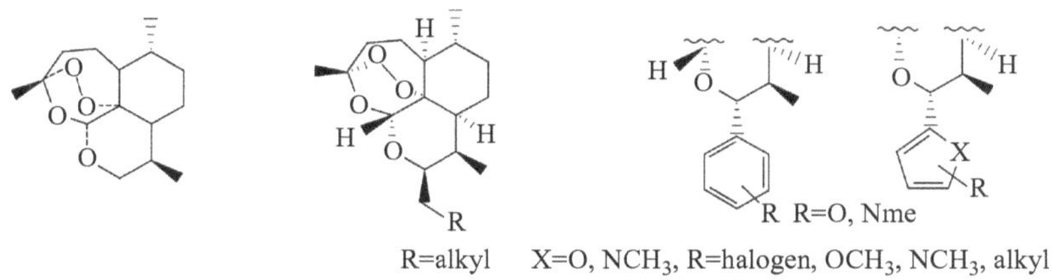

Figure 25.11: Structures of Decarbonylated Artemisinin Structure-Related Compounds (I).

(Fig. 25.11). In the same year, Ye al. [27] reported that compound A was derived from arteannuic acids via an eight-step reaction by means of total synthesis (Fig. 25.11A). The results of in vitro anti-*Plasmodium falciparum* activity test showed that the antimalarial activity of the compound was about 8 times higher than that of artemisinin. The results suggested that the stability of the molecular structure was improved after reduction of 12-carbonyl group that might not be essential to antimalarial activity. Over the past 10 years, the decarbonylated artemisinin structure-related compounds attached the attention of many research groups. A large number of such compounds were synthesized by Haynes group [28], Posner group [29–32], O'Neill group [33–35], Jung group [36–38], Ziffer group [39–41], and so on.

Avery [42] and Jung [43] groups have gained a number of decarbonylated artemisinin C13 derivatives (Fig. 25.12) by total synthesis or semisynthetic method, and conducted the efficacy evaluation and exploration of structure–activity relationship.

TDR 40292 (Fig. 25.13) is a decarbonylated artemisinin-like antimalarial compound in WHO collaborative study [28,34].

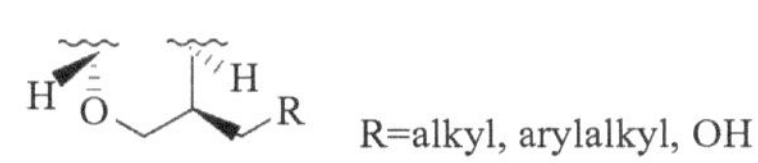

Figure 25.12: Structures of Decarbonylated Artemisinin C13 Derivatives.

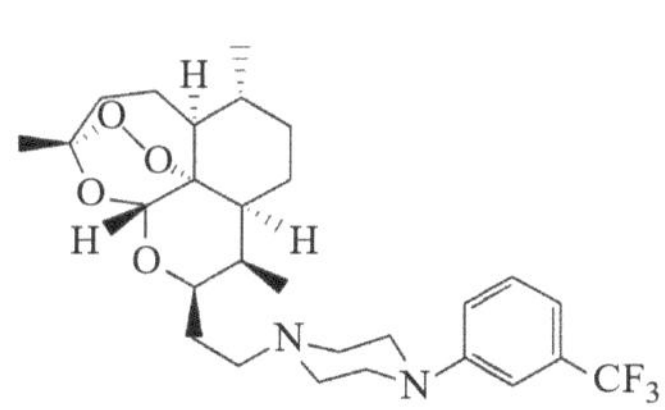

Figure 25.13: The Structure of TDR 40292.

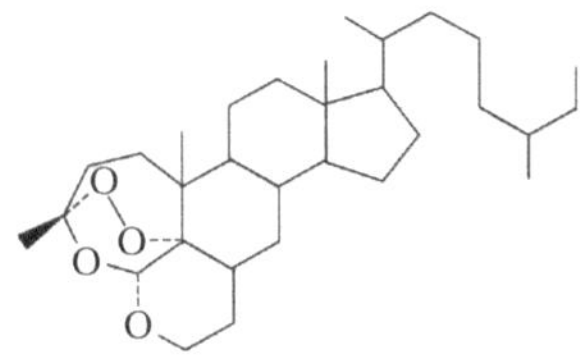

Figure 25.14: The Structure of Steroid Artemisinin Structure-Related Compound.

2.4 Steroid Artemisinin Structure-Related Compounds

In 1993, Rong and Wu [44] synthesized steroid artemisinin structure-related compounds (Fig. 25.14) by introducing cholesterol structure into the molecular structure of artemisinin to increase the affinity of artemisinin in the study, aiming to explore the introduction of lipophilic steroid groups on the antimalarial activity. In addition, some research groups simply connect artemisinin molecules with steroidal structure.

2.5 Lactam Artemisinin Structure-Related Compounds

In 1995, Torok et al. [45] transformed the lactone structure into lactam structure, which was more stable under acid environment by introducing the nitrogen into the artemisinin molecular structure to find more effective artemisinin drugs and increase their stability (Fig. 25.15). A series of lactam artemisinin analogs were synthesized with corresponding amine under the catalyzation of H_2SO_4/SO_2. They have drawn the conclusion that the transformation of lactone structure into lactam structure can increase the stability rather than reducing its physiological activity.

2.6 Ring-Opening Artemisinin Structure-Related Compounds

Posner et al. [46–48] synthesized several analogs of artemisinin without Ring D. Although the antimalarial mechanism was completely different from that of quinine, these compounds

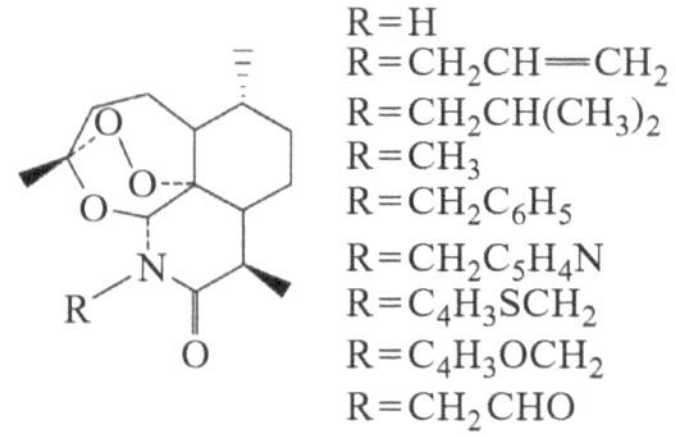

Figure 25.15: Structures of Lactam Artemisinin Structure-Related Compounds.

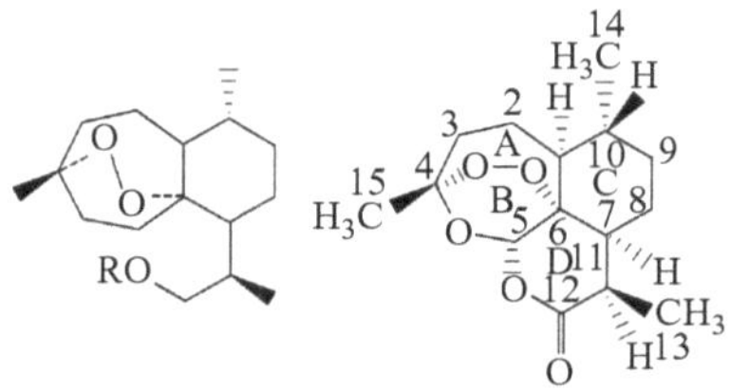

Figure 25.16: Structures of 4,5-Epoxy-Carbon-Substituted–Artemisinin Structure-Related Compounds.

had antimalarial effect similar to artemisinin. In this chapter, Ring D is not considered as an essential group to antimalarial activity.

In 1994, Avery et al. [49] synthesized artemisinin analogs with opened Ring A, with higher antimalarial activity than artemisinin.

In 2000, Yi et al. [50] reported artemisinin analogs without Ring C synthesized from natural(−)-carvone, but this compound with lower yield had no antimalarial activity.

2.7 4,5-Epoxy-Carbon-Substituted–Artemisinin Structure-Related Compounds

Avery et al. [51] synthesized a series of 4,5-epoxy-carbon-substituted–artemisinin structure-related compounds with artemisinin as raw material (Fig. 25.16), with enhanced antimalarial activity.

2.8 Artemisinin Structure Simplifier

Peroxide groups in molecular structure of artemisinin are essential to its antimalarial effect, so simplifiers of artemisinin structure often have similar structure, such as dioxane, tetraalkoxysilane, ozonide, and so on. The structure of some simplifiers is shown in Fig. 25.17.

In particular, the research and development of Compounds G are actively conducted by The Medicines for Malaria Venture MMV jointly with Ranbaxy (India).

2.9 Other Artemisinin Structure-Related Compounds

Other artemisinin structure-related compounds include thioethers, sulfones, artemisinin multimers, and other hybrids that are formed by interconnection of active pharmaceutical molecules (Fig. 25.18).

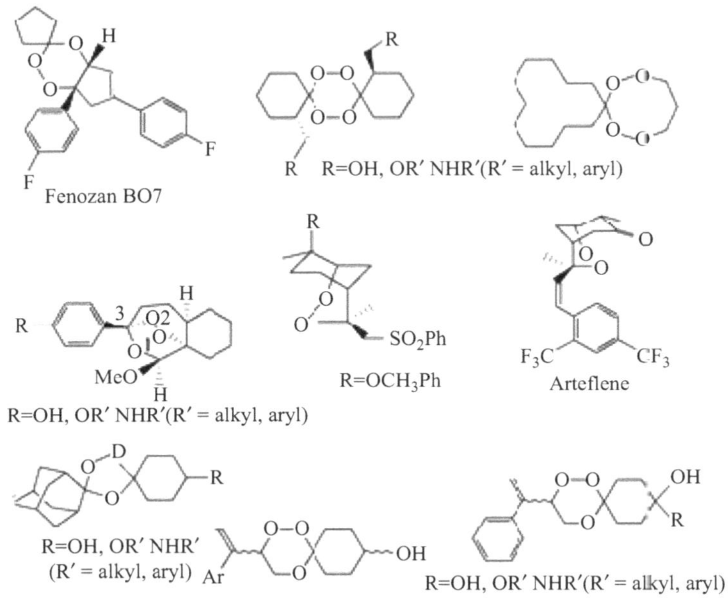

Figure 25.17: Structures of Some Artemisinin Structure Simplifiers.

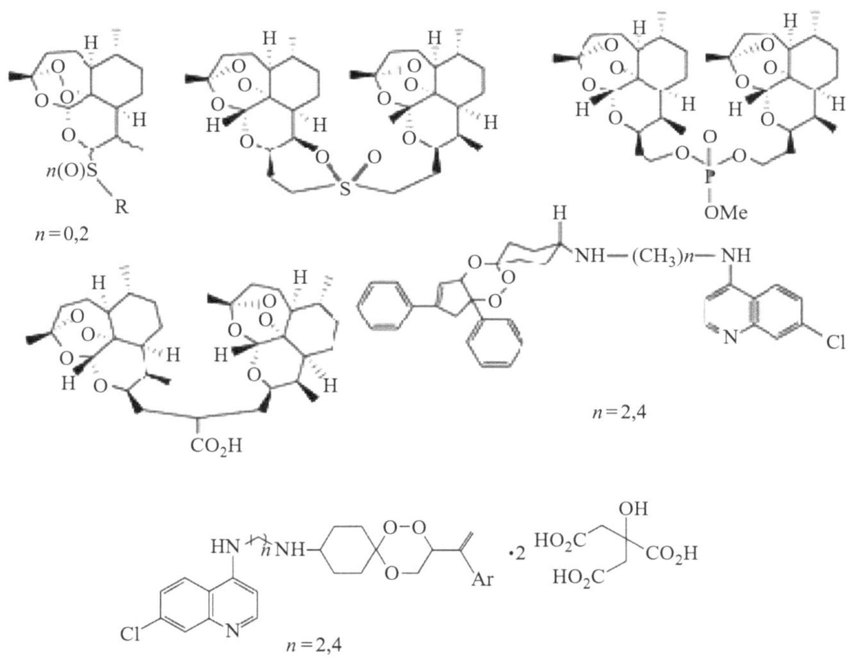

Figure 25.18: Structures of Some Other Artemisinin Structure-Related Compounds.

References

[1] Traditional Chinese medicine research institution. Artemisinin-based antimalarial album (1971–1978). Traditional Chinese medicine research data. 1978;3:54–55.

[2] Tu YY. Chemical studies on qinghaosu. J Tradit Chin Med 1982;2(1):3–8.

[3] Tu YY, Ni MY, Zhong YR, Li LN. Studies on the constituents of artemisinin and its derivatives. China J Chin Mater Med 1981;6(2):31.

[4] Li Y, Yu PL, Chen YX, et al. The synthesis of artemisinin derivatives. Chin Sci Bull 1979;24(14):667–9.

[5] Li Y, Yu PL, Chen YX, Liang Q, Gai YZ, Wang DS, Zheng YP. Studies on analogs of artemisinine I. The synthesis of ethers, carboxylic esters and carbonates of dihydroartemisinine. Acta Pharm Sin 1981;16(6):429–39.

[6] Li Y, Yu PL, Chen YX, Ji RY. Studies on analogs of artemisinine. II. The synthesis of some carboxylic esters and carbonates of dihydroarteannuin by using 4-(*N,N*-dimethylamino) pyridine as an active acylation catalyst. Acta Chim Sin 1982;40(6):557–61.

[7] Liang J, Li Y. Synthesis of the aryl ether derivatives of artemisinin. Chin J Med Chem 1996;6(1):22–5.

[8] Haynes PK, Chan HW, Cheung MK, et al. C-10-ester and ether derivatives of dihydroartemisinin-10α-artesunate, preparation of authentic-10-β-artesunate, and of other ester and ether derivatives bearing potential aromatic intercalating groups at C-10. Eur J Org Chem 2002;(1):113–32.

[9] O'Neil PM, Miller A, Ward SA, et al. Application of the TMSOTf-AgClO4 activator system to the synthesis of novel, potent, C-10-phenoxy derivatives of dihydroartemisinin. Tetrahedron Lett 1999;40(51):9129–32.

[10] O'Neil PM, Miller A, Bishop LPD, et al. Synthesis, antimalarial activity, biomimetic iron(II) chemistry, and in vivo metabolism of novel, potent C-10-phenoxy derivatives of dihydroartemisinin. J Med Chem 2001;44(1):58–68.

[11] Venugopalan B, Karnik PJ, Bapat CP, et al. Antimalarial activity of new ethers and thioethers of dihydroartemisinin. Eur J Med Chem 1995;30(9):697–706.

[12] Delhaes L, Biot C, Berry L, et al. Novel ferrocenic artemisinin derivatives: synthesis, in vitro antimalarial activity and affinity of binding with ferroprotoporphyrin IX. Bioorg Med Chem 2000;8(12):2739–45.

[13] Li Y, Yang ZS, Zhang H, et al. Artemisinin derivatives bearing Mannich base group: synthesis and antimalarial activity. Bioorg Med Chem 2003;11(20):4363–8.

[14] Li Y, Zhang HB, Ye YP. Synthesis of esters of dihydroartemisinin and 11,12-dihydroxyartemisinin. Chin J Med Chem 1995;5(2):127–30.

[15] Pu YM, Boris Y. Stereoselective oxidations of 13-methylglycal, anhydrodihydroartemisinin. Tetrahedron Lett 1994;35(14):2129–32.

[16] Lin AJ, Li LQ, Klayman DL, et al. Antimalarial activity of new water-soluble dihydroartemisinin derivatives 3. Aromatic amine analogs. J Med Chem 1990;33(9):2610–4.

[17] Yang YH, Li Y, Shi YL, et al. Artemisinin derivatives with 12-aniline substitution: synthesis and antimalarial activity. Bioorg Med Chem Lett 1995;5(16):1791–4.

[18] Paitayatat S, Tarnchompoo B, Thebtaranonth Y, et al. Correlation of antimalarial activity of artemisinin derivatives with binding amity with ferroorotoporphyrin IX. J Med Chem 1997;40(5):633–8.

[19] Grellepois F, Chorki F, Ourevitch M, et al. Allylic bromination of anhydrodihydroartemisinin and of its 10-trfluoromethyl analogue: a new access to 16-substituted artemisinin derivatives. Tetrahedron Lett 2002;43(43):7837–40.

[20] Rong YJ, Wu YL. Synthesis of C-4-substituted qinghaosu analogues. J Chem Soc Perkin Trans 1993;(18):2147–8.

[21] Jung M, Li X, Bustos DA, et al. Synthesis and antimalarial activity of (+)deoxoartemisinin. J Med Chem 1990;33(5):1516–8.

[22] Ye B, Wu YL, Li GF, Jiao YQ. Antimalarial activity of deoxoqinghaosu. Acta Pharm Sin 1991;26(3):228–30.

[23] Avery MA, Mehrotra S, Bonk JD, et al. Structure-activity relationships of the antimalarial agent artemisinin. 4. Effect of substitution at C-3. J Med Chem 1996;39(15):2900–6.

[24] Ye B, Wu YL. Synthesis of carba-analogues of qinghaosu. Tetrahedron 1989;45(23):7287–90.

[25] Jung M, Li X, Bustos DA. Synthesis and antimalarial activity of (+)-deoxoartemisinin. J Med Chem 1990;33(5):1516–8.

[26] Avery MA, Jennings-White C, Chong WKM. Simplified analogues of the antimalarial artemisininaynthesis of 6,9-desethylartemisinin. J Org Chem 1989;54(8):1792–5.

[27] Ye B, Wu YL. An efficient synthesis of qinghaosu and deoxoqinghaosu from arteanuic acid. J Chem Soc Commun 1990;726.

[28] O'Neill PM, Posner GH. A medicinal chemistry perspective on artemisinin and related endoperoxides. J Med Chem 2004;47(12):2945–64.

[29] Borstnik K, Paik IH, Shapiro TA, et al. Antimalarial chemotherapeutic peroxides: artemisinin, yingzhaosu A and related compounds. Int J Parastiol 2002;32(13):1661–7.

[30] O'Dowd H, Polypradith P, Xie SJ, et al. Antimalarial artemisinin analogs. synthesis via chemoselective C-C bond formation and preliminary biological evaluation. Tetrahedron 1999;55(12):3625–36.

[31] Posner GH, Paik IH, Sur S, et al. Orally active, antimalarial, anticancer, artemisinin-derived trioxane dimers with high stability and efficacy. J Med Chem 2003;46(6):1060–5.

[32] Posner GH, Parker MH, Northrop J, et al. Orally active, hydrolytically stable, esmisynethetic, antimalarial trioxanes in the artemisinin family. J Med Chem 1999;42(2):300–4.

[33] O'Neill PM, Pugh M, Stachulski AV, et al. Optimisation of the allylsiane approach to C-10 edoxo carba anaolgues of dihydroartemisinin: synthesis and in vitro antimalarial activity of new, metabolically stable C-10 analogues. J Chem Soc Perkin Trans 2001;2682–9.

[34] Hindley S, Ward SA, Storr RC, et al. Mechanism-based design of parasite-targeted artemisinin derivatives: synthesis and antimalarial activity of new diamine containing analogues. J Med Chem 2002;45(5):1052–63.

[35] O'Neill PM, Searle NL, Kan KW, et al. Novel, potent, semisynthetic antimalarial carba analogues of the first-generation 1,2,4-trioxane artemether. J Med Chem 1999;42(26):5487–93.

[36] Jung M, Bae J. An efficient synthesis of new analogs of water-soluble and hydrolytically stable deoxoartemisinin. Heterocycles 2000;53:261.

[37] Jung M, Yu D, Bustos D, et al. A concise synthesis of 12-(3′-hydroxy-*n*-propyl) deoxoartemisinin. Bioorg Med Chem Lett 1991;1(12):741–4.

[38] Jung M, Lee S, Ham J, et al. Antitumor activity of novel deoxoartemisinin monomers, dimers, and trimer. J Med Chem 2003;46(6):987–94.

[39] Ma JY, Katz E, Kyle DE, et al. Synthesis and antimalarial activities of 10-substituted deoxoartemisinins. J Med Chem 2000;43(22):4228–32.

[40] Ma JY, Katz E, Ziffer H. A new synthetic route to 10-β-alkyldeoxoartemisinins. Tetrahedron Lett 1999;40(49):8543–5.

[41] Pu YM, Ziffer H. Synthesis and antimalarial activities of 12-β-allyldeoxoartemisinin and its derivatives. J Med Chem 1995;38(4):613–6.

[42] Avery MA, Alvim-Gaston M, Vroman JA, et al. Structure-activity relationships of the antimalarial agent artemisinin. 7. Direct modification of (+)-artemisinin and in vivo antimalarial screening of new, potential preclinical antimalarial candidates. J Med Chem 2002;45(19):4321–35.

[43] Jung M, Lee K, Jung H. First synthesis of (+)-deoxoartemisitene and its novel C-11 derivatives. Tetrahedron Lett 2001;42(24):3997–4000.

[44] Rong YJ, Wu YL. Synthesis of steroidal 1,2,4-trioxane as potential antimalarial agent. J Chem Soc Perkin Trans 1993;21(18):2149–50.

[45] Torok DS, Ziffer H, Meshnick SR, et al. Synthesis and antimalarial activity of *N*-substituted 11-azaartemisinins. J Med Chem 1995;38(26):5045–50.

[46] Posner GH, Oh CH, Milhous WK. Olefin oxidative cleavage and dioxetane formation using triethylsilyl hydrotrioxide: applications to preparation of potent antimalarial 1,2,4-trioxanes. Tetrahedron Lett 1991;32(34):4235–8.

[47] Posner GH, Oh CH, Milhous WK. Extraordinarily potent antimalarial compounds: new, structurally simple, easily synthesized, tricyicl 1,2,4-trioxanes. J Med Chem 1992;35(13):2459–67.

[48] Posner GH, Cumming JN, Woo SH, et al. Orally active antimalarial 3-substituted trioxanes: new synthetic methodology and biological evaluation. J Med Chem 1998;41(6):940–51.

[49] Avery MA, Gao FL, Chong WKM, et al. Synthesis, conformational analysis, and antimalarial activity of tricycilc analogs of artemisinin. Tetrahedron 1994;50(4):957–72.
[50] Yi T, Shi ZD, Qin DG, Zhang YF, Wu YK, Li Y, Wu YL. Study on the synthesis of C-Nor-qinghaosu analogues. Acta Chim Sin 2000;58(4):448–53.
[51] Avery MA, Fan PC, Karle JM, et al. Structure-activity relationships of antimalarial agent artemisinin. 3. Total synthesis of (+)-13-carbaartemisinin and related tetra- and tricyclic structures. J Med Chem 1996;39(9):1885–97.

CHAPTER 26

Study Progress in Other Pharmacological Effects of Artemisinins

Chapter Outline

Malaria is the most serious insect-borne infectious disease in the world. In all insect-borne infectious diseases, malaria is one of the diseases with the highest morbidity and mortality.

In 1950s, scientists were very optimistic that malaria could be eliminated because at that time, by using dichlorophenyl trichloroethane and other mosquito biocides, the incidence of malaria and related diseases was significantly decreased. However, due to the long survival history of mosquitos, they have a strong ability to adapt to the environment, and quickly developed resistance to all used pesticides.

It has been believed that malaria has no other reservoir host than humans; therefore, it was thought that as long as the plasmodia in humans were killed, even if malaria cannot be completely eliminated, at least the spread could be significantly reduced. Based on this view, collective prophylaxis was made in many parts of the world, including regions in Thailand and South America, but it was just in these two regions that plasmodium resistance to chloroquine was first reported.

From *Artemisia annua* L. to Artemisinins
http://dx.doi.org/10.1016/B978-0-12-811655-5.00026-X

Mosquito resistance and plasmodia resistance made malaria more difficult to control, which resulted in the incidence of malaria and death toll in the 1990s were much higher than 30 years ago [1].

Plasmodium falciparum currently has developed resistance to almost all the original antimalarial drugs. It was found to be resistant to quinine in early 1900s, to proguanil and cycloguanil in the late 1940s, to pyrimethamine in mid-1950s, to chloroquine in late 1950s, and then it developed multidrug resistance. Suspected chloroquine resistance cases were found in Thailand in 1957, and chloroquine-resistant cases have been officially reported in Colombia and Thailand in 1960. Thereafter the resistance to chloroquine was widely spread. At present, except Central America, the chloroquine-resistant *P. falciparum* has been spread to all *P. falciparum* malaria epidemic areas throughout the world.

Artemisinins were developed in the process of looking for new structural types of antimalarial drugs to treat resistant malaria. This type of drugs is highly efficient, with quick onset of actions and low toxicity. These remarkable effects play an important role in today's global malaria treatment. We must protect the sensitivity of plasmodia to the artemisinins and prevent artemisinins abuse and occurrence of drug resistance. Therefore, it should be noted that these drugs should not be arbitrarily used for malaria prevention [2]. The unique and novel chemical structure of artemisinin has been attracting much attention. Pharmacologists are constantly exploring other pharmacological effects of these compounds in addition to antimalarial effect. This chapter provides an overview of other pharmacological studies on artemisinins.

1 Antitumor Effects

1.1 The Antitumor Effects of Artemisinins

The anticancer effects of artemisinin, dihydroartemisinin, artesunate, artemether, and other artemisinins are all significant.

Foreign scholars have carried out a wide range of antitumor studies on artemisinin-based compounds. Woerdenbag et al. [3] found that the aforementioned drugs have significant actions against *Ehrlich ascites* carcinoma cells in vitro in 1993. Efferth et al. [4] found that a variety of leukemia and colon cancer cell lines were very sensitive to these drugs, and the proliferation of nonsmall cell lung cancer cell lines, melanoma cells, breast cancer cells, ovarian cancer cells, prostate cancer cells, and renal cancer cells could be significantly inhibited in vitro. They also found that multidrug-resistant human leukemia CORE-CEM cells were sensitive to artemisinins; there were no cross-tolerance effects. Sadava et al. [5] also found that artemisinins could inhibit the growth of multidrug-resistant small cell lung cancer.

Chinese scholars [6–8] also found that the antitumor spectrum of artemisinins is broad in vitro. It could inhibit human hepatoma cell lines BEL-7402, SMMC-7721, H22 tumor strain,

HepC12 cell lines, human cervical carcinoma cells, human low differentiated squamous nasopharyngeal carcinoma CNE2 and SUME-1 cell lines, sarcoma cell S180, salivary-like cystic carcinoma cell line SACC-83 cells, and induced the apoptosis of these cells. In animal experiments, artemisinin and its derivatives could inhibit the tumor growth of liver cancer BEL-7403, ascites tumor, human low differentiated squamous nasopharyngeal carcinoma CNE2 and SUNE-1 transplanted in nude mice. They also have significant inhibiting effects on the growth of tumor S-180, liver cancer, *E. ascites* carcinoma LII and L615 transplanted in hybrid mice.

Overall, artemisinin of therapeutic concentration and its derivatives almost have no effects on normal cells. Artesunate (100 μmol/L) has no effect on the growth of human fibroblasts [9]. The proliferation of normal breast cells did not change after being treated with dihydrotestosterone [10,11].

Artemisinin and its derivatives have prominent antitumor activity [12,13], especially against multidrug-resistant tumor cells [14,15]. A lot of studies have been carried out on the antiproliferation, angiogenesis inhibition, and induction of apoptosis of artemisinin and its derivatives, and it has been found that artemisinin and its derivatives not only could act on tumor cells alone, but also have synergistic antitumor effect with 5-fluorouracil [14], pirarubicin, adriamycin [16], and other antitumor drugs.

1.2 The Antitumor Mechanism of Artemisinins

1.2.1 The inhibition of tumor cell proliferation

Artemisinin derivatives can inhibit the proliferation of tumor cells by affecting cell cycle at any stage, the most common is the G0/G1~S phase inhibition, and its mechanism of action is achieved through changes of the expression and activity of adjustment enzymes in the cell cycle [17–19]. Fang et al. [20] treated human breast cancer MCF7 cells with dihydroartemisinin; 24 h later, the intercellular arrangement became loose, which seemed to be separated, and 35.6% of the cells were dead, IC50 was 0.31 μmol/L.

1.2.2 Inhibition of angiogenesis

Angiogenesis refers to the process of generating new blood vessels (capillaries and venules) in a budding or division manner. Angiogenesis involves in many physiological responses, such as embryogenesis, wound healing in female reproductive cycle; it also involves in pathological processes, such as tumor, rheumatoid arthritis, and diabetes. Moreover, both the growth and metastasis of solid tumor are dependent on angiogenesis. Therefore, the antiangiogenesis has become a new strategy for cancer therapy [21]. Chen et al. [22, 23] found that artesunate can significantly inhibit the in vitro proliferation of four human tumor cell lines and the VEGF expression of human umbilical vein endothelial cells, and its inhibition rate on vascular endothelial growth factor receptor KDR/flk-1 is 80%. **Artemisinin**

can also inhibit the activity of glutathione *S*-transferase, reverse multidrug resistance of tumor cells against anthracyclines, platinum, and other chemotherapy drugs [24]. Dihydroartemisinin can induce the apoptosis of human umbilical vein endothelial cells in a dose- and time-dependent manner, and significantly reduces the expression of two major growth factor receptors Flt-1 and KDR/flk-1 of vascular endothelial cells [25]. **In addition, artesunate, artemisinin, and dihydroartemisinin can play a role by mitochondria and endoplasmic reticulum to reverse MDR** [26, 27].

1.2.3 Induction of apoptosis

Apoptotic study found that 24 h after artesunate and human umbilical vein endothelial cells were coincubated, the cells shrunk, the chromosome condensed, and apoptotic bodies appeared [28]. Artesunate not only can inhibits the proliferation of human hepatoma cell H22, but also can induce tumor cell apoptosis. It is detected that the expression of Bcl-2 gene and PCNA gene decreased, and Bax protein expression increased [29]. Heng et al. [30] found in their studies that artesunate can inhibit the proliferation of fibroblasts derived from lung and skin, but has no significant impacts on the proliferation of umbilical vein endothelial cells. **Dihydroartemisinin-induced apoptosis may be not depend on P-53, and more likely to depend on the activation of P-38MAPK** [31]. **Dihydroartemisinin can effectively improve the cell proapoptotic protein Bax and downregulate the antiapoptotic protein Bcl-2** [32] **while increasing the fas and activating the caspase-8** [33]. Whether artemisinins can inhibit vascular endothelial cell growth by inducing umbilical vein endothelial cell apoptosis still needs further investigation.

2 Immunomoregulation Effects [34]

2.1 Action of Artemisinins on Immune-Related Diseases

2.1.1 Immunoregulatory effects on systemic lupus erythematosus

Systemic lupus erythematosus is an autoimmune disease that compromises multisystem, which is characterized by the emergence of a series of autoantibodies, such as anti-ds-DNA and ANA antibodies within the body. In recent years, studies have shown that artemisinins can regulate the immune system in different ways to reduce the production of autoantibodies, thereby achieving target organ protection effects. Youyou Tu presided over the research work in the field of dihydroatemisinin, carried out the studies in collaboration with relevant institutions, and received invention patents. Limin et al [35] reported that dihydroatemisinin can increase the quantity of T lymphocytes (especially CD8+ T cells) in BXSB mice while reducing the high reactivity of B lymphocytes, thereby decreasing immunoglobulin generation and effectively suppressing the further development of systemic lupus erythematosus conditions. Yanjun et al. [36] conducted a series of studies on these drugs for their actions on lupus nephritis, and then reported dihydroatemisinin can significantly inhibit the anti-ds-DNA antibodies and TNF-α levels in the serum of BXSB mice, meanwhile dihydroartemisinin at the doses of 125 and 25 mg/kg can significantly improve the kidney lesions in BXSB mice. For example, the proliferation of glomerular mesangial cells and leukocyte infiltration was significantly improved, and there was no platinum ear-like structure generation. On the basis of this, Yanjun et al [37] also conducted in-depth studies on the molecular basis of immune regulation mechanism of dihydroartemisinin for the treatment of lupus nephritis. NF-κB is a protein factor capable of specifically binding to the enhancerκB sequence of immunoglobulin κ chain gene. When activated it can regulate the transcription of a variety of immune-related genes, including the gene expression of IL-1β, TNF-α, ICAM-1, IL-6, and other cytckines. Yanjun et al. have found in their studies that the levels of NF-κBp65 in the renal tissues of mice were lower in varying degrees in the experimental groups with 125, 25, and 5 mg/kg of dihydroartemisinin than those of lupus model group, which indicated the NF-κB levels in the target organ are suppressed by dihydroartemisinin, resulting in the downregulation of the gene expression of related cytokines, so that the deposition of immunoglobulins and complements in the kidney are decreased, thereby achieving the effect of delaying disease development. More recently, using spleen cells from lupus-prone MRL/lpr mice, Huang et al. [38] reported that dihydroartemisinin inhibits activation of the Toll-like receptor 4/interferon regulatory factor signaling pathway in spleen cell triggered by lipopolysaccharide (LPS). Wu et al. [39] reported the therapeutic effects of an artemisinin analog, termed SM934. A twice daily oral administration of SM934 ameliorated the lymphadenopathy symptoms of MRL/lpr mice by suppressing B cell activation and plasma cell formation.

2.1.2 *Immune regulation on* Pneumocystis carinii *pneumonia*

Pneumocystis carinii pneumonia (PCP) is a pulmonary complication susceptible in patients with immune deficiencies. Currently, the animal PCP models established by subcutaneously injecting cortisone acetate into Wistar rats are used to explore the role of drugs. Yatang et al. [40] used artemether to treat PCP in rats for the first time, and found that artemether is a promising anti-PCP drug. Bin et al. [41] used dihydroartemisinin and artesunate in PCP rat models, and then observed the survival rate, survival days, lung weight, lung weight/body weight, and mean number of cysts per visual field in lung printed images, and found both drugs have effects for this disease. Wengui et al. [42–44] further found PCP infection can cause high level of TNF-α secretion in rat's alveolar macrophages, while dihydroartemisinin treatment can reduce TNF-α level and improve the inflammatory reactions in host tissues, so that the host immune response can reach near normal state. Wengui et al. also found that dihydroatemisinin can reduce alveolar macrophage apoptosis, and also reduces the apoptosis of spleen cells in rat PCP models.

In addition, Noori et al. [45] found in their studies that artemisinin has immunosuppressive activity for delayed hypersensitivity. Some relevant scholars have found that the treatment with artemisinins for cerebral malaria is related not only to the actions of directly killing plasmodia, but also to its immunoregulatory mechanism [46]. Artemisinins also play an immunoregulatory role in tumor and inflammatory responses.

2.2 The Action Mechanism Study on the Immunoregulation of Artemisinins

Xiuzhen et al. [47] conducted a study by injecting different doses of artemisinin and dihydroartemisinin to mice, and found that these drugs can inhibit the humoral and cellular immunity in mice. When mice were intraperitoneally injected with 25–100 mg/kg of artemisinin or dihydroartemisinin, the quantity of IgG and IgM antibodies secreting cells (PFC) in their spleen cells were significantly inhibited. When the dose of artemisinin or dihydroartemisinin was 100 mg/kg, the delayed hypersensitivity reaction in mice to bovine serum albumin was significantly inhibited; the ability of mouse spleen cells in producing IL-2 and the proliferative responses of T cell mitogen ConA and B cell mitogen LPS were inhibited. Li et al. [48] reported that artesunate can inhibit IL-1 production. Ping et al. [49] found that large doses of artesunate can significantly reduce the weight of thymus and decrease serum lysozyme concentration in mice. Lysozyme is a nonspecific immune substance that is derived from mononuclear cells, and can specifically bind to Gram-positive bacteria and dissolve their cell walls. Due to decreased serum levels in mice, Ping et al. [49] speculated that artesunate can inhibit monocyte's activity; meanwhile, they cocultured different concentrations of artesunate with suboptimal concentrations of ConA and spleen lymphocytes, and found that artesunate can suppress IL-2 secretion in spleen lymphocytes when their supernatant's concentration was 6.5×10^{-5} mmol/L. By inhibiting IL-1 and IL-2 production, artesunate can prevent lymphocytes from entering S phase from G1 phase, so that the proliferation reaction was reduced, and DNA synthesis was inhibited. This may be the important mechanism for its immunosuppressive effect.

Yang et al. [50] reported that artemisinin and its derivatives can enhance mouse splenocytes DNA synthesis activated by alloantigen or ConA, but have no effects on LPS-activated mouse splenocytes DNA synthesis; meanwhile, in vivo experiments showed that sodium artesunate can amplify animals' DTH responses to sheep erythrocytes and antibodies generating reaction, and promote ConA-induced IL-2 expression in mouse splenocytes. Therefore, they thought that this type of drugs may have a role in the reconstruction of immune function. Since then, further study using cells culture and molecular hybridization technique have found that dihydroatemisinin can significantly promote ConA-induced mouse splenocytes proliferation and promote IL-2 production and their mRNA expression in mouse splenocytes and T cell lines LBRM. The study also found that dihydroartemisinin of the same concentration has no impacts on converted T cells and LPS-induced proliferation of lymphocyte, and thus proposed dihydroatemisinin may promote IL-2 production and lymphocytes proliferation through different pathways [51].

Peiying et al. [52] found that artesunate can downregulate sheep erythrocytes-induced mouse plasma IgG and hemolysin level, so that the plasma C3 level in plasmodia-infected mice was significantly increased. Meanwhile, these studies also showed that artesunate not only can increase the conversion rate of PHA-induced mouse lymphocytes and increase mouse spleen weight, but also can enhance DNFB-induced delayed hypersensitivity. Thus, they believed artesunate has the effects of inhibiting humoral immunity and enhancing cellular immunity.

3 Antischistosomal Effect [53–56]

In early 1980s, people have found that artemisinin has antischistosomal effect, and then further studies confirmed that artemisinin and its various derivatives all have antischistosomal effects. They can kill larvae of schistosome during the whole administration period, so it has good killing effect. Artemisinin has a protective effect on people who contacted the contaminated water. When they are used for early treatment of schistosoma japonicum cercariae infection, they can reduce the schistosome infection rate and the degree of infection, and they can prevent the occurrence of schistosomiasis. The antischistosomal active group of such substances is peroxide bridge, and the mechanism of action is interfering with glucose metabolism. The studies found that after liver phase schistosome is treated with artemether, glycogens were reduced, the activity of alkaline phosphatase was inhibited, body surface intestine tubes and gonads had obvious degenerative changes, schistosome body became thin, and protein content decreased, especially in female schistosome. The antischistosomal activity of artemisinin derivatives, such as artemether, artesunate, and arteether, was well reviewed in a newly published review [57].

4 Antiinflammatory Effect

The lipopolysaccharide/endotoxin (LPS) in the outer membrane of Gram-negative bacteria is the major causative factor that causes systemic infection, which can induce excessive release of inflammatory mediators [58]. There have been reports about studies on artemisinin antagonizing LPS-induced inflammation, and artemisinin can inhibit the synthesis of inducible NO synthase and NF-κB activation [59]. Aihua et al. [60] found that artesunate has significant inhibitive effects against LPS and combined interferon-stimulated NO synthesis of mouse peritoneal macrophages; artesunate has the same protective effect on LPS-stimulated mouse peritoneal macrophages RAW264.7; and with the increase of artesunate concentrations, artesunate inhibition on NO production also increased. Artesunate has significant inhibitive effect on LPS-induced TNF-α production, and compared with LPS alone the inhibition rate was 58% [61]. Yuqing et al. [62] also found that artemisinin can reduce the concentrations of lipid peroxide, alkaline phosphatase, endotoxin, TNF-α, and P450 of mice with endotoxic shock, increase the activity of SOD, reduce mice mortality, prolong the mean survival time of mice, and has protective effects on liver and lung tissues of mice with endotoxic shock.

In a BV2 microglia/HT22 neuron coculture system, artemether, a lipid-soluble derivative of artemisinin, has been reported to suppress LPS-induced proinflammatory mediators, including

NO/iNOS, PGE_2/COX-2/mPGES-1, TNFα, and IL-6. Artemether inhibits nueroinflammation via Nrf2-dependent mechnisms [63].The antiinflammatory and immunoregulatory functions of artemisinin and its derivatives were comprehensively discussed recently [64].

5 Liver Protection Effect

Chongyang et al. [65] have found in animal experiments that artesunate can antagonize liver reductive glutathione (GSH) depletion caused by acetaminophen and phosphorus, where larger doses can increase GSH content in the liver of normal hungry mice, suggesting that the antiliver injury effect may be associated with increasing liver GSH reserves and promoting detoxification of toxic metabolites. Xinsheng et al. [66] observed the role of artemisinin on reducing the formation of lipid peroxidation in liver microsomes, and found that artemisinin can significantly reduce the formation of lipid peroxide in liver microsomes.

6 Antiarrhythmic Effect

Baofeng et al. [67–71] reported that artemisinin has antiarrhythmic effect. They used whole-cell voltage-clamp technique to investigate artemisinin's effect on potassium ion current of isolated guinea pig ventricular myocytes and dog Purkinje fibers. The results showed the antiarrhythmic effect is related to potassium current inhibition. The study using patch clamp has found that artemisinin has a concentration-dependent inhibitive effect on IK of rabbit ventricular myocytes. The study using gene clamp has demonstrated that after injecting Kir2.1cRNA in African frog oocytes, artemisinin has significant inhibitive effect on Kir2.1/IK1. IK1 and Kir2.1 are related to the late repolarization phase of myocardial action potential, and are closely related to the occurrence and development of arrhythmias, so they are the important antiarrhythmic drug targets. In animal models, artemisinin has significant antagonistic effect on rat and guinea pig cardiac arrhythmias models induced by ouabain and aconitine. Artemisinin may have antiarrhythmic effect in clinical practice. Studying with nodose ganglion neurons of adult rats using whole-cell patch, artemisinin was demonstrated to inhibit both depolarization and repolarization of action potentials recorded from either A- or C-type nodose ganglion neurons in a concentration- and time-dependent manner [72].

7 Antiviral Effect

Artemisinins possess strong inhibitory effects against double-stranded DNA viruses [73–75], such as Epstein-Barr virus, cytomegalovirus, and human herpesvirus 6A. They are some effective against hepatitis B virus and hepatitis C virus [76, 77]. Ruisheng et al. [78] found in chicken embryo experiments that artemisinin has antiinfluenza virus effect, the nature of which is inhibition rather than killing.

In 2005, Romero et al. [79] reported that artemisinin has antihepatitis B virus effect. Peilin et al. [80] investigated the in vitro anti-Coxsackie virus B 3 (CVB3) effect of artemisinin, and found artemisinins have in vitro anti-CVB3 effect, and the antiviral mechanism is achieved by virus adsorption blocking and virus copy inhibition.

8 Toxoplasmosis Treatment Effect

Artemisinins have reported to inhibit *Toxoplasmagondii* replication, growth, and attachment to and invasion of host cells [81]. Li et al. [82] reported that dihydroatemisinin can inhibit the nucleic acid metabolism of toxoplasma, and dihydroartemisinin combined with sodium sulfadiazine can produce a synergistic effect in the treatment of acute toxoplasma infection in mice, which can more effectively prevent relapse after drug withdrawal, and have better effect than using dihydroartemisinin alone. Chonggong [83] used dihydroatemisinin to treat 18 patients with recent active infection and 7 cases of toxoplasma infection, and believed that it has some clinical effect on toxoplasma, but whether it can cure toxoplasma infection completely to achieve a curative purpose or not still needs more long-term clinical tracking observations to accumulate more data. In vitro experiments showed that artemisinins can inhibit toxoplasma from invading cells. Artemisinin mainly acts on cell membranes, mitochondria, and nuclei, and then extensively damages their membrane system, resulting in such phenomenon as nuclear membrane rupture, mitochondrial swelling, vacuolar degeneration, endoplasmic reticulum extension, and even nuclear fragmentation and nuclear fusion [55]. These reports demonstrate the therapeutic potential of artemisinins to prevent and treat human *Toxoplasmosis*.

9 Anticanine Eperythrozoon Effect [84]

It was found in the study of treating canine eperythrozoonosis with artemisinin and allicin that artemisinin not only has significantly better effect for the treatment of canine eperythrozoonosis than tetracycline and neoarsphenamine, but also cured more than 10 cases of eperythrozoonosis animals that respond poorly to tetracycline and neoarsphenamine. In addition to inhibiting and killing eperythrozoon, artemisinin also has antipyretic effect, which is good for both clinical symptoms improvement and symptomatic therapy for most of acute eperythrozoonosis animals with fever.

10 Anticoccidial Effect

Allen et al. [85] and Dhingra et al. [86] reported that low concentration of artemisinin as a feed additive can control avian coccidiosis. Changshen et al. [87] reported that they found in their chicken Eimeria prevention experiments by using artemisinin that different concentrations of artemisinin all can treat artificially Eimeria-infected chickens to prevent them from dying.

11 Antiasthmatic Effect

Maosheng et al. [88] demonstrated through animal experiments that artesunate has relaxant effect on the smooth muscles of trachea and bronchia of whole animals, indicating that through inhalation or oral administration, artesunate has better inhibitive effect on asthma in guinea pigs. The mechanism may be related to the blocking of calcium inflow and the activity of adenylate cyclase in activated tracheal tissues.

12 Antipregnancy Effect

Jihong and Yuanpei [89] reported that artesunate has antipregnancy effect in animal trials. The results showed that artesunate has very high selective toxicity on embryos. Even low dose of artesunate can cause embryonic death leading to abortion, but has insignificant impact on maternal uterus and ovaries. Therefore, they raised a possibility that artesunate could be developed as an artificial abortion drug. And they pointed out that the present clinical dose of artesunate is 112 mg/kg × 4 days (i.v.), which is close to the antipregnancy ED50 in golden hamster, so they recommended that the drug should be used with caution in pregnant women.

So far it has been more than 30 years since artemisinin was found, and artemisinin has become a new antimalarial drug that plays an important role in the treatment of resistant malaria following quinine and chloroquine and other original quinoline drugs. Beyond antimalarial effect, scholars from various countries also carried out other pharmacological exploration, such as antitumor, antiviral, immune regulation, and antischistosome effects as described earlier. Artemisinins have novel and unique chemical and structural characteristics, and with the progress of science and technology, the application of different active test methods and systems [90–92] will create conditions for the scientific and rational use of artemisinins and further research and development of new drugs for the benefit of mankind.

References

[1] Foote SJ, Cowman AF. The mode of action and the mechanism of resistance to antimalarial drugs. Acta Trop 1994;56(2–3):157.

[2] Peihui C, Youyou T, Fengyun W, et al. The effects of dihydroartemisinin on the development of plasmodium yoelii in mosquitoes. Chin J Parasitol Parasit Dis 1998;16(6):421–4.

[3] Woerdenbag HJ, Moskal TA, Pras N, et al. Cytotoxicity of artemisinin-related endoperoxicles to Ehrlich ascites tumor cells. J Nat Prod 1993;56(6):849–56.

[4] Efferth T, Davey M, Olbrich A, et al. Activity of drugs from traditional Chinese medicine towards sensitive and MDRI or MRPI-overerpressing multidrug-resistant human CORE-CEM Leukenna cells. Blood Cells Mol Dis 2002;28(2):160–8.

[5] Sadava D, Philips T, Lin C, et al. Transferrin overcomes drug resistance to artemisinin in human small-cell lung carcinoma cells. Cancer Lett 2002;179(2):151–6.

[6] Xiaoping Y, Qichao P, Yongju L, et al. The anti-tumor research of artesunate sodium. Cancers 1997;16(3). 186–187, 190.

[7] Qin W, Limao W, Aiyuan L, et al. Experimental study on the anti-hepatoma effect of artesunate. Chin Mater Med 2001;26(10):707–9.
[8] Liuying X, Peiying L, Dan Z, et al. The inhibition of artesunate on mice hepatic carcinomaand S180 solid tumor. Pract J Cancer 2001;16(3):245–7.
[9] Efferth T, Marschall M, Wang X, et al. Antiviral activity of artesunate towards wild-type, recombinantand ganciclovir-resistant human cytomegaloviruses. J Mol Med 2002;80(4):233–42.
[10] Singh NP, Lai H. Selective toxicity of dihydroartemisinin and holotransferrin toward human breast cancer cells. Life Sci 2001;70(1):49–56.
[11] Lai H, Singh NP. Oral artemisinin prevents and delays the development of 7, 12-dimethylbenz [α] anthracene (DMBA)-induced breast cancer in the rat. Cancer Lett 2006;231(1):43–8.
[12] Beekman AC, Woerdenbag HJ, Van Uden W, et al. Stability of artemisinin in aqueous environments: impact on its cytotoxic action to Ehrlich ascites tumour cells. J Pharm Pharmacol 1997;49(12):1254–8.
[13] Yang XP, Liang YJ. Study on anti-tumor effect of sodium artesunate. Cancer 1997;16(3):186–90.
[14] Zhang J, Wang S, Zhang F, et al. The effect of artesunate on the in vitro proliferation of human cancer cells Hela and SACC-83 cells. Chin Tradit Herb Drugs 2001;32(4):345–7.
[15] Sadava D, Phillips T, Lin C, et al. Transferrin overcomes drug resistance to artemisinin in human small-cell lung carcinoma cells. Cancer Lett 2002;179(2):151–6.
[16] Reungpathanaphong P, Mankhetkorn S. Modulation of multidrug resistance by artemisinin, artesunate and dihydroartemisinin in K562/adr and GLC4/adr resistant cell lines. Biol Pharm Bull 2002;25(12):1555–61.
[17] Willoughby JA Sr, Sundar SN, Cheung M, Tin AS, Modiano J, Firestone GL. Artemisinin blocks prostate cancer growth and cell cycle progression by disrupting Sp1 interactions with the cyclin-dependent kinase-4 (CDK4) promoter and inhibiting CDK4 gene expression. J Biol Chem 2009;284(4):2203–13.
[18] Wang SJ, Gao Y, Chen H, Kong R, Jiang HC, Pan SH, Xue DB, Bai XW, Sun B. Dihydroartemisinin inactivates NF-kappaB and potentiates the anti-tumor effect of gemcitabine on pancreatic cancer both in vitro and in vivo. Cancer Lett 2010;293(1):99–108.
[19] Uren AG, Wong L, Pakusch M, Fowler KJ, Burrows FJ, Vaux DL, Choo KH. Survivin and the inner centromere protein INCENP show similar cell-cycle localization and gene knockout phenotype. Curr Biol 2000;10(21):1319–28.
[20] Fang L, Jian D, Liping L. The in vitro inhibition of dihydroartemisinin on human breast cancer MCF7 cells. Chin J New Drugs 2002;11(12):934–6.
[21] Kim KJ, Li B, Winer J, et al. Inhibition of vascular endothelial growth factor-induced angiogenesis suppresses tumour growth in vivo. Nature 1993;362(6423):841–4.
[22] Chen HH, Zhou HJ, Wang WQ, et al. Antimalarial dihydroartemisinin also inhibits angiogenesis. Cancer Chemother Pharmacol 2004;53(5):423–32.
[23] Chen HH, Zhou HJ, Fang X, et al. Inhibition of human cancer cell line growth and human umbilical vein endothelial cell angiogenesis by artemisinin derivatives in vitro. Pharmacol Res 2003;48(3):231–6.
[24] Mukanganyama S, Widersten M, Naik YS, Mannervik B, Hasler JA. Inhibition of glutathione S-transferases by antimalarial drugs possible implications for circumventing anticancer drug resistance. Int J Cancer 2002;97(5):700–5.
[25] Dell Eva R, Pfeffer U, Vené R, et al. Inhibition of angiogenesis in vivo and growth of Kaposi's sarcoma xenograft tumors by the anti-malarial artesunate. Biochem Pharmacol 2004;68(12):2359–66.
[26] Reungpatthanaphong P, Mankhetkorn S. Modulation of multidrug resistance by artemisinin, artesunate and dihydroartemisinin in K562/Adr and GLC4/Adr resistant cell lines. Biol Pharm Bull 2002;25(12):1555–61.
[27] Efferth T, Sauerbrey A, Olbrich A, et al. Molecular modes of action of artesunate in tumor cell lines. Mol Pharmacol 2003;64(2):382–94.
[28] Efferth T, Oesch F. Oxidative stress response of tumor cells: microarray-based comparison between artemisinins and anthracyclines. Biochem Pharmacol 2004;68(1):3–10.
[29] Qin W, Limao W, Yi Z, et al. Research on the mechanism of artesunate's antitumor action. Acta Pharm Sin 2002;37(6):477–8.
[30] Heng K, Qingsheng Y. Research on the selectivity of artesunate on fibroblasts and endothelial cells. Chin J Clin Pharm Therapeut 2005;10(6):687–90.

[31] Mu D, Zhang W, Chu D, Liu T, Xie Y, Fu E, Jin F. The role of calcium, P38 MAPK in dihydroartemisinin-induced apoptosis of lung cancer PC-14 cells. Cancer Chemother Pharmacol 2008;61(4):639–45.
[32] Chen H, Sun B, Wang S, Pan S, Gao Y, Bai X, Xue D. Growth inhibitory effects of dihydroartemisinin on pancreatic cancer cells: involvement of cell cycle arrest and inactivation of nuclear factor-kappa B. J Cancer Res Clin Oncol 2010;136(6):897–903.
[33] Jiao Y, Ge CM, Meng QH, Cao JP, Tong J, Fan SJ. Dihydroartemisinin is an inhibitor of ovarian cancer cell growth. Acta Pharmacol Sin 2007;28(7):1045–56.
[34] Bei S, Xingyi M. The immunoregulatory effects of artemisinin and its derivatives. Chin J Integr Tradit West Nephrol 2005;6(3):176–8.
[35] Limin X, Xuerong C, Youyou T, et al. The effects of dihydroatemisinin on Lupus BXSB mice. Chin J Dermatovenereol Integr Tradit West Med 2002;1(1):19–20.
[36] Yanjun D, Weidong L, Youyou T, et al. The effects of dihydroarteannuin on the autoantibodies generation, TNF-α secretion and Lupus nephritis pathologic changes in BXSB lupus mice. Chin J Integr Tradit West Med 2003;23(9):676–9.
[37] Yanjun D, Weidong L, Youyou T, et al. Research on the effect and its mechanism of dihydroarteannuin on lupus nephritis in BXSB mice. Chin Pharmacol Bull 2003;19(10):1125–8.
[38] Huang X, Xie Z, Liu F, Han C, Zhang D, Wang D, Bao X, Sun J, Wen C, Fan Y. Dihydroartemisinin inhibits activation of the Toll-like receptor 4 signaling pathway and production of type I interferon in spleen cells from lupus-prone MRL/lpr mice. Int Immunopharmacol 2014;22(1):266–72.
[39] Wu Y, He S, Bai B, Zhang L, Xue L, Lin Z, Yang X, Zhu F, He P, Tang W, Zuo J. Therapeutic effects of the artemisinin analog SM934 on lupus-prone MRL/lpr mice via inhibition of TLR-triggered B-cell activation and plasma cell formation. Cell Mol Immunol 2016;13(3):379–90.
[40] Yatang C, Liang M, Qin M, et al. The preliminary observations on the treatment of pneumocystis carinii pneumonia in rats with artemether. Chin J Parasit Dis Control 1994;7(4):278–81.
[41] Bin Y, Yatang C, Chengwei L, et al. The efficacy study on dihydroartemisinin and artesunate for the treatment pneumocystis carinii pneumonia in rats. Chin J Zoonoses 2001;17(4). 43–45, 48.
[42] Wengui L, Yatang C, Chengwei L, et al. The effects of dihydroartemisinin on TNF-α levels in rats with pneumocystis carinii pneumonia. Immunol J 2001;17(4). 274–276, 294.
[43] Wengui L, Yatang C, Chengwei L, et al. The effects of dihydroartemisinin on alveolar macrophages apoptosis in rats with pneumocystis carinii pneumonia. Immunol J 2002;18(3):207–10.
[44] Wengui L, Yatang C, Chengwei L, et al. The effects of dihydroartemisinin on splenocytes apoptosis in rats with pneumocystis carinii pneumonia. Chin J Zoonoses 2003;19(2):55–9.
[45] Noori S, Naderi GA, Hassan ZM, et al. Immunosuppressive activity of a molecule isolated from *Artemisia annua* on DTH responses compared with cyclosporin A. Int Immunopharmacol 2004;4(10–11):1301–6.
[46] Xiaoyu L, Huizhen L. The effects of artemether on the immune functions of erythrocytes. Acta Pharmacol Sin 1986;7(5):471.
[47] Xiuzhen S, Shusheng X, Youyou T, et al. The experimental study on the immunosuppressive effects of artemisinin and its derivatives. J Chin Integr Med 1991;11(1):37–8.
[48] Li W, Yuxiang G. The effects of artesunate on mice allergic contact dermatitis and interleukin 1. Chin J Dermatol 1992;25(3):165–7.
[49] Ping Z, Yuxiang G. The effects of artesunate on immune functioni in mice. J Bengbu Med Coll 1996;21(1):5–7.
[50] Yang SX, Xie SS, Gao HL, et al. Artemisinin and its derivatives enhance T lymphocyte-mediated immune responses in normal mice and accelerate immunoreconstitution of mice with syngeneic bone marrow transplantation. Clin Immunol Immunopathol 1993;69(2):143–8.
[51] Sixun Y, Shusheng X, Dalong M, et al. Dihydroartemisinin promoting IL-2 production and mRNA expression. Acta Pharmacol Sin 1994;15(6):515–20.
[52] Peiying L, Zhaoming F, Jingqiang P, et al. The effects of artesunate on immune functions in mice. Acta Pharmacol Sin 1995;16(5):441–4.
[53] Wenju Y, Genfa W, Jiqing Y, et al. The experimental study on artemisinin derivatives for the treatment of animal schistosomiasis. Chin Pharm Bull 1980;15(4):182.

[54] Institute of Chinese Materia Medica, China Academy of Chinese Medical Sciences. Variety sorting and quality research of commonly used Chinese herbal medicine—*Artemisia annua*. 1990:6–7.
[55] Mingsheng X. The research progress in schistosomiasis prevention with artemisinin, artemether and artesunate. Chin J Schisto Control 1998;10(4):251–3.
[56] Xiao SH. Development of antischistosomal drugs in China, with particular consideration to praziquantel and the artemisinins. Acta Trop 2005;96(2–3):153–67.
[57] Saeed ME, Krishna S, Greten HJ, Kremsner PG, Efferth T. Antischistosomal activity of artemisinin derivatives in vivo and in patients. Pharmacol Res 2016;110:216–26.
[58] Rietschel ET, Kirikae T, Schade FU, et al. The chemical structure of bacterial endotoxin in relation to bioactivity. Immunobiology 1993;187(3–5):169–90.
[59] Aldieri E, Atragene D, Bergandi L, et al. Artemisinin inhibits inducible nitric oxide synthase and nuclear factor NF-κB activation. FEBS Lett 2003;552(2–3):141–4.
[60] Aihua L, Baoyun X, Chunying L, et al. The inhibition of artesunate on endotoxin induced nitric oxide synthesis. Foreign Medical Sciences TCM volumes 2001;23(3):156.
[61] Aihua L, Baoyun X, Jinhua W, et al. The study on artesunate's inhibition on endotoxin-induced inflammatory cytokine synthesis. Chin Mater Med 2001;26(11):770.
[62] Yuqing T, Yi Z. The experimental study on artemisia annua extract against endotoxin. Chin Mater Med 1999;24(3):166–71.
[63] Okorji UP, Velagapudi R, El-Bakoush A, Fiebich BL, Olajide OA. Antimalarial drug artemether inhibits neuroinflammation in BV2 microglia through Nrf2-dependent mechanisms. Mol Neurobiol 2016;53(9):6426–43.
[64] Shi C, Li H, Yang Y, Hou L. Anti-inflammatory and immunoregulatory functions of artemisinin and its derivatives. Mediators Inflamm 2015;2015:435713.
[65] Chongyang C, Zunian T, Ronggan L. The effects of artesunate on glutathione contents in normal and liver toxic poisoned mouse liver. Chin Pharmacol Bull 1993;9(1):52–4.
[66] Xinsheng F, Quan Z, Hong M. The effects of artemisinin on the peroxidation injury of rat liver microsomes. Chin J Biochem Pharm 1999;20(6):295–6.
[67] Huizhen W, Baofeng Y, Dali L, et al. The study on the anti-arrhythmic effects of artemisinin. Chin Pharmacol Bull 1998;14(1):94.
[68] Baofeng Y, Yurong L, Changqing X, et al. Mechanisms of artemisinin antiarrhythmic action. Chin J Pharmacol Toxicol 1999;13(3):169–75.
[69] Jin Z, Zhimin D, Xiaohong Q, et al. The antiarrhythmic mechanism of antimalarial artemisinin. Acta Pharm Sin 1999;34(8):569–72.
[70] Baoxin L, Baofeng Y, Yurong L, et al. The antiarrhythmic effect and mechanism of artemisinin. Chin Pharmacol Bull 1999;15(5):449–52.
[71] Zhimin D, Ying L. The study on anti-arrhythmic effect of artemisinin. Chin Pharm J 2003;38(5):372.
[72] Qiao G, Li S, Yang B, Li B. Inhibitory effects of artemisinin on voltage-gated ion channels in intact nodose ganglion neurones of adult rats. Basic Clin Pharmacol Toxicol 2007;100(4):217–24.
[73] Efferth T, Romero MR, Wolf DG, Stamminger T, Marin JJ, Marschall M. Theantiviral activities of artemisinin and artesunate. Clin Infect Dis 2008;47:804–11.
[74] Milbradt J, Auerochs S, Korn K, Marschall M. Sensitivity of human herpesvirus 6 and other human herpesviruses to the broad-spectrum antiinfective drug artesunate. J Clin Virol 2009;46:24–8.
[75] Schreiber A, Harter G, Schubert A, Bunjes D, Mertens T, Michel D. Antiviral treatment of cytomegalovirus infection and resistant strains. Expert Opin Pharmacother 2009;10:191–209.
[76] Cui X, Wang Y, Kokudo N, Fang D, Tang W. Traditional Chinese medicine and related active compounds against hepatitis B virus infection. Biosci Trends 2010;4:39–47.
[77] Paeshuyse J, Coelmont L, Vliegen I, Van heme J, Vandenkerckhove J, Peys E, et al. Hemin potentiates the anti-hepatitis C virus activity of the antimalarial drug artemisinin. Biochem Biophys Res Commun 2006;348:139–44.
[78] Ruisheng Q, Zhuliang L, Jianliang Y, et al. The immune and antiviral effects of artemisinin. J Tradit Chin Med 1981;22(6):463–6.

[79] Romero MR, Efferth T, Serrano MA, et al. Effect of artemisinin/artesunate as inhibitors of hepatitis B virus production in an "in vitro" replicative system. Antiviral Res 2005;68(2):75–83.
[80] Peilin M, Fengmin Z, Weihua S, et al. The experimental observations on artemisinin's anti-Coxsackie B virus infection. Chin J Endemiol 2004;23(5):403–5.
[81] D'Angelo JG, Bordon C, Posner GH, Yolken R, Jones-Brando L. Artemisinin derivatives inhibit *Toxoplasma gondii* in vitro at multiple steps in the lytic cycle. J Antimicrob Chemother 2009;63:146–50.
[82] Li Y, Shaobo G, Zhiqun Q, et al. Further observations on the efficacy of dihydroartemisinin for the treatment of mice with acute toxoplasmosis. Acta Parasitol Med Entomol Sin 2002;7(2):70–6.
[83] Chonggong W. The study on artemisinin treatment for toxoplasma and preliminary observation on clinical treatment efficacy. Chin J Zoonoses 1997;13(1):79–80.
[84] Xiuguo H, Hongquan L, Hua Yang, et al. The study on artemisinin and allicin for treatment of canine eperythrozoonosis. J Shanghai Agric Coll 1997;15(3):218–23.
[85] Allen PC, Danforth HD, Augustine PC. Dietary modulation of avian coccidiosis. Int J Parasitol 1998;28(7):1131–40.
[86] Dhingra V, Pakki SR, Narasu ML. Antimicrobial activity of artemisinin and its precursors. Curr Sci 2000;78(6):709–13.
[87] Changshen N, Longxian Z, Min L, et al. The prevention experiment of synthetic artemisinin on Eimeria tenella. Henan Agric Sci 2001;1(3):29–30.
[88] Maosheng Y, Kang H, Weiming X. The study on artesunate's inhibition on guinea pig asthma model. Chin Pharm J 1999;34(3):160.
[89] Jihong X, Yuanpei T. Anti-pregnancy effect of dihydroartemisinin and artesunate. Acta Pharm Sin 1996;31(9):657.
[90] Bailey NJC, Wang YL, Sampson J, et al. Prediction of anti-plasmodial activity of *Artemisia annua* extracts: application of H-1 NMR spectroscopy and chemometrics. J Pharm Biomed Anal 2004;35(1):117–26.
[91] Bhonsle JB. Role of array of oxygen atoms of artemisinin in the anttmalarial activity: Part 1—molecular modeling studies to investigate compounds with artemisinin-like array of oxygen atoms. Indian J Chem 1995;34B(5):380–8.
[92] Ranasinghe A, Sweatlock JD, Cooks RG. A rapid screening methodfor artemisinin and its congeners using MS/MS: search for new analogues in *Artemisia annua*. J Nat Prod 1993;56(4):552–63.

Appendix

Curriculum Vitae

Name: Youyou TU 屠呦呦
Date of birth: December 30, 1930
Nationality: Chinese
Present professional positions: Lifelong Professor, Chief Professor, Supervisor of Doctorate and the Director of the Qinghaosu Research Center, China Academy of Chinese Medical Sciences, Beijing, China

Education

1951–55	Studied and graduated from the School of Pharmacy, Beijing Medical College
1959–62	Studied and graduated from the training course on theories of Traditional Chinese Medicine (TCM) for personnel with a background in Western Medicine (2.5 years)

Brief Chronology of Employment

2000–	Chief Professor, China Academy of Chinese Medical Sciences (previously known as China Academy of Traditional Chinese Medicine, before 2005)
1997–	Director, Qinghaosu Research Center, Institute of Chinese Materia Medica, China Academy of Traditional Chinese Medicine
1992–	Lifelong Professor, China Academy of Traditional Chinese Medicine
1985–	Professor and Member of the Experts Committee, China Academy of Chinese Medical Sciences
1985–90	Director, Department of Chemistry, Institute of Chinese Materia Medica, China Academy of Chinese Medical Sciences
1979–84	Associate Professor, China Academy of Chinese Medical Sciences
1965–78	Practice Researcher and Assistant Professor, Department of Chemistry, Institute of Chinese Materia Medica, China Academy of Chinese Medical Sciences
1955–64	Practice Researcher, Department of Pharmacognosy, Institute of Chinese Materia Medica, China Academy of Chinese Medical Sciences

From *Artemisia annua* L. to Artemisinins
http://dx.doi.org/10.1016/B978-0-12-811655-5.00036-2

Professional Experiences

With the help of the integrated medical knowledge of Western Medicine and TCM, in the past 50 years, Professor Tu gained outstanding achievements in the chemical analysis and identification for Chinese Materia Medica. She firmly believes that TCM is a treasure house, and efforts should be made to explore its essence. In 1969, she devoted herself to antimalarial studies and was in charge of the research group later on. On the basis of collection and analysis of traditional prescriptions, over 200 herbs and 380 abstracts from them were analyzed and screened by the research group. Enlightened by the ancient medical book *A Handbook of Prescriptions for Emergencies*, by Ge Hong, she and her research group made extensive chemical analysis for *Artemisia annua* L. Eventually, she discovered a completely new antimalarial substance, artemisinin (*qinghaosu*). Also, the team and she evaluated the efficacy of extracts of *A. annua* L. and *qinghaosu* for the treatment of patients suffering from malaria in the Hainan Province in 1972–73; the clinical findings of which were very promising. The structure of *qinghaosu* was then determined as a new sesquiterpene lactone in the mid-1970s. *Qinghaosu* and its derivatives are now administered to patients suffering from malaria worldwide. Her research group and she have had six "New Drugs Certificates" ratified by CFDA since the discovery of *qinghaosu*.

Grant Proposals in Recent Years

1. Research and application of the new antimalarial drug, dihydroartemisinin (*dihydroqinghaosu*); funded by the World Health Organization and Ministry of Health.
2. Series studies on dihydroartemisinin; funded by the State Administration of TCM.
3. Modulation of dihydroartemisinin on single transduction of NFκB/IκB; funded by the Natural Science Foundation of China.
4. Chemical and pharmacological studies on active ingredients from *A. annua* L.; funded by the Natural Science Foundation of China.

Memberships and Social Positions

1. Council Member and Founding Member, Chinese Association of Invention.
2. Member, Executive Committee, All-China Women's Federation (1988–93).
3. Committee Member, Beijing Committee of Natural Science Foundation (1995–97).

Awards

1. Award for Progress in Antimalarial Research by the National Congress of Science and Technology (1978).
2. National Scientific Discovery Award for the Antimalarial Drug-Qinghaosu by the Ministry of Science and Technology (1979).

3. Invention Award (as the first inventor) by the National Congress for Awards in Science and Technology (1982).
4. Young and Middle-Aged Experts With Outstanding Contribution by the National Ministry of Personnel (1984).
5. Highest Honorary Award by the China Academy of Traditional Chinese Medicine (1992).
6. Top Ten National Achievements for Progress in Science and Technology awarded for dihydroartemisinin and its tablet by the National Committee of Science and Technology (1992).
7. Quality Studies on Commonly Used Chinese Materia Medica (*A. annua* L. included), First-Rate Award of National Achievements in Science and Technology, and National Award Committee for Advances in Science and Technology (1992).
8. National Model Worker by the State Council (1995).
9. Award for Outstanding Achievement in Traditional Chinese Medicine by the Guangzhou Zhongjing Award Foundation for Traditional Chinese Medicine (1995).
10. Outstanding Scientific Achievement Award by the Qiu Shi Science and Technologies Foundation of Hong Kong (1996).
11. Top Ten Health Achievements Award in New China, awarded for dihydroartemisinin and its tablet by the Ministry of Health (1997).
12. Woman Inventor of the New Century by the National Bureau of Intellectual Property (2002).
13. Golden Medal of the Fourteenth National Invention Exhibition by the National Bureau of Intellectual Property (2003).
14. Prince Mahidol Award by the Prince Mahidol Award Foundation, Thailand (2003).
15. Development of Chinese Materia Medica by the Cyrus Chung Ying Tang Foundation (2009).
16. Lasker–DeBakey Clinical Medical Research Award by the Albert and Mary Lasker Foundation (2011).
17. GlaxoSmithKline Award for Outstanding Achievements in Life Science by the Royal Society of London (2011).
18. National Woman Red-Banner Pacesetter by the All-China Women's Federation (2012).
19. Outstanding Schoolfellow by the Medical College of Peking University (2012).
20. Warren Alpert Foundation Prize (2015).
21. Nobel Prize in Physiology or Medicine by the Nobel Foundation (2015).

Meeting Presentations and Invited Presentations

1. Fourth Meeting of the Scientific Working Group on the Chemotherapy of Malaria, United Nations Development Programme/World Bank/World Health Organization Special Programme for Research and Training in Tropical Diseases; Beijing, China (1981).

2. Fourth Women Congress, United Nations; Beijing, China (1995).
3. US–China Conference Symposium on Chinese Herbal Medicine; Chicago, IL, United States (1997).
4. First World Congress on Integrated Traditional Chinese Medicine With Western Medicine; Beijing, China (1997).
5. Second International Forum on Creativity and Invention; Beijing, China (2002).
6. Sino–African Forum on Traditional Medicine; Beijing, China (2002).

Index

H

I

K

L

T

U

V

W

X

Y

Z

Kolibri 360 GmbH
Ferdinand-Jühlke-Straße 7
99095 Erfurt, Deutschland
produktsicherheit@kolibri360.de